D0141762

1979

W. B. SAUNDERS COMPANY/PHILADELPHIA/LONDON/TORONTO

R. C. GULEZIAN

ELEMENTS OF BUSINESS STATISTICS

W. B. Saunders Company: West Washington Square
Philadelphia, PA 19105

1 St. Anne's Road
Eastbourne, East Sussex BN21 3UN, England

1 Goldthorne Avenue
Toronto, Ontario M8Z 5T9, Canada

ELEMENTS OF BUSINESS STATISTICS ISBN 0-7216-4351-5

 Made in the United States of America. Press of W. B. Saunders Company. Library of Congress catalog card number 79-63262

Last digit is the print number: 9 8 7 6 5 4 3 2 1

To
LONGINA MARIA ELIZABETH ZIENKIEWICZ
CHRIS
HARRY
and
AL

PREFACE

The primary goal of this text is to provide the basic concepts of applied statistics suited to the needs of courses shorter in length than two semesters. This is accomplished in terms of the topic areas covered which emphasize the main concepts needed to obtain an understanding of the field of statistics without considering more specialized topics that often are treated as optional.

The format of the chapters has been designed to overcome many of the student learning difficulties encountered in courses in basic statistics. Many of the chapters are shorter than most texts and have a consistent format. These chapters begin with an exposition that introduces the material to be covered together with definitions and background material needed in the chapter. This section concludes with a completely defined summary of symbols, which serves as a glossary for the entire chapter. Symbols are provided at this point since they supply the student with a preliminary exposure to the terms and concepts covered throughout the remainder of the chapter. A distinct section follows that provides a step-by-step outline of the main procedure considered in the chapter. Numerical examples illustrating the main procedure follow immediately.

A discussion section appears near the end of every chapter. Here, material considered previously is elaborated upon, and this section includes interpretations of results, extensions, limitations, and possible applications. In some cases, related methods and procedures are included in the discussion. When appropriate, considerations associated with the use of modern calculating devices and computers as well as shortcut formulas are presented in an identified section at the end of the discussion. By placing the discussion section at the end of each chapter, the student is allowed to obtain a deeper understanding after acquiring knowledge of the fundamentals. In this way, the frequent diffusion of concept, formulas, and computation is avoided. From the teacher's point of view the format is flexible and can accommodate many classroom styles of presentation.

Questions and problems appear at the end of every chapter. These sections are divided into three distinct parts: (1) definitional questions, (2) computational problems, and (3) conceptual questions and problems. By structuring the problem section in this way, it no longer is necessary to hunt for the type of problem that one wants to assign. From the student's point of view, identification of the type of problem allows the student an opportunity to assess his or her performance in terms of accomplishment at the different levels provided.

The definitional questions are directed exclusively toward defining terms and concepts presented in the chapter and provide a list of important points for the student's awareness. The computational problems include only problems requiring the application of computational procedures. The third section includes computational problems that are coupled with questions of interpretation in addition to other types of conceptual questions aimed at a deeper understanding of the material. Combinations of questions and problems from any of the sections can be assigned depending upon the demands of a particular course. A selection of answers and solutions is provided for the student at the end of the text. In some cases a full solution is provided, whereas in others only the final answer is given. Some answers to conceptual questions are provided in order to make a point that is difficult for the student to understand. Mainly, however, it is felt that conceptual questions are directed toward developing the students' ability to verbalize conceptual material on their own. No answers to the definitional questions are given since the student can find these directly in the chapter.

The text is divided into four parts. Descriptive statistics has been included in Part One since it is felt that this material provides more motivation for the student to understand the material that follows. Part Two covers prerequisite concepts of probability required in the chapters on inference. Chapter 6 introduces the basic concept of a probability distribution without considering special distributions. This chapter can be used to connect the ideas of probability presented earlier to the binomial and normal distributions; however, the chapters concerned with these distributions have been written so that they may be taught without the introductory material in Chapter 6.

Statistical inference is presented in Part Three and is introduced in terms of four core chapters at the beginning in terms of means. The format of Chapter 16, which covers regression and correlation, varies from that of the other chapters. Each of the main sections within this chapter, however, has a consistent format. Part Four deals exclusively with problems involving time series data exclusively in terms of business and economic problems.

The number of people that have contributed to this text are numerous. Each in his own way provided different elements that have enhanced the quality of the book and deserves much thanks. I am grateful to the reviewers who made many helpful suggestions. Those who provided reviews were Thomas Hawk of the Community College of Philadelphia, Henry Tingey of the University of Delaware, Judith Stoyle of Temple University, Dick Howe of Orange Coast College, the late Irving Hirsh of the Bronx Community College of the City University of New York, John Potts of Peirce Junior College, and John deCani of the Wharton School of the University of Pennsylvania. Of those students who contributed to the text, special thanks are deserved by Tranda Cummins who methodically drudged through the material a second time as a final check on the continuity of the book. Finally, the entire staff at Saunders deserves a great deal of credit for their continued support and tireless efforts toward publication of the text.

RONALD GULEZIAN

CONTENTS

Introduction

Statistical thinking will one day be as necessary for efficient citizenship as the ability to read and write.

H. G. Wells

The ever-increasing role played by numbers in our lives is staggering. Quantified information in one form or another seems to penetrate into almost every aspect of modern existence. Daily, computers consume vast amounts of data and produce more and more input into modern society. Never before have we been able to carry pocket calculating devices that, in an instant, provide the results of complex numerical calculations. Data are collected; numbers are manipulated; information is presented; probabilities are quoted; conclusions are reached and decisions are made. Our efforts to understand our world by collecting and analyzing information of all kinds appears to be an unending process. These efforts often are guided by the use of methods from the field of statistics. The views commonly held about the field of statistics, however, are numerous and often incomplete. Frequently, statistics is confused with the field of accounting, and many exposed to statistical methods in their work or studies lack a full appreciation of the field.

Primarily during the last half century, statistical methods have been developed and applied to a wide variety of problems. For example, the effectiveness of various medicines is determined with the aid of statistical techniques, as is the quality and safety of other consumer products. Such diverse problems as economic stabilization, weather forecasting and control, population dynamics, market research, auditing, inventory control, and disputed authorship in literature are being solved with the aid of statistics. In addition, more and more legal disputes are resolved by applying statistical methods. The list is virtually endless. Actually, the advent of modern computing equipment has opened the way for using statistics to solve problems that were too difficult or impossible to solve with previously existing techniques.

During the years to come the use of statistical methods will be even greater. Vast amounts of money currently are being allocated for the solution of many social problems. Coupled with this expenditure is a new emphasis on measurement that is directed toward determining the impact of programs for social reform. For example, by law, the health care industry is to undergo greater scrutiny through the development of an interrelated set of local planning and coordinating units. Collection and analysis of data using statistical methods has been built-in as a key component of the system. The socioeconomic impact of these expenditures not only will be influenced by the results based upon statistics, but the magnitude of the impact also will be established by its use.

DEFINING STATISTICS

There are two distinct, but related, areas of statistics; mathematical and applied. Mathematical statistics is a branch of mathematics and is theoretical in nature. Although the results of statistical theory will be utilized, mathematical statistics is *not* the subject matter of this text. Our interests lie in concepts and methods that are the *result* of the theory that can be used to formulate and solve problems in other subject areas. Thus, our work will focus upon the area of *applied statistics.*

The field of applied statistics is neither pure art nor pure science. It is a mixture of activities connecting the results of other disciplines and numerous tasks directed toward the solution of many substantive problems. The field is complicated since the rules necessary to solve a particular problem are not always obvious. Although guiding principles exist, experience and judgment are necessary in order to execute an actual "statistical investigation." As a result, it is difficult to define the field of statistics. We shall, however, provide a definition that can be used to obtain some insight into the contents of this book and the workings of the field. Statistics can be defined in the following manner.

> **Statistics** is a special discipline, or body of principles and methods, useful as an aid in the collection and analysis of numerical information for the purpose of drawing conclusions and making decisions.

Certainly, we all make decisions and draw conclusions using quantified or numerical data, and have done so most of our lives. In many instances, our experience alone is sufficient; there is no apparent need to use some formal method. Moreover, professionals regularly make decisions that are based upon numerical information without using statistical methods. In other words, like most definitions, the given definition of statistics is correct, but not extremely informative; it does not supply features that clearly distinguish the field from much of our everyday activity.

Actually, in our daily lives we utilize thinking that in many ways resembles the kind employed by professional statisticians. The key distinction lies in the fact that statistical methods are *explicit* and possess clearly defined measures of error, whereas this is not true in our everyday activities. In order to elaborate upon our definition of statistics, it is convenient to isolate certain components that characterize the field. The breakdown that follows is consistent with the main subdivisions of the text.

Descriptive Statistics. Whenever statistical methods are employed, a problem always is formulated in terms of a **population** or **universe,** which is defined as all of the elements about which a conclusion or decision is made. For example, suppose we are faced with a problem where we must determine the percentage defective in a shipment of machine parts. The entire shipment of parts represents the population or universe of elements about which the conclusion (percentage defective) is drawn.

In cases where numerical information or data is taken from each element of the universe and used to reach a conclusion about the *entire* population, we are dealing with **descriptive statistics.** That is, we are describing certain aspects of the entire population based upon a complete enumeration. In survey language, this process is referred to as a *census.* With respect to the machine parts illustration, we would be dealing with a "100 percent" inspection of the shipment. Descriptive statistics, therefore, is a specific set of techniques used to describe characteristics of a body of data.

In some cases, data are collected from a population other than the one about which a conclusion is drawn. This occurs when it is either impossible to identify the elements of the population of interest or when it is too costly to obtain data from the population. In such cases, a "substitute" universe that closely resembles the one of interest is used, and the two universes can be identified separately. The population of interest is referred to as the *target population* or *ideal universe.* The so-called substitute universe is referred to as the *statistical universe,* since it is the one from which the data actually are collected.

Consider the problem of predicting the outcome of an election. The target population is the set of all people who actually will vote in the election, since it is this group that will determine the outcome. Unfortunately, actual voters cannot be identified until after the election. Consequently, another group resembling the target must be considered in order to make the prediction. One obvious choice frequently made in voting problems is to use registered voters as the statistical universe. This group can be identified prior to the election and contains those who can vote. There is no guarantee, however, that registered voters will supply the correct outcome of the election since they all may not go to the polls. In similar cases, as long as we use data from all members of the statistical universe, we are dealing with a problem of descriptive statistics. The corresponding methods, however, do not provide a way to select the substitute or statistical universe. Strictly speaking, this is a non-statistical problem that is based upon individual judgment and experience.

Statistical Inference. Another way that conclusions or decisions are made is through the use of a portion or **sample** of elements from the universe. The sample data are analyzed, and the corresponding conclusions are generalized to the target population.

Much of the data collected today is sampled. Obviously, a sample is cheaper to collect than census data. Also, a complete enumeration of certain populations is impractical or virtually impossible. A national election poll is a case in point. Not only is the cost of a census prohibitive, but the necessary time and personnel are unavailable; the required timeliness of the results may be lost as well by undertaking a census. In other instances, a complete enumeration of the population is possible, but unrealistic. For example, in cases where the quality of a shipment, say, of light bulbs is to be determined, a destructive test is required; if the entire shipment is tested, no product will be available for sale since it will be destroyed in testing. For many reasons, sampling is the most reasonable and efficient means of determining the nature of various populations. To the amazement of many, samples which are small relative to the size of the population provide useful information.

In cases where sampling is used, there is an element of uncertainty surrounding the final conclusion; a particular form of error is associated with the final result. Methods for collecting and analyzing sample data have been developed in order to account for error or uncertainty due to sampling. This body of methods is referred to as statistical inference. In other words, **statistical inference** is a body of principles and methods that can be used to make meaningful generalizations about populations on the basis of sample information.

Many of the principles and methods of statistical inference are based upon results that have been taken from the field of probability theory. These probability concepts provide the basis for coping with sampling uncertainty or error. Procedures taken from descriptive statistics also are needed when solving sampling problems.

Concepts of Probability. The field of probability theory, a branch of mathematics, had its real beginning in the 17th Century and was initially applied to games of chance or gambling. Later, it was found that the behavior of other phenomena was similar to these ordinary games of chance. After much theorizing and experimentation, a formal theory was developed that provided explanations for phenomena in numerous fields of study. As already mentioned, the results of probability theory represent an important component of modern methods of statistical inference. Consequently some understanding of probability is vital when dealing with problems in sampling.

Statistics in Business and Economics. The areas of descriptive and inferential statistics together with concepts of probability make up the entire field of applied statistics. The corresponding methods are applied to many areas of study and decision-making. Frequently, particular types of statistical procedures are used more extensively in one problem area than in another owing to the special nature of the subject matter. Of these problem areas, business and economics are ones that directly reach us most on a day to day basis, whether it is as a student of business or as a citizen in general. We constantly are deluged with marketing campaigns making numerous claims, massive amounts of government data, and forecasts concerning interest rates, stock prices, and many other aspects of economic activity. As the nations of the world become increasingly interdependent, more reports and forecasts will become concerned with other nations beside our own.

Virtually all aspects of business and economics depend upon data and statistical analysis in one form or another. The need for students as well as the general public to understand the way business and economic data are analyzed and manipulated is becoming increasingly evident. The ability to critically evaluate claims, reports, and forecasts is important as well.

Many of the statistical methods used in business, economics and government are similar to those applied to other areas, but in a modified form. This results from the nature of most economic data that are generated by government sources, and the manner of data collection of businesses that follows along accounting lines.

ABOUT THIS BOOK

This textbook focuses upon the concepts and methods of applied statistics, which are used in problem-solving, research, and decision-making. Remember, every study utilizing statistical methods begins with a problem in another applied field or discipline. Sometimes, much thought and effort is necessary to formulate a problem properly before statistical methods can be applied. Not infrequently, it is more difficult to define a problem than to solve it. This book, however, is directed toward a solution (once the problem is defined), although some insight into problem formulation is supplied.

The book is subdivided into four main parts:

Part One. Descriptive Statistics
Part Two. Concepts of Probability
Part Three. Statistical Inference
Part Four. Analyzing Business and Economic Data

Each part of the book corresponds to the four topic areas introduced in the last section. Part One presents ways of describing data and also provides some background for the material presented later. Whenever possible, the material in Part Two is related to decision-making in general and anticipates certain topics covered in Part Three. Actually, many topics that are dealt with as isolated concepts in earlier chapters reoccur in another form at later points in the book. Business illustrations are interspersed among others in the first three sections. In many cases, examples are used that convey concepts better by appealing to our everyday understanding. Part Four concentrates on methods commonly associated with business and economics.

Most of the chapters in the book have the same format. Each of these begins with an *exposition* providing background material and chapter goals. A *procedure* section outlining the steps necessary to perform required computations follows the exposition. Numerical *examples* illustrating the procedure appear next in a separate section. Generally, understanding of the procedure cannot be obtained until these examples are covered. Interpretations, extensions, and limitations of earlier ideas appear in a *discussion* section near the end of the chapter. Where applicable, a section presenting *shortcuts* for performing computations has been added. In cases where "canned" computer programs are available, a sample computer output is presented and discussed. Isolated chapters do exist throughout the book where the format is modified. This modification occurs mainly when more than one case of a related procedure is presented or when two or more closely related methods are considered in one chapter.

The *problems and questions* provided at the end of each chapter are designed to supplement the chapter material. A full appreciation of the material can be obtained only by diligently working the problems. A selection of answers and solutions is located at the back of the book allowing you to determine how well you understand the material presented. A number of problems require the use of *tables* in order to arrive at a solution. All necessary tables appear in the Appendix along with an explanation of their use. A *bibliography* of current statistics textbooks for reference and extended reading is provided before the Appendix.

DESCRIPTIVE STATISTICS

CHAPTER 1

BUILDING A FREQUENCY DISTRIBUTION

The Nature of Data

It's not the figures themselves . . . it's what you do with them that matters.

K. A. C. Manderville

If you torture the data long enough, it will confess.

Ronald Coase

Virtually every phase of life is subjected to recorded observation. Data of many kinds are gathered on a daily basis in order to meet a variety of needs. The type of data and the way it is observed differ by the nature of an underlying problem. Using a ruler to measure heights of individuals obviously differs from using a questionnaire to determine attitudes about a political issue. There is one feature, however, that is common to all data: variability. For example, reported stock prices for various stocks are different; the price of an individual stock generally varies from day to day. Heights among individuals, personal attitudes, air temperatures, and annual incomes among families all exhibit differences or variation. The output of the economy and the productivity of a single firm rarely remain the same from year to year. Whenever measurements are taken, differences among individual observations exist. In some cases it is necessary to use an extremely precise measuring instrument to detect these differences. This would be the case when dealing with measurements of finely tooled machine parts.

Due to variability among measurements, it is difficult to know much until we determine special features that are present in the data; unless something is done to a set of numbers, they tell us little. In other words, we must summarize and analyze a body of data in order for it to be informative. Actually, we can think of the field of statistics simply as a special body of methods designed to "make sense" out of variability in data.

Remarkably, measurements associated with most phenomena follow recognizable patterns. Understanding these patterns and knowing how to describe them is basic to our development of statistical reasoning. In this chapter we deal with the concept of a frequency distribution. This is used as a general summary of a body of data and provides the basis for understanding possible patterns that exist. Moreover, the concept of a frequency distribution is of fundamental importance in statistical work. The following two chapters present measures to describe particular characteristics of a set of numerical observations: central tendency and dispersion. Central tendency is treated in terms of averages, whereas dispersion is measured in terms of variability about these averages. The three chapters cover the main features of descriptive statistics and provide a basis for presenting findings or results associated with a set of observations.

EXHIBIT 1
EXAMPLE OF RAW DATA

135	105	200	93	124
93	124	105	150	135
124	135	93	124	124
105	150	124	105	135
124	105	124	150	124

Raw Data and Frequency Distributions. The preceding discussion referred to the concept of numerical data very loosely and did not use any numbers. Consider, as an example, measurements representing the weekly wage rate of a group of 25 workers (Exhibit 1). The collection of the 25 worker wage rates is referred to as raw data. In general, **raw data** represents the originally recorded measurements of any quantity of interest.

A useful way of summarizing raw data is in the form of a **frequency distribution.** This is a table that presents the values of some quantity together with the frequency, or number of times, that each of the values appears in the set of original observations. By counting the number of times each weekly wage rate appears in the above list and tabulating, we can construct the table or frequency distribution shown in Exhibit 2.

Each distinct wage rate appearing in the exhibit is associated with a frequency (i.e., number of workers) representing the number of times a wage rate occurs in the original set of observations. The long list of wages has been summarized in compact tabular form. The longer the list of raw observations, the more useful is a summary of this kind. If the order of recording the observations is unimportant, no information about the raw data is lost by tabulating in the form of a frequency distribution like the one illustrated. Such distributions are referred to as *ungrouped* frequency distributions.

EXHIBIT 2
AN EXAMPLE OF A FREQUENCY DISTRIBUTION

Weekly Wage Rate ($)	*Number of Workers*
93	3
105	5
124	9
135	4
150	3
200	1
Total	25

Frequency distributions are easily graphed, providing a visual description of the way measurements behave. For example, a graph of the above distribution appears in Exhibit 3. Different sets of data do not have shapes exactly the same as the one shown in the exhibit, although many display similar tendencies. That is, low frequencies generally correspond to relatively low and relatively high magnitudes, whereas higher frequencies correspond to values within the center of the distribution. In other words, when viewed graphically, frequency distributions generally begin on the left with lower frequencies, rise to a single peak, and then "tail-off" to lower frequencies again.

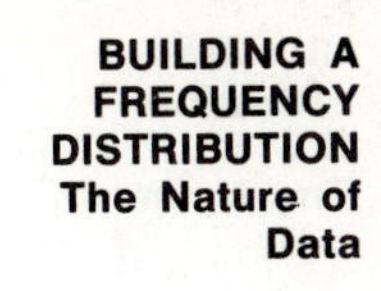

EXHIBIT 3
AN EXAMPLE OF A FREQUENCY DISTRIBUTION WHEN GRAPHED

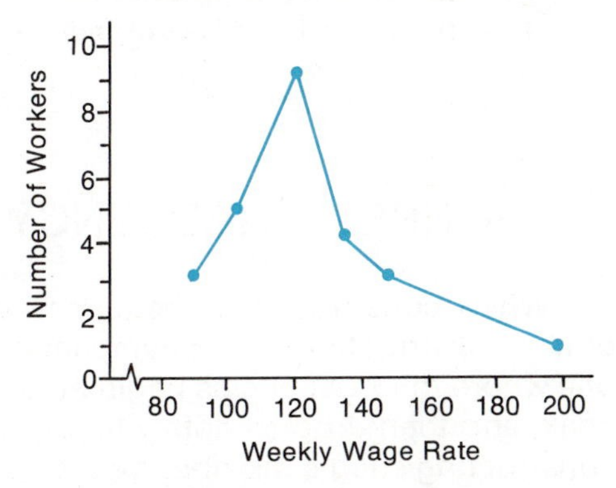

When working with masses of data containing many distinct observations, ungrouped frequency distributions do not provide a useful description of the data. However, by grouping the observations into categories or classes corresponding to ranges of values, a *grouped* frequency distribution can be constructed that provides a better description. An example of a grouped distribution is shown in Exhibit 4. In this case, we see that exact values of the individual observations are not presented and that frequencies correspond to wage rate intervals. For example, the first interval in the table is 75 to 99 with a frequency of 5. This tells us that 5 workers have wages somewhere between the *class limits* of $75 and $99. In general, intervals of this kind are referred to as *class intervals,* or classes.

EXHIBIT 4
AN EXAMPLE OF A GROUPED FREQUENCY DISTRIBUTION

Weekly Wage Rate ($)	*Number of Workers*
75– 99	5
100–124	25
125–149	35
150–174	20
175–200	15
Total	100

It is important to consider the size of class intervals when constructing grouped distributions. The **size** of a class interval is defined as the difference between the upper and lower limit of a class. For example, again consider the first class shown in Exhibit 4. The upper limit is 99 and the lower limit is 75. The difference between these values equals 24 (i.e., $99 - 75 = 24$). Therefore the size of the first class is $24. All of the classes for the distribution shown in the exhibit have the same size. This is not true for all grouped distributions, although equal intervals should be used whenever possible.

When summarizing a set of raw data with a grouped distribution, all the information contained in the original observations is not available. If constructed properly, however, a grouped frequency distribution can provide a useful description of the original data. In the next two sections we outline and illustrate the steps needed to construct a grouped frequency distribution.

The symbols to be used in this chapter are defined as follows:

- X = a numerical value of the quantity measured
- f = the frequency of occurrence, or the number of times that a value or values of X occur
- N = the total number of observations or measurements; N equals the sum of the f-values
- c = the size of the class interval when all class intervals are of equal size

R = range of the observations, or the difference between the largest and smallest observed values
p = the relative frequency
F = the cumulative frequency

CONSTRUCTING A GROUPED FREQUENCY DISTRIBUTION

When constructing a frequency distribution the basic goal is to obtain a distribution that is compact in appearance and, when possible, one where the frequencies begin at low values, rise smoothly to a single peak, and then drop smoothly to lower values. In practice, the rules for constructing such a distribution are general; the steps presented below serve as a guide but do not provide a complete set of rules for constructing a grouped distribution. Not infrequently, we may construct more than one distribution on a trial-and-error basis before being satisfied with the way a grouped distribution describes the original data.

The general procedure for constructing a grouped frequency distribution is as follows:

Step 1. *Decide whether class intervals are to be equal or unequal.* Whenever possible, class intervals should be equal in size. In cases where frequencies change substantially from interval to interval or when a small number of extremely high or low values occur, unequal intervals should be used.

Step 2. Determine the number and size of class intervals. This is the most difficult and arbitrary step. Whether the number or size of the intervals is chosen first varies from person to person and with each problem. Since the two are related, number and size somehow are determined in terms of the usefulness of the final summary that is obtained. We treat the number of classes first.

(a) *Select the number of class intervals.* The general rule for choosing the number of classes is the number of class intervals should neither be "too few" nor "too many." In practical terms, this means that the number of intervals should be somewhere between 5 and 15. Where the range of observations *or* the total number of observations is relatively small, the number of intervals should be closer to 5 than to 15.

(b) *Determine the size of the class interval.* When *equal* class intervals are used, the size of the interval is determined conveniently by using the following formula as a guide:

$$c = \frac{R}{\text{No. of Classes}}$$

That is, the interval size can be determined by dividing the range, or the difference between the largest and smallest values observed, by the number of class intervals. The value obtained from the formula should be modified to a convenient round number if necessary. In cases where *unequal* intervals are used, interval sizes are based upon the nature of the problem solved and the particular set of raw observations.

Step 3. *Establish the actual class intervals* based upon the choices made in Step 2. Sufficient intervals should be established so all of the original observations can be placed into the intervals. The limits of the class intervals should not overlap so there is no confusion about the interval into which a particular observation is placed.

Step 4. *Tally* the raw data in order to determine the number of observations that fall within each class interval.

Step 5. *Count the number of tally bars* and assign the corresponding number, or frequency, to each interval. Add the frequencies to obtain the required total. Compare the total with the original number of observations as a *check* on the tallying procedure; these should be equal.

NUMERICAL EXAMPLE

The following set of numbers represents the highest mutual fund prices reported at the end of a week for a selection of 85 nationally sold funds.

Mutual Fund Prices

4.31	7.52	10.29	8.40	8.71	16.32	2.36	4.58
7.18	8.17	3.05	3.19	10.97	8.69	6.43	7.15
5.87	9.18	2.35	5.63	9.75	10.22	13.98	7.12
9.88	10.87	9.03	7.41	4.10	4.10	7.48	4.05
7.43	10.38	8.62	6.94	7.40	16.94	3.41	8.21
10.39	8.97	5.51	19.84	17.78	*20.06*	7.07	13.37
13.06	10.35	7.53	12.39	*1.00*	3.63	1.75	3.48
15.74	11.43	14.24	4.87	3.72	12.71	12.27	10.61
7.96	10.34	5.22	13.36	6.39	5.42	17.25	9.97
10.99	5.24	8.41	8.48	8.96	2.63	4.26	11.91
7.35	5.40	3.25	19.66	10.05			

In order to summarize the data in a form that describes the general behavior of the mutual fund prices, we are to construct a grouped frequency distribution of the raw data.

Step 1. Unequal vs. Equal Class Intervals.

Equal intervals are to be considered first. If a "smooth progression" of frequencies does not result, unequal classes then may be considered.

Step 2. (a) Number of Class Intervals.

Without prior experience in handling data, it is difficult to automatically choose the appropriate number of classes. Since the observations vary over a relatively small range (i.e., from 1.00 to 20.06) the number of classes should be closer to 5 than to 15. Arbitrarily, let us use 6 class intervals.

(b) Size of the Class Interval.

$$\text{Compute } c = \frac{R}{\text{No. of Classes}}$$

$$= \frac{\text{Largest } \textit{minus} \text{ Smallest}}{\text{No. of Classes}}$$

$$= \frac{20.06 - 1.00}{6}$$

$$= \frac{19.06}{6}$$

$$= 3.17$$

3.17 is an awkward number, although it can be used. More reasonable possibilities are 3.25 and 3.50. 3.00 could be used but the number of intervals must be changed to 7. Let us consider 3.50.

Step 3. Establishing Actual Intervals.

The starting number (lower limit of the first class) can be the lowest observation appearing in the raw data, or some number less than, but close to, this number. Use of "round numbers" is desirable. For the given body of data, 0.00 or 1.00 can be used conveniently; we shall choose 0.00. The class intervals are

$0.00 but less than $3.50
$3.50 but less than $7.00
$7.00 but less than $10.50
$10.50 but less than $14.00
$14.00 but less than $17.50
$17.50 but less than $21.00

Step 4. Tallying the Raw Data

0.00 but less than 3.50	~~||||~~ ~~||||~~
3.50 but less than 7.00	~~||||~~ ~~||||~~ ~~||||~~ ||||
7.00 but less than 10.50	~~||||~~ ~~||||~~ ~~||||~~ ~~||||~~ ~~||||~~ ~~||||~~ ||||
10.50 but less than 14.00	~~||||~~ ~~||||~~ |||
14.00 but less than 17.50	~~||||~~
17.50 but less than 21.00	||||

Step 5. The Completed Distribution is obtained by counting the tally bars and assigning frequencies.

X *Mutual Fund Price*	f *No. of Funds*
$0.00 but less than $3.50	10
$3.50 but less than $7.00	19
$7.00 but less than $10.50	34
$10.50 but less than $14.00	13
$14.00 but less than $17.50	5
$17.50 but less than $21.00	4
Total	85

Note. The progression of frequencies (f) appearing in the table is sufficiently smooth, so the constructed distribution is acceptable. No one choice, however, is correct; others exist. Consider, as examples, the following two based upon the same set of raw data.

X	f
1.00– 4.49	17
4.50– 7.99	24
8.00–11.49	27
11.50–14.99	9
15.00–18.49	5
18.50–21.99	3
	85

X	f
0.00– 2.49	4
2.50– 4.99	14
5.00– 7.49	20
7.50– 9.99	18
10.00–12.49	15
12.50–14.99	6
15.00–17.49	4
17.50–19.99	3
20.00–22.49	1
	85

These two cases are acceptable alternatives in which the starting value, the size of the interval, and the number of classes vary. Although the values of the frequencies are different, all exhibit the same basic tendency regarding the frequencies. The second of the two alternatives provides more detail since the size of the intervals is smaller.

DISCUSSION

The preceding sections furnish background on the nature of data and the way data can be summarized by a frequency distribution. This section will highlight some details about the construction and interpretation of frequency distributions.

Constructing Grouped Distributions. By grouping, we can describe the overall pattern present in a body of data. The pattern is made up of individual components representing different characteristics of the data. When constructing a grouped distribution, we are interested in preserving the characteristic relating to different parts of the overall distribution. These characteristics are described later in this section, and discussed in detail in the following two chapters.

Advantages of equal class intervals.

By employing equal class intervals when constructing a grouped distribution, the objectives stated above generally are met. There are additional advantages associated with equally spaced intervals. Interpretation of the frequencies is simpler since it is not necessary to concentrate on the size of the class intervals. The graph of a grouped distribution does not contain misleading visual distortions that can result from the use of unequal classes. Computations made from frequency distributions with equal intervals also are easier to perform. Unequal class intervals are preferable, however, if the frequencies corresponding to equal intervals vary substantially or are very low or zero in a number of classes.

Unequal and open-ended intervals.

When heavy concentrations of frequencies occur in the center of the distribution accompanied by frequencies that are very low or zero at the extremes, a different type of interval sometimes is employed: an *open-ended interval.* This is a class interval where one of the limits is not specified and is used either at the beginning or at the end of a frequency distribution. Typical examples of open-ended intervals take the following form:

Less than 5000
5000 or less
More than 10,000
10,000 or more

The first two examples would appear at the beginning of a distribution, whereas the last two would appear at the end. Notice that the first two illustrations do not provide the same information. A raw observation of 5000 would not be placed in the first interval, but would be placed in the second. Similar considerations apply to the last two examples. In cases where open-ended intervals are used, it is good practice to indicate something about the magnitudes of the values in the interval by a footnote. Typically, the average value of the observations in the open interval is used.

Consider the precision of measurements when establishing class limits.

The precision of the measurements also should be considered when establishing class limits. Thus, if we are interested in grouping weekly wage rates, they would be measured in dollars and cents or just dollar amounts. If, for example, we use a classification in terms of "dollars only," but measure a wage as $149.53 it will lie between two classes. If rounding is permitted, the "dollars only" classification would be satisfactory; there would, however, be no way to determine directly from the distribution whether measurements were taken in dollar amounts or to the nearest cent and rounded unless this was specifically noted. Two points are apparent: (1) more than one way exists to specify a given class interval, and (2) the precision of the measurements must be considered in the limits.

The number of classes in an important consideration.

In order to obtain a useful description of the raw data when grouping, it is necessary to use a sufficient number of classes. If too few classes are employed, most of the information about the raw data is lost, and the distribution becomes an ineffective summary. If too many classes are used, much information is provided; however, the ultimate purpose of grouping is defeated, since the distribution would nearly be the same as the set of raw observations. To a great extent, the number (and size) of class intervals is based upon the judgment of the investigator. Both the visual summary and the achievement of a relatively smooth progression of frequencies should be used as a guide. Sometimes, additional guidance is provided by considerations associated with a particular problem.

Plotting Frequency Distributions. All frequency distributions can be graphed. Graphing provides a useful way to obtain a quick understanding of the underlying *shape* of the distribution and its characteristics. In the case of an ungrouped distribution, plotting is a simple matter. By convention, values of the variable under study (X) are scaled along the horizontal axis; frequencies (f) are scaled along the vertical axis. Adjacent plotted points usually are connected by straight lines in order to enhance the "shape" of the distribution visually. The same basic form also can be used for grouped distributions.

The two basic ways to graph a grouped distribution are in terms of a frequency polygon and a histogram.

Actually, grouped distributions can be graphed in two ways: by means of a *frequency polygon* or a *histogram*. As an example, consider the distribution of mutual fund prices. Alternative graphs of this distribution appear in Exhibit 5. When plotting a grouped frequency distribution, either in the form of a frequency polygon or a histogram, the endpoints of the class intervals are specified on the horizontal axis. In the case of a polygon, a dot representing the coordinates of a specific frequency and value of X is placed directly across from the value of the frequency and above the *mid-point* of the appropriate class interval; successive dots are connected by straight lines. For a histogram, each frequency is represented by a rectangle whose height is equal to the frequency and whose width corresponds to the size of the class interval. In both cases, we can see that the general "shape" or form of the original distribution is described graphically.

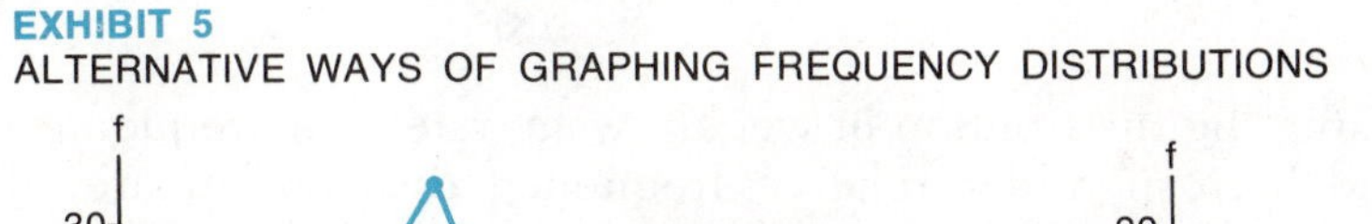

EXHIBIT 5
ALTERNATIVE WAYS OF GRAPHING FREQUENCY DISTRIBUTIONS

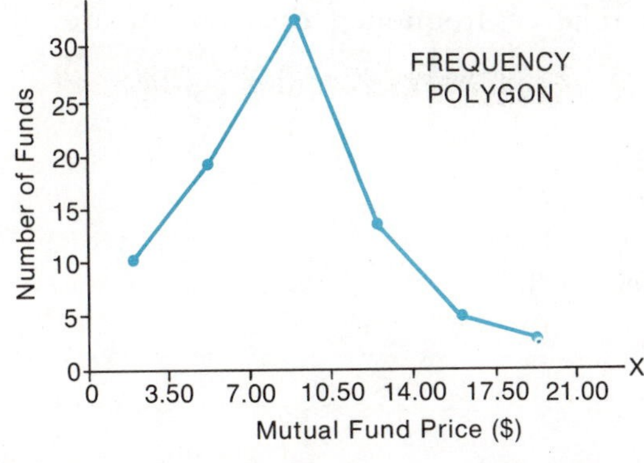

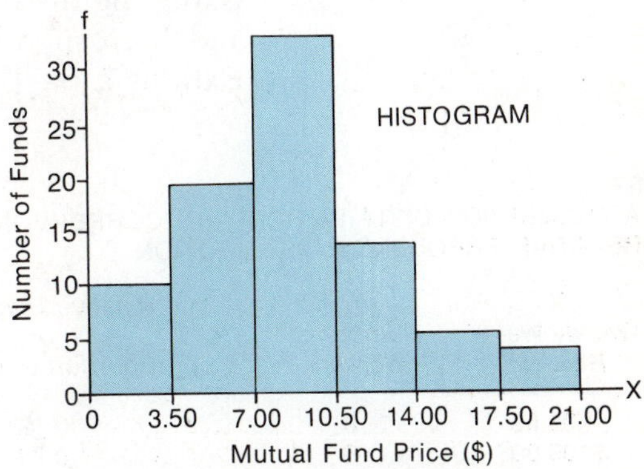

An alternative way to graph a grouped distribution is with a frequency curve.

The graphs shown above can be streamlined by drawing a smooth curve through the rectangles of the histogram to form another kind of graphic description called a *frequency curve*. An example of a frequency curve using the same illustration appears in Exhibit 6. Visually, it is

EXHIBIT 6
AN EXAMPLE OF A SMOOTHED HISTOGRAM OR FREQUENCY CURVE

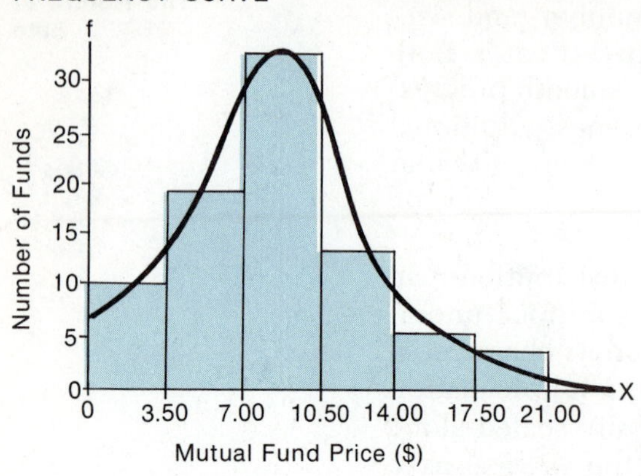

more appealing than the other two graphs. By using the smoothed frequency curve we can obtain a clearer picture of the underlying form or shape of the data. The relationship between the histogram and the frequency curve becomes especially important in later chapters when the concepts of data description and probability are used for the purpose of analyzing data statistically.

Other Forms of Frequency Distributions. Two useful distributions that are closely related to the frequency distribution are the relative frequency distribution and the cumulative frequency distribution. These apply to both the ungrouped and grouped form. The **relative frequency distribution** is obtained by dividing the total number of observations, N, into the absolute frequency, f, for *each* value or group of values of X. Thus, the relative frequency, p, is found by the formula

$$p = \frac{f}{N}$$

Relative frequency.

Using the distribution of weekly wage rates, the frequency distribution and corresponding relative frequency distribution are illustrated in Exhibit 7.

EXHIBIT 7
A FREQUENCY DISTRIBUTION AND CORRESPONDING RELATIVE FREQUENCY DISTRIBUTION

X *Weekly Wage Rate ($)*	*f* *Number of Workers*	*p* *Relative Frequency or Proportion of Workers*
$93.00	3	0.12
$105.00	5	0.20
$124.00	9	0.36
$135.00	4	0.16
$150.00	3	0.12
$200.00	1	0.04
Total	25	1.00

In order to understand how the relative frequencies in the last column of the exhibit are obtained, consider the first value of 0.12 corresponding to a wage rate of $93.00. This is obtained by dividing the frequency of 3 by the total of 25 (i.e., 3/25 = 0.12). Hence, 0.12 or 12 percent of the workers earned $93.00 weekly. Similar calculations and interpretations apply to the remaining values in the exhibit.

The *sum* of all relative frequencies of any distribution must equal one. Relative frequencies also can appear as percentages rather than proportions without altering the distribution. *Percent relative frequencies* are obtained by multiplying each relative frequency by 100. When in percent form, the sum of the relative frequencies equals 100 percent.

Relative frequencies are used to compare the shapes of two distributions with differing total observations. Also, by representing frequencies in relative form, a greater degree of generality is obtained. In other words, relative frequencies can be viewed as estimates applying to any number of observations in addition to the total upon which they were computed. This concept closely relates to the problem of obtaining empirical probability estimates which is discussed in Part Two.

The **cumulative frequency distribution** is obtained by successively adding or cumulating the absolute frequencies, and placing each new sum next to the corresponding value or group of values of X. In order to illustrate the concept of a cumulative distribution, consider the distribution of mutual fund prices again. This distribution together with the cumulative frequencies appears in Exhibit 8. Each of the cumulative frequencies, F, in the last column of the table provides the total number of mutual funds falling in the corresponding category *and* all of the preceding categories. For example, for the third category, $7.00 to $10.49, the cumulative frequency of 63 tells us that 63 mutual funds have dollar values that are less than or equal to $10.49; or *at most* $10.49. Cumulative frequency distributions also can be constructed when the original distribution is in relative or percent form.

EXHIBIT 8

A FREQUENCY DISTRIBUTION AND CORRESPONDING CUMULATIVE FREQUENCY DISTRIBUTION

X *Mutual Fund Price*	*f* *Number of Funds*	*F* *Cumulative Frequency*
$0.00– $3.49	10	10
$3.50– $6.99	19	29
$7.00–$10.49	34	63
$10.50–$13.99	13	76
$14.00–$17.49	5	81
$17.50–$20.99	4	85
	85	

The Nature of Data. The concept of a frequency distribution is of fundamental importance in the area of applied statistics. With it, we are able to describe or summarize the way a body of data behaves. Although most bodies of data exhibit the same basic tendencies, differences exist.

Frequency distributions have different shapes.

EXHIBIT 9
FREQUENCY DISTRIBUTIONS WITH DIFFERENT SHAPES

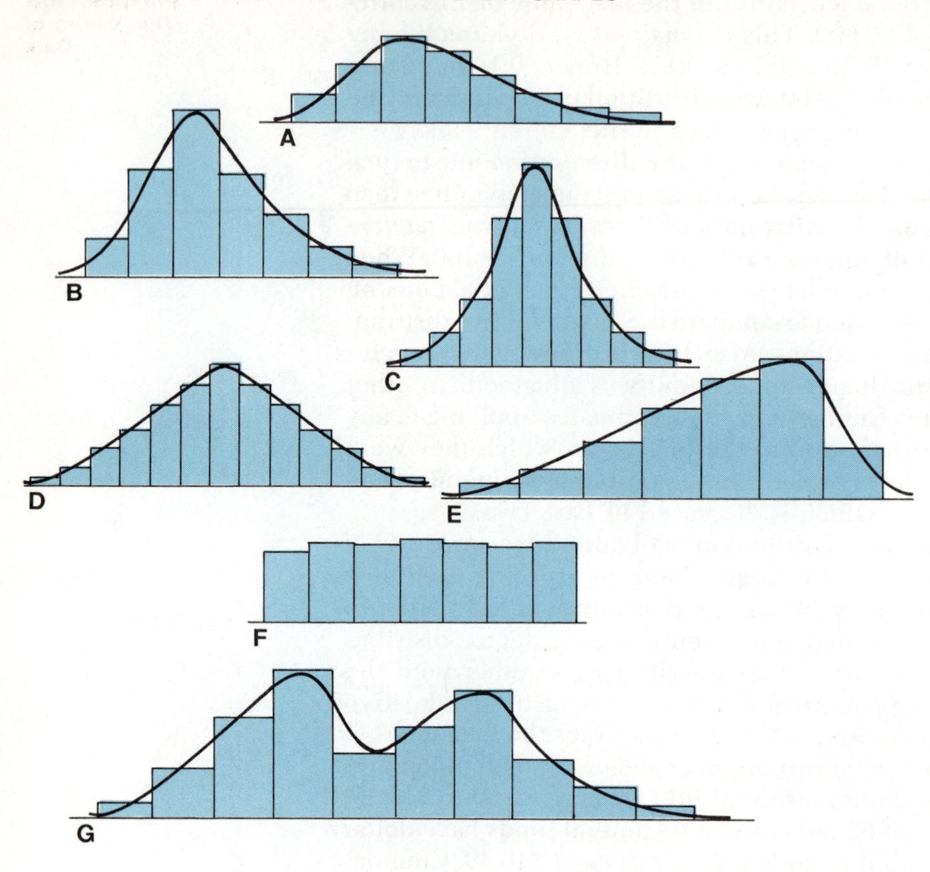

Some distributions are "taller" than others, some are "wider," while others have a concentration of observations about different values.

Consider seven distributions depicted in Exhibit 9. Most observed distributions have a single peak as depicted in diagrams A through E, whereas the exceptions have either no peak (F) or two (G) or more peaks. A peak exists when a value has a frequency which is somewhat higher than the other frequencies. When more than one peak occurs, the data generally must be reanalyzed before they can yield meaningful descriptions. Consequently G will be treated separately from the other forms.

The shape of a frequency distribution can be described in terms of different properties.

Distributions possess properties referred to as symmetry, kurtosis, dispersion, and central tendency. *Symmetry* occurs when each half of the distribution is the "mirror image" of the other half. The distributions labeled as C and D are symmetric, and F is roughly symmetric. Measurements such as the length and weight of assembly-line items often are distributed symmetrically similar to C and D. Distributions that are not symmetric are said to be *skewed;* examples are depicted in A, B, and E. A and B are said to be skewed *to the right,* since the "tail" of the distribution points to the right side. More specifically, a distribution is skewed to the right when relatively small frequencies are attached to large values.

We already have examined the distribution of mutual fund prices, which was skewed to the right. The distribution of family income represents another case where a variable is skewed to the right; we know that relatively few families are very rich, the bulk of the population (the "middle class") have incomes somewhere between the poor and the rich, and there are more poor people than rich people.

When a distribution has relatively low frequencies on the left, or has a "tail" on the left, we say that the distribution is skewed *to the left;* this is shown in the diagram labeled as E. Phenomena exist that are distributed in this way, but obvious examples such as income and mutual fund prices are not as easy to find. One possible example is a distribution of numerical exam grades where few students receive lower or failing grades and there are a lot of A's and B's.

Diagrammatically, we can think of dispersion in terms of the "width" of the distribution. For example, A appears "wider" than B. In such a case, we say that the characteristic described in A possesses greater *variability* than in B. The same type of comparison can be made between C and D. More variability exists in D than in C. As the variability becomes greater in both cases, the peak or highest frequency falls. The degree of "peakedness" in a distribution is referred to as *kurtosis.* Little attention is given to kurtosis in practical work.

As the variability in a set of observations or in a distribution becomes smaller, the values of the observations concentrate about some centrally located value or values. If all the observations had the same value, there would be no variability and consequently no distribution. Since virtually all sets of data exhibit some variation, it is convenient to think in terms of a single value about which the other observations tend to cluster or concentrate. This value is said to be representative of the distribution or of the observations; in statistical terms this phenomenon is referred to as *central tendency.* The concept of central tendency is more meaningful when distributions with the same degree of variability and skewness are compared.

Phenomena described by F do not occur frequently when dealing with observed data. In this case, the frequencies are somewhat *uniformly* assigned; that is, the various values occur with basically the same frequency. Phenomena described in this manner are of importance in statistical work, however, and will be discussed when we reach the problem of sampling.

Although the last diagram (G) can be observed in practice, we seldom deal with it directly. The double peak generally indicates one of two situations: (1) either a single peaked distribution has been distorted by using unequal class intervals, or (2) the double peak may result from situations where a second characteristic is related to the one being measured and concentrations of values occur at two levels. For example, if a distribution of wage rates is constructed for workers including both skilled and unskilled individuals, concentrations of values would occur about different wage levels, reflecting the disparity in wages paid to the two groups. In such instances, if two separate frequency distributions were constructed, one for skilled and one for unskilled workers, we would obtain two single peaked distributions. When drawing conclusions about wages, we then could treat the two frequency distributions separately as descriptions of the wages of skilled and unskilled workers.

The two distributions also could be compared to determine the extent of the differences between the two groups. In general, data behaving in the way described in this paragraph are said to be *non-homogeneous*.

In summary, most bodies of data have certain tendencies that are represented by different properties of frequency distributions, whether in grouped or ungrouped form. The frequency distributions shown above were used to illustrate these properties and the possible patterns typically encountered. Recognize that actual data will exhibit irregularities. Hence, distributions observed in practice do not appear exactly as depicted, but do possess similar tendencies. The specific properties of data, namely, variability, symmetry, peakedness, and central tendency, can be quantified in terms of specific measures. For purely descriptive purposes, these measures give useful summaries for drawing conclusions about a set of data. When used with other statistical procedures, they are indispensable tools. Due to the nature of these procedures, the properties of central tendency and variability are the most important and are given special treatment in the following two chapters.

The properties of central tendency and variability are most important.

USING THE COMPUTER

Computers can be helpful when performing tallies in order to construct frequency distributions. In cases where it is of interest to construct a single distribution and perform no accompanying calculations, the advantage of the computer is questionable. In such a case, the time required to keypunch the original observations is nearly the same or greater than that of a direct hand-tally. However, in cases where additional calculations are desired, or where different distributions with varying sets of class intervals are to be tried, time and effort will be saved by using a computer.

More and more public and private institutions have data bases already stored in their computer files. When this is the case, it is a simple matter to use the computer to develop frequency distributions and perform other types of computations.

Programmed routines are available for constructing frequency distributions. For grouped distributions, generally you must decide upon the class interval size to be used. In addition, programs generally print simple numerical codes for the categories or classes, which you must supply as well. Some programs will print the actual class limits if they are supplied. In the case of an ungrouped distribution, actual numerical values of the variable may be used. Routines for constructing grouped distributions on handheld and desk-top programmable calculators also are available.

The output from a computer program can take one of two basic forms: either the form used in this chapter or expressed horizontally across the page. The latter is illustrated since there is little difference in the computer format for the other case. Assume we are interested in a distribution of incomes for which we have 15 equal class intervals of $2500; the distribution starts at zero. We shall code the categories simply with the numbers 1 through 15. The output of a "canned" program may appear as follows (Exhibit 10):

EXHIBIT 10
AN EXAMPLE OF COMPUTER OUTPUT OF A FREQUENCY DISTRIBUTION

VARIABLE: INCOME

CODE	1.	2.	3.	4.	5.
ABSOLUTE FREQUENCY	5	15	5	45	25
RELATIVE FREQUENCY (PERCENT)	2.9	8.6	2.9	25.7	14.3
CUMULATIVE FREQUENCY (PERCENT)	2.9	11.5	14.4	40.1	54.4
CODE	6.	7.	8.	9.	10.
ABSOLUTE FREQUENCY	10	30	15	10	7
RELATIVE FREQUENCY (PERCENT)	5.7	17.1	8.6	5.7	4.0
CUMULATIVE FREQUENCY (PERCENT)	60.1	77.2	85.8	91.5	95.5
CODE	11.	12.	13.	15.	
ABSOLUTE FREQUENCY	5	1	1	1	
RELATIVE FREQUENCY (PERCENT)	2.9	0.6	0.6	0.6	
CUMULATIVE FREQUENCY (PERCENT)	98.4	99.0	99.6	100.2	

TOTAL NUMBER OF OBSERVATIONS 175
NUMBER OF MISSING OBSERVATIONS 0

Printouts from canned programs typically provide the absolute frequency as well as the relative and cumulative frequencies in percentage form. Based upon the initial output, for example, we see that the second category (CODE 2) has a frequency of 15. This tells us that 15 families have incomes between \$2500 and \$5000, that these families represent 8.6 percent of the total number of families, and that 11.5 percent of the families have an income that does not exceed \$5000. Similar interpretations apply to the other codes. Notice code number 14 is missing. This means the frequency in this category is *zero*, and therefore is not printed. Also the cumulative frequency for the last category exceeds 100.0; this is due to rounding error and cannot be avoided in this case unless more decimal places are employed.

Appearing at the end of the printout is the total number of observations and the number of observations missing from the tally. In some cases the information may actually be unavailable, and in others the information may be key-punched incorrectly; as a result, such data are not included in the tally. The information regarding missing observations provides the number of cases not included. With respect to this example, no observations are missing.

QUESTIONS AND PROBLEMS

DEFINITIONAL QUESTIONS

1. What is the meaning of the following terms?

(a) Raw data
(b) Range
(c) Frequency distribution
(d) Shape of a distribution
(e) Ungrouped frequency distribution
(f) Grouped frequency distribution

2. What is meant by the following terms relating to a grouped frequency distribution?

   (a) Class interval
   (b) Class limit
   (c) Class size
   (d) Equal class intervals
   (e) Unequal class intervals
   (f) Open-ended interval

3. Explain what is meant by a frequency polygon, a histogram, and a frequency curve.

4. Define the terms *relative frequency* and *cumulative frequency.* How are these related to a frequency distribution?

5. Explain what is meant by the terms *central tendency, variability* or *dispersion, skewness,* and *peakedness.*

NUMERICAL PROBLEMS

6. Based upon the following 30 numbers, construct an ungrouped frequency distribution. Use the symbols X and f as labels to indicate the quantity being summarized and the frequency of these values.

1	0	1	2	0	3	1	2	3	0
2	0	4	1	0	3	1	1	2	1
1	4	1	2	2	0	1	3	2	2

7. Appearing below are the number of years of service of 20 faculty members of a department of business administration.

Years of Service

10	7	4	1	7	5	2	10	7	2
5	10	1	7	4	10	7	5	7	2

Use the above values to construct an ungrouped frequency distribution of the number of years of service. Be sure to identify the columns of the distribution appropriately.

8. Given below are a set of 35 values.

5.2	5.6	7.3	5.4	5.7	3.2	3.3
3.7	5.4	3.3	5.9	3.4	3.0	4.5
3.5	1.7	1.4	3.9	3.6	5.0	2.6
5.3	3.8	5.6	7.6	5.7	3.1	5.2
5.5	5.0	4.1	3.2	5.1	4.4	6.0

Summarize these values in the form of a grouped frequency distribution. Use equal class intervals of 1.5 where the lower limit of the first class equals 1.0. Identify the components of the distribution appropriately with the symbols X and f.

9. Examination grades for a class of 25 students appear below.

Exam Grades

61	55	87	83	75
67	81	71	59	71
79	71	51	55	59
73	73	99	61	57
69	49	97	55	77

Construct a grouped frequency distribution of these grades using equal class intervals of 10 and a lower limit of the first class equal to 40. Label the components of the distribution appropriately.

10. A portfolio contains 50 stocks whose prices are given as follows:

Stock Prices

109	17	$50\frac{5}{8}$	$16\frac{3}{8}$	$28\frac{3}{8}$	$1\frac{5}{8}$	$2\frac{3}{8}$	$9\frac{7}{8}$	$2\frac{1}{2}$	$6\frac{3}{4}$
$58\frac{3}{4}$	$21\frac{1}{2}$	$3\frac{1}{4}$	$33\frac{5}{8}$	$9\frac{1}{8}$	$20\frac{5}{8}$	$29\frac{3}{4}$	44	$7\frac{5}{8}$	$13\frac{5}{8}$
$19\frac{1}{2}$	$20\frac{1}{8}$	$55\frac{1}{2}$	$20\frac{3}{4}$	$6\frac{1}{2}$	8	$27\frac{3}{8}$	$23\frac{1}{4}$	$23\frac{3}{4}$	$18\frac{1}{4}$
$90\frac{1}{4}$	$11\frac{7}{8}$	129	$19\frac{5}{8}$	$20\frac{1}{4}$	$25\frac{5}{8}$	95	$30\frac{5}{8}$	$3\frac{5}{8}$	$6\frac{5}{8}$
$8\frac{5}{8}$	30	$12\frac{1}{2}$	$10\frac{7}{8}$	$27\frac{1}{2}$	$22\frac{3}{8}$	$10\frac{1}{4}$	$28\frac{1}{4}$	$7\frac{3}{8}$	$24\frac{3}{4}$

Summarize these stock prices in the form of a frequency distribution.

11. The length of life, measured in hours, of 40 light bulbs appears below. Construct a frequency distribution of these values.

Light Bulb Life (No. of hours)

381	225	482	580	506	284	340	329	341	400
110	403	180	406	499	387	300	462	280	475
264	483	380	531	430	427	420	509	416	622
419	564	401	220	698	554	540	336	106	418

12. The distribution of yearly incomes of 175 families appears below.

Family Income	*Number of Families*
\$ 0–\$ 4999	20
\$ 5000–\$ 9999	36
\$10000–\$14999	50
\$15000–\$19999	42
\$20000–\$24999	17
\$25000–\$29999	5
\$30000–\$34999	3
\$35000–\$39999	1
\$40000–\$45000	1
Total	175

Plot the distribution of incomes in the form of a frequency polygon and a histogram. Smooth the histogram to obtain a frequency curve.

13. The distribution of ages of 500 readers of a nationally distributed magazine is given as follows:

Age in Years	*Number of Readers*
14 & Under	20
15–19	125
20–24	25
25–29	35
30–34	80
35–39	140
40–44	30
45 & Over	45
Total	500

For this distribution, find the corresponding relative frequency distribution and cumulative frequency distribution.

14. The following frequency distribution shows the distribution of the inventory to sales ratio of 200 retail outlets.

Inventory to Sales Ratio	*Number of Retail Outlets*
1.0 but less than 1.2	20
1.2 but less than 1.4	30
1.4 but less than 1.6	60
1.6 but less than 1.8	40
1.8 but less than 2.0	30
2.0 but less than 2.2	15
2.2 but less than 2.4	5
Total	200

Find the relative frequency distribution and cumulative frequency distribution of the inventory to sales ratio.

CONCEPTUAL QUESTIONS AND PROBLEMS

15. What is the basic property of virtually all data that leads to methods of describing and analyzing data? How is a frequency distribution related to this property of data?

16. What advantages are there for using a frequency distribution to describe a body of raw data? What are the disadvantages?

17. What is meant by the phrase "the way data are distributed"? Is this phrase related to the "shape" of a frequency distribution? Explain.

18. Why should one use a grouped frequency distribution instead of an ungrouped distribution?

19. When constructing a grouped frequency distribution, should equal intervals always be used? Under what circumstances should unequal intervals be used instead?

20. What are the advantages and disadvantages of using open-end intervals when constructing a grouped frequency distribution?

21. Consider the distribution of the inventory to sales ratio presented in Problem 14. What is the size of the class interval used? If the class intervals were kept equal but changed to 1.00 to 1.19, 1.20 to 1.39, and so on, would the frequencies change? Would a change in the intervals as suggested alter the size of the class intervals? Does the suggested change in the class intervals possibly convey any different information about the precision of the original data? Explain.

22. When constructing a grouped frequency distribution, is it necessary that the resulting distribution be symmetric? Explain.

23. Consider the age distribution of readers in Problem 13. Plot the distribution in the form of a histogram. What difficulties did you find? Explain. Would the same problem be associated with the frequency polygon and the frequency curve?

24. What are the advantages of using a graph to describe a frequency distribution?

25. Consider the distribution of family incomes presented in Problem 12. For the third income class, $10,000 to $14,999, what is the value of the absolute frequency, the relative frequency, and the cumulative frequency? Interpret these values in terms of the actual data presented.

26. Consider a firm engaged in the business of conducting educational meetings for business executives. Participants pay the firm a fee to attend meetings. Proceeds received by the firm are used to conduct the meetings and to make a profit. Attendance at the meetings generally ranges between 6 and 55 participants. It has been determined that the break-even number of participants equals 16. Should this information be considered when constructing a grouped frequency distribution of the number of participants? Discuss.

27. Problem 9 presents a set of exam grades for a class of 25 students. The problem requires that you construct a grouped distribution based upon equal class intervals of 10. Can you suggest a reason why this choice may not be satisfactory? Would it help to know that letter grades are awarded on the basis of the following classification scheme: A, 94–100; B, 85–93; C, 75–84;

D, 65–74; and F, less than 65? What is suggested by this question about working with data and frequency distributions?

28. What is meant by the term *non-homogeneous*? Prepare an original example of a situation where data might be non-homogeneous.

29. The distribution of heights of all college students has two peaks or is bimodal. The distribution of the IQ's of the same students, however, has only one peak. How is this possible, since the same students are considered in both cases? Explain.

CHAPTER 2

REPRESENTING DATA

On the Average

We can easily represent things as we wish them to be.
Aesop

If at first you don't succeed, you're just about average.
Bill Cosby

The word *average* should be familiar to us all. As students we must contend with grade point averages. As consumers we are concerned with price changes that are reported in terms of index numbers, another type of average. It is now standard practice to post gas mileage on new cars; the posted figure also is supposed to be an average. Daily news reports include statements about stock prices; a popularly quoted figure is the Dow-Jones industrial stock average. The nation's daily oil requirement is specified as a single number. Since the amount of oil consumed varies from day to day, an average of some sort is reported.

In these instances and many others, there is a need to describe a situation with a number, or average, that is *representative* of a body of data without looking at every observation. Obviously, more information would be provided if an entire distribution were presented. However, useful information can be obtained from averages in many cases. In others, averages should be accompanied by additional information in order to provide a meaningful description of a body of data.

Unfortunately, when averages are reported, we are not told much about them. In cases where we are told that an average is used, the kind of average generally is not specified. In other words, more than one type of average exists but we rarely know which one is being used. Each is defined differently and generally does not have the same value. Consequently, different averages are used in different situations.

The most commonly used averages are the *arithmetic mean,* briefly referred to as the mean, the *median,* and the *mode.* The mean is found by adding all the observations and dividing by the total number. The median is defined as the "middle" value. That is, fifty percent of the observations have numerical values less than the median and fifty percent have values greater than it. The mode is the value that occurs most frequently and is the value that lies under the peak of a frequency distribution. A distribution can have more than one mode; when this occurs we refer to it as *multi-modal.*

Both the median and mode are based on certain values rather than on all of the observations. The mean, however, is based upon all of the observations. All three averages are used to represent a body of data; we shall discuss the meaning of a representative measure after we present the procedures for computing each type of average. Unlike other chapters, there is a separate section for each of the three averages. Since raw data and ungrouped and grouped frequency distributions are presented separately, the procedure for each case is illustrated before another case is considered. The meaning and application of the averages are presented in the discussion section near the end of the chapter.

The symbols to be used in this chapter are defined as follows:

X = a numerical value of the quantity measured
f = the absolute frequency, or the number of times a value or values of X occur
F = the cumulative frequency
p = the relative frequency, or the proportion of the total number of observations corresponding to a value or values of X
N = the total number of observations
m = the midpoint of a class interval of a grouped distribution
c = the size of the class interval when all classes are of equal size
$\overline{X}$ = the arithmetic mean (read as "X-bar")
X_{50} = the median, or fiftieth percentile
X_P = the P-th percentile, where P is a number between zero and 100
m_a = the midpoint of an arbitrarily chosen class interval used in the shortcut for finding the mean
d = the number of class intervals that a particular interval is from the arbitrarily chosen interval containing m_a
Σ = summation (read as Sigma)
= an instruction meaning that all terms following this symbol should be added

COMPUTING THE ARITHMETIC MEAN

Raw Data. In order to find the arithmetic mean of a set of raw observations, substitute into the following formula and solve.

$$\overline{X} = \frac{\Sigma X}{N}$$

Mean of raw data.

This formula tells us to add the observations and divide by the total number of observations.

Example 1 ***Mean of Raw Data.*** The weekly wage rates of 12 workers are given below.

124	150	124	105	135	150
105	124	135	124	135	124

Since these values are in the form of raw data, the arithmetic mean weekly wage rate per worker is found by substituting into the following formula as

$$\bar{X} = \frac{\Sigma X}{N}$$

$$= \frac{124 + 150 + 124 + 105 + 135 + 150 + 105 + 124 + 135 + 124 + 135 + 124}{12}$$

$$= \frac{1535}{12}$$

$$= 127.92 \text{ or } \$127.92 \text{ per worker}$$

The final answer equal to $127.92 is obtained by adding the raw observations and dividing by the total of 12.

Ungrouped Frequency Distribution. The arithmetic mean for an ungrouped distribution is found using the formula

$$\bar{X} = \frac{\Sigma fX}{N}$$

or

$$\bar{X} = \frac{\Sigma fX}{\Sigma f}$$

Mean of ungrouped frequency distribution.

In this case, the mean is found by adding the product of each X-value multiplied by its corresponding frequency, and then dividing by the total number of observations. The two expressions given above are equivalent, since the total number of observations (N) equals the sum of the frequencies (Σf). The second demonstrates that you may see alternative formulas that lead to the same result.

> *Note.* If relative frequencies in the form of proportions are used, there is no need to divide by N since the total frequency equals one; the formula for the mean is simply ΣpX. This formula tells us simply to multiply each X by the corresponding relative frequency, p, and add the results. The sum must be divided by 100 if percents are used instead of proportions.

Example 2 ***Mean of an Ungrouped Frequency Distribution.*** Consider the same set of data given in Example 1 but presented in tabular form.

X *Weekly Wage Rate ($)*	f *Number of Workers*
105	2
124	5
135	3
150	2
	12

These data are in the form of an ungrouped distribution so the mean is found by substituting into the following formula.

$$\overline{X} = \frac{\Sigma fX}{N}$$

$$= \frac{(2 \times 105) + (5 \times 124) + (3 \times 135) + (2 \times 150)}{12}$$

$$= \frac{210 + 620 + 405 + 300}{12}$$

$$= \frac{1535}{12}$$

$$= 127.92$$

Here, each wage rate is multiplied by its frequency, these products are added, and the sum is divided by the total of 12 to obtain the final answer.

Note. An orderly and convenient way to perform the above operation when using hand computation is to continue in tabular form:

X	f	fX
105	2	210
124	5	620
135	3	405
150	2	300
Total	12	1535

The mean is obtained by dividing the totals: $\frac{1535}{12} = 127.92$.

Grouped Frequency Distribution. The formula for finding the mean of a grouped distribution is similar to the one used for the ungrouped case. That is,

$$\bar{X} = \frac{\Sigma fm}{N}$$

or

$$\bar{X} = \frac{\Sigma fm}{\Sigma f}$$

Mean of grouped frequency distribution.

where the midpoint of each class interval is used instead of values of X. In other words, add the products of the frequencies and the class midpoints and then divide by the total of the frequencies.

Example 3 ***Mean of a Grouped Frequency Distribution.*** Consider the grouped frequency distribution of mutual fund prices from Chapter 1. We shall find the totals in the form of a table to determine the mean. Before performing the calculations, the midpoints of each class must be found. These are found as the values in the *middle* of the class intervals. The midpoints and preliminary calculations to find the totals appear in the following table.

X *Mutual Fund Price*	f *No. of Funds*	m *Midpoint*	fm
\$0.00 but less than \$3.50	10	1.75	17.50
\$3.50 but less than \$7.00	19	5.25	99.75
\$7.00 but less than \$10.50	34	8.75	297.50
\$10.50 but less than \$14.00	13	12.25	159.25
\$14.00 but less than \$17.50	5	15.75	78.75
\$17.50 but less than \$21.00	4	19.25	77.00
Total	85		729.75

The totals from the table are substituted in the formula for the mean of a grouped distribution as follows:

$$\bar{X} = \frac{\Sigma fm}{N}$$

$$= \frac{729.75}{85}$$

$$= 8.59 \text{ or } \$8.59$$

Example 4 ***A Related Problem.*** Suppose we are told that the average cost per unit to produce a particular item is \$11.23, and we are interested in determining the *total* cost of producing 1500 items. Based upon the formula for the arithmetic mean, the total (ΣX) can be found by multiplying the mean ($\bar{X}$) by the total number of observations (N).

$$\Sigma X = N\bar{X}$$
$$= 1500 \times 11.23$$
$$= 16845 \text{ or } \$16,845.00$$

Find the total based upon the mean.

The total cost of producing 1500 items therefore is $16,845.00.

FINDING THE MEDIAN

Raw Data. In order to find the median of a set of raw observations, *array* the observations in order of increasing magnitude and choose the *middle* value. For an even number of observations there is no middle value, but the median can be found by taking the arithmetic mean of the *middle two* values in the array. That is, the median is found by dividing the sum of the middle two values by 2 when the number of observations is even.

Example 5 ***Median of Raw Data.*** Odd Number of Observations

Given the following *seven* observations:

13 9 16 12 7 8 3

Arrayed in order of increasing magnitude, the observations appear as follows:

3 7 8 9 12 13 16

The middle observation is 9; therefore, the median, $X_{50} = 9$.

Example 6 ***Median of Raw Data.*** Even Number of Observations

Given the following *six* observations:

20 19 17 27 24 30

Arrayed in order of increasing magnitude, we have

17 19 20 24 27 30

The middle two values are 20 and 24; the arithmetic mean of these two values is found as $(20 + 24)/2 = 22$. Therefore $X_{50} = 22$.

Ungrouped Frequency Distribution. In order to find the median for an ungrouped distribution, perform the following three steps.

Step 1. Compute $\frac{N}{2}$, the total number of observations divided by 2.

Step 2. Find values of the cumulative frequency, F, until F equals or first exceeds $\frac{N}{2}$. In other words, successively add the frequencies from the top until the sum equals the value found in Step 1 or is greater than this value for the first time.

Step 3. The median is chosen as the value of X corresponding to the value of F found in Step 2.

> *Note.* If the frequency distribution is given in terms of relative frequencies, p, the median is the value of X that corresponds to the cumulative frequency that equals or first exceeds 0.50 or 50 percent.

Example 7 ***Median of an Ungrouped Frequency Distribution.*** Given the following frequency distribution:

X	f
107.3	10
110.7	20
155.2	25
173.9	15
194.6	5
	75

The median is found by performing the following three steps:

Step 1. Divide the total number of observations by 2.

$$\frac{N}{2} = \frac{75}{2}$$
$$= 37.5$$

Step 2. Compute cumulative frequencies until they reach the value $\frac{N}{2} = 37.5$ or exceed it for the first time.

X	f	F Cumulative Frequency	
107.3	10	10	
110.7	20	30	
155.2	25	55	STOP
173.9	15		
194.6	5		
	75		

Notice, we stop cumulating frequencies when F = 55 since the preceding value of 30 is less than 37.5, and 55 exceeds it for the first time.

Step 3. Choose the median as the value of X corresponding to the last value of F found in the preceding step. The median is 155.2, or $X_{50} = 155.2$

Grouped Frequency Distribution. The following steps are required to find the median of a grouped distribution.

Step 1. Compute $\frac{N}{2}$, the total number of observations divided by 2.

Step 2. Find the value of F until F equals or first exceeds $\frac{N}{2}$. The median is in the class corresponding to this value of F, since this class contains the middle observation. Since the identity of the observations in this class is not available, the exact value of the median cannot be determined. We know it equals the lower limit of this class *plus* some quantity.

Step 3. The quantity that must be added to the lower limit of the class containing the median is found by *interpolation.* This is accomplished in the following way:

(a) Find the difference between $\frac{N}{2}$ and the value of F corresponding to the class *preceding* the one containing the median.
(b) Divide the above difference by the frequency of the class containing the median.
(c) Multiply the resulting quotient by the size of the class interval containing the median.

(d) The median is found by adding the result found in (c) to the lower limit of the class corresponding to the cumulative frequency equal to $\frac{N}{2}$.

Note. The procedure for finding the median of a grouped distribution is presented above in terms of the method of interpolation since this form can be generalized easily to other measures, called percentiles, that are similar to the median or the 50th percentile. These measures are referred to in the discussion and problem sections. There is, however, a formula that can be used to find the median which takes the following form:

$$X_{50} = L_{X_{50}} + \frac{\frac{N}{2} - F_{precede}}{f_{X_{50}}} \times c_{X_{50}}$$

The symbols used in the formula are defined as follows:

$L_{X_{50}}$ = the lower limit of the class containing the median
N = the total number of observations
$F_{precede}$ = the cumulative frequency corresponding to the class preceding the one containing the median
$f_{X_{50}}$ = the frequency of the class containing the median
$c_{X_{50}}$ = the size of the class interval containing the median

The formula leads to the same result as the interpolation method.

Example 8 ***Grouped Frequency Distribution.*** Let us find the median for the distribution of 85 mutual fund prices appearing on page 33.

Step 1. Divide the total number of observations by 2, which yields

$$\frac{N}{2} = \frac{85}{2} = 42.5$$

Hence, the median would occupy the "42.5th" position if the actual observations could be arrayed in order of increasing magnitude.

Step 2. Compute cumulative frequencies, F, until they reach the value of $\frac{N}{2} = 42.5$ or exceed it for the first time. The distribution of mutual fund prices appears below together with the calculation of the cumulative frequencies.

X Mutual Fund Price	f Number of Funds	F Cumulative Frequency	
\$0.00 but less than \$3.50	10	10	
\$3.50 but less than \$7.00	19	29	
\$7.00 but less than \$10.50	34	63	STOP
\$10.50 but less than \$14.00	13		
\$14.00 but less than \$17.50	5		
\$17.50 but less than \$21.00	4		
	85		

We stop calculating expected frequencies when F = 63 since the previous value of F equal to 29 is less than 42.5 and 63 exceeds it for the first time. Therefore, the median must fall in the interval "\$7.00 but less than \$10.50" and must equal the *lower limit* of this class *plus* a quantity computed by interpolating into the interval.

Step 3. The quantity added to the lower limit is based upon the following values:

$$\frac{N}{2} = 42.5 \text{ (from Step 1)}$$

The cumulative frequency of the class preceding the one containing the median = 29.
The frequency of the class containing the median = 34.
The class interval size for the class containing the median = 10.50 − 7.00 = 3.50. The value added to the lower limit is

$$\frac{42.5 - 29}{34} \times 3.50 = 1.39$$

This value is added to the lower limit of the class containing the median. Therefore, the median is

$$X_{50} = 7.00 + 1.39$$
$$= 8.39$$

Note. Based upon the formula for the median presented in the previous note, the median is computed as

$$X_{50} = L_{X_{50}} + \frac{\frac{N}{2} - F_{precede}}{f_{X_{50}}} c_{X_{50}}$$

$$= 7.00 + \frac{\frac{85}{2} - 29}{34} \times 3.50$$

$$= 8.39$$

FINDING THE MODE

The determination of the mode is a simple matter. When dealing with raw data, we must construct a frequency distribution of the observations. In *ungrouped* form, the mode is found as the value of X corresponding to the highest value of f. In *grouped* form, the value of the mode generally used is the midpoint of the class with the highest frequency. In cases where a distribution has the same numerical frequency for all values of X or for all classes, the distribution has *no* mode.

If two or more values of X, or classes, have frequencies that are somewhat higher than the others, more than one mode exists. In other words, if a distribution has two or more peaks, the distribution is multimodal.

DISCUSSION

The procedures for computing the most common averages were presented in the last two sections. In order to obtain a full appreciation of their usefulness, the interpretation and application of these and related measures can now be discussed in greater detail.

An average is a single number used to represent a body of data.

The Meaning of an Average. When a set of measurements are to be represented or summarized with a single number, averages are used. No matter which average we compute, its value *must* lie between the highest and lowest raw observations. Generally, an average will fall somewhere in the vicinity of the center of a frequency distribution; consequently, we may think of it as a measure of *central tendency.* Actually, the extent to which an average is representative or adequately describes central tendency is related to the amount of variability present in the data; we shall discuss this in more detail later.

One way to understand averages is to compare frequency distributions describing different kinds of data. Consider the frequency curves presented in Exhibit 1 associated with two sets of data measured on the same scale.

EXHIBIT 1
AVERAGES AS MEASURES OF CENTRAL TENDENCY

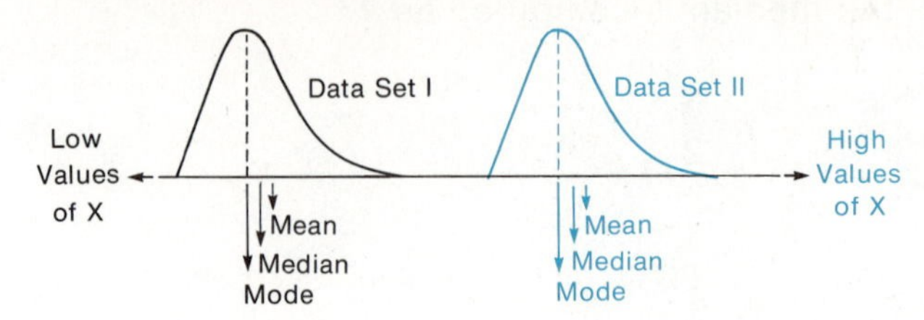

For simplicity, the distributions have been drawn with the same shape. Immediately, we can see that the observations described by distribution II lie above, or are higher in value than, those summarized by distribution I. By comparing a similarly computed average, whether it is the mean, median, or mode, we are provided with the same information: the mode of II is greater than the mode of I; the median of II is greater than the median of I; and the mean of II is greater than the mean of I. The two values of a similarly computed average bear the same relationship to each other as do the individual values of the observations corresponding to the two distributions. In other words, an average describes the position or *location* of the "bulk" of the values of the observations.

When averages are compared, the distributions should have the same basic shape.

When comparing two sets of data using averages only, the distributions should be of the same general shape. Otherwise, the meaning of the averages is unclear. It is true that averages reflect the location of the bulk of the observations; however, a complete picture regarding the nature of the observations generally is not presented with the use of an average alone. In order to shed some light on the meaning of an average in relation to the shape of the distribution, consider the frequency curves presented in Exhibit 2.

EXHIBIT 2
LOCATION OF AVERAGES RELATED TO DISTRIBUTION SHAPE

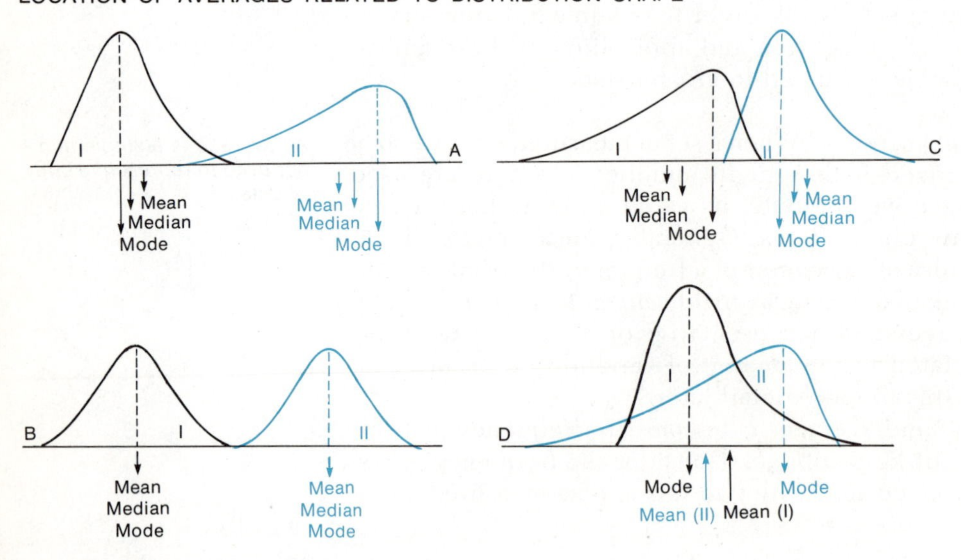

Although the situations appear different in cases A, B, and C, each of the averages corresponding to the two distributions in each case have the same relationship to one another as in the earlier case discussed. With respect to D, which is similar to A regarding distribution shape, we can see the mode gives the greatest indication of the relative positions of the bulk of the observations corresponding to I and II. If the means are examined, however, they could provide a misleading impression about the location of the bulk of the observations, due to the degree of overlap of the two distributions, the relative amounts of variability, and the direction and magnitude of skewness; in other words, the mean of II is shown to be less than the mean of I, whereas the opposite relationship is true of the mode. Whether this occurs in other cases depends upon the actual data used in a particular problem.

For symmetric distributions the mean, median, and mode are equal.

After examining the diagrams, we can see that the relationship among the three averages for any distribution is related to skewness. In the case of a symmetric distribution (B), the mean, median, and mode have the same numerical value; for skewed distributions (A, C, and D), however, the three measures are not the same. The mode, which is the value of X lying under the peak of the distribution, is not influenced by extreme observations and is not affected by skewness. The median, which is based upon position but not all the data, is influenced by extremes, or skewness, to some extent. As a result, the median is pulled somewhat in the direction of the tail of the distribution. The mean, which is based upon all the observations, is influenced most by the extremes, and is pulled more in the direction of skewness than the median.

The mode is not affected by skewness, whereas the mean is affected the most by skewness.

Using Averages to Describe Measurements. Because each average is defined differently, each can be used under different circumstances. In some cases they may apply to the same set of data, whereas in others there may be a question regarding which is the appropriate measure to use. Obviously, when a set of data is distributed symmetrically, it does not matter which average is used as a *descriptive* measure, since they all possess the same numerical value. When using averages in sampling problems or inference, the distinction among the averages is important, even in the case of symmetric distributions.

When dealing with measurements in practice, the mode is rarely used. In fact, it may not even exist, especially in small data sets. When working with skewed distributions the median is generally preferred, since it lies closer to the point of maximum concentration than the mean. A good illustration of this is found in the use of the median by the Bureau of the Census when summarizing income information; median income is the reported average since income distributions are typically skewed to the right. Another advantage of the median is that it can be computed for frequency distributions with open-ended intervals, as long as it does not fall in an open class. Unless additional information about the values in the open-ended intervals accompanies a distribution, the same is not true of the mean; either the actual observations or the arithmetic mean of the observations in the open-ended interval must be available in order to compute the mean of the entire distribution.

The most commonly used average in statistical work is the arithmetic mean. This is due more to its current popularity and to certain mathematical properties that it possesses than to its ability to describe a set of observations. The implications of this should become clear when we reach statistical inference. One property of the mean worth noting is that we can *combine* the means of several sets of data to obtain the overall mean of the data if considered as one group. In other words, it is not necessary to know the values of the original observations of each set of data in order to compute the overall mean. In the case of the mode and median we must work directly with the original data. The procedure for finding the combined mean is precisely the same as the one used to find the mean of an ungrouped frequency distribution, where the total number of observations of each group represents the frequencies, and the means of each group correspond to the individual values of the characteristic measured.*

Frequencies represent weights that reflect the importance of values in determining the mean.

The Weighted Arithmetic Mean. Special names occasionally are used to distinguish the formula for the mean of raw data from the one used for a frequency distribution, whether it is in ungrouped or grouped form. In the case of raw data, the formula is referred to as the *unweighted arithmetic mean;* the one corresponding to frequency distributions, therefore, is called the *weighted arithmetic mean.* Since the difference in the formulas lies in the frequencies, the frequencies can be thought of as weights, since they reflect the impact or weight that a particular value has upon the value of the mean. In other words, frequencies represent the number of times observations occur in the original set of raw data. Although an obvious inconsistency exists, the term *weighted average* is popularly used in place of the phrase "weighted arithmetic mean."

Aside from the terminology, it is important to understand the role frequencies play in determining the value of the mean. Suppose we take 60 measurements corresponding to some characteristic and find one of the measurements assumes a value of 1.0, five a value of 2.0, eleven a value of 3.0, thirteen a value of 4.0, and thirty a value of 5.0. If we ignore the number of times each value occurs and compute the mean, we would obtain the unweighted mean as

$$\text{Unweighted Mean} = \frac{\Sigma X}{N}$$

$$= \frac{1.0 + 2.0 + 3.0 + 4.0 + 5.0}{5}$$

$$= 3.0$$

which is *not* correct. The *correct* answer is obtained as the weighted

*The formula for the combined mean in the case of two groups is

$$\bar{X}\text{ Combined} = \frac{N_1\bar{X}_1 + N_2\bar{X}_2}{N_1 + N_2}$$

where the subscripts denote the individual groups. The formula generalizes easily to more than two groups.

mean computed as

$$\text{Weighted Mean} = \frac{\Sigma fX}{N} \text{ or } \frac{\Sigma fX}{\Sigma f}$$

$$= \frac{(1 \times 1.0) + (5 \times 2.0) + (11 \times 3.0) + (13 \times 4.0) + (30 \times 5.0)}{1 + 5 + 11 + 13 + 30}$$

$$= 4.1$$

In order to explore the idea of weights further, consider Exhibit 3 where we compare four cases having different values of the frequencies; all of the cases are based upon the same number of observations and values of X. In the first case (1), all the frequencies are equal, and the mean has the same value (i.e., 3.0) as the unweighted case computed earlier. Whenever all frequencies are identical, the weighted and unweighted means have the same numerical value; conceptually, the two situations are alike since they correspond to distributions that are uniform.

The distribution given as (2) also has a mean of 3.0 because it is symmetric. Although this was true of the first case, case (2) is different with respect to the way the observations are distributed about the center of the distribution. The distribution "tails-off" to lower values on both sides of the peak; the pattern of variability is different.

In the last two cases, (3) and (4), the distributions are skewed but in opposite directions. The means are not equal to 3.0, but lie close to the points of highest frequency, and are influenced by the lower frequencies representing the extremes; consequently the means are pulled away from the modes in the direction of skewness. Notice in (3) and (4) that the modes are 1.0 and 5.0, respectively.

EXHIBIT 3
MEAN VALUES CORRESPONDING TO VARYING WEIGHTS OR FREQUENCIES

Case	*(1)*	*(2)*	*(3)*	*(4)*
X	f_1	f_2	f_3	f_4
1.0	12	3	30	1
2.0	12	12	13	5
3.0	12	30	11	11
4.0	12	12	5	13
5.0	12	3	1	30
Total	60	60	60	60
MEAN	3.0	3.0	1.9	4.1

The Median and Other Percentiles. Early in the chapter the median was presented as a value with 50 percent of the observations less than and 50 percent greater than it. As such, we can designate the median as the *fiftieth percentile*, reflected by the symbol X_{50}. Just as we define the median as the fiftieth percentile, we can define percentiles corresponding to other percentages. For example, there is the ninetieth percentile (X_{90}), below which 90 percent and above which 10 percent of the observations fall. Another term for the ninetieth percentile is the ninth decile, which views the data in terms of ten groupings, each containing 10 percent of the observations. Another commonly used percentile is the seventy-fifth (X_{75}), the value below which 75 and above which 25 percent of the observations fall. Another name for this is the third quartile. Similarly, terms such as the fifth decile or the second quartile can be used for the median.

A percentile is a number such that a certain percentage of the observations lies below it and the remaining observations lie above it.

The procedures for locating percentiles are similar to those used to find the median, but with a modification for the percentage used. For example, the median corresponds to 50 percent, so the position of the median of a frequency distribution is $\frac{1}{2}N$ or $\frac{N}{2}$. Hence, the 75th percentile corresponds to 75 percent, so the position is found as $\frac{3}{4}N$. When finding any percentile, the general procedure for the median can be used with an appropriate modification for the position of the percentile. Those interested in working with percentiles may refer to the last part of the problem section in order to develop an understanding of these measures.

Percentiles are used selectively in practice. The most familiar use is in establishing student position or class standing. Another application of percentiles is to compare two frequency distributions that are markedly different in shape. For example, if two distributions have different amounts of variability and are skewed in opposite directions, single values in the form of an average may be inappropriate. Instead, the first and third quartile, or the first and ninth decile, may describe them more meaningfully.

Conclusions Drawn From Averages. Due to the common acceptance of averages and the seeming ease with which they are understood, averages alone are often quoted in order to convey an impression or transmit a message. One should not, however, blindly accept them at face value, since unqualified averages have little meaning as descriptive measures. The simplest question to ask when confronted by an average is "which average?" Since we now know different averages can assume different values, it is important to know which of the averages is being quoted; in the case of a comparison, obviously similarly computed averages must be used. In addition, it is important to know something about the shape of the underlying frequency distribution. This is of special importance when averages corresponding to two bodies of data are compared, since the meaning of the comparison is questionable unless the underlying distributions are similar in shape.

Averages can be misleading and are not always used correctly.

Sometimes averages are presented with an implied conclusion that is really never stated. Often, the conclusion may be misleading, since it may relate to the wrong problem. For example, if we are told that on the average a particular disinfectant kills more germs than all leading brands, we are led to believe that the disinfectant is of superior quality. The real question, however, is whether the number of germs killed is sufficient to provide adequate protection.

Frequently, we read statements about "the average consumer" or "the average driver." The word average is used rather loosely, with no information regarding who it is that is being considered. What specific characteristics are considered when speaking of the average consumer? Is an average driver one who drives at a particular speed, or one who gets five parking tickets a year! In many cases, the word average is used to convey an idea without specific information to support it.

In some instances, averages can be used when they shouldn't. An amusing story related by the actor Anthony Quinn in his book *The Original Sin* illustrates this point. When he was somewhat younger, he and his friends were interested in ways to make money. They decided that since their talents concentrated on dancing, they would enter a series of dance contests in order to win the prize money. Unfortunately they each needed a pair of decent shoes but had sufficient funds for only one pair. As a decision problem, a question arises regarding the choice of a representative shoe size to serve everyone's need. Stated another way, would an average of some kind solve the problem? Since only the largest shoe size of the group would accommodate the others, a representative measure in the form of an average is not appropriate. Incidentally, the largest shoe size of the group represents the "100th" percentile.

When we speak of an average as a representative measure or a measure of central tendency, it is important to consider the amount of variability about the average. If the observations differ greatly, there is little meaning in talking about a representative value about which the observations tend to cluster. For example, two localities may exhibit similar average daily temperatures for an entire year, but experience entirely different climates during different seasons. If in one region summers are excessively hot and winters are extremely cold, the average temperature could be similar to the average of another region with a moderate climate all year-round. The first region experiences extreme temperatures resulting in much greater variability than the second. Variability, in addition to central tendency, is essential to consider in order to provide an adequate assessment of the climates in the two regions. In general, it is good practice to report the amount of variability in addition to any average value when using summary or descriptive measures to describe a body of data. The next chapter is devoted to the concept of variability in detail.

When using averages it is important to consider the pattern of variability.

EASING COMPUTATIONS

Although computation of averages does not present any real difficulties, savings in time and effort can be realized by using various calculating devices. A number of pocket calculators are pre-programmed to compute the arithmetic mean and possess a separate mean key. Computer programs are available to compute all of the averages. These generally accompany programs for constructing grouped distributions and computing other descriptive measures. Furthermore, shortcut formulas for computing the arithmetic mean are available that are helpful when hand computation is necessary. Of the shortcuts available, the one associated with grouped frequency distributions with equal class intervals is the most useful.

It should be noted that in cases where the raw data are available, we should compute the mean directly rather than grouping and then computing the mean, since the raw data provide the exact value. Only when a grouped distribution is provided from a secondary source, without the raw data, is it necessary to compute the mean from the distribution. If, however, one wishes to present a summary of the raw data with a grouped distribution, a partial *check* on the ability of the distribution to describe the data can be obtained by computing the mean (as well as other measures) for both the raw data and the grouped distribution and then comparing the results.

Shortcut for the Mean-Grouped Distribution With Equal Intervals. For grouped distributions in the case of equal intervals only, the following shortcut formula can be used to compute the arithmetic mean.

$$\bar{X} = m_a + \frac{\Sigma fd}{N} \times c$$

Mean for grouped frequency distribution with equal intervals.

m_a is the midpoint of a class interval chosen arbitrarily; d represents the number of class intervals that each interval is from the arbitrarily chosen interval; and c represents the size of the class intervals. By definition, the arbitrarily chosen interval is always assigned a value of d equal to zero.

In order to illustrate the use of the formula, consider the following distribution of mutual fund prices.

X *Mutual Fund Price*	f *No. of Funds*	d	fd
\$0.00 but less than \$3.50	10	−2	−20
\$3.50 but less than \$7.00	19	−1	−19
\$7.00 but less than \$10.50	34	0	0
\$10.50 but less than \$14.00	13	1	13
\$14.00 but less than \$17.50	5	2	10
\$17.50 but less than \$21.00	4	3	12
Total	85		−4

If the third class interval is arbitrarily chosen, $m_a = 8.75$. It happens that the mean of the distribution falls in the third interval. Any other interval could have been chosen, however, and the values of d would differ correspondingly. After placing a value of d equal to zero alongside the "chosen" interval, the remaining values are assigned as successive integers. The values above zero are assigned a minus sign and those below a plus sign. Based upon the information given, we find the mean by substituting into the above formula as

$$\begin{aligned}\overline{X} &= m_a + \frac{\Sigma fd}{N} \times c \\ &= 8.75 + \frac{(-4)}{85} \times 3.50 \\ &= 8.59\end{aligned}$$

where the size of the class interval is 3.50. The result is the same as the one obtained in Example 3 using the longer procedure.

QUESTIONS AND PROBLEMS

DEFINITIONAL QUESTIONS

1. What is the meaning of the following terms?

   (a) Location
   (b) Central tendency
   (c) Representative

2. Without using formulas, define the following terms.

(a)	Average	(e)	Multi-modal
(b)	Mean	(f)	Unweighted mean
(c)	Median	(g)	Weighted mean
(d)	Mode	(h)	Weighted average

3. What is the meaning of the following terms?

(a)	Percentile	(e)	Twenty-fifth percentile
(b)	Fiftieth percentile	(f)	Seventy-fifth percentile
(c)	Tenth percentile	(g)	Quartile
(d)	Ninetieth percentile	(h)	Decile

COMPUTATIONAL PROBLEMS

4. Find the mean, median, and mode of the following 20 values.

7	3	9	7	12	19	7	3	3	9
3	7	7	12	3	7	9	7	7	7

5. The following figures represent the number of packs of cigarettes smoked by 15 adults during a period of one year.

182	1000	330	215	440
200	435	746	519	943
325	815	250	400	354

Find the arithmetic mean and median number of packs smoked per adult for the year. Does a mode for the data exist?

6. For the frequency distribution given below, find the arithmetic mean, median, and mode.

X	f
3.0	5
7.0	9
9.0	3
12.0	2
19.0	1
Total	20

7. Appearing below is the frequency distribution of the number of parcels delivered on a given day by the 64 trucks of a delivery service company.

Number of Parcels	*Number of Trucks*
20	10
25	12
35	23
40	9
55	7
70	3
Total	64

Find the mean, median, and modal number of parcels delivered per truck on that day.

8. Find the mean, median, and mode of the following frequency distribution.

X	f
100–199	3
200–299	5
300–399	8
400–499	15
500–599	7
600–699	2
Total	40

9. The distribution of examination grades for a class of 25 students appears below.

Examination Grade	*Number of Students*
40 but less than 50	1
50 but less than 60	7
60 but less than 70	4
70 but less than 80	8
80 but less than 90	3
90 but less than 100	2
Total	25

Find the arithmetic mean, median, and mode of the given grade distribution.

10. For the following frequency distribution given in terms of relative frequencies, find the mean, median, and mode.

X	p
1.0	.10
2.0	.15
4.0	.10
5.0	.15
7.0	.30
10.0	.20
Total	1.00

11. Find the mean, median, and mode of the ages of readers described by the following frequency distribution.

Age in Years	*Proportion of Readers*
20 but less than 30	.09
30 but less than 40	.44
40 but less than 50	.20
50 but less than 60	.15
60 and over*	.12
Total	1.00

*The mean age of the readers in this class equals 68

12. If the average daily oil consumption of the nation is 3 million barrels, what is the total amount of yearly consumption?

13. The mean exam grades for two classes of students are 82 and 74. If the first class has 40 and the second has 27 students, what is the arithmetic mean grade of the two classes combined?

CONCEPTUAL QUESTIONS AND PROBLEMS

14. What information about a body of data is provided by an average? How are averages useful as a descriptive measure?

15. How is an average considered as a representative measure or a measure of central tendency? Explain. Is the extent to which an average is representative the same for all bodies of data? How is the ability of an average to measure central tendency related to other characteristics of data?

16. It has been said that the less variability that exists, the more an average is representative of a set of data. Comment on the meaning of this statement.

17. What are the advantages and disadvantages of the three common averages: the mean, the median, and the mode?

18. It was stated in the chapter that the weighted arithmetic mean is commonly referred to as a "weighted average." How is the use of this phrase inconsistent with the definition of an average?

19. What is the relationship between the mean, median, and mode? Under what circumstances are they equal?

20. Based upon reasoning underlying the computation of the median, find the twenty-fifth and seventy-fifth percentiles for the values presented in Problem 4. Interpret your results.

21. Using the same kind of reasoning to find the median, find the values of the tenth and ninetieth percentiles for the cigarette data presented in Problem 5. What is the meaning of these values?

22. Find the eightieth percentile for the distribution of parcels delivered presented in Problem 7.

23. Why is it necessary to interpolate in order to find the median of a grouped frequency distribution?

24. Based upon the procedure used to find the median, find the ninth decile for the grade distribution appearing in Problem 9 and interpret the result. *Hint:* The ninth decile corresponds

to the value that would occupy the position $\frac{9N}{10}$ in the array of observations if the array were available.

25. It has been said that the same percentage of frequencies falls between the first and ninth decile for symmetric and skewed distributions. Criticize or explain this statement. Generalize your answer to other percentiles.

26. For the situations listed below, indicate whether an average can be used to aid in solving the problem posed. If an average is useful, specify the kind of average, the quantity measured, and any other characteristics that may be helpful. If an average is not appropriate, indicate what other measures could be used.
    (a) A clothing manufacturer needs to determine a method for establishing sizes of jeans to be produced.
    (b) The height of a bridge to be built over a large river must be determined to accommodate ships passing under it.
    (c) A tourist bureau needs to issue a report regarding seasonal climatic conditions in a particular resort area.
    (d) A testing bureau needs to make a report to consumers regarding the durability of various brands of auto tires.
    (e) Wages of workers of a firm must be compared with those of the industry to determine whether the wages paid by the firm are fair.
    (f) A dam is to be built to protect a small town from potential flooding conditions.
    (g) The total cost of producing various quantities of a manufactured item is to be established periodically.

CHAPTER 3

MEASURING DISPERSION

Variability About the Average

There never was in the world two opinions alike, no more than two hairs or two grains; the most universal quality is diversity.

Michel De Montaigne

I feel like a fugitive from th' law of averages.

Bill Mauldin

The importance of the concept of variability was discussed in the first two chapters. Although most of us were familiar with averages before reading the last chapter, measurement of variability is not as common. Just as central tendency can be measured by a number in the form of an average, the amount of variation, or differences among observations in a set of data, can be measured.

In order to understand what is meant by the phrase "measuring variability," consider the two sets of numbers appearing below.

Set 1						Set 2				
7	8	9	10	11		3	6	9	12	15

Both sets of numbers have the same mean and median, equal to 9; however, the differences among the values are greater in Set 2. An easy way to see this is to compare the two sets of values on a scale as follows:

	3	4	5	6	7	8	9	10	11	12	13	14	15
Set 1					☆	☆	☆	☆	☆				
Set 2	☆			☆			☆			☆			☆

The asterisks represent the position of the values for the two sets of numbers. We can see that the values in Set 2 are spread more than those in Set 1. In other words, there is more variability or dispersion in Set 2.

The simplest way to measure the variability described above is with the range. We used the range in Chapter 1 to find the size of class intervals for a grouped frequency distribution. Recall, the **range** is defined as the difference between the largest and the smallest observations. For the two sets of values given earlier, the range is found as

Set 1	Set 2
Range = 11 − 7	Range = 15 − 3
= 4	= 12

The differences in the amount of variability between the two sets of numbers is summarized by the calculated ranges: the range of 12 for Set 2 is greater than the range of 4 for Set 1. By calculating the range, we have described the variability in terms of a single summary or descriptive measure.

Although the range is useful in particular problems, it is limited in its ability to describe variability since it is based solely upon the extreme observations and does not consider the pattern of variability within these extremes. This point can best be understood by comparing two more sets of values. This is done in the following diagram.

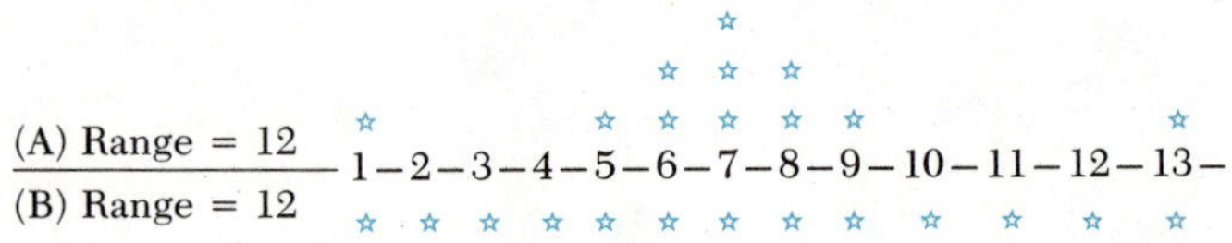

```
                                      *
                                   *  *  *
                    *           *  *  *  *  *           *
(A) Range = 12
------------------- 1–2–3–4–5–6–7–8–9–10–11–12–13–
(B) Range = 12
                    *  *  *  *  *  *  *  *  *  *  *  *  *
```

Because both sets of values described in the diagram have the same extremes, their ranges are identical. It is obvious, however, that the pattern of variability in the points above the scale (A) is much different from those below (B). The points above the scale cluster between the values of 5 and 9, while those below are evenly distributed between 1 and 13. The differences in these patterns of variability are not considered by the range.

Other measures beside the range are available. Of these, the most important measures are the *variance* and the *standard deviation*. These measures consider each value in the set of data and have many properties that are useful for solving problems in statistics. Both are based upon *deviations* or differences between the individual observations and the arithmetic mean. These deviations are squared and the average of the squares is defined as the variance. The standard deviation is defined as the positive square root of the variance.

The variance and the standard deviation are appropriate measures to use when the mean is employed as the measure of central tendency. These measures gain their importance because of their relation to the mean and because of certain properties not associated with other measures of variation. In addition, they enjoy a unique role in problems of statistical inference where samples are used to draw conclusions about universe characteristics. Since the variance and standard deviation are the most widely used measures of variability, we shall deal with them exclusively in the procedure and example sections. After exploring them in greater detail in the discussion section, other measures of variability are considered.

The symbols to be used in this chapter are:

X = a numerical value of the quantity measured
f = frequency of occurrence
N = the total number of observations
$\overline{X}$ = the arithmetic mean of a set of observations
S^2 = the variance of a set of observations
S = the standard deviation of a set of observations
m = the midpoint of a class interval
R = the range of a set of observations
X_{50} = the median of a set of observations
MD = the mean deviation of a set of observations
V^2 = the rel-variance
V = the coefficient of variation
c = the size of the class interval
m_a = the midpoint of an arbitrarily chosen class interval
d = the number of class intervals that a given interval is from an arbitrarily chosen interval
Σ = summation (read as Sigma)

COMPUTING THE VARIANCE AND STANDARD DEVIATION

The formulas for finding the variance and standard deviation are presented separately in this section for raw data and for ungrouped and grouped frequency distributions. These formulas represent the definitional form and provide some insight into the underlying meaning of the variance. Shortcut expressions that save time and effort appear later in the computation section at the end of the chapter. Since three separate cases are presented, the basic format is slightly different. Each formula is illustrated immediately after it is presented. Consequently, there is no separate example section as in most of the other chapters.

Raw Data. The variance of a set of raw observations is defined by the formula

$$S^2 = \frac{\Sigma(X - \overline{X})^2}{N}$$

Variance of raw data.

In order to find the variance the formula tells us to (a) subtract the mean from each observation, (b) square each of these differences or deviations from the mean, (c) sum them, and (d) divide by the total number of observations. Before applying the formula, it is necessary to compute the arithmetic mean.

The standard deviation is found by the expression

$$S = \sqrt{S^2}$$ *Standard deviation.*

which tells us to take the square root of the value found as the variance.

Example 1 ***Raw Data.*** Given four observations (N = 4): 3, 7, 5, and 9. In order to compute the variance and standard deviation, first find the mean.

$$\bar{X} = \frac{\Sigma X}{N} = \frac{3 + 7 + 5 + 9}{4} = \frac{24}{4} = 6$$

Then, the variance is found by substituting into the given formula as follows:

$$S^2 = \frac{\Sigma(X - \bar{X})^2}{N}$$

$$= \frac{(3-6)^2 + (7-6)^2 + (5-6)^2 + (9-6)^2}{4}$$

$$= \frac{(-3)^2 + (1)^2 + (-1)^2 + (3)^2}{4}$$

$$= \frac{9 + 1 + 1 + 9}{4}$$

$$S^2 = \frac{20}{4} = 5$$

The variance is equal to 5. The standard deviation is found as the square root of the variance as follows:

$$S = \sqrt{S^2} = \sqrt{5}$$

$$S = 2.236068 \text{ or } 2.24 \text{ (approximate)}$$

The standard deviation, rounded to two decimal places, is 2.24.

Note. The simplest way to find a square root is by using the "$\sqrt{\ }$ key" that appears on many pocket and desk calculators. If this is not available, **Table 12** in the Appendix can be used. To find $\sqrt{5}$ (see above), look up 500 in Table 12 in the "n" column and find the square root in the $\sqrt{n}$ column by moving the decimal point of the result one place to the left. A complete explanation of the use of the table is provided in the Appendix along with the table.

Example 2 Consider another set of raw data representing the exam grades of 10 students with a mean grade of 72.4.

61 55 87 83 75
73 73 99 61 57

It is convenient to organize the calculations for the variance in the form of a table instead of substituting directly into the formula as in Example 1. This is done as follows:

PRELIMINARY CALCULATIONS FOR THE VARIANCE OF RAW DATA

X *Exam Grade*	X − 72.4	(X − 72.4)²
61	−11.4	129.96
55	−17.4	302.76
87	14.6	213.16
83	10.6	112.36
75	2.6	6.76
73	.6	.36
73	.6	.36
99	26.6	707.56
61	−11.4	129.96
57	−15.4	237.16
Total		1840.40

The total (1840.40) appearing in the last column is the sum of the squared deviations about the mean (72.4) appearing in the numerator of the formula for the variance. Hence, the variance of the 10 grades is found as

$$S^2 = \frac{\Sigma(X - \overline{X})^2}{N}$$

$$= \frac{1840.40}{10}$$

$$= 184.04$$

The variance of the grades is 184.04. The standard deviation is equal to the square root and is found as

$$S = \sqrt{184.04}$$

$$= 13.57 \text{ (approximate)}$$

Ungrouped Frequency Distribution. The formulas for the variance and standard deviation of an ungrouped distribution are similar to those for raw data, except each "squared deviation" must be weighted or multiplied by the corresponding frequency of occurrence. The formula for the variance is

$$S^2 = \frac{\Sigma f(X - \overline{X})^2}{N} \quad or \quad \frac{\Sigma f(X - \overline{X})^2}{\Sigma f}$$

Variance of ungrouped frequency distribution.

Notice each squared deviation, $(X - \overline{X})^2$, is multiplied by the frequency f before summing. The sum is divided by the total number of observations to obtain the variance. The two expressions given yield the same result and are presented only to indicate that the total number of observations can be expressed in two ways, N or Σf.

The standard deviation of an ungrouped distribution also is found by taking the square root of the variance. That is,

$$S = \sqrt{S^2}$$

Standard deviation.

Example 3 ***Ungrouped Frequency Distribution.*** Given the following frequency distribution

X	f
10	2
12	5
13	12
17	8
20	3
Total	30

It is simpler to find the variance and standard deviation of a frequency distribution with the aid of a table of preliminary calculations.

1. Find the Mean

X	f	fX
10	2	20
12	5	60
13	12	156
17	8	136
20	3	60
Totals	30	432

2. Preliminary Calculations for the Variance

(X − 14.4)	(X − 14.4)²	f(X − 14.4)²
−4.4	19.36	38.72
−2.4	5.76	28.80
−1.4	1.96	23.52
2.6	6.76	54.08
5.6	31.36	94.08
		239.20

3. Use Total of Last Column to Find the Variance

$$\overline{X} = \frac{\Sigma fX}{\Sigma f} = \frac{432}{30} = 14.4$$

$$S^2 = \frac{\Sigma f(X - \overline{X})^2}{\Sigma f} = \frac{239.20}{30} = 7.97 \text{ (rounded)}$$

The variance of the ungrouped distribution is 7.97. The standard deviation is found as

$$S = \sqrt{7.97} = 2.82 \text{ (approximate)}$$

Grouped Frequency Distribution. For a grouped distribution, the same expression as in the ungrouped case is used, but substitute the mid-point of a class interval in place of the value of X. Hence, the variance for a grouped distribution is given as

$$S^2 = \frac{\Sigma f(m - \overline{X})^2}{N}$$

or

$$\frac{\Sigma f(m - \overline{X})^2}{\Sigma f}$$

Variance of grouped frequency distribution.

where m is the class midpoint. The standard deviation is found as the square root.

$$S = \sqrt{S^2}$$

Standard deviation.

Example 4. ***Grouped Frequency Distribution.*** Consider the distribution of mutual fund prices presented in the previous two chapters with *mean* equal to 8.59. The original distribution together with the intermediate calculations necessary to find the variance is given in tabular form as

X *Mutual Fund Price*	f *No. of Funds*	m *Midpoint*	(m − 8.59)	$(m - 8.59)^2$	$f(m - 8.59)^2$
\$0.00 but less than \$3.50	10	1.75	−6.84	46.7856	467.856
\$3.50 but less than \$7.00	19	5.25	−3.34	11.1556	211.956
\$7.00 but less than \$10.50	34	8.75	0.16	0.0256	0.870
\$10.50 but less than \$14.00	13	12.25	3.66	13.3956	174.143
\$14.00 but less than \$17.50	5	15.75	7.16	51.2656	256.328
\$17.50 but less than \$21.00	4	19.25	10.66	113.6356	454.542
Total	85				1565.695

$$S^2 = \frac{\Sigma f(m - \overline{X})^2}{\Sigma f}$$

$$= \frac{1565.695}{85}$$

$$S^2 = 18.42$$

$$S = \sqrt{18.42} = 4.29 \text{ (approximately)}$$

The variance and standard deviation for the distribution of mutual fund prices are \$18.42 and \$4.29. In this example, the mean was given. If not, it would be found using the procedure outlined in Chapter 1.

DISCUSSION

Any measure of variability should assume large values when observations are highly dispersed and should equal zero when all the observations have the same value.

It is difficult to gain a full appreciation of the meaning and usefulness of the variance and standard deviation by applying the formulas. Some insight into the meaning of the variance and standard deviation is provided at this point. A clearer view is presented in Part Three.

Any measure of variability must assume large values for a set of observations that is highly dispersed; at the other extreme, when a set of observations possesses the same numerical values, a measure of variability should equal zero. By inspecting the formulas for the variance, we can see that the more dispersed the observations are from one another, the larger will be the individual differences of the observations from the mean. As a result, the square of the deviations from the mean will be larger, and so will the variance. When all observations are identical, the variance is zero since all deviations from the mean and the corresponding squared values are zero.

Consider the graphs of frequency distributions for three sets of data with the same mean of six appearing in Exhibit 1. By examining the three diagrams we obtain an impression about the variance as a descriptive measure. The first case deals with data where all fifteen observations are identical. When graphed, the data plot in the form of a frequency distribution as a single point, which indicates there is no variation. Consequently, the variance is zero.

The second case also deals with fifteen observations. Although the most frequently occurring value remains at six, some observations are different. This results in a frequency distribution which is somewhat peaked and narrow, indicating a concentration of observations about six. The corresponding variance equals one.

As the observations become more dispersed or differ more from six, we see that the distribution that is described in the third diagram is less peaked, flatter, and wider or more spread-out than the previous case. The variance, equal to eight, is greater than the variance of the observations in the second situation. In general, when one set of observations is more dispersed than another, evidenced by a wider frequency distribution, the variance will be greater; the wider the distri-

EXHIBIT 1
COMPARISON OF THE VARIANCE FOR DIFFERENT PATTERNS OF VARIABILITY

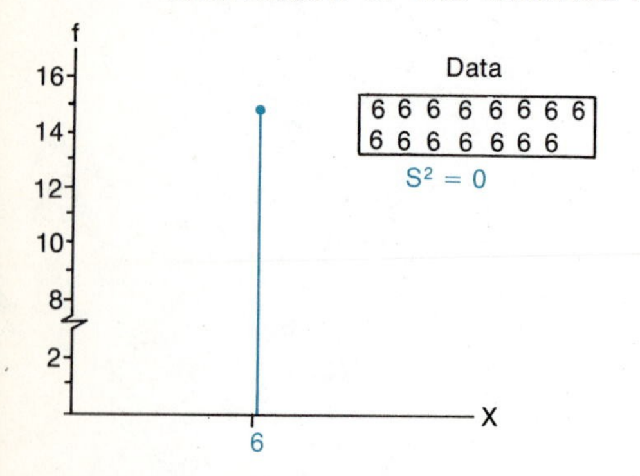

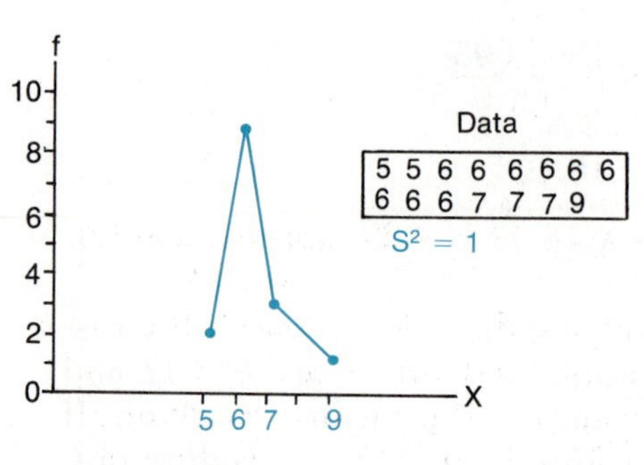

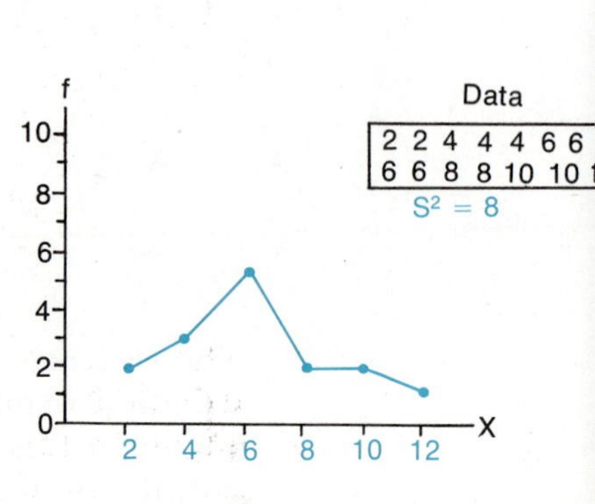

bution, the less concentrated are the observations about some centrally located value.

Recognize that the variance is interpreted on a comparative basis; we speak of *wider, more* dispersed, *less* concentrated when attaching a meaning to the variance. Without a basis for comparison, or a standard, we cannot speak of a variance as large or as small in an absolute sense, but only in terms of the way one computed value compares with another.

Interpreting variances and standard deviations must be done on a comparative basis.

Although our discussion has focused on the variance, it should be noted that as a descriptive measure the same interpretation can be given to the standard deviation: when one variance is greater than another variance, the corresponding standard deviation will be greater. The variance assumes squared units, whereas the standard deviation is expressed in units of the original observations. For descriptive purposes it does not matter which of the two measures we use, since they both impart similar information. When making comparisons between two sets of data, the corresponding measure must be in the same units. It is not meaningful to compare the variance of one set of data with the standard deviation of another set.

Other Measures. As previously mentioned, there are other measures of variability beside the variance and standard deviation. The simplest measure is the range, which was introduced at the beginning of the chapter. For descriptive purposes, the range can be used like any other measure, and can be compared with the range of another set of data to determine which is more variable. Remember, however, the range does not consider the pattern of variability within the extremes. Consequently, many sets of data could possess the same value of the range and have markedly different patterns of values within the extremes. Furthermore, when sample data are used, the range is affected by the number of observations. Therefore comparisons of two samples are limited to cases where the number of observations are equal. The chief advantage of the range is that it is the easiest measure to compute and is easily understood.

There are other measures of variability besides the variance and standard deviation.

Another measure that does consider all of the observations is the mean deviation. The mean deviation, designated by the symbol MD, can be computed about any average but is generally computed in terms of the median. The formula for the mean deviation about the median is

$$MD = \frac{\Sigma|X - X_{50}|}{N}$$

Mean deviation about the median.

This formula tells us that the mean deviation is the average (mean) of the *absolute* deviations of the observations about the median. In order to find an absolute deviation, compute the differences between the observations and the median, always considering the sign to be *positive.* This is indicated by the vertical lines in the formula which represent an "absolute value." As an example, consider the mean deviation of the following five values.

2 5 7 10 12

The median for this set of values is 7. Therefore,

$$MD = \frac{|2-7| + |5-7| + |7-7| + |10-7| + |12-7|}{5}$$

$$= \frac{|-5| + |-2| + |0| + |3| + |5|}{5}$$

$$= \frac{5 + 2 + 0 + 3 + 5}{5}$$

$$= \frac{15}{5} = 3$$

The mean deviation can be used like any other measure of variability when making comparisons among different bodies of data. It can be used to determine whether there is more variability about the average in one set of data relative to another. The mean deviation generally is used when the median is the appropriate measure of central tendency since it assumes a smaller value about the median than about other averages. The same property applies to the variance and standard deviation computed about the arithmetic mean, rather than other averages. Actually, a complete understanding of the properties of various measures of variability is complex and not appropriate for this text. The reason for presenting the mean deviation here is for reference purposes since it is not to be used later. Due to their importance, the variance and standard deviation are used in the remaining parts of the text.

Relative measures of variability are used to compare data in different units and to reflect the importance of the variability in data.

Relative Measures. When describing variability in data, there are times when we are interested in comparing two sets of data expressed in different units. In such cases, a comparison of absolute measures of variability would have little meaning unless there was a common basis for comparison. For example, we may be interested in determining whether employment or sales among firms in a particular industry exhibit greater variability. In other situations absolute measures of variation do not reflect the importance of the actual variability described. An obvious example is associated with stock prices. Equivalent price increases of two stocks would have different meanings, depending upon the basic price of each stock. As a result, there is a need to employ a measure of *relative variability.*

All measures of relative variability are expressed as the ratio of an absolute measure of variability to some form of average. When the variance is used, the appropriate relative measure is the **rel-variance.** This is defined as the variance divided by the square of the arithmetic mean. That is,

$$V^2 = \frac{S^2}{\bar{X}^2}$$

Formula for the rel-variance.

Closely related to the rel-variance is the **coefficient of variation.** This is defined as the ratio of the standard deviation to the mean, or the square root of the rel-variance. Hence, we have

$$V = \frac{S}{\bar{X}}$$

or

$$\sqrt{V^2}$$

Formula for the coefficient of variation

Both formulas for the coefficient of variation yield the same numerical answer.

In order to illustrate the formulas for V^2 and V, consider the mutual fund price distribution with a mean of $8.59 and variance of 18.42. The rel-variance is found as

$$V^2 = \frac{S^2}{\bar{X}^2}$$

$$= \frac{18.42}{(8.59)^2}$$

$$= \frac{18.42}{73.79}$$

$$= 0.25$$

Hence, the rel-variance equals 0.25. This means the variance of mutual fund prices is 0.25 or 25 percent of the square of the mean. The coefficient of variation equals the square root of 0.25, or 0.5.

When the mean deviation is computed, the relative measure is the ratio of the mean deviation to the median. In cases where the range is computed, a number of relative measures are appropriate. Sometimes the ratio of the range to the median is used, while in other cases the ratio of the range to the midrange is calculated. The midrange is the value midway between the largest and the smallest observations.

EASING COMPUTATIONS

As in the case for means, time and effort can be saved by using calculating devices to compute the variance and standard deviation. Some hand-held and desk-top calculators are preprogrammed to compute the standard deviation in addition to the mean. A special key sometimes is provided. Furthermore, the variance and standard deviation are included in computer programs that are used to compute averages and to construct frequency distributions.

The formulas for the variance presented in the body of the chapter are definitional forms; they represent the way the variance is defined. They were presented to provide understanding about the meaning of the variance. There exist, however, simpler ways to compute the variance, whether one does it "long-hand" or with the aid of a computing device. Whenever the original observations are available, they should be used rather than data from a grouped distribution since these may yield an approximate result. As a partial check on the grouped distribution, the variances of the raw data and the grouped distribution can be compared along with the means and other measures.

Shortcut Formulas. Of the many shortcut formulas for the variance available three are presented here. These represent the most convenient to use for each of the three forms of data.

Raw Data. When dealing with the original observations, the shortcut formula for the variance is

$$S^2 = \frac{\Sigma X^2}{N} - \bar{X}^2$$

Shortcut for the variance of raw data.

Notice it is not necessary to calculate each deviation of an observation from the mean before squaring. We have only to square the original observations, divide by the total number, and then subtract the square of the mean.

Example Consider four observations (N = 4): 3, 7, 5, and 9 with mean equal to 6. Recall, the mean is found with the formula $\bar{X} = \Sigma X/N$.

$$S^2 = \frac{3^2 + 7^2 + 5^2 + 9^2}{4} - (6)^2$$

$$= \frac{9 + 49 + 25 + 81}{4} - 36$$

$$= \frac{164}{4} - 36$$

$$= 41 - 36 = 5$$

Ungrouped Frequency Distributions. The shortcut formula for an ungrouped frequency distribution is basically the same as that of the previous case except each squared value must be weighted by the corresponding frequency before summing. That is,

$$S^2 = \frac{\Sigma f X^2}{N} - \bar{X}^2$$

Shortcut for variance of ungrouped frequency distribution.

> *Note.* Recall that $N = \Sigma f$ and $\bar{X} = \Sigma fX/N$. In order to avoid confusion, it is wise to compute $\bar{X}$ separately and then insert it into the formula.

Example Based upon Example 2 illustrated earlier, perform the preliminary computations in tabular form

X	f	X^2	fX^2
10	2	100	200
12	5	144	720
13	12	169	2028
17	8	289	2312
20	3	400	1200
Total	30		6460

$\bar{X} = 14.4$

$$S^2 = \frac{6460}{30} - (14.4)^2$$
$$= 215.33 - 207.36$$
$$= 7.97$$

Grouped Frequency Distribution. The formula used for an ungrouped frequency distribution can be adapted to grouped distributions by substituting the class midpoint (m) for X. It should be used in cases where the class intervals are *not* equal. A simpler method is available for cases in which we have *equal* class intervals. The formula is

$$S^2 = c^2\left[\frac{\Sigma fd^2}{N} - \left(\frac{\Sigma fd}{N}\right)^2\right]$$

Formula for finding the variance of a grouped distribution with equal intervals.

This expression appears awkward, but is simple to use. It is based upon the coding procedure used in the previous chapter as a shortcut for the mean (i.e., d represents the number of class intervals that each interval is from the arbitrarily chosen interval, where a value of d = 0 is automatically attached to the chosen interval). The size of the class intervals is represented by c.

Example Use of the formula is illustrated with the mutual fund price distribution used previously.

X *Mutual Fund Price*	f *No. of Funds*	d	fd	d^2	fd^2
\$0.00 but less than \$3.50	10	−2	−20	4	40
\$3.50 but less than \$7.00	19	−1	−19	1	19
\$7.00 but less than \$10.50	34	0	0	0	0
\$10.50 but less than \$14.00	13	1	13	1	13
\$14.00 but less than \$17.50	5	2	10	4	20
\$17.50 but less than \$21.00	4	3	12	9	36
Total	85		−4		128

$$S^2 = c^2\left[\frac{\Sigma fd^2}{N} - \left(\frac{\Sigma fd}{N}\right)^2\right]$$

$$= (3.50)^2\left[\frac{128}{85} - \left(\frac{-4}{85}\right)^2\right]$$

$$= 12.25\ [1.5059 - (-.0471)^2]$$

$$= 12.25\ [1.5059 - 0.0022]$$

$$= 12.25\ [1.5037] = 18.42 \text{ (rounded)}$$

The result is the same as the one obtained earlier, but with an obvious saving in computational effort.

> *Note.* In general, when the variance is computed, the mean also is found. The term $\Sigma fd/N$ also is used in the shortcut for the mean, and therefore no additional preliminary work needs to be done to obtain both results.

The *standard deviation* always can be found as the square root of the variance, regardless of the method used to find the variance.

QUESTIONS AND PROBLEMS

DEFINITIONAL QUESTIONS

1. What is the meaning of the following terms and concepts?
   (a) Variability
   (b) Dispersion
   (c) Deviation
   (d) Squared deviation
   (e) Deviation from the mean

2. What is meant by the following terms?
   (a) Range
   (b) Variance
   (c) Standard deviation
   (d) Rel-variance
   (e) Coefficient of variation
   (f) Mean deviation

COMPUTATIONAL PROBLEMS

3. Find the variance and standard deviation of the following 15 values.

13	3	7	18	20
5	16	9	14	19
6	9	13	7	4

4. The number of items ordered by the last 20 customers of a mail order house appear below.

3	1	7	2	3	12	4	8	1	5
2	4	1	6	4	1	5	3	7	1

Compute the variance and standard deviation of the number of items ordered.

5. Find the variance and standard deviation of the ungrouped frequency distribution given below.

X	f
2.0	6
3.0	10
5.0	4
9.0	3
13.0	2
Total	25

6. The distribution of the amounts of the day's credit sales for 100 customers of a small retailer appear below.

Credit Amount	*Number of Customers*
$6.50	13
$10.30	24
$15.65	33
$17.90	16
$23.50	9
$26.75	5
Total	100

Compute the variance and standard deviation of credit sales.

7. Appearing below is the distribution of outstanding balances on mortgages of a small savings and loan association.

Outstanding Balance	*Number of Mortgages*
$ 0 – $9999	21
$10000 – 19999	32
$20000 – 29999	43
$30000 – 39999	24
$40000 – 49999	17
$50000 – 59999	13
Total	150

Find the variance and standard deviation of outstanding mortgage balances based upon the above distribution.

8. Find the rel-variance and coefficient of variation for the data in Problem 4 regarding the number of items ordered.

9. The mean exam grade for a class of 20 students is 75.3 with a standard deviation of 13.2. Find the corresponding rel-variance and coefficient of variation.

10. Find the range and the mean deviation about the median for the values given in Problem 3.

11. For the data given in Problem 4, compute the range and the mean deviation about the median.

CONCEPTUAL QUESTIONS AND PROBLEMS

12. What is the purpose of computing a measure of variability?

13. What additional information about a body of data is provided by a measure of variability that is not obtained from an average? What information is not provided by a measure of variability?

14. What advantages are associated with the variance and standard deviation relative to the range as measures of variability? Are the same advantages associated with the mean deviation?

15. What information is provided by the variance or standard deviation? When all observations in a set of data are the same, what is the value of the variance and the standard deviation?

16. At the beginning of this chapter, two diagrams were used (with asterisks) together with the concept of the range in order to introduce the concept of measuring variability. For each of the two sets of values, determine the values indicated by the asterisks and compute the variance for each set of values. Relate the results to the points made about variability in the introductory remarks.

17. Suppose you read a published statement that the average amount of food consumption of families in this country is adequate; the overall conclusion based upon the statement is that everyone is properly fed. Criticize the conclusion in terms of the concept of variability as it relates to the use of averages.

18. Is it necessarily true that being above average indicates that someone is superior? Explain.

19. Of the two sets of 8 values given below, which is more variable? Why?

Set 1	5	8	7	1	6	4	7	5
Set 2	9	8	3	6	2	8	5	3

20. Appearing below are the prices of two mutual funds for five consecutive days. The average price for each fund is the same.

Day	*Price* *Fund 1*	*Fund 2*
1	17.51	10.18
2	13.40	30.81
3	15.26	5.26
4	19.79	15.31
5	16.04	20.44

Find the variance for each fund. Based upon the results, what conclusion can you draw about the two mutual funds?

21. A distributor of electrical supplies has two loading docks. Because of their location, trucks delivering shipments favor one dock more than another. Appearing below are the number of deliveries made at each dock for each day of a typical week.

Number of Deliveries	
Dock 1	*Dock 2*
16	34
13	37
15	33
17	36
14	35

(a) Compute the mean and standard deviation number of deliveries for each loading dock.
(b) Compute the coefficient of variation for each loading dock.
(c) Which loading dock has more variability in deliveries? Explain.
(d) Assume overtime employment is needed in order to handle more than 25 deliveries a day. Is variability important to consider in order to remedy the problem of overtime? If the same basic delivery patterns observed apply generally to other weeks, how can the problem of overtime be corrected?
(e) Suppose 10 trucks a day were diverted from Dock 2 to Dock 1. What would happen to the values of the standard deviation and coefficient of variation corresponding to the two loading docks? What does your answer suggest about the standard deviation and coefficient of variation as measures of variability?

22. The mean grade point average of a group of college students after their first year of school is 2.8 with a standard deviation of 0.6. The mean SAT score for the same group is 640 with a standard deviation of 75. Which is more variable, grade points or SAT scores?

23. Suppose you are interested in buying tires for your car. Considerations such as tire appearance are not important and only length of tire life is to be considered.
(a) If your choice were narrowed down to two equally priced

tires, Brand A and Brand B, and you were told the average life of Brand B is greater than Brand A, which tire would you buy? What additional information would you need to make a better decision? Explain.

(b) Suppose the lengths of life of the two brands actually are distributed according to the two frequency curves given below. Which tire would you purchase? Consider both central tendency and variability in your answer.

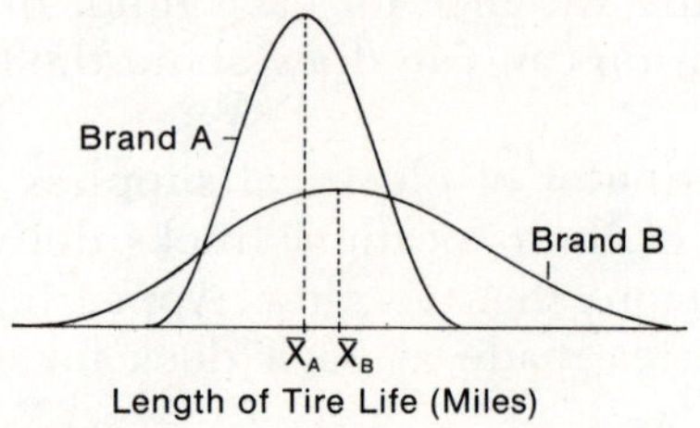

Brand A
Brand B
$\bar{X}_A$ $\bar{X}_B$
Length of Tire Life (Miles)

CONCEPTS OF PROBABILITY

CHAPTER 4

FUNDAMENTAL PROBABILITY

. . . is touched by that dark miracle of chance which makes new magic in a dusty world.

Thomas Wolfe

We have insulted you as Lady Luck.

Phelps Putnam

The words *chance* and *random* are quite common. We know they have something to do with probability, which we generally associate with gambling and lotteries, and maybe with the way premiums are determined on insurance policies. The daily weather forecast customarily ends with a statement about the probability of bad weather. We rarely stop to think what this number means or, for that matter, how it was obtained or computed.

Typically, when told the probability of rain is 80 percent we may assume that it will rain and plan accordingly; when told it is 20 percent, we are inclined to act as though it will not rain. Even though we may not have a complete understanding of the meaning of these numbers or the way they are computed, probabilities provide information that frequently becomes a *basis for action.* Based upon our intuition alone, we can and do use probability. A deeper understanding of the concept can provide us with a basis for drawing conclusions and making decisions more effectively.

In order to obtain a deeper understanding of probability, it is necessary to employ certain terms and definitions more precisely than the way we normally use them. Of fundamental importance is a special type of phenomenon known as *randomness* or *random variation.* Based upon situations where randomness is present, we can define particular types of occurrences or *events.* The main purpose of this chapter is to provide a definition of *the probability of an event* that will be used throughout this book.

The Concept of Randomness. The easiest way to introduce the concept of randomness is to define a *random experiment.* This is a phenomenon having the property that (1) all possible outcomes can be specified in advance, (2) it can be repeated, and (3) the same outcome does not necessarily occur on various repetitions so that the actual outcome is not known beforehand. For example, measurement of the blood pressure of a group of individuals constitutes a series of random experiments. Checking a car's gas mileage is another example of such an experiment. The simplest and most popular illustration is flipping a coin. Examples of situations that do not constitute a random experiment are recording the amount of an individual's weekly paycheck or the amount or rent someone pays to a realtor or landlord. Here, the contracted amounts are known in advance and generally are the same between payment periods. It should be clear that the word experiment is used in a very broad sense. Typically an experiment is thought of in terms of a test or set of trials under controlled laboratory conditions. The term is used here to refer to any type of situation that can be experienced on a repeated basis.

The essential feature of a random experiment is that the outcome of a single repetition is *not predictable* in advance. In other words, there is uncertainty surrounding the outcome. Although an individual occurrence associated with a random experiment cannot be predicted exactly, it can be shown that something can be said about the frequency of occurrence in a *large number* of repetitions. It is through the use of probability that this can be accomplished, and becomes the basis for resolving problems dealing with uncertainty.

In order to develop our understanding of probability, the following concepts are defined:

> A **sample space** is a set or collection of all possible outcomes associated with a random experiment.
>
> An **event** is defined as any collection of elements or outcomes within the sample space.
>
> An event is said to **occur** if the outcome of a random experiment, once performed, is contained within the given event.

Consider a very simple case in order to illustrate the three definitions. Suppose we are interested in an experiment in which a single coin is flipped twice. The sample space can be designated in the following way:

{HH, HT, TH, TT}

We can obtain two heads in a row, or a head on the first flip and a tail on the second, and so on. In this example each possible outcome or *sample point* is considered separately in characterizing the sample space.

Sample spaces may be characterized in ways that conceal the identity of the sample points. For example, another legitimate way to specify the space corresponding to the two flips of a coin would be in terms of tails as

{No Tails, At Least One Tail}

In terms of the number of heads, the same experiment is represented by the sample space

$$\{0, 1, 2\}$$

The essential points to keep in mind are that a properly designated sample space considers or exhausts *all* possible outcomes, and that there is no overlap among the elements within the space; namely, that they are *mutually exclusive.* Moreover, there is no single correct way of specifying the sample space. The specification depends upon the way it is to be used in a particular problem.

Based upon the sample space, {HH, HT, TH, TT}, there are a number of events that can be considered. For example, two possibilities are {HT, TT} and {HH, HT, TH}. Events also can be characterized in other ways, depending upon the way the sample space is specified.

Once an experiment is performed, the outcome always will be represented by an individual possibility or single sample point. An event is said to occur if the outcome is contained within the event, regardless of the way the event is characterized or the number of sample points contained in the event. For example, if we obtain two tails after flipping a coin twice, we can say that the event {HT, TH, TT} has occurred since it contains the outcome, TT. Other ways of designating this same event are {At Least One Tail} and in terms of the number of heads, {0}. Furthermore, any other event that contains the outcome, TT, also is said to occur.

The Probability Of An Event. Now that random experiment, sample space, and events have been defined, we can define the probability of an event.

> The **probability of an event** is a number, between zero and one, that represents the proportion of times the event occurs in an infinite sequence of trials of the same random experiment.

In order to understand this definition, imagine repeating a random experiment endlessly. You define a particular event, and each time the experiment is performed you record whether the event has occurred or not. As the number of trials of the experiment becomes large, begin computing the ratio of the number of times the event occurs to the total number of trials, or compute the relative frequency of occurrence. Loosely speaking, the larger the number of trials the closer will the ratio or relative frequency approach the probability of the event. Strictly, the computed ratio approaches the probability in "the limit" as the number of trials approaches infinity.

If, for example, we are told that the probability of selecting "a spade" on a single draw from a well-shuffled card deck is 0.25, how should we interpret this number? Based upon the definition, we can say that 0.25 or one-fourth of an "infinite" number of draws would yield a spade. In more practical terms, we can say that if a large number of trials were performed, "close" to 25 percent would yield a spade. Recognize that the probability statement tells us nothing about which of the trials comprise the 25 percent!

Consider another situation in which an assembly line item is produced. We are told the probability that a defective item will be produced is 0.05. Using the given definition, we can say that 0.05 or 5 percent of all the items produced will be defective, under the assumption that the manufacturing process continues endlessly into the future. More practically, we can say that large shipments taken from this process will contain nearly 5 percent defective items.

The definition states that the probability of an event is a number between zero and one. A zero probability implies that an event is impossible, whereas a probability of one signifies that an event will occur with certainty, or 100 percent of the time. Relative to the entire scale of probability values, probabilities of zero and one are the only two values that provide complete information about a particular outcome of a random experiment. If zero, we know the event will not occur; if one, we know it must occur. Any other value provides information only about the relative frequency of the occurrence over a long sequence of trials, and it is toward these values that we concentrate our interest.

The symbols to be used are defined as follows:

S = sample space
E = an event
N(S) = the number of possible outcomes or sample points in the sample space
N(E) = the number of outcomes or sample points corresponding to any event E
P(E) = the probability of any event E
N(A&B) = the number of sample points that the two events A and B have in common
P(A&B) = the joint probability of A and B, or the probability that the two events A and B occur simultaneously
P(A|B) = the conditional probability of an event A given another event B
P(B|A) = the conditional probability of B given A

A METHOD OF COMPUTATION

The simplest method for computing the probability of an event follows directly from its definition. In order to apply the definition, however, it is necessary to introduce two assumptions. For the present, we shall consider cases where the sample space is *finite*, or spaces where we actually can count the exact number of outcomes or sample points in the space. On the other hand, we shall assume that the random experiment is such that each of the sample points is *equally likely*, which means that each possible outcome has the same probability of occurrence attached to it.

Based upon the two assumptions provided, we can compute the probability of an event according to the formula

$$P(E) = \frac{N(E)}{N(S)}$$

Probability of an event.

This formula tells us to compute the ratio of the number of ways an event can occur to the total number of possible outcomes.

NUMERICAL EXAMPLES

Example 1 To find the probability of drawing "a spade" on a single draw from a well-shuffled card deck.

There are 52 cards in the deck or 52 distinct possible outcomes or sample points, of which 13 are spades. Therefore

$$P(\text{Spade}) = \frac{\text{Number of spades}}{\text{Number of cards}}$$
$$= \frac{13}{52}$$
$$= 0.25 \text{ or } \frac{1}{4}$$

Example 2 To find the probability of drawing "an ace or a spade" on a single draw from a well-shuffled card deck.

Of the 52 possibilities, there are 13 spades and 4 aces, one of which is both a spade and an ace. Therefore, there are 16 *distinct* cards corresponding to the given event, and

$$P(\text{Ace or spade}) = \frac{16}{52}$$
$$= 0.31$$

Example 3 Suppose you face a shelf containing five books: a novel, an economics text, a math book, a history book, and a dictionary. What is the probability that you will select the "novel or the history book" at random?

Two out of 5 possibilities correspond to the event specified, and therefore

$$P(\text{Novel or history book}) = \frac{2}{5}$$
$$= 0.40$$

Example 4 Suppose a single coin is flipped twice, and we are interested in finding the probability of obtaining "at least one tail."

The sample space for the experiment is {HH, HT, TH, TT}. "At least one tail" can be obtained in one of three ways; HT, TH, or TT. The probability then becomes

$$P(\text{At least one tail}) = \frac{3}{4} = 0.75$$

Example 5 Suppose a single coin is flipped once, and we are to find the probability of "a head."

The sample space is simply {H, T}, and therefore

$$P(\text{Head}) = \frac{1}{2} = 0.50$$

Example 6 A shipment of 5000 auto parts contains 250 that are defective. If one part is drawn from the shipment at random, what is the probability that it will be defective? The answer is found as

$$P(\text{Defective}) = \frac{\text{Number of defectives}}{\text{Number of parts}} = \frac{250}{5000} = 0.05$$

DISCUSSION

The preceding treatment of probability was quite simple, especially the numerical illustrations. Traditionally, "games of chance" are used to illustrate the way probabilities are obtained. To some, this is not very motivating since the problems do not seem realistic. It should be recognized that games of chance more closely resemble *idealized* circumstances dictated by the stated assumptions. Shuffling cards, rolling dice, and flipping coins are situations that conform to the condition of randomness and, as a result, provide a better basis for understanding the basic concepts.

Ultimately, our objective is to move away from these idealized experiments, but we first must obtain the basis for dealing with more complicated and realistic problems. What we need are ways of establishing *theoretical* probabilities that describe different kinds of problems. For example, consider the case of flipping a single coin once.

Using the "equally likely assumption," the theoretical probability of obtaining a head is one half. If you were to rely on the outcome of the flip, wouldn't you feel better if you knew that this theoretical number applied to the particular coin being used?

The last statement suggests that, to be useful, the theoretical probability should describe the *relative frequency* that would be observed if we were to actually flip the coin a "very large number of times." In other words, the relative frequency observed by repeatedly flipping the coin should equal one half before the theoretical value is useful. This is not true for all coins. Coins with a probability of a head not equal to one half exist.

Probabilities can be obtained theoretically or empirically as a relative frequency.

As an alternative to determining the probability of an event theoretically, we may estimate it *empirically* by actually performing a large number of trials and computing the observed relative frequency. In the case of flipping coins, it is obvious that we are not able to flip a coin forever. Hence, the flips we do make constitute a sample of all possible trials. How many flips or what size sample do you think we should take in order to estimate the probability of a head properly? This is a problem that can be treated with the aid of statistical methods that are discussed in the next part of the text.

Earlier we cited the use of a probability statement reported at the end of broadcasted weather forecasts. When a statement of this sort is made, we really do not know how the value was obtained. The method of using equally likely outcomes is inappropriate. It is possible that the probability of rain, say, is computed as an observed relative frequency by dividing the total number of days on record with the given date into the number of those days when it rained, based upon past weather data. One would need additional information about rainfall behavior, however, to determine whether such an empirical estimate is valid, since other factors such as changing weather patterns over time, weather on adjacent days, and related patterns in other regions may have an effect.

Two types of probabilities, theoretical and observed relative frequencies, have been cited. Another alternative exists that is worth noting. The probability scale of values may be used to convey a message in a simple and concise way; in other instances, probability statements may be used to relate one's "degree of belief" in the occurrence of an event. In other words, *subjective* assessments regarding a particular situation sometimes are made, and then translated into a number between zero and one and used as a probability. For example, if it is cloudy and certain indicators suggest that it may rain, this may be reported as a high probability, or as a number "close" to one, without the aid of any numerical information or calculations. Actually, information in this form can be useful and is considered in particular branches of statistics.

Probabilities can be established subjectively.

What are the essential points that can be made on the basis of the above discussion? Initially, we should note that different views exist about probability and the way probabilities are established. A second point is that regardless of the method of computation we shall use the definition of probability presented in this chapter throughout the remaining portions of the text. As the kind of problem varies, different methods of computation are required. One procedure has been given earlier. Others will be considered. Some will apply to sample spaces

The relative frequency interpretation of probability is used throughout the text although different ways of finding probabilities are used.

that assume different forms. Finally, the results of probability theory can be used to solve many types of problems. Methods exist for establishing theoretical probabilities that must correspond to the behavior of real situations in order to be useful. Ultimately, we shall develop an understanding of the application of probability to problems of sampling and statistical inference.

Extending the Concept. The material presented to this point in the chapter is concerned with a single event. The basic ideas can be extended to problems where more than one event is considered. In order to understand how this is accomplished, consider the table of data appearing in Exhibit 1. The data correspond to a pool of 100 workers offering their services for temporary employment. Interest is focussed upon two characteristics, area of residence and qualification for the job considered. The frequencies in the body of the table represent the number of workers that simultaneously or jointly fall into particular categories associated with the two characteristics. These frequencies are called *joint frequencies.* Tables like the one below that provide joint frequencies are referred to as crossclassified tables or *contingency tables.* Contingency tables are used to describe relationships among characteristics or variables.

Total figures appear in the margins of the table corresponding to each column and each row. These totals are referred to as *marginal frequencies.* For example, the sum of the joint frequencies in the center column tells us that 30 workers from the entire group are urban residents. Likewise, a total of 55 workers do not qualify for the job. Notice the marginal frequencies correspond to *one* characteristic only, while the joint frequencies in the body of the table consider two characteristics simultaneously.

In order to relate the information in the table to the concept of probability, consider the pool of 100 workers as a sample space from which one worker is to be selected at random. Each worker or element in the space is now characterized in two ways; they fall into categories corresponding to each of the two characteristics, area of residence and job qualification. Each of the individual categories can be thought of as events. When a worker is chosen, *two events* occur simultaneously. The possible combinations of events that can occur are "qualified and suburban," "qualified and urban," "qualified and rural," "unqualified and suburban," "unqualified and urban," and "unqualified and rural." Recognize that certain events cannot occur at the same time. For example, a worker cannot be "qualified and unqualified" simultaneously.

EXHIBIT 1
AN EXAMPLE OF A CONTINGENCY TABLE

NUMBER OF TEMPORARY WORKERS

Job Qualification	***Area of Residence*** *Suburban*	*Urban*	*Rural*	***Total***
Qualified	14	20	11	45
Unqualified	21	10	24	55
Total	35	30	35	100

In such cases where two events do not have any sample points in common, we refer to the events as *mutually exclusive.*

When events do not have any common sample points, they are referred to as mutually exclusive.

Generally speaking, whenever we deal with two events we can extend the concept of probability in a number of ways. In many applications, it is useful to determine the **joint probability** based upon the joint frequency or the probability that two events will occur simultaneously. In other instances, we may be interested in the **marginal probabilities,** based upon the marginal totals, which simply are the probabilities of each of the events considered individually. Another related probability is referred to as a **conditional probability.** This is the probability that an event will occur *given* that another event is known or assumed to occur. Situations arise where information given about the occurrence of one event in terms of conditional probabilities is helpful in providing additional information about another event.

In general, under the assumption that the sample space is finite and individual outcomes are equally likely, the joint probability that any two events, A and B, occur simultaneously is given by the following formula:

$$P(A\&B) = \frac{N(A\&B)}{N(S)}$$

Probability of events A and B occurring simultaneously assuming equally likely outcomes.

The joint probability is found by dividing the number of points corresponding to both events, N(A&B), by the total number of points, N(S). This is equivalent to dividing a joint frequency in a contingency table by the overall total.

In order to illustrate the formula for a joint probability, consider the employment pool data in Exhibit 1. Suppose we want to find the probability that a worker drawn at random from the pool is an "unqualified rural resident." This is found as

$$\begin{aligned} P(\text{Unqualified } \textit{and} \text{ Rural}) &= \frac{\text{Number of unqualified rural residents}}{\text{Total number of workers}} \\ &= \frac{24}{100} \\ &= 0.24 \end{aligned}$$

This probability is found by dividing the number of workers that jointly fall into the two categories, unqualified and rural, by the total number of workers. The result, 0.24, tells us that 0.24 or 24 percent of the workers are unqualified rural residents. Consequently, repeated random drawings from the employment pool will result in a worker with these two characteristics. Joint probabilities for other possible combinations are found and interpreted in the same way.

Interpreting joint probabilities.

A marginal probability describes the behavior of one event without considering any of the other events. Hence, a marginal probability is no different from the probabilities presented earlier in the chapter but is given a special label since other types of probabilities now are being considered. For example, we could find the probability that a randomly drawn worker is unqualified as

$$P(\text{Unqualified}) = \frac{\text{Number of unqualified workers}}{\text{Total number of workers}}$$

$$= \frac{55}{100}$$

$$= 0.55$$

Interpreting marginal probabilities.

In other words, 0.55 or 55 percent of the workers are unqualified. The result is a marginal probability since it is associated with one event without considering any of the others that are possible.

A conditional probability is based upon a joint and a marginal frequency. Assuming a finite sample space and equally likely outcomes, the conditional probability of any event A given another event B is found with the formula

$$P(A|B) = \frac{N(A\&B)}{N(B)}$$

Conditional probability of the event A given B assuming equally likely outcomes.

The conditional probability is found by dividing the number of points corresponding to both events, N(A&B), by the number of points corresponding to the event B, N(B). This is equivalent to dividing a joint frequency in a contingency table by a marginal frequency. The conditional probability of the event B given A, P(B|A), is found by using N(A) in the denominator of the formula; the numerator is the same.

Suppose we want to find the conditional probability that a worker is a rural resident given that the worker is unqualified. Using the above formula, we have

$$P(\text{Rural } \textit{given} \text{ Unqualified}) = \frac{\text{Number of unqualified rural residents}}{\text{Number of unqualified workers}}$$

$$= \frac{24}{55}$$

$$= 0.44$$

Interpreting conditional probabilities.

We can interpret this number to mean that 0.44 or 44 percent of the unqualified workers live in rural areas.

More About Conditional Probabilities. Conditional probabilities require special attention since they summarize relationships existing among characteristics. In order to understand the importance of a conditional probability, examine the table appearing in Exhibit 2 that presents conditional and marginal probabilities based upon the table of workers presented in the earlier exhibit. The probabilities in Exhibit 2 are associated with worker qualification only. The marginal probabilities describe the behavior of job qualification without reference to area of residence. The conditional probabilities provide information about job qualification given area of residence. A similar table can be constructed

EXHIBIT 2

COMPARISON OF CONDITIONAL AND MARGINAL PROBABILITIES OF WORKER QUALIFICATION

Job Qualification	*Conditional Probabilities* Suburban	*Urban*	*Rural*	*Marginal Probabilities*
Qualified	0.40	0.67	0.31	0.45
Unqualified	0.60	0.33	0.69	0.55
Total	1.00	1.00	1.00	1.00

for area of residence; however, the overall conclusions would be the same.

To understand the contents of the table, consider the marginal probability of 0.45 in the last column. This tells us that 0.45 or 45 percent of the worker pool is qualified, regardless of area of residence. The conditional probability of 0.67 at the top of the center column tells us that 0.67 or 67 percent of urban residents in the pool are qualified; in other words, the probability is restricted to urban residents only. Similar interpretations apply to the other marginal and conditional probabilities in the table.

By focusing upon the top row of the table, we see that the probabilities are different. This means that the conditional probability that a worker is qualified varies with the worker's area of residence. Consequently, the conditional probabilities provide different information about worker qualification, *depending* upon area of residence. If area of residence were not considered, only the marginal probabilities of worker qualification would be available.

A reasonable question to ask as a result of the above comments is how is such information useful? Assume you represent an organization interested in the services of a worker from the pool, and the worker received virtually is selected at random. In this case, 55 percent of the time a worker would not qualify for the job. If the additional information regarding area of residence is available, we could apply at the urban regional office and reduce the chance of receiving an unqualified worker to 33 percent. Consequently, the additional information about area of residence increases one's chance of receiving a qualified worker for the job. Knowledge of the relationship between the characteristics in terms of probability provides us with a better basis for action in this particular situation.

Independent and Dependent Events. We can summarize our discussion of conditional probabilities in another way in terms of the concept of statistical independence. Two events, A and B, are said to be **independent** if information about B tells nothing about the occurrence of A. In terms of probability, this occurs when

$$P(A|B) = P(A)$$

Independence of events A and B.

In other words, two events are independent when the conditional probability of one event, A, given another event, B, is equal to the marginal probability of the first event. In the worker pool example, independence between job and area of residence would occur if the probabilities in each row of the table in Exhibit 2 were equal to the marginal probability. If this were the case, the probability of selecting a qualified worker would be the same regardless of the area of residence.

Since we saw that the conditional probabilities in the exhibit vary and do not equal the marginal probabilities, job qualification and area are said to be dependent. For example, consider the conditional probability that a worker is qualified given urban residence, $P(Q|U) = 0.67$. This is not equal to the marginal probability that a worker is qualified, $P(Q) = 0.45$. Therefore qualification and area of residence are dependent.

When events are dependent, knowledge about one event provides different information about the probability of occurrence of the other event. When events are independent, knowledge about one event provides no more information than the marginal probability of the other event.

In general, two events, A and B, are said to be **dependent** if information about B tells something about A. In terms of probability, this means

$$P(A|B) \neq P(A)$$

Dependence of events A and B.

In other words, the events A and B are dependent when the conditional probability of A given B is *not* equal to the marginal probability of A. Dependence between characteristics implies that a relationship exists and, therefore, knowledge of one characteristic is useful in assessing the probability of the other.

In this part of the discussion we have seen how the basic computational formula for finding probabilities can be applied to different types of probabilities. The ideas that we have developed are useful when dealing with further topics in probability and statistics. In order to apply these ideas to other topic areas, it is useful to generalize them into a set of rules that are presented in the next chapter.

QUESTIONS AND PROBLEMS

DEFINITIONAL QUESTIONS

1. What is meant by the terms *random, random variation,* and *random experiment?*

2. Define the terms *sample space* and *sample point.*

3. How is an event defined? What do we mean by the phrase "an event has occurred"?

4. In terms of the definition of a sample space and events, what is meant by the term *mutually exclusive?*

5. What is the meaning of the probability of an event as it is used throughout this text?

6. Define the terms *theoretical probability, observed relative frequency,* and *subjective probability.*

7. What is meant by the term *equally likely?*

8. Briefly explain what is meant by a contingency table. Why is it appropriate to use the term *crossclassified table* instead?

9. What is meant by the terms *joint frequency* and *marginal frequency?*

10. Define the terms *joint probability, marginal probability,* and *conditional probability.*

11. What is meant by the terms *independence* and *dependence* of events?

12. In terms of probabilities, explain the meaning of the phrase "a relationship exists between two characteristics."

COMPUTATIONAL PROBLEMS

13. A single card is drawn from a well-shuffled card deck.
    (a) Find the probability of drawing a nine.
    (b) What is the probability of drawing the nine of hearts?
    (c) Find the probability that the card drawn is a king or a six.
    (d) What is the probability of drawing the king of spades or the six of diamonds?
    (e) Find the probability of drawing a six or a diamond.

14. A single die is rolled once. Find the probability of the following events.
    (a) A five.
    (b) A three.
    (c) A three or a five.
    (d) At least a 4.
    (e) More than a 4.

15. Two dice are rolled once.
    (a) What is the probability of getting two sixes?
    (b) Find the probability of obtaining a six on either die, or at least one six.
    (c) What is the probability that the sum of the result on the two dice equals 7?

16. What is the probability that three heads will result from three flips of a coin?

17. Recipients of leases to federal oil lands are selected monthly by a lottery administered by the Bureau of Land Management. Suppose there are 650 applicants for a certain parcel of land in Wyoming. Of the applicants, 17 are oil companies; the remaining are private citizens. The recipient of the lease is drawn at random. All applicants have an equal chance of being selected.
    (a) If you are one of the applicants, what is the probability that you will receive the lease for the Wyoming parcel?

(b) What is the probability that the lease will be given to: (1) an oil company, (2) a private citizen?

18. A shipment of 1000 light bulbs contains 150 defective bulbs. If you buy one of these 1000 bulbs, what is the probability that the bulb will not work?

19. Given the following frequency distribution.

X	f
2.0	3
2.5	7
3.0	11
4.3	9
5.0	5
Total	35

Find the following probabilities.

(a) $P(X = 2.5)$
(b) $P(X = 4.3)$
(c) $P(2.5 \leq X \leq 4.3)$
(d) $P(X \leq 3.0)$
(e) $P(X > 2.5)$

20. Given below is the distribution of weekly wage rates for 25 workers.

Weekly Wage Rate ($)	*Number of Workers*
93	3
105	5
124	9
135	4
150	3
200	1
Total	25

(a) If one worker is drawn at random from the group, what is the probability that the worker's wage rate is $124.
(b) What is the probability that the wage rate of a worker drawn at random does not exceed $135?

21. The distribution of closing prices for 85 mutual funds appears below.

Mutual Fund Price	*Number of Funds*
$1.00–$4.49	17
$4.50–$7.99	24
$8.00–$11.49	27
$11.50–$14.99	9
$15.00–$18.49	5
$18.50–$21.99	3
Total	85

(a) If one fund were selected at random, what is the probability that its price is between $8.00 and $11.49?
(b) What is the probability that the price of a randomly drawn stock exceeds $14.99?

22. The table below presents the results of extensive testing of 20 brands of television sets. Each of the sets is classified in terms of quality of reception and whether the television set requires major repair within the six-month period following expiration of the warranty.

NUMBER OF TELEVISION SETS

	Major Repair Required	*Major Repair Not Required*	*Total*
Acceptable Reception	4	10	14
Reception Not Acceptable	4	2	6
Total	8	12	20

Assume the results presented in the table reflect the actual performance of the twenty brands. If a television set were chosen at random, find the probability that the brand chosen

(a) Needs major repair within six months after expiration of the warranty.
(b) Has acceptable reception.
(c) Does not need major repair but has unacceptable reception.
(d) Does not need major repair given that it is rated with acceptable reception.

23. Records of a university for the previous year indicate that 1600 students applied for admission. Of these, 850 applicants were admitted and 750 were denied admission. Of those admitted 550 were men, whereas 400 of those denied were women. Assume one person is chosen at random from the list of applicants.

(a) Find the probability that the one chosen is male.
(b) What is the probability that the one chosen is denied admission?
(c) What is the probability of selecting a woman granted admission?
(d) Given that the person chosen is female, what is the probability that she was denied admission?

24. A medical study of ten thousand people over a long period yielded the following results regarding a particular disease and its symptoms.

NUMBER OF PEOPLE

	Symptoms Evident	*No Symptoms*	*Total*
Contracted Disease	200	50	250
No Disease	100	9650	9750
Total	300	9700	10000

Based upon the data in the table:

(a) What is the probability of contracting the disease?
(b) What is the probability that a person has no symptoms and contracts the disease?
(c) Find the probability of having the disease given no symptoms present.
(d) Find the probability of having the disease given that symptoms are present.

25. Applicants for credit cards to a major department store are classified below in terms of three income categories and whether they received a credit line with the store.

NUMBER OF APPLICANTS

Income	*Received Credit*	*Credit Not Received*	*Total*
Low	70	230	300
Middle	260	240	500
High	165	35	200
Total	495	505	1000

Assuming the figures in the table reflect the store's policy for extending credit, find the probability that (a) a person in a low income bracket will not receive credit from the store, (b) an applicant is in a high income bracket who does not receive credit (c) an applicant receives credit, and (d) a person receiving credit is from a middle income bracket.

CONCEPTUAL QUESTIONS AND PROBLEMS

26. When given the probability of an event, what information is provided about that event?

27. What information is provided about an event with a probability of zero? Of one? How is this related to the probability of an event in general?

28. If you are told the probability of getting a six on a roll of a die is 1/6, or 0.17, how is this number interpreted? If, instead, you are told the probability of the same event is 0.25, how is this

value interpreted? Can it be possible that the probability of a six is 0.25? Explain. Aside from the numerical difference between the two probability values cited, what is the difference between them?

29. Problem 17 describes the way recipients of oil leases are selected by the federal government. Assuming the answer to part (a) of the problem is 0.002, how is this probability interpreted?

30. Suppose an entire shipment of 1000 items is inspected and 50 items are found to be defective. Assume the defective items are not removed from the shipment before being sent to a retail outlet for sale.

    (a) If you purchase one item from this shipment, what is the probability that it will be one of the defective items?
    (b) What is the meaning of the value found in part (a)?
    (c) What assumption must you make in order to compute the value in part (a)?

31. Suppose you are told that the price of a particular stock will increase with a probability of 0.7.

    (a) How is this probability interpreted?
    (b) Did you have difficulty interpreting the probability? Would it be easier to interpret the probability of a head resulting from the flip of a coin? Explain.
    (c) If you were told the probability represented the opinion of a particular broker, would you change the interpretation in part (a)? Explain.
    (d) Assuming the definition of probability in terms of long run relative frequencies adopted in this chapter, how would you find the probability that a stock price will increase? In realistic terms, what are the difficulties in finding such a probability?

32. The marketing vice president of a company producing packaged snack foods estimates the probability of success of a newly developed product to be 0.75.

    (a) How can this probability be interpreted?
    (b) Since the introduction of a new product cannot be considered as a repeatable event, does it make sense to interpret the probability of success as a relative frequency in a large number of trials? Explain.
    (c) By using data about the success of similar new products from the past is it possible to establish a probability of success for a new product not yet on the market? Explain. What problems are associated with this approach?

33. Criticize or explain the following statement. Of the 50,000 newborn babies that died within one month of birth during the last year, 35,000 were white and 15,000 were black. Consequently, whites have a poorer chance of survival.

34. Shuffling a deck of cards is a way of producing randomness.
    (a) What is meant by the above statement?
    (b) What is the relationship between card shuffling and other types of random experiments that are not related to gambling?

35. Some people do not take trips during holidays because their chances of having an accident are greater. What is meant by this statement? If your chances are greater, does it mean that you will have an accident? Explain.

36. The concept of randomness is commonly associated with fairness. In terms of the following three situations, explain the meaning of this statement.
    (a) Playing poker or bridge.
    (b) Betting at a roulette wheel.
    (c) Selection for the armed forces through a draft lottery.

37. Suppose you are to undergo surgery for which there is a chance of 1 in 25 of survival. Your surgeon wants to reassure you and tells you there is no reason to worry since the last 24 patients undergoing similar surgery died so that you will be the one to survive. In terms of the definition of probability, what is wrong with the surgeon's reasoning?

38. Closely related to the probability of an event is the concept of *odds*, or the odds ratio. The odds associated with an event is defined as the probability the event will occur divided by the probability that the event will not occur. The probability that the event will not occur equals one minus the probability that it will occur.
    (a) Suppose you are told the probability that the price of a particular stock will increase is 0.75. What are the odds that the stock will increase in price? What are the odds "against" the price increase?
    (b) Suppose the odds in favor of winning a bet are 4 to 1. What is the probability of winning? Of losing?

39. By comparing the three kinds of probabilities (joint, conditional, and marginal), explain what information is provided by each.

40. Refer to Problem 24. Identify each of the probabilities in parts (a) through (d) as joint, marginal, or conditional. Express each of the probabilities in terms of symbols.

41. Refer to Problem 25. Identify each of the probabilities in parts (a) through (d) as joint, marginal, or conditional. Express each of the probabilities in terms of symbols.

42. Before working on the job, 525 salesmen were given an aptitude test. At the end of the first year of employment, these same sales-

men received performance ratings assigned by their regional sales managers. The results of the test and performance ratings are summarized in the following table.

NUMBER OF SALESMEN

Test Score	Performance Rating *Acceptable*	*Unacceptable*	Total
High	150	25	175
Medium	160	90	250
Low	35	65	100
Total	345	180	525

Using the figures in the table, find the probability that (a) a salesman will receive an acceptable performance rating after one year of employment, (b) a salesman receives a medium test score and is rated as acceptable, and (c) a salesman will receive an acceptable rating given that he received a medium test score prior to the first year of employment. Interpret each of these probabilities in terms of the problem.

43. Under what circumstances are conditional probabilities informative? Explain.

44. Based upon the information given in Problem 24, construct a table of conditional probabilities associated with disease categories conditional upon symptom categories and similar to the table appearing in the discussion section. What information is provided by this table? Based upon the marginal totals in the original table, what should be the values of the joint frequencies for the disease and symptoms to be independent? Explain how you arrived at these results.

45. Using the data in Problem 25, construct a table of conditional probabilities associated with the categories "received credit" and "credit not received" given each income category similar to the table of conditional probabilities in the discussion section. What information is supplied by this table. Assuming the marginal totals are the same as given in the original table, what values of the joint frequencies would lead to the conclusion that receipt of credit is independent of income level? Explain how you arrived at these results?

CHAPTER 5

SOME GENERAL RULES

Everything in life is six to five against.
Damon Runyon

. . . with the serene confidence which a Christian feels in four aces.
Mark Twain

Many results from the field of probability can be reduced to simple rules and formulas. Emphasis in this chapter is upon some commonly used rules that apply in cases where *two* events are considered. Although the material draws upon definitions from the previous chapter, it is presented in general terms. By this we mean that each of the rules not only applies to finite sample spaces with which we already are familiar, but to other kinds of spaces as well. The strength of the rules lies in their usefulness in computing probabilities that are too complicated or impossible to find with the direct counting method used so far. Three basic rules are considered: the *addition rule,* a rule for finding *conditional* probabilities, and the *multiplication rule.* Extensions of the addition and multiplication rules to more than two events are presented in the discussion section. A way of using the three rules together also is presented in the discussion and is summarized in the form of another rule called Bayes' Theorem.

The addition rule for two events is employed where it is of interest to find the probability that either one of two events or both events will occur. In other words, it applies when one is interested in finding the probability that one event *or* another event will occur. A probability using the addition rule is found by adding probabilities associated with individual events defined within the sample space.

The procedure used to determine conditional probabilities is very similar to the one considered in the previous chapter. That is, they are found by dividing a joint probability by a marginal probability; in the case of finite sample spaces, we employed the number of sample points instead of probabilities.

The multiplication rule is nothing more than a general procedure to find the joint probability of two events; it applies when one is interested in finding the probability that the events occur at the same time and is based upon the product of probabilities. An alternative way of defining the concept of independence of events is directly related to a special case of the multiplication rule.

The symbols to be used are summarized as follows:

P(A) = the marginal probability of the event A, or the probability of A
P(B) = the marginal probability of the event B, or the probability of B
P(A & B) = the joint probability of the events A and B
P(A *or* B) = the probability that event A or B will occur
P(A|B) = the conditional probability of the event A given B
P(B|A) = the conditional probability of the event B given A

PROCEDURE

The procedures for finding probabilities presented in this section are concerned with cases where any two events are involved. For convenience, we designate these as A and B. Each of the rules considered are general and require *no* assumptions about the way the individual probabilities are established.

Addition Rule. In order to find the probability that either event A *or* B (or both A and B) will occur use the addition rule, which takes the form

$$P(A \text{ or } B) = P(A) + P(B) - P(A \,\&\, B)$$

Probability of the event A or B occurring.

In other words, the probability that A or B will occur is found by adding the marginal probabilities and subtracting the joint probability from the sum of the marginals of the two events.

Special Case. When A and B are mutually exclusive, the addition rule reduces to the "straightforward addition" of the marginal probabilities. That is,

$$P(A \,\&\, B) = P(A) + P(B)$$

Addition rule for two mutually exclusive events.

Recall, two events are mutually exclusive when they have no common elements or sample points; in other words, they cannot occur at the same time, or P(A & B) = 0.

Conditional Probability. The conditional probability of an event A *given* the event B is found using the formula

$$P(A|B) = \frac{P(A \& B)}{P(B)}$$

Probability of event A given B.

In other words, the conditional probability of an event given another event is found by dividing the joint probability by the marginal probability. [The conditional of B is found analogously by dividing the joint probability by the marginal of A, P(A)].

Multiplication Rule. In order to find the probability that events A *and* B occur simultaneously, the multiplication rule for joint probabilities is used. The formula is

$$P(A \& B) = P(A)P(B|A)$$

Probability of events A and B occurring simultaneously.

In other words, the joint probability can be found by multiplying the marginal probability of one event by the conditional probability of the other event. By interchanging A and B in the above expression, an equivalent result can be obtained. That is, $P(A \& B) = P(B)P(A|B)$.

Special Case. When A and B are *independent,* the multiplication rule reduces to the "straightforward multiplication" of the marginal probabilities:

$$P(A \& B) = P(A)P(B)$$

Multiplication rule for two independent events.

Recall, two events are independent when the conditional probability equals the marginal.

NUMERICAL ILLUSTRATIONS

Example 1 ***Addition Rule.*** Suppose two events A and B occur with probabilities, 0.15 and 0.35, respectively. The probability that they will occur simultaneously is 0.06. The probability that either "A or B" will occur is found by the addition rule as

$$\begin{aligned} P(A \textit{ or } B) &= P(A) + P(B) - P(A \& B) \\ &= 0.15 + 0.35 - 0.06 \\ &= 0.44 \end{aligned}$$

In other words, the probability that A or B will occur is found by adding the marginal probabilities and subtracting the joint probability of A and B.

Example 2

Addition Rule. Suppose a single card is to be drawn from a well-shuffled straight deck of cards. What is the probability of drawing "an ace *or* a spade"?

$$P(\text{Ace}) = \frac{4}{52}$$

$$P(\text{Spade}) = \frac{13}{52}$$

$$P(\text{Ace \& Spade}) = P(\text{Ace of Spades}) = \frac{1}{52}$$

Therefore, using the addition rule based upon the above probabilities:

$$P(\text{Ace } or \text{ Spade}) = P(\text{Ace}) + P(\text{Spade}) - P(\text{Ace of Spades})$$

$$= \frac{4}{52} + \frac{13}{52} - \frac{1}{52}$$

$$= \frac{16}{52}$$

$$= \frac{4}{13} \text{ or } 0.31$$

Example 3

Special Case of Addition Rule. What is the probability of drawing an "ace or a nine" on a single draw from a well-shuffled card deck?

Since an ace and a nine cannot occur together and are *mutually exclusive* events,

$$P(\text{Ace } or \text{ Nine}) = P(\text{Ace}) + P(\text{Nine})$$

$$= \frac{4}{52} + \frac{4}{52}$$

$$= \frac{8}{52}$$

$$= \frac{2}{13} \text{ or } 0.15$$

For mutually exclusive events the addition rule reduces to the sum of the marginal probability of an ace and of a nine.

Example 4

A Related Problem. Suppose some event, A, will occur with probability equal to 0.65 and you are interested in the probability that A will *not* occur.

Since the events "A" and "not A" are mutually exclusive, the sum of the two corresponding probabilities equals one, or P(A) + P(not A) = 1. By the special case of the addition rule, therefore,

$$\begin{aligned} P(\text{A will not occur}) &= 1 - P(\text{A will occur}) \\ &= 1 - 0.65 \\ &= 0.35 \end{aligned}$$

In other words, the probability that an event will not occur equals one minus the probability that it will occur.

Example 5

Conditional Probability. Given two events A and B with marginal probabilities of 0.20 and 0.65, respectively. The joint probability of A and B is 0.15. The conditional probability of A given B is

$$\begin{aligned} P(A|B) &= \frac{P(A \,\&\, B)}{P(B)} \\ &= \frac{0.15}{0.65} \\ &= 0.23 \end{aligned}$$

Analogously, the conditional probability of B given A is

$$\begin{aligned} P(B|A) &= \frac{P(A \,\&\, B)}{P(A)} \\ &= \frac{0.15}{0.20} \\ &= 0.75 \end{aligned}$$

The conditional probabilities are found by dividing the joint probability by the appropriate marginal.

Example 6

Conditional Probability. Given the probability that an employer will receive a qualified worker who lives in the suburbs from a pool of temporary workers is 0.14. Thirty-five percent of the workers in the pool are suburban residents. Find the conditional probability that the employer will receive a qualified worker given the worker is drawn from the suburban office.

$$P(\text{Qualified Worker and Suburban Resident}) = P(Q \,\&\, SU) = 0.14$$

$$P(\text{Suburban Resident}) = P(SU) = 0.35$$

Therefore, the conditional probability of receiving a qualified worker given suburban residence is found as

$$P(Q|SU) = \frac{P(Q \,\&\, SU)}{P(SU)} = \frac{0.14}{0.35} = 0.40$$

Example 7 ***Multiplication Rule.*** Find the probability of drawing two queens by drawing *two* cards *without replacement* from a well-shuffled card deck.

$$P(\text{Queen on first draw}) = P(Q_1) = \frac{4}{52}$$

$$P(\text{Queen on second draw}|\text{Queen on first draw}) = P(Q_2|Q_1) = \frac{3}{51}$$

Therefore, the joint probability of drawing two queens is obtained by the multiplication rule as

$$P(\text{Queen on first draw } \textit{and} \text{ Queen on second draw}) = P(Q_1 \,\&\, Q_2) = P(Q_1)P(Q_2|Q_1) = \frac{4}{52} \times \frac{3}{51} = \frac{12}{2652} = \frac{1}{221} \text{ or } 0.005$$

The joint probability of two queens is obtained by multiplying the marginal probability of a queen on the first draw by the conditional probability of drawing a queen on the second given the queen drawn on the first.

Example 8 ***Special Case of the Multiplication Rule.*** If the above problem were considered *with replacement,* the composition of the deck on the second draw would be the same as on the first, and the events would be *independent.* The probability of two queens is

$$\begin{aligned} P(Q_1 \& Q_2) &= P(Q_1)P(Q_2) \\ &= \frac{4}{52} \times \frac{4}{52} \\ &= \frac{16}{2704} \\ &= \frac{1}{169} \text{ or } 0.006 \end{aligned}$$

Since the two events are independent, the joint probability reduces to the product of the marginal probabilities.

DISCUSSION

Many problems arise where probabilities are given rather than actual frequencies. In such cases, rules like the ones presented in this chapter are useful to find other probabilities based upon the ones given. For example, suppose we are told that two key components of an electronic device fail with probabilities $P(F_1) = 0.1$ and $P(F_2) = 0.2$. The symbol F indicates failure and the subscript denotes the component number. Further, assume we know that the second component fails 70 percent of the time when the first component fails, or $P(F_2|F_1) = 0.7$. The original data regarding failures is not available, and only the above probabilities are known from reported specifications.

Suppose the device to be built with the two components experiences a total breakdown only if *both* parts fail. The probability of total breakdown, therefore, is found as a joint probability of failure of both components using the multiplication rule.

$$\begin{aligned} P(\text{Device breakdown}) &= P(F_1 \& F_2) \\ &= P(F_1)P(F_2|F_1) \\ &= (0.1)(0.7) \\ &= 0.07 \end{aligned}$$

Hence, the chance that the device will fail is 0.07, or 7 times in 100.

On the other hand, suppose the device will break down if *either* of the components fails. There the probability of total device failure is given by the addition rule as

$$\begin{aligned} P(\text{Device breakdown}) &= P(F_1 \textit{ or } F_2) \\ &= P(F_1) + P(F_2) - P(F_1 \,\&\, F_2) \\ &\;\; .1 + .2 - .07 \\ &= 0.23 \end{aligned}$$

Consequently, there is a higher chance, equal to 0.23, that the device will fail if a breakdown can occur as a result of either part failing rather than both.

We could, of course, introduce different types of assumptions about the device used in the above example to show how other probabilities can be obtained from the ones given. For example, we could calculate the conditional probability that one part fails given that the second fails, or we could calculate the probability of device failure based upon simultaneous failure of both components assuming independence. Overall, the example was introduced to illustrate how rules in terms of probabilities can be useful without the direct counting procedure for finding probabilities discussed in the previous chapter.

Rules of probability are useful to find probabilities when the relative frequency method is difficult or impossible to use.

Extensions to More than Two Events. Problems exist where extensions of the addition and multiplication rules to more than two events are required. For our purposes, it is sufficient to consider the extension to the special case of the addition rule for mutually exclusive events. In this case the probability that any one of a number of events occurs is obtained by adding the marginal probabilities of the events considered. For example, in the case of three mutually exclusive events, A, B, and C. the probability that either "A or B or C" occurs is found as

$$P(A \textit{ or } B \textit{ or } C) = P(A) + P(B) + P(C)$$

Addition rule for three mutually exclusive events.

In order to illustrate the rule, suppose a certain portion of enrollments at a large university is made up of students from three poor rural areas, A, B, and C. The probability that a student comes from (a) area A is 0.03, (b) area B is 0.04, and (c) area C is 0.02. The probability that a student comes from any of the three areas is found using the addition rule for three mutually exclusive events:

$$\begin{aligned} P(A \textit{ or } B \textit{ or } C) &= P(A) + P(B) + P(C) \\ &= .03 + .04 + .02 \\ &= 0.09 \end{aligned}$$

Consequently, the probability that a student comes from any poor rural area is 0.09, or 9 percent. If other characteristics beside area were considered such that a student possesses more than one characteristic simultaneously, the general form of the addition rule would be used. The formula is more complicated and is not considered here.

With respect to three events, A, B, and C, the multiplication rule takes the form

$$P(A \& B \& C) = P(A)P(B|A)P(C|A \& B)$$

Multiplication rule for three events.

In other words, the probability that three events, A, B, and C, occur simultaneously is obtained by multiplying the marginal probability of the event A, P(A), by the conditional probability of the event B given A, P(B|A), and the conditional probability of the event C given A and B, P(C|A & B).

As an example, consider drawing 3 items at random from a shipment of 50 items containing 15 defectives; we want to find the probability of selecting 3 defective items in a row. This is found by using the multiplication rule as follows:

$$\begin{aligned}
P(\text{3 Defectives}) &= P(D_1 \& D_2 \& D_3) \\
&= P(D_1)P(D_2|D_1)P(D_3|D_1 \& D_2) \\
&= \frac{15}{50}\left(\frac{14}{49}\right)\left(\frac{13}{48}\right) \\
&= \frac{2730}{117600} \\
&= 0.02
\end{aligned}$$

The probability of three defectives in a row, equal to 0.02, is obtained by multiplying the probability of a defective on the first draw, $P(D_1)$ by the probability of a defective on the second draw given a defective on the first, $P(D_2|D_1)$, and the probability of a defective on the third draw given defectives on the first and second draws, $P(D_3|D_1 \& D_2)$.

Notice the numerator and denominator of the individual probabilities in the above example are different. This results from the sampling process assumed where an item is not replaced before the next draw. The size of the shipment and the number of defectives are reduced by one on each draw. Consequently, the draws are not independent. If, on the other hand, each item were replaced before the next draw, the draws would be independent and each of the conditional probabilities would equal the marginals. For the sampling problem above, the probability of selecting a defective on all draws would be equal and the probability of three defectives in a row becomes

$$\begin{aligned}
P(\text{3 Defectives}) &= P(D_1)P(D_2|D_1)P(D_3|D_1 \& D_2) \\
&= P(D_1)P(D_2)P(D_3) \\
&= \frac{15}{50}\left(\frac{15}{50}\right)\left(\frac{15}{50}\right) \\
&= \left(\frac{15}{50}\right)^3 \\
&= \frac{3375}{125000} \\
&= 0.03
\end{aligned}$$

This last example is an illustration of the multiplication rule for three independent events. For any three independent events, A, B, and C, the multiplication rule takes the form

$$P(A \& B \& C) = P(A)P(B)P(C)$$

Multiplication rule for three independent events.

In other words, the probability that three independent events occur simultaneously is found as the product of their marginal probabilities.

Combining the Rules—An Example. Suppose we are to conduct an experiment where one of two urns containing blue and red chips is chosen, and a chip is drawn from the chosen urn at random. Assume the probability of choosing the first urn is 0.33. The probability of drawing a blue chip from the first urn is 0.8, and the probability of drawing a blue chip from the second urn is 0.4. Based upon the information given, we automatically can find the probability of selecting the second urn and the conditional probabilities of selecting a red chip from each urn. A summary of the available information is presented in the following table where a I is used to denote the first urn and a II is used for the second.

Urn I *P(I) = 0.33*	**Urn II** *P(II) = 0.67*
P(Blue|I) = 0.8	P(Blue|II) = 0.4
P(Red|I) = 0.2	P(Red|II) = 0.6

With this information, let us find the probability that the experiment results in a blue chip, or the marginal probability of a blue chip.

The probability of choosing Urn I *and* a blue chip is obtained using the multiplication rule as

$$\begin{aligned} P(\text{Blue Chip \& Urn I}) &= P(I)P(\text{Blue}|I) \\ &= 0.33(0.8) \\ &= 0.264 \end{aligned}$$

This is found by multiplying the marginal probability of Urn I by the conditional probability of a blue chip given Urn I. In a similar way, we can find the probability of chosing Urn II and a blue chip:

$$\begin{aligned} P(\text{Blue Chip \& Urn II}) &= P(II)P(\text{Blue}|II) \\ &= 0.67(0.4) \\ &= 0.268 \end{aligned}$$

The marginal probability of a blue chip is found using the addition theorem for mutually exclusive events since a blue chip can result from Urn I *or* Urn II.

$$\begin{aligned} P(\text{Blue Chip}) &= P(\text{Blue Chip \& Urn I}) + P(\text{Blue Chip \& Urn II}) \\ &= P(I)P(\text{Blue}|I) + P(II)P(\text{Blue}|II) \\ &= .264 + .268 \\ &= .532 \end{aligned}$$

Hence, the marginal probability of a blue chip equal to 0.532 is obtained by adding the two conditional probabilities obtained earlier.

Now suppose we are interested in determining the probability that a blue chip came from the first urn, or the conditional probability of Urn I *given* a blue chip. From the definition of a conditional probability, we have

$$P(\text{Urn I}|\text{Blue Chip}) = \frac{P(\text{Blue Chip \& Urn I})}{P(\text{Blue Chip})} = \frac{0.264}{0.532} = 0.496$$

This probability is obtained by dividing the joint probability of a blue chip and Urn I by the marginal probability of a blue chip, where each of these values has already been determined above. Instead of finding this probability in parts, we could use one formula that includes all of the steps:

$$P(\text{Urn I}|\text{Blue Chip}) = \frac{P(\text{I})P(\text{Blue}|\text{I})}{P(\text{I})P(\text{Blue}|\text{I}) + P(\text{II})P(\text{Blue}|\text{II})}$$

In general, rules for finding probabilities can be combined to find other probabilities. A simple example is used here to illustrate the point. By characterizing the events in a special way, the result of this example represents another rule called Bayes' Theorem.

By substituting the individual values into this expression the same result is obtained. The result is based upon the combined use of the addition rule, the multiplication rule, and the rule for conditional probabilities.

Bayes' Theorem. When faced with a decision-making problem we must make a judgment regarding a particular aspect of reality, since we do not possess full knowledge of what actually exists or will happen. Frequently, decisions are based upon experience or knowledge gained from similar problems dealt with in the past. In other cases, we seek evidence or additional information and combine it with our experience in order to reach a conclusion. Consequently, we can formulate problems in terms of events either in the form of existing or future conditions and possible indications or evidence of such conditions. When viewed in this way, the existing conditions can be termed *states* or *states of nature,* whereas evidence can be considered in terms of *observable outcomes*.

The table appearing in Exhibit 1 presents examples of situations

EXHIBIT 1
EXAMPLES DISTINGUISHING BETWEEN STATES OF NATURE AND OBSERVED OUTCOMES

State of Nature	*Observed Outcome*
Unknown Disease	Symptoms
Quality of a Shipment of Items	Quality of a Sample
Unknown Author	Piece of Literature
Success of a New Product	Survey Attitudes
Winning Election Candidate	Election Poll Results
Job Performance of a New Employee	Aptitude Test Score

where we can distinguish between events in terms of states of nature and observable outcomes. For each case presented in the exhibit the state of nature represents an existing situation where there is incomplete knowledge. In each case information in the form of an observable outcome can be collected that may provide more knowledge regarding what actually exists.

There are many cases where it is meaningful to characterize events in terms of states and observable outcomes. This section presents a way to combine prior experience about states of nature and observed information in the form of a conditional probability. Computation of these probabilities is based upon the same reasoning that we used in the previous section where we found the conditional probability of a blue chip given Urn I. In that problem the urns represent what we now call states, and the color of the chip corresponds to an observed outcome. The result in these terms is given special consideration because of its impact upon statistical thinking and formal decision-making. The computational procedure is summarized in the form of a formula and is referred to as Bayes' Rule or *Bayes' Theorem.*

If we assume that states of nature are expressed as capital letters and observed outcomes in terms of lower case letters, Bayes' Theorem for two states is given by the formula

$$P(A|a) = \frac{P(A)P(a|A)}{P(A)P(a|A) + P(B)P(a|B)}$$

Bayes' Theorem for two states.

where A and B represent states and "a" is an observed outcome.

In order to illustrate the formula, suppose a rare disease occurs with a probability of 0.0001. Symptoms of this disease occur with a probability of 0.75 when a patient has the disease and with probability of 0.10 when a patient does not have the disease. The following symbols are used:

D = patient has the disease

ND = patient does not have disease

d = patient has symptoms of disease

nd = patient does not have symptoms

In terms of these symbols, we are supplied with the following information:

$$P(D) = 0.0001$$

$$P(ND) = 1 - P(D) = 0.9999$$

$$P(d|D) = 0.75$$

$$P(nd|D) = 1 - P(d|D) = 0.25$$

$$P(d|ND) = 0.10$$

$$P(nd|ND) = 1 - P(d|ND) = 0.90$$

With this information, let us use Bayes' Theorem to find the probability that a patient *has* the disease *given* symptoms are present:

$$P(D|d) = \frac{P(D)P(d|D)}{P(D)P(d|D) + P(ND)P(d|ND)}$$

$$= \frac{.0001(.75)}{.0001(0.75) + .9999(.10)}$$

$$= 0.0008$$

When symptoms are present, the probability of having the disease is 0.0008 or 8 in 10,000.

We also can use Bayes' Theorem to find the probability that the patient has the disease given that *no* symptoms are present. That is,

$$P(D|nd) = \frac{P(D)P(nd|D)}{P(D)P(nd|D) + P(ND)P(nd|ND)}$$

$$= \frac{.0001(.25)}{.0001(.25) + .9999(.9)}$$

$$= 0.00003$$

The chance of having the disease given no symptoms is .00003 or 3 in 100,000.

Note. The conditional probability of having the disease given symptoms is roughly 8 times the marginal probability of having the disease without considering symptoms (i.e., $P(D|d)/P(D) = 8$). The conditional probability of having the disease given the symptoms is 27 times the conditional probability of having the disease given *no* symptoms [i.e., $P(D|d)/P(D|nd) = 27$]. These results are consistent with what we believe to occur. Hence, Bayes' Theorem demonstrates that the conditional probability either is greater if the evidence supports a given state or is less if the evidence does not support that state.

The formula provided for Bayes' Theorem considers two states of nature. This is the simplest case to work with and is sufficient to convey the main concept. The formula can be extended to more than two states; the difference lies in the denominator of the formula. That is, a term similar to the ones appearing in the formula for two states must be added to the denominator for every additional state of nature. For example, in the case of three states, A, B, and C, we have

$$P(A|a) = \frac{P(A)P(a|A)}{P(A)P(a|A) + P(B)P(a|B) + P(C)P(a|C)}$$

Bayes' Theorem for three states.

where P(A|a) represents the conditional probability of State A given the observed outcome "a." Probabilities for the other states are found by replacing the numerator with the appropriate term; the denominator remains the same. In general, the number of terms to be added in the denominator equals the number of states of nature.*

It is interesting to note the origin of Bayes' Rule. It seems that during the middle of the 18th Century Reverend Thomas Bayes, a Presbyterian minister, sought to prove the existence of God with the aid of mathematics. His approach centered upon the probability that the universe resulted from an "intelligent cause," based upon evidence of the orderly phenomena constituting our world. His work was published posthumously and was not considered very seriously for two hundred years. Then reasoning similar to Bayes' was reconsidered and has become the basis of an advancing field of statistics known as Bayesian decision theory.

QUESTIONS AND PROBLEMS

DEFINITIONAL QUESTIONS

1. What is meant by the terms *marginal, joint,* and *conditional probability? Note.* If you have trouble defining these terms, refer to the material presented in Chapter 4.

2. Define the terms *mutually exclusive* and *independent* as they relate to events.

3. What is the addition rule? Identify the individual components of the rule.

4. What is the addition rule for mutually exclusive events?

5. What is the multiplication rule? Identify the individual components of this rule.

6. What is the multiplication rule for independent events?

7. What is meant by the term *state* or *state of nature?*

---

*The formula for the general case of Bayes' Theorem, which applies to any number of states, takes the form

$$P(A_i|a) = \frac{P(A_i)P(a|A_i)}{\sum_{i}^{k} P(A_i)P(a|A_i)}$$

where A_i represents the ith state of nature, "a" represents the observed outcome, and k is the number of states.

8. What is meant by the terms *observed outcome* or *evidence?*

9. Define the following terms:

   (a) Marginal probability of a state.
   (b) Conditional probability of a state given an observed outcome.

10. In words, explain what is meant by Bayes' Rule or Bayes' Theorem.

COMPUTATIONAL PROBLEMS

11. Given two events A and B. Event A occurs with a probability of 0.2 and B occurs with a probability of 0.15. The probability that they will occur simultaneously is 0.03. Find the following probabilities:

    (a) The probability that either A or B will occur.
    (b) The probability of A given that B occurs.
    (c) The conditional probability of B given A.

    What is the joint probability of A and B?

12. Given two mutually exclusive events A and B, where P(A) = 0.3 and P(B) = 0.45.

    (a) What is the probability that either A or B will occur?
    (b) What is the probability that A and B will occur simultaneously?

13. Two events A and B occur with probabilities of 0.25 and 0.30, respectively. Event B occurs 40 percent of the time when A occurs.

    (a) What is the probability that A and B occur simultaneously?
    (b) Using your answer to part (a), find the conditional probability of A given B.

14. Given two independent events A and B where P(A) = 0.6 and P(B) = 0.4. Find the probability that A and B will occur simultaneously. What are the conditional probabilities of A and B?

15. Suppose members of the Senate are to be polled regarding their views toward legislation unfavorable to the oil industry. Sixty percent of the senators are Democrats and eighteen percent of them have opposed legislation against the oil industry in the past. Twelve percent of the Senate is comprised of Republicans who opposed unfavorable oil legislation in the past. Assume members of the Senate are either Republicans or Democrats.

    (a) If one senator is drawn at random, what is the probability that the person is a Republican?
    (b) If one senator is drawn at random, what is the probability

that the person either is a Republican or a senator who previously opposed unfavorable oil legislation?

16. A device is made up of two components. Each component can fail with probabilities of failure equal to 0.06 and 0.08. When the first component fails, the second component fails with a probability of 0.40. The device will not operate if either component fails.

    (a) What is the probability that both components will fail at the same time?
    (b) What is the probability that the device will fail to operate?
    (c) If the second component fails, what is the probability that the first will fail?

17. Two diseases, D_1 and D_2, occur with probabilities 0.01 and 0.02, respectively. Both diseases can occur at the same time with a probability of 0.001. What is the probability that an individual can contract either disease?

18. A large group of individuals are classified in terms of income according to five categories ranging between low and high. The probability that a person falls in each of the categories is given in the following table:

Income Level	Probability
Low	.10
Low Middle	.20
Middle	.40
Upper Middle	.25
High	.05
	1.00

    (a) What is the probability that a randomly drawn individual falls in the low or high income category?
    (b) Find the probability that a person drawn at random is not from a low or high income category.

19. A manufacturing process produces steel rods that are rejected if they are too long or too short. The probability that a rod is too long is 0.02 and the probability that a rod is too short is 0.03.

    (a) What is the probability that a rod is rejected?
    (b) What is the probability that a rod is accepted?

20. Assume that television sets are rated in terms of two characteristics, reception and the need for major repairs within 6 months after expiration of the warranty. If the probability that a TV set will have unacceptable reception is 0.2 and the probability that it needs major repairs within 6 months given acceptable reception is 0.3, what is the probability that a randomly drawn TV

set will have acceptable reception and needs major repairs within 6 months following expiration of the warranty?

21. A shipment of 20 items contains 3 defectives. If two items are drawn from the shipment at random, what is the probability of drawing 2 defectives?

22. Each of two stocks has a probability of increasing in price equal to 0.4. What is the probability that both stocks will increase in price? Assume price movements of the stocks are independent.

23. A baseball player has a batting average of .400, which can be used as an estimate of the probability of getting a hit. What is the probability of two hits in a row for the next two times at bat?

24. Given two events or states A and B with probabilities $P(A) = 0.3$ and $P(B) = 0.7$. Another event or observable outcome, b, occurs with conditional probabilities for each state equal to $P(b|A) = 0.8$ and $P(b|B) = 0.1$. Using Bayes' Theorem, find the probability of state A given b, or $P(A|b)$.

25. A Democratic and a Republican candidate are running for office in a local election. In the past Democrats have won 55 percent of the time. The results of a recent poll indicate the Republican will win. With respect to previous polls of this kind, survey results indicate a Republican victory 25 percent of the time when a Democrat wins and 75 percent of the time when a Republican wins.

    (a) Find the probability that the Democratic candidate will win the election given the results of the poll.
    (b) Find the probability that the Republican candidate will win the election given the election poll result.

26. Sixty-two percent of the sales force of a large insurance company have been rated as acceptable after one year of service with the firm. Of these, 11 percent received a low score on an aptitude test administered prior to actual employment. Of the sales personnel rated as unacceptable after one year of service, 47 percent received a low test score. What is the probability that a sales trainee who receives a low aptitude test score will receive an acceptable performance rating after one year of actual employment?

27. It is known that 90 percent of a large group of suspected criminals actually are guilty and 10 percent are innocent. A particular type lie detector has been shown to be 99 percent reliable when a suspect is innocent and 95 percent reliable when a suspect is guilty. In other words, 99 percent of innocent suspects are correctly evaluated by the lie detector test as innocent and 95 percent of guilty suspects are correctly evaluated as guilty. What is the probability that a suspect actually is guilty given that the lie detector indicates that he or she is innocent?

CONCEPTUAL QUESTIONS AND PROBLEMS

28. (a) What is the relationship between the material presented in this chapter and the concepts presented in Chapter 4 concerning contingency tables?
    (b) Why are rules like the ones provided in this chapter useful?

29. Refer to Problem 22. Is it reasonable to assume that price movements of the two stocks are independent? Explain. What additional information is needed in order to calculate the appropriate probability that the two stocks will increase in price?

30. Refer to Problem 23. In order to find the probability that a player will get two hits for the next two times at bat, it is necessary to assume independence. Is this assumption justified? Explain.

31. Suppose a device made of three components is constructed in such a way that it fails to operate only if all three components fail. Each of the components has the same probability of failure, equal to 0.10. Component failures are independent. Use the multiplication rule for three independent events to find the probability that the device will fail to operate.

32. A batch of 10 items contains 4 defectives. (a) Use the multiplication rule to find the probability that 3 defectives will be randomly drawn in a sample of 3 assuming the item on each draw is replaced before the next draw. (b) Repeat (a) assuming items are not replaced before each draw.

33. Suppose three stocks each have a probability of 0.2 of increasing in price, a probability of 0.5 of staying the same, and a probability of 0.3 of decreasing in price. Find the probability that 2 of the three stocks will decrease in price. Assume price movements are independent. *Hint.* Find the probability that a particular sequence of two stocks will decrease in price and the third will not decrease. Do the same for all possible sequences and apply the addition theorem to obtain the final result.

34. A batch of 20 items contains 5 defectives. Find the probability that a sample of 3 items will contain at least 2 defectives, assuming (a) sampling with replacement, and (b) sampling without replacement.

35. (a) Using the multiplication rule, find the probability of obtaining 3 heads in 3 flips of a fair coin.
    (b) Using the multiplication and addition rules, find the probability of 3 heads in 4 flips of a fair coin.
    (c) Is it possible to find the probabilities in parts (a) and (b) with the procedure presented in Chapter 4? Explain how this would be done. What is the advantage of using the rules of this chapter instead of the basic method presented earlier?

36. New products marketed by a firm vary in terms of success. Product success can be measured in terms of 3 levels of sales: high, moderate, and low volume. Before marketing a new product, a survey is conducted to determine whether the product should be placed on the market. Assume such a survey is conducted for a new product that indicates a moderate sales volume. Marginal probabilities of success and probabilities of moderate volume survey indications conditional upon success levels are given in the following table:

Sales Level	*Probability of Success*	*Conditional Probability of Moderate Survey Result*
High Volume	.25	.1
Moderate Volume	.40	.5
Low Volume	.35	.1

Using Bayes' Theorem for 3 states of nature calculate the conditional probability for *each* state or level of sales based upon a moderate volume survey indication.

37. Suppose a nationwide screening program instituted through the schools is being considered to uncover child abuse. It is estimated that 2 percent of all children are subject to abuse. Further, existing screening programs are able to determine correctly that abuse occurs 92 percent of the time and that abuse is incorrectly suspected 5 percent of the time.

    (a) What is the probability that the results of screening indicating abuse are associated with children that actually are not abused?
    (b) Interpret the result found in part (a).
    (c) What is the probability that the screening program will result in suspicions of child abuse?
    (d) Based upon every 100,000 children screened, how many screenings will be expected to lead to an accusation of abuse?
    (e) Based upon every 100,000 children screened, how many screenings will be expected to lead to a false accusation of abuse?
    (f) Do your answers to the above questions suggest any problems with a national screening policy? Explain.
    (g) Based upon your answer to part (a), is it valid to conclude that 73 percent of the families not abusing children would be falsely accused? Why or why not?

CHAPTER 6

FINDING PROBABILITIES EASILY

Probability Distributions

The public. . . .demands certainties; it must be told definitely. . . .that this is true and that is false! But there are no certainties.

H. L. Mencken

And did not, for he could not, cheat the grave.

Ambrose Bierce

Our treatment of statistics began with the frequency distribution, a basic tool for describing the way data behave. Early in the history of probability and statistics it was discovered that frequency distributions could be represented mathematically and explained as a consequence of the laws of probability. It was this discovery that led to the development of modern statistical methods. The mathematical model used to represent frequency distributions is called a probability distribution. Once we understand the concept and know which distribution to use in a particular situation, finding probabilities with a distribution is easier than applying techniques used in earlier chapters to find probabilities of individual events. The probability distribution is the main device underlying statistical inference and is used often throughout the remainder of the text.

A **probability distribution** can be defined as a rule that assigns probabilities to *every* value of a variable, x, where the values of x correspond to individual outcomes related to some random phenomenon; probability distributions can appear either in the form of a table or as a formula. This chapter provides a general introduction to the concept of a probability distribution and the way it is used to find probabilities.

Instead of dealing with individual events corresponding to random phenomena as before, we shall focus on the entire sample space and work with numerical values attached to *all* possible outcomes. In order to develop the idea of a probability distribution, consider the table appearing in Exhibit 1. The table is comprised of x-values together with a set of values designated as f(x). The f(x)-values are positive numbers that lie between *zero and one* where their *sum* equals one. The f(x)-values possess the same properties as probabilities of events described in Chapter 4. Therefore, the x's can be viewed as points in a sample space that assume *numerical* values. For example, the x-values could be grade point averages, accident rates, or numerically scaled attitudes relating to some social or political issue. Hence, the table presented in the exhibit can be considered a probability distribution of the variable x since the table serves as a rule or way of assigning probability values to each value of x.

EXHIBIT 1

AN EXAMPLE OF A PROBABILITY DISTRIBUTION

x	*f(x)*
1.0	0.05
2.0	0.45
3.0	0.25
4.0	0.15
5.0	0.10
	1.00

In appearance, the table looks very much like a relative frequency distribution, except f(x) is used instead of "p." This is done to distinguish between theoretical probabilities [i.e., f(x)] and observed relative frequencies (i.e., p). The resemblance between the two types of distributions is important, since ultimately the probability distribution is used as a model of empirical results or of frequency distributions.

When dealing with chance phenomena using probability distributions, special terminology is used when referring to the numerical valued quantity, x, whose probabilities are provided by the distribution. We call x a random variable. A **random variable** is defined as a quantity whose numerical value is determined by a random experiment. In other words, the value of x is unknown before observing the result of a random experiment and its numerical value is determined by chance. The probability associated with an individual value of x is obtained from the probability distribution. Another term for a random variable is a *chance variable.*

Some examples of probability distributions are presented in the next section. The way these distributions are used to find probabilities also is illustrated. Related concepts are considered in the discussion section that follows.

The symbols used in this chapter are summarized as follows:

x	= an unknown quantity whose value is determined by chance
	= a random variable
f(x)	= a function or rule used to assign probabilities to values of x
P(x = a)	= the probability that the random variable assumes a numerical value equal to a; sometimes abbreviated as P(a)
F(x)	= the cumulative probability
μ *or* E(x)	= the mean or expected value of a probability distribution
σ^2	= the variance of a probability distribution
σ	= the standard deviation of a probability distribution

EXAMPLES OF PROBABILITY DISTRIBUTIONS

Example 1 Given the following table:

x	f(x)
1	0.1
2	0.2
3	0.3
4	0.4
Total	1.0

Since each of the f(x)-values are non-negative, less than one, and sum to one, the table constitutes a probability distribution. Each f(x)-value represents a probability. For instance, if we want to find the probability that 3 will occur, it is read directly from the table as 0.3. That is,

$$P(x = 3) = f(3) = 0.3$$

Probabilities for the remaining values of x are found in the same manner.

Example 2 Given the following function in *formula* form:

$$f(x) = \frac{x}{10};\ x = 1,2,3,4$$

The values of x following the formula represent the values for which the function is defined.

Since all the x-values are positive, substitution into x/10 will yield positive f(x)-values. The sum of the f(x)'s is determined as

$$\Sigma f(x) = \Sigma \frac{x}{10} = \frac{1}{10} + \frac{2}{10} + \frac{3}{10} + \frac{4}{10} = \frac{10}{10} = 1$$

Therefore, the above function is a probability distribution.

Probabilities corresponding to the x's are found by *substituting* into the formula for f(x). For *all* values of x, the probabilities are found as

$$P(x = 1) = f(1) = \frac{1}{10} = 0.1$$

$$P(x = 2) = f(2) = \frac{2}{10} = 0.2$$

$$P(x = 3) = f(3) = \frac{3}{10} = 0.3$$

$$P(x = 4) = f(4) = \frac{4}{10} = 0.4$$

Note. The distribution in this example is precisely the same as the one given in Example 1; in other words, this particular distribution can be represented either tabularly or algebraically.

Example 3 Given the following probability distribution:

$$f(x) = \frac{x^2}{90}; \quad x = 2,3,4,5,6$$

In order to find the probability that the value 5 will occur, we substitute 5 into the formula as follows:

$$P(x = 5) = f(5)$$
$$= \frac{(5)^2}{90}$$
$$= \frac{25}{90} = 0.28$$

Probabilities for the remaining values of x are found by substituting and solving in the same way.

Note. Consider the distribution given in Example 3 and suppose we are interested in finding the probability that x assumes a value that does *not exceed* 4. This is found by applying the special case of the *addition theorem* of Chapter 5. That is, we add the probabilities of the x-values less than and equal to 4 as follows:

$$P(x \leq 4) = \sum_{x=2}^{4} f(x)$$

$$\sum_{x=2}^{4} \frac{x^2}{90}$$

$$= \frac{(2)^2}{90} + \frac{(3)^2}{90} + \frac{(4)^2}{90}$$

$$= \frac{29}{90} = 0.32$$

In general, the probability that x will assume a value within some range is found by *adding* the probabilities of the x-values within that range.

Example 4 Consider the probability distribution

$$f(x) = \frac{1}{5}; \quad x = 1,2,3,4,5$$

In this case, each value of x is assigned the same value of f(x), equal to 1/5. Hence *all* values of x have a probability of 1/5, or 0.20. For example, the probability that x equals 3 is

$$P(x = 3) = f(3)$$

$$= \frac{1}{5}$$

DISCUSSION

There are a number of ideas related to probability distributions that are similar to frequency distributions presented in earlier chapters. Some of these ideas are discussed in this section.

Graphing Probability Distributions. By graphing a probability distribution we can visualize its underlying shape, just as we did in the case of a frequency distribution. By convention, probability distributions are plotted a little differently than empirical distributions. For example, consider graphs of two distributions illustrated earlier that appear in Exhibit 2. Instead of connecting the dots representing the coordinates of x and f(x), vertical lines connecting the coordinates with

Graphs of probability distributions provide information about their shape similar to that provided by graphs of frequency distributions.

EXHIBIT 2
GRAPHS OF SELECTED PROBABILITY DISTRIBUTIONS

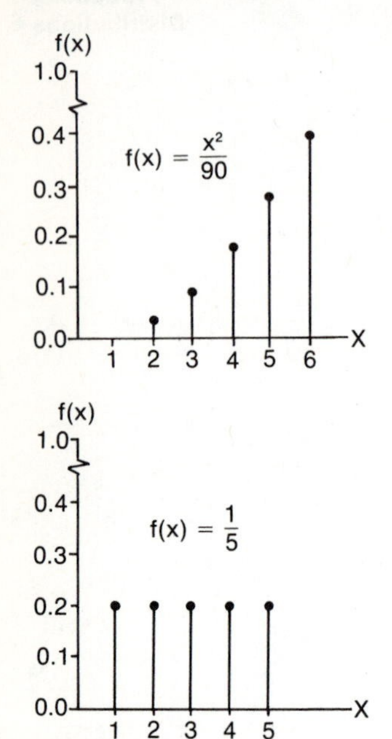

the x-axis generally are used. However, information similar to the case of frequency distributions regarding shape is provided. For the cases diagrammed, we see that they are skewed to the left and uniform. Other probability distributions that are bell-shaped, symmetric, and skewed to the right also exist. Specific examples will be provided in later chapters.

The Cumulative Probability Distribution. Cumulative probability distributions similar to cumulative frequency distributions also can be defined. The **cumulative probability distribution,** designated as F(x), provides the probability that a value of the random variable will not exceed each value of x of the original distribution. Symbolically, the cumulative distribution is defined in the following way:

$$F(a) = P(x \leq a)$$

Definition of a cumulative probability.

where "a" represents any numerical value of x. F(a) represents the probability that x assumes a value that does not exceed any number "a." Consider the example appearing in Exhibit 3.

A cumulative probability or F(x)-value automatically provides the probability of not exceeding a given value of x.

Each value of F(x) in the exhibit is obtained by successively cumulating the individual f(x)-values. An F(x)-value automatically provides the probability of not exceeding each given x-value. For example, the value of F(x) corresponding to x equal to 3 is 0.75. Therefore, we can write

$$\begin{aligned} \text{Probability that x does not exceed } 3 &= P(x \leq 3) \\ &= 0.05 + 0.45 + 0.25 \\ &= F(3) \\ &= 0.75 \end{aligned}$$

The concept of a cumulative probability is important when using tables expressed in terms of a cumulative probability distribution.

The cumulative distribution is important because tabulated probabilities in some instances only are available in cumulative form. In some problems, these are used directly. If, however, we require the probability of an individual value in such instances, we must *subtract* two cumulative probabilities to obtain the required result. Based upon the previous table, suppose we want the probability of x *equal* to 3. This is found as

$$\begin{aligned} P(x = 3) &= P(x \leq 3) - P(x \leq 2) \\ &= F(3) - F(2) \\ &= 0.75 - 0.50 \\ &= 0.25 \end{aligned}$$

Using cumulative probabilities, the probability that x equals 3 is obtained by subtracting the cumulative probability of x equal to 2 from the cumulative probability of x equal to 3.

EXHIBIT 3
PROBABILITY DISTRIBUTION AND CUMULATIVE PROBABILITY DISTRIBUTION

x	*f(x)*	*F(x)*
1.0	0.05	0.05
2.0	0.45	0.50
3.0	0.25	0.75
4.0	0.15	0.90
5.0	0.10	1.00

Characteristics of Probability Distributions. Just as we defined measures describing properties of frequency distributions, we can do the same for probability distributions. The most important of these measures are the mean, the variance, and the standard deviation. Procedures for finding each are the *same* as those for a relative frequency distribution. Their interpretations in terms of location and distribution spread also are similar. The mean and variance (and standard deviation) also can be interpreted in terms of probabilities. This is especially important when explaining the way chance phenomena occur.

Probability distributions also have a mean, a variance, and a standard deviation that provide similar information about location and dispersion.

The mean of a probability distribution is referred to as the mathematical expectation, or the *expected value.* The expected value is designated by one of two symbols, μ (read Mu) or E(X), and is defined by the following formula:

$$\mu \text{ or } E(X) = \Sigma x f(x)$$

Mean or expected value.

The expected value is found by multiplying every value of x by its probability of occurrence and summing. Procedurally, computation is the *same* as in the case of a relative frequency distribution, except the weights are probabilities rather than frequencies.

Suppose we have a probability distribution associated with a particular random phenomenon. Visualize the random phenomenon repeated endlessly. Each repetition will yield values that reoccur according to the relative frequency or probability of occurrence of each value. The expected value of a random variable is the value obtained as the arithmetic mean of the observations resulting from the endless repetition of the random phenomenon. The larger the number of repetitions, the closer the computed mean will be to the expected value or theoretical mean. In other words, we "expect" the average result of a large number of repetitions of a random phenomenon will be *close* to the expected value or mean.

Interpreting the mean or expected value in terms of probabilities.

Consider a simple case in order to illustrate the ideas presented in the last two paragraphs. Suppose you are confronted with a situation where you win \$500 with probability equal to 0.35 and lose \$100 with probability equal to 0.75. The mean or expected value is found as

$$\begin{aligned} E(x) &= \Sigma x f(x) \\ &= 500(.25) + (-100)(.75) \\ &= 125 - 75 \\ &= 50 \text{ or } \$50 \end{aligned}$$

The mean value of \$50 was obtained by multiplying each possible value by its probability of occurrence and adding the results. A minus sign was attached to \$100 since this figure represents a loss.

The result of \$50 can be interpreted as the average amount of money received if you were confronted with the above situation or gamble a large number of times. In other words, various repetitions of the above gamble would yield a return of \$500 or a loss of \$100 resulting in an average return of \$50 over a large number of repetitions.

The variance of a probability distribution is designated by the symbol σ^2 (read Sigma-squared), and is defined by the formula

$$\sigma^2 = \Sigma(x - \mu)^2 f(x)$$

Variance of a probability distribution.

The variance is defined as the sum of the squared deviations of the x-values from the mean weighted by their probability of occurrence. Note once again that this procedure is the *same* as in the case of a frequency distribution, where relative frequencies are replaced by probabilities of occurrence. The *standard deviation,* designated as σ (or Sigma), is obtained in the usual way as the square root of the variance. For our purposes, however, it is not necessary to compute a variance or standard deviation of a probability distribution. It is important to understand what they mean.

As a measure of the spread or width of a probability distribution, the variance indicates the amount of concentration or variability of the x-values within the vicinity of the mean or expected value. This is similar to the interpretation of the variance of a frequency distribution presented in Chapter 3. The concentration of x-values about the mean also is reflected in the probability associated with these x-values. Hence, *the smaller the variance of a probability distribution, the larger will be the probability that an x-value falls within some fixed interval centered about the mean.* Correspondingly, the larger the variance, the smaller the probability that an x-value will fall within a specified interval about the mean. Relatively large variances are associated with greater uncertainty, whereas small variances are associated with less uncertainty. A variance of zero corresponds to absolute certainty.

Interpreting the variance and standard deviation in terms of probabilities.

As an example, consider the variances of the probability distributions of the yields of two investment portfolios. Suppose one has a standard deviation of 4 percent and the other a standard deviation of 7 percent. Since the second has a larger standard deviation, there is greater risk attached to this portfolio. In other words, more uncertainty is associated with the yield of the second portfolio since more variability exists.

Continuous Probability Distributions. The probability distributions illustrated in this chapter all exhibit a certain property; they are associated with x-values that are not connected when considered on a numerical scale. This becomes evident when the distributions are graphed, since spaces exist between the plotted points. In other words, the distributions are defined for a distinct set of values rather than everywhere along an interval. Distributions exhibiting this property are called **discrete.**

There are two types of probability distributions: discrete and continuous.

While the concept of a probability distribution is understood by examining the discrete case, many methods used in statistics rely upon another type of distribution, referred to as a **continuous** probability distribution. A continuous distribution is associated with x-values that may assume *any* numerical value in an interval. Probabilities are obtained as *areas* under the distribution that are associated with a set of values, rather than by direct substitution. Examples of some continuous distributions are depicted in Exhibit 4. Each of the distributions shown in the exhibit is represented by a smooth, unbroken curve associated with a set of x-values plotted on the horizontal axis. This is true of all continuous probability distributions regardless of their shape.

EXHIBIT 4
GRAPHIC ILLUSTRATIONS OF SOME CONTINUOUS PROBABILITY DISTRIBUTIONS

Shaded Areas Represent a Probability, $P(a \leq x \leq b)$

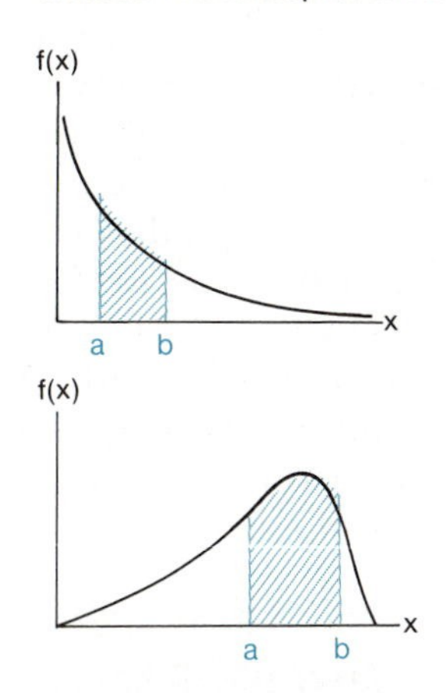

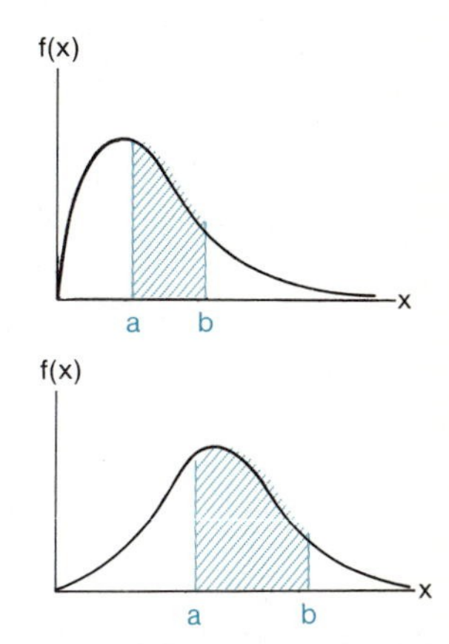

The area under all continuous probability distributions *must* equal one. Consequently, the area under each curve associated with a subinterval of x-values is less than one, and therefore can represent a probability. The shaded regions are included in each diagram appearing in Exhibit 4 to indicate that a probability associated with any interval is represented by an area under the curve. For example, the probability that a value of x falls between any two numbers "a" and "b" is equal to $P(a \leq x \leq b)$.

Whenever continuous distributions are used, probabilities *always* must be found as areas; the probability associated with a single x-value is equal to *zero* and has no meaning. It should be emphasized that, unlike the discrete case, substitution directly into the formula for f(x) is *not* appropriate in order to find probabilities for continuous variables. In general, probabilities are more difficult to calculate when using continuous probability distributions, and rely upon mathematical methods derived from calculus. Fortunately, the distributions we need are available in the form of tables. Hence, there is no need for us to deal with the mathematical considerations at all.

Probabilities for discrete distributions are found by direct substitution, whereas they are found as areas for continuous distributions.

Continuous probability distributions are used widely in statistical work. Surprisingly, they can be used to approximate discrete distributions, and frequently save calculating time and effort. In addition, continuous probability distributions are applied directly to the behavior of various measurements such as sales, consumption, machine tolerances, and so on. In such cases we can think of continuous distributions as models of smoothed histograms, or frequency curves, similar to those presented in Chapter 1.

Some Final Remarks. Probability distributions represent a basic tool that is used as a model of a set of empirical observations. In this chapter we were interested in introducing the concept of a theoretical distribution, which we illustrated with artificial examples. A question arises, however, about the source of distributions that apply to real problems. Much work has been done in the area of applied probability that has resulted in a number of distributions that apply to various situations, based upon given assumptions. When dealing with a particular problem we first look at our assumptions, and, in many cases, we are able to find an established model or distribution that applies.

Other problems exist in which actual data must be collected in order to uncover an appropriate distribution to use. This requires the construction of a frequency distribution or histogram and finding a theoretical distribution that adequately represents the observed distribution.

In the next two chapters we shall become familiar with two commonly used distributions. Once this is accomplished, we can apply the distributions to problems involving data. Other distributions also are presented in later chapters. Regardless of the probability distribution used in a particular situation, it should be emphasized that probabilities determined with the aid of a distribution are given the *same interpretation* as the ones provided in Chapter 4 where we introduced probability for the first time.

QUESTIONS AND PROBLEMS

DEFINITIONAL QUESTIONS

1. Define the terms *probability distribution* and *random variable.*

2. Distinguish between a discrete and a continuous probability distribution.

3. What is a cumulative probability and a cumulative probability distribution?

4. What is meant by the mean or expected value, variance, and standard deviation of a probability distribution?

NUMERICAL PROBLEMS

5. Given the following probability distribution in the form of a table:

x	$f(x)$
6.0	.10
7.0	.40
8.0	.30
9.0	.15
10.0	.05

   (a) Explain why the distribution is a probability distribution.
   (b) What is the probability that x equals 8.0?
   (c) Based upon the distribution, find the probability that x is less than 8.0.
   (d) Find the probability that x does not exceed 8.0 and the probability that x exceeds 8.0.

6. Given the following probability distribution in the form of a formula:

$$f(x) = \frac{x}{18}; \ x = 3,4,5,6$$

(a) Explain why the distribution is a probability distribution.
(b) What is the probability that x equals 5?
(c) What is the probability that x is less than 5? At most 5? Greater than 5?

7. The probability distribution of the number of heads, x, resulting from 3 flips of a fair coin is given by the following table.

Number of Heads, x	*Probability*
0	.125
1	.375
2	.375
3	.125
	1.000

(a) What is the probability of getting exactly 2 heads in 3 flips?
(b) What is the probability that 3 flips will result in a least 1 head?

8. The number of trucks, x, arriving at the loading dock of a distributor of electrical appliances during a typical hour is given by the following table.

Number of Trucks, x	*Probability*
0	.14
1	.27
2	.27
3	.18
4	.09
More than 4	.05
	1.00

(a) What is the probability that no trucks will arrive during a typical hour?
(b) What is the probability that more than 4 trucks will arrive?
(c) What is the probability that exactly 3 trucks arrive during a typical hour? At least 3 trucks? More than 3 trucks?

9. After two years of operation a particular automobile engine requires between 1 and 4 tries before the engine will start. The probability distribution of the number of tries, x, necessary to start the engine is given by the formula on the following page:

$$f(x) = \frac{x^2}{30};\ x = 1,2,3,4$$

(a) What is the probability that the engine will start on the first try?
(b) Construct the cumulative probability distribution in table form of the number of tries necessary to start the engine.
(c) Use the cumulative distribution found in part (b) to find the probability that it takes at least 3 tries to start the engine.
(d) Use the cumulative distribution found in part (b) to find the probability that it takes 2 or 3 tries to start the engine.
(e) Find the mean, or expected value, of the number of tries necessary to start the engine.

10. Refer to the probability distribution in Problem 5 and construct a cumulative probability distribution.

    (a) Based upon the cumulative distribution, find the probabilities specified in parts (b), (c), and (d) of Problem 5.
    (b) Use the cumulative distribution to find the probability that x falls between 7.0 and 9.0.

11. Refer to the probability distribution given in Problem 6 and construct, in the form of a table, a cumulative probability distribution. Based upon the cumulative distribution, find the probabilities specified in parts (b) and (c).

12. Find the mean, or expected value, of the distribution given in Problem 5.

13. Find the mean, or expected value, of the distribution given in Problem 6.

14. Find the mean, or expected value, of the distribution of the number of heads appearing in Problem 7.

15. (a) Plot the probability distribution given in Problem 5 on a graph.
    (b) Plot the distribution given in Problem 6 on a graph.
    (c) Plot the probability distribution of the number of heads given in Problem 7 on a graph.
    (d) Plot the distribution of the number of tries to start the automobile engine in Problem 9 on a graph.

    State which of the distributions are skewed and which are symmetric.

CONCEPTUAL QUESTIONS AND PROBLEMS

16. What is the relationship between a probability distribution and the probabilities computed in earlier chapters?

17. State which of the following variables are discrete and which are continuous.

    (a) Annual family incomes
    (b) Air temperature
    (c) The number of passengers boarding a train
    (d) The weights of a packaged item
    (e) The number of defective light bulbs in the daily output of a manufacturer
    (f) The number of heads resulting from 3 flips of a coin

18. How is a cumulative probability distribution useful?

19. Given below are expressions for the probability associated with a variable x and particular numbers "a" and "b."

    (a) $P(x = a)$
    (b) $P(x < a)$
    (c) $P(x \leq a)$
    (d) $P(x > a)$
    (e) $P(x \geq a)$
    (f) $P(a < x < b)$
    (g) $P(a \leq x \leq b)$

    In words, explain the meaning of each of the above expressions. Does the expression in (a) have any meaning for a continuous variable? Under what circumstances are the expressions in (b) and (c) the same? Under what circumstances are they different?

20. In words, explain how probabilities are found for discrete and continuous random variables.

21. Consider the random experiment where a single die is rolled once and the random variable, x, represents the number resulting from the roll.

    (a) Write an expression for the probability distribution of x.
    (b) What is the probability that the result of a single roll will be a 2? At most 2? Use the expression from part (a) to find these values.
    (c) Find the mean or expected value of the distribution found in part (a). Interpret the result.

22. Consider the random experiment where a coin is flipped 3 times.

    (a) Write down the set of all possible sequences or elements in the sample space that can result from the experiment.
    (b) Using the multiplication rule, find the probability associated with each sequence possible.
    (c) For each sequence, assign a number between 0 and 3 representing the number of heads. Using the addition rule when necessary, find the probability of obtaining each of the possible number of heads.

(d) By letting the symbol x represent the number of heads in 3 flips, construct a table representing the probability distribution of the number of heads. How does the result compare with the distribution given in Problem 7?

(e) What is the relationship between this problem and the material on probability presented in earlier chapters? Does this problem help you to understand the nature of probability distributions in general?

23. Refer to Problem 9, which provides the probability distribution necessary to start a car after two years of operation.

(a) By examining the distribution, what happens to the magnitude of the probabilities for larger numbers of tries? Does your answer make sense when applied to some cars as they get older? Explain.

(b) The probability that the engine will start on the second try is 4/30 or 0.13. Interpret this result in terms of the problem.

(c) The mean, or expected, number of tries equals 10/3 or 3.33. How should this number be interpreted in terms of the problem? How is it related to your answer to part (a)?

24. Suppose you are confronted with an investment opportunity where you can make a profit of $5000 with probability 0.3 or suffer a loss of $2000 with a probability 0.7.

(a) What is the probability distribution associated with this problem?

(b) Find the mean, or expected, profit associated with the investment opportunity. How is this value interpreted?

25. A *fair bet* or gamble is defined as one in which the expected value is equal to zero.

(a) In terms of the meaning of an expected value, what is meant by a fair bet?

(b) Consider the investment opportunity presented in Problem 24. Can this be considered a fair gamble? Explain.

(c) What probabilities associated with the profit and loss figures in Problem 24 would make the investment fair?

26. (a) What is the relationship between probability distributions and frequency distributions presented in Chapter 1?

(b) What information is provided by the mean, variance, and standard deviation of a probability distribution? What are the similarities and differences between these measures and the mean, variance, and standard deviation presented in Chapters 2 and 3 concerned with descriptive statistics?

CHAPTER 7

THE BINOMIAL DISTRIBUTION

Anybody can win unless there happens to be a second entry.

George Alda

There are many situations where the result of a random phenomenon is represented by one of two possibilities. For example, in the area of quality control certain assembly-line items are characterized as acceptable *or* defective. The main question of interest in a poll associated with a two-party election is whether an individual is for *or* against a particular candidate. In medicine, problems sometimes are reduced to determining whether a patient has a particular disease *or* not. Although there are many considerations when introducing a new product, the ultimate question is whether it will *or* will not be successful.

Frequently in situations like those above, it is helpful to know the probability of the *number* of times one of the outcomes will occur in a fixed number of repetitions, where the number of repetitions constitutes a sample of a given size. For example, we may want to know the probability that a certain number of defectives will be observed in a sample of assembly-line output, or we may be interested in the probability that a certain number of respondents prefer a firm's product, based upon a survey of consumers.

If we *assume* that on every repetition the probability of each outcome remains constant and that repetitions are independent, we can use a formula to find the probability of the number of times a particular outcome will occur. Formulas of this kind that are used to generate probabilities are called *probability distributions*. The appropriate formula to be used here is called the *binomial probability distribution*, or simply the binomial distribution, which takes the form

$$f(x) = \binom{n}{x} P^x Q^{n-x};\ x = 0,1,2,\ .\ .\ .,\ n$$

The binomial distribution.

The symbol, x, is used to represent the number of times an outcome will occur, and is referred to as a binomial random variable. That is, the number of times an outcome occurs will vary from sample to sample at random, or by chance, between the values of zero and n.

Since integer values or counts can occur rather than any value between zero and n, the binomial is referred to as a *discrete* probability distribution. This information accompanies the distribution and is found alongside the above formula. The symbol n represents the sample size, or the total number of repetitions. P and Q correspond to the probabilities that an outcome will and will not occur on any of the n repetitions. The *sum* of P and Q is one.

Special attention should be given to the symbol $\binom{n}{x}$, referred to as the *binomial coefficient*, which is given by the formula

$$\binom{n}{x} = \frac{n!}{n!(n-x)!}$$

Formula for the binomial coefficient.

The exclamation point following each term indicates that each is a *factorial;* the term n!, for instance, is read as "n-factorial."

The factorial symbol is a convenient way of representing a more complicated expression that can best be explained with a numerical example; consider "five-factorial." This is given as

$$5! = 5 \times 4 \times 3 \times 2 \times 1$$

Five-factorial.

In general, the *factorial* of a positive integer represents the product of the integer and all integers less than it. By definition, "zero factorial" equals one. That is,

$$0! = 1$$

Zero-factorial.

The binomial probability distribution is important in many areas of application. This chapter provides exposure to the binomial distribution and begins to develop an understanding of special distributions used in practice. After showing how to find probabilities with the formula in the next two sections, other properties of the binomial distribution are discussed.

A summary of symbols used in this chapter is as follows:

x	= a random quantity representing the number of elements in a sample possessing a particular characteristic
	= the number of times an outcome occurs
n	= the number of repetitions, or the sample size
n − x	= the number of times an outcome does *not* occur
P	= the probability that an outcome will occur on a single repetition
Q	= the probability that an outcome will *not* occur on a single repetition
	= 1 − P
f(x)	= a general expression representing the formula or rule for assigning probabilities to values of x
P(x = a)	= the probability that a binomial variable, x, assumes a numerical value of "a" in n repetitions
	= f(a)

$P(x|n,P)$ *is* used sometimes to denote a binomial probability for specific values of n and P

μ = the general symbol for the mean, or expected value of a probability distribution (read as "Mu")

σ^2 = the general symbol for the variance of a probability distribution (read as "Sigma-squared")

σ = the general symbol for the standard deviation of a probability distribution (read as "Sigma")

PROCEDURE

For situations where n independent random trials are performed where one of two possible outcomes can occur on each trial, the probability that one particular outcome will occur "x-times" is found by *substituting* into the formula for the binomial distribution:

$$f(x) = \binom{n}{x} P^x Q^{n-x}$$

or

$$\frac{n!}{x!(n-x)!} P^x (1-P)^{n-x}$$

where P is the probability that the particular outcome will occur on any of the n trials and Q = 1 − P.

> *Note.* The problems in the following examples are solved by substituting into the above formula. For selected values of P and values of n between 1 and 20, **Table 1** in the Appendix can be used as an alternative. When it applies, using the table is the most convenient way of finding binomial probabilities. When reading the examples, it is instructive to compare the solutions with the tabulated values. Instructions for using Table 1 accompany the table in the Appendix.

NUMERICAL EXAMPLES

Example 1 Given a binomial distribution with n = 5, P = 0.3, Q = 0.7. The distribution takes the form

$$f(x) = \binom{n}{x} P^x Q^{n-x}$$

$$= \binom{5}{x} (.3)^x (.7)^{5-x}$$

$$x = 0,1,2,3,4,5$$

(a) Find the probability that x *equals* 2. The answer is obtained by substituting 2 into f(x) and solving.

$$
\begin{aligned}
P(x = 2) &= f(2) \\
&= \binom{5}{2}(.3)^2(.7)^{5-2} \\
&= \frac{5!}{2!(5-2)!}(.3)^2(.7)^3 \\
&= \frac{5!}{2!3!}(.3)^2(.7)^3 \\
&= \frac{5 \times 4 \times 3 \times 2 \times 1}{(2 \times 1) \times (3 \times 2 \times 1)}(.3)^2(.7)^3 \\
&= 10(.09)(.343) \\
&= 0.3087
\end{aligned}
$$

Note. When working with an expression like $\binom{5}{2}$ in terms of factorials, some terms can be cancelled without writing out each factorial completely. For example,

$$
\begin{aligned}
\binom{5}{2} &= \frac{5!}{2!3!} = \frac{5 \times 4 \times 3!}{2!3!} \\
&= \frac{5 \times 4}{2!} = \frac{20}{2} \\
&= 10
\end{aligned}
$$

Since $5! = 5 \times 4 \times 3!$, the two values of $3!$ cancel, which simplifies the work.

(b) Using the same distribution, find the probability that x does not exceed 2, or is *at most* 2. Since x does not exceed 2 in three ways (i.e., x = 0,1, and 2), we must substitute 0,1 and 2 into f(x) and *add* the terms. The basis for adding the three terms is the special case of the addition theorem for mutually exclusive events.

$$
\begin{aligned}
P(x \leq 2) &= \sum_{x=0}^{2} f(x) \\
&= f(0) + f(1) + f(2) \\
&= \binom{5}{0}(.3)^0(.7)^5 + \binom{5}{1}(.3)^1(.7)^4 + \binom{5}{2}(.3)^2(.7)^3 \\
&= \frac{5!}{0!5!}(1)(.7)^5 + \frac{5!}{1!4!}(.3)(.7)^4 + \frac{5!}{2!3!}(.3)^2(.7)^3 \\
&= (1)(1)(.1681) + (5)(.3)(.2401) + (10)(.09)(.343) \\
&= .1681 + .3602 + .3087 \\
&= 0.8370
\end{aligned}
$$

(c) Using the same distribution again, find the probability that x is *greater* than 2. If done from scratch, this problem can be solved by substituting into f(x) three times for the values, x = 3,4, and 5. Since we computed the probability that x does not exceed 2 in part (b), our work is simplified by subtracting that result from 1. That is,

$$\begin{aligned} P(x > 2) &= P(x \geq 3) \\ &= 1 - P(x \leq 2) \\ &= 1 - 0.8370 \\ &= 0.1630 \end{aligned}$$

Note. Part (c) of Example 1 illustrates a basic point. Instead of substituting into the formula for the binomial distribution directly, the probability that x is greater than 2 was found by subtracting the probability that x is at most 2 from 1. This is possible since each f(x)-value represents a probability and the sum of the f(x)-values for all values of x must equal 1. This condition is necessary for the binomial to be a probability distribution. The point is illustrated further in the next example.

Example 2 A balanced or unbiased coin (i.e., P = Q = 0.5) is to be flipped three times (i.e., n = 3). Find the probability of *all* possible number of heads, where x represents the number of heads.

The underlying binomial distribution takes the form

$$\begin{aligned} f(x) &= \binom{n}{x} P^x Q^{n-x} \\ &= \binom{3}{x} (.5)^x (.5)^{3-x}, \\ x &= 0,1,2,3 \end{aligned}$$

where P = Q = 0.5 and n = 3. Because P and Q are equal, the distribution simplifies to

$$f(x) = \binom{3}{x} (.5)^3$$

Since probabilities for all values of x are desired, it is convenient to present the work in the form of a table as follows:

x	f(x)		$\frac{3!}{x!(3-x)!}(.5)^3$				P(x|3,0.5)
0	$\binom{3}{0}(.5)^3$	=	$\frac{3!}{0!3!}(.5)^3$	=	(1) (.125)	=	.125
1	$\binom{3}{1}(.5)^3$	=	$\frac{3!}{1!2!}(.5)^3$	=	(3) (.125)	=	.375
2	$\binom{3}{2}(.5)^3$	=	$\frac{3!}{2!1!}(.5)^3$	=	(3) (.125)	=	.375
3	$\binom{3}{3}(.5)^3$	=	$\frac{3!}{3!0!}(.5)^3$	=	(1) (.125)	=	.125
	Total						1.000

The answers are given in the last column labeled P(x|3,0.5). For example, the probability of 2 heads in 3 flips, P(x = 2|3,0.5), equals 0.375. Notice the sum of the probabilities in the last column is one. This is the case since the sum of the probabilities for all possible outcomes must equal one.

Example 3 Suppose a random sample of four items (i.e., n = 4) is drawn from a continually operating manufacturing process that has been demonstrated to produce 10 percent defective items (i.e., P = 0.10). Find the probability that exactly *two* of the four sample items will be defective and interpret the final result obtained.

The underlying binomial distribution is of the form

$$f(x) = \binom{n}{x}P^xQ^{n-x}$$

$$= \binom{4}{x}(0.1)^x(0.9)^{4-x},$$

$$x = 0,1,2,3,4$$

Substituting into f(x) and solving, we have

$$P(x = 2) = f(2)$$

$$= \binom{4}{2}(.1)^2(.9)^{4-2}$$

$$= \frac{4!}{2!2!}(.1)^2(.9)^2$$

$$= 6 \times 0.01 \times 0.81$$

$$= 0.0486 \text{ or } 0.05 \text{ (rounded)}$$

Interpretation. Using the rounded figure, we can say that roughly 0.05 or 5 percent of all possible samples of size four taken from *this* process will have exactly 2 defectives. Since 2 defectives in a sample of four represents a 50 percent defective *sample* (i.e., 2/4 = 0.5 or 50 percent), we also can say that 5 percent of all possible samples of size four taken from a 10 percent defective process will yield samples that are 50 percent defective.

Example 4 Suppose a firm consistently has a success rate of 25 percent with new products put on the market. If we assume new products marketed by the firm are independent of one another, what is the probability that *one* of the next seven new products will be successful?

Based upon the given information, the binomial distribution of the number of successful products, x, is written as

$$f(x) = \binom{n}{x} P^x Q^{n-x}$$

$$= \binom{7}{x} (.25)^x (.75)^{7-x},$$

$$x = 0,1,2,3,4,5,6,7$$

where n = 7, P = 0.25, and Q = 0.75. Hence,

$$P(x = 1) = \binom{7}{1} (.25)^1 (.75)^6$$

$$= \frac{7!}{1!6!} (.25)(.75)^6$$

$$= 7 \times .25 \times .178$$

$$= 0.3115$$

Example 5 This example is directed toward understanding the binomial and why the formula works. Consider the coin flipping problem of Example 2 where 3 coins are flipped. Using the symbols H and T to represent a head and a tail, let us write down the possible sequences associated with the event "2 heads." Two heads can be obtained in the following three ways:

HHT HTH THH

In other words, we can get 2 heads in 3 flips if 2 heads and then a tail occurs, or a head first, a tail second, and then a head, or a tail first and then 2 heads. The probability of each of the above sequences can be found using the multiplication rule for independent events. If we let the probability of a head equal P and the probability of a tail equal Q, the probability of each sequence can be written as

$$PPQ \quad PQP \quad QPP$$

Each of these probabilities can be written as P^2Q; they are all equal. Since the three sequences correspond to the same event "2 heads," the probability of 2 heads in 3 flips is obtained by using the addition rule for mutually exclusive events. Since the probability of each sequence is the same, we can multiply P^2Q by 3. Hence,

$$P(2 \text{ heads}) = 3P^2Q$$

Notice the number of possible sequences, equal to 3, equals the binomial coefficient $\binom{3}{2}$. This coefficient, therefore, represents the number of ways that three symbols of which two are alike (i.e., 2H's and one T) can be ordered or arranged. In general, the binomial coefficient $\binom{n}{x}$ represents the number of ways that n symbols of which x are alike can be ordered.

If, instead of 2 heads in 3 flips, we want the probability of x heads in n flips we multiply the probability of one sequence where this occurs, P^xQ^{n-x}, by the number of possible sequences or orderings of x heads and n − x tails, $\binom{n}{x}$. The result is

$$P(x \text{ heads}) = \binom{n}{x}P^xQ^{n-x}$$

which is the general term of the binomial distribution. The reasoning used above applies to any two outcomes beside heads and tails.

DISCUSSION

Now that we have been exposed to the binomial distribution and have seen how to compute probabilities using the formula, we shall examine the properties of the distribution in greater detail.

Characteristics of the Binomial. Once we are told a distribution is binomial, the form of the particular binomial distribution is identified by a specific set of values of n and P (where Q is determined once P is known). Quantities such as these that identify a particular form of a specific distribution are referred to as *parameters*. Hence, the parameters of the binomial distribution are n and P. Different values of n and P identify different binomial distributions which lead to different probabilities of the x-values.

A particular binomial distribution is identified by specific values of the parameters n and P. The mean, variance, and standard deviation can be expressed in terms of these parameters.

Just as a frequency distribution can be described in terms of a mean and variance (and standard deviation), so can the binomial probability distribution. For the binomial, however, these can be expressed in terms of its parameters, and are given as

$$\mu = nP$$ *Mean of the binomial.*

$$\sigma^2 = nPQ$$ *Variance of the binomial.*

$$\sigma = \sqrt{nPQ}$$ *Standard deviation of the binomial.*

where μ, σ^2, and σ represent the general symbols used in statistics for the mean, variance, and standard deviation of a probability distribution. For example, if a binomial distribution has parameter values of $n = 20$ and $P = 0.6$, where Q equals 1-P or 0.4, then

$$\mu = nP$$
$$= 20(0.6)$$
$$= 12$$

$$\sigma^2 = nPQ$$
$$= 20(0.6)(0.4)$$
$$= 4.8$$

$$\sigma = \sqrt{4.8}$$
$$= 2.19 \text{ (rounded)}$$

Plotting the Binomial. In order to understand the binomial and its characteristics, we can plot the distribution on a graph. For example, examine the three cases shown in Exhibit 1, all with a value of n equal to 6. In the three cases illustrated, from left to right, the value of P is increased. Notice the skewness varies with the value of P; when P is small, the distribution is skewed to the right and when P is large the distribution is skewed to the left. When P and Q are equal (i.e., $P = 0.5$),

Binomial distributions can be skewed to the right, symmetric, or skewed to the left. They are symmetric when P and Q are equal.

EXHIBIT 1

COMPARISON OF THREE BINOMIAL DISTRIBUTIONS WITH THE SAME SAMPLE SIZE, n = 6

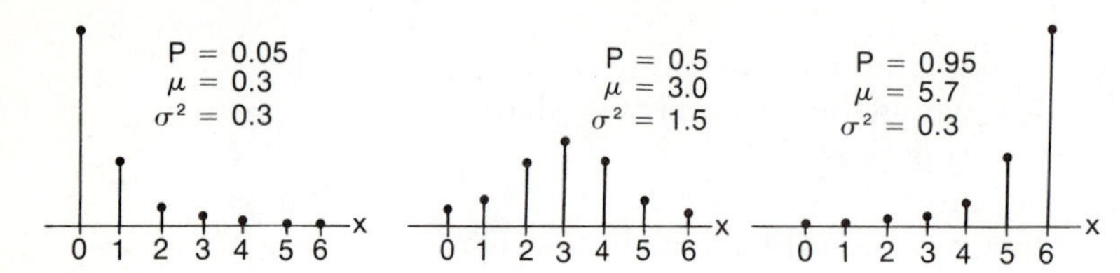

the binomial distribution is symmetric. The point of greatest concentration of probability always is in the vicinity of the mean, which increases as P is increased. The variance is largest when P and Q are equal, which is evidenced by the flatness of the second diagram with higher probabilities in both tails. The smaller the variance, the larger the probability that a value of x falls within the vicinity of the mean.

Regardless of the value of P, the binomial approaches a symmetric distribution as the sample size, n, becomes larger.

The diagrams above depict distributions corresponding to the same sample size. For any value of P other than 0.5 *the binomial approaches a symmetric distribution as the sample size,* n, *is increased.* For example, consider the two cases shown in Exhibit 2. Both distributions describe binomial distributions with the same value of P, equal to 0.25. Although both have "tails to the right," the bulk of the probability tends to be distributed more symmetrically about the mean for n equal to 20 rather than for n equal to 6. Recognize that the means of the two distributions are different, since the mean is larger as n is increased. For values of n greater than 20, the distribution will appear closer to symmetry. This illustration considers two distributions with a specific value of P. If a different value of P were used, the same patterns would appear; however, the degree of symmetry corresponding to the larger sample size would not be the same. In general, the smaller the value of P, the larger the sample size necessary for near symmetry to occur.

EXHIBIT 2

ILLUSTRATION OF A CASE WHERE THE BINOMIAL BECOMES SYMMETRIC AS THE SAMPLE SIZE INCREASES

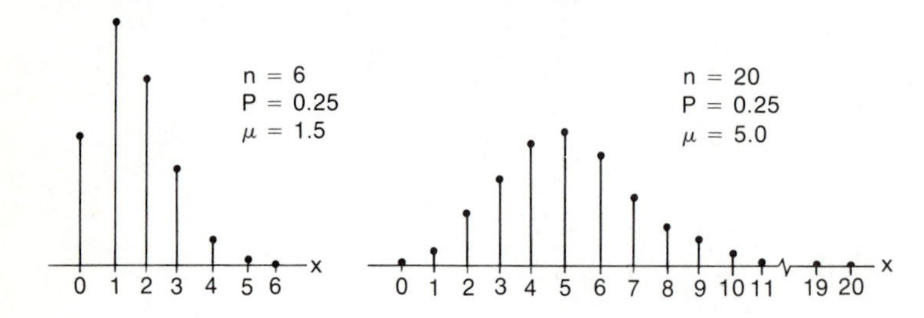

Interpreting the Mean. The mean of a probability distribution describes the central tendency in a distribution similar to the mean of a frequency distribution. This was just illustrated for particular binomial distributions. The mean of the binomial also can be interpreted as the *expected value* or average resulting from a large number of samples. For example, consider a manufacturing problem where a sample of 40 items is drawn from a process producing 10 percent defectives. The mean or expected number of defectives equals 4 (i.e., $40 \times 0.1 = 4$).

We can interpret the mean in the example as the average of the number of defectives observed in a large number of samples drawn at random from the process. Each time we sample 40 items from the process a different number of defectives is observed. The arithmetic mean of these values will approach 4 as the number of samples is increased. Hence, we would "expect" to observe 4 defectives per sample on the average.

An Interesting Application. Some time ago a Las Vegas hotel issued the following newspaper advertisement.*

FIESTA HOLIDAY
2 DAYS/1 NIGHT FOR ONE OR TWO PERSONS
You Pay $15
WE GIVE YOU $10 in FREE Play Chips!
Your NET Cost is $5

Apparently, when a chip was played on a gamble, the player received $1 for winning. Once played, the chip was relinquished whether the player won or lost.

Let us examine the claim made in the advertisement regarding the stated net cost to a purchaser of the plan. We can assume the number of times a player will win in ten tries is binomially distributed. Under the assumption that we are dealing with fair odds, the probability of winning on a single play is 0.5. By fair odds we mean that the chances of winning and losing are equal. In order to obtain a net cost of $5 as advertised, the purchaser would have to receive $10 from gambling (i.e., purchase price minus winnings = $15 − $10 = $5), which means one must win all ten plays. Using the binomial, the probability of winning ten times in ten plays is approximately 0.001 [i.e., $(.5)^{10} = 0.001$]. Since it is quite unlikely to win $10, which will occur one time in one thousand, it also is unlikely that one's net cost will be as low as the $5 claimed in the hotel ad.

A second way to look at the problem is based upon the expected cost, using the mean of the binomial distribution. The expected *number* of wins in ten tries is equal to 5 (i.e., $nP = 10 \times 0.5 = 5$). The expected *winnings*, therefore, equal the expected number of wins multiplied by $1, or $5. The expected *net* cost of the vacation plan is the difference between the purchase price and the expected winnings, or $10 (i.e., $15 − $5 = $10). Using the mean of the binomial we can see that the advertised claim is, on the average, invalid since the claim states the net cost to be $5. Use of the phrase "on the average" implies that the expected winnings of $10 would be realized as the arithmetic mean based on a large number of similar vacation plans purchased over the long-run.

The second form of reasoning was used as the basis for a court

*_Los Angeles Times_, September 2, 1970, Classified Advertisement Section, p. 1. Attention to the problem was originally given by A. R. Sampson in Letters to the Editor, *The American Statistician*, Vol. 28, No. 2, May, 1974.

settlement of a class action suit against the hotel resulting in damages totalling $1,250,000, awarded in the form of unconditionally redeemable certificates to all guests associated with the advertised vacation plan.* This is a very simple application of the use of probability for settling court disputes. The importance of probability and statistics in the courtroom is growing, although more sophisticated techniques than the one cited above often are employed.

It is interesting to note that the settlement in this case was based on the assumption of fair odds. Under this assumption, we saw that the likelihood of the hotel claim was 1 in a 1000. If the gamble was not fair, the probability of winning 10 times in 10 tries would be considerably less. For example, if the probability of winning on one play were 0.4, the probability of winning 10 times reduces to 1 in 10,000, and for 0.3 the probability drops dramatically to 1 chance in 169,000.

A Related Problem Involving Data. The problem in the previous section relied solely upon a theoretical distribution for its solution. No data was used to reach a conclusion. Let us consider another application of the binomial distribution that does require data. The illustration serves as a preview of the kind of reasoning introduced later in statistical inference. When confronted with a gamble, we may want to know whether the probability of winning is equal to 0.5, or whether the odds are fair. This is similar to determing whether the probability of obtaining a head on a flip of a coin is "one half."

Preview of statistical inference.

In order to make this determination, we must collect evidence, or data, by simply repeating the gamble a number of times and recording the number of wins and losses. Suppose we actually undertake 10 repetitions and win 10 times. *If* the probability of winning on a single repetition is 0.5, then the probability of winning all ten, based upon the binomial, is

$$\begin{aligned} P(x = 10) &= \binom{10}{10}(.5)^{10}(.5)^{0} \\ &= (.5)^{10} \\ &= 0.00098 \text{ or } 0.001 \text{ (rounded)} \end{aligned}$$

An event that occurs with a probability of 0.001 can be considered rather unlikely. Consequently, we can reason that such an event occurs so infrequently *under the assumption* that P equals 0.5, that the assumption is not consistent with the observed outcome. In the language of statistics, we say that this difference between the observed result and the hypothesized value of P, equal to 0.5, *is greater than one attributed to chance*, or that the difference is significant. Hence, it is reasonable to conclude that we are dealing with a gamble with a probability of winning that is *greater* than 0.5; that is, it is more likely to observe 10 wins out of 10 if the probability of winning is greater than 0.5.

If one is dealing with a gambling problem, it is more realistic to worry about a bias that is not in our favor. Using similar reasoning, an

**Los Angeles Times*, "Vegas Hotel Payoff—$1.25 Million in Chips," January 17, 1973, Sec. 1, p. 3.

observed outcome that is small (e.g., 0 or 1 in 10 tries) may lead to the conclusion that the probability of winning is *less* than 0.5. In general, this type of reasoning applies to a variety of problems where interest centers upon *testing hypotheses* concerning probabilities and proportions, or percentages.

The Assumptions. In order to use the binomial distribution we assume the probability, P, associated with one of two possible outcomes is constant among repetitions and that the repetitions are independent. These assumptions imply that when we sample, it is done *with replacement*. That is, each item sampled is returned to the population before the next repetition of the sampling process. By replacing each item before the next selection the probability, P, stays the same. In cases where the sample size is small relative to the size of the universe from which it is drawn, the binomial provides a satisfactory *approximation* in cases where sampling is done without replacement. The generally accepted classroom rule to follow is that the approximation is satisfactory when the sample size is *less than 10 percent* of the universe. In practice, the decision regarding acceptability of the approximation should be based upon the degree of precision required of the final result.

The binomial distribution is based upon certain assumptions:

(a) A single trial can result in one of two possible outcomes.

(b) The probability of each of the outcomes is constant from trial to trial.

(c) There are a fixed number of trials that are independent.

Consider again the problem of selecting a sample of items from a continuing manufacturing process. If the items were not selected at random, but in sequence as they were produced, it is possible that process conditions at the time of the selection are influenced by some factor that affects the quality of the items. In such a case the independence assumption would not hold. The same would be true in the case dealing with marketing a firm's new products. If the promotion of an earlier marketed product has a sustained effect on the market environment for the firm's products in general, the assumption of independence can be questioned. Hence, the binomial would not be applicable for computing the probability of a given number of successful products.

QUESTIONS AND PROBLEMS

DEFINITIONAL QUESTIONS

1. What is a binomial probability distribution? A binomial random variable?

2. What is meant by the term *parameter of a probability distribution?* Relate the concept to the binomial distribution.

3. What assumptions must be met for a binomial distribution to be applied to a real life situation?

4. How are probabilities found using the binomial distribution?

5. What are the general expressions for the mean, variance, and standard deviation of the binomial distribution?

6. What is a binomial coefficient? A factorial?

7. What is meant by the terms *sampling with replacement* and *sampling without replacement*?

COMPUTATIONAL PROBLEMS

8. Given the binomial distribution $f(x) = \binom{4}{x}(0.35)^x(0.65)^{4-x}$, find the probability that x equals 3.

9. Given the binomial distribution $f(x) = \binom{5}{x}(0.15)^x(0.85)^{5-x}$, find the probability that x equals 2.

10. Using the formula for the binomial distribution, find the probability that x equals 2 when $n = 3$ and $P = 0.8$.

11. Using the formula for the binomial distribution, find the probability that x equals 4 when $n = 6$ and $P = 0.3$.

12. For the binomial distribution with $n = 4$ and $P = 0.35$, find
    (a) the probability that x does not exceed 2, or $P(x \leq 2)$.
    (b) the probability that x is greater than 2, or $P(x > 2)$.
    (c) $P(1 \leq x \leq 3)$.
    (d) $P(1 < x < 4)$.

13. For the binomial with $n = 5$ and $P = 0.15$, find
    (a) the probability that x does not exceed 3, or $P(x \leq 3)$.
    (b) the probability that x is greater than 3, or $P(x > 3)$.
    (c) $P(2 \leq x \leq 5)$.
    (d) $P(2 < x < 5)$.

14. Using Table 1 entitled "Binomial Probabilities" in the Appendix, find the following probabilities:
    (a) $P(x = 3|n = 4, P = 0.05)$
    (b) $P(x \geq 4|n = 5, P = 0.95)$
    (c) $P(x < 2|n = 6, P = 0.10)$

15. Using Table 1 in the Appendix, find
    (a) $P(x = 4|n = 5, P = 0.85)$
    (b) $P(x \geq 3|n = 4, P = 0.90)$
    (c) $P(x < 2|n = 8, P = 0.10)$

16. What are the values of the mean and standard deviation of a binomial distribution with $n = 50$ and $P = 0.2$?

17. What are the values of the mean and standard deviation of a binomial distribution with $n = 175$ and $P = 0.05$?

18. A manufacturer of small motors knows from past experience that 5 percent of all motors produced do not start on the first try after production. Assuming the binomial is applicable, what is the probability that two motors in a sample of ten selected from the production line will not start on the first try?

19. It was discovered that a bookkeeper incorrectly records entries 3 times in 100, on the average. Assuming incorrect entries are recorded independently, what is the probability that the bookkeeper will make no recording errors in the next 5 entries?

20. A panel of 16 people is to be selected at random from a very large group of consumers consisting of an equal number of men and women. Find the probability that
    (a) the panel will consist of an equal number of men and women.
    (b) the panel will contain at least 50 percent women.

21. If 1 out of 10 smokers eventually develops lung cancer and 1 out of 100 nonsmokers develops the disease, what is the probability that
    (a) 2 persons in 20 randomly drawn smokers develop lung cancer?
    (b) 2 persons in 20 randomly drawn nonsmokers develop lung cancer?

CONCEPTUAL QUESTIONS AND PROBLEMS

22. Cite two original examples where the binomial might apply; specifically identify the basic random variable and the meaning of the parameters in each case.

23. Demonstrate that the binomial coefficient $\binom{n}{x}$ equals $\binom{n}{n-x}$ and illustrate this with a specific numerical example.

24. What information is provided by the mean, variance, and standard deviation of the binomial distribution?

25. Comment upon the validity of the independence assumption associated with the following survey selection methods aimed at determining opinions about a particular moral issue.
    (a) A sample of individuals is chosen randomly from an entire city.
    (b) One person is chosen randomly and an opinion is sought ten separate times during the course of one year.
    (c) A sample of the city consisting of ten neighbors of the same socioeconomic bracket living in adjacent houses and belonging to the same community organization are chosen.

26. Of the following situations listed, which can reasonably be assumed to conform to the assumption of independence?

    (a) A sample of 10 items chosen consecutively from an assembly line in order to determine the percentage defective.
    (b) A random sample of 10 items chosen from a large shipment in order to estimate the shipment quality.
    (c) Five consecutive bookkeeping entries are selected in order to determine a clerk's rate of committing errors.

27. A portfolio consists of 10 stocks. The investor feels the probability that a stock will go down in price is 0.35 and that the price movements are independent. What is the probability that exactly 4 will decline? That 4 or more will decline? Does the assumption of independence here seem reasonable? If not, is the binomial distribution the appropriate probability distribution for this problem?

28. A quality control inspector is concerned about the number of defectives being produced by three (identical) machines. It is assumed that "in the long run" all three produce the same proportion, P, of defectives. Ten items are made on each machine representing independent trials.

    (a) What is the probability distribution of the number of defectives produced on machine one?
    (b) What is the probability distribution of the total number of defectives produced?

29. When sampling without replacement from a shipment of items in order to determine shipment quality, in which of the following two circumstances is it safe to assume that the proportion defective is relatively constant from draw to draw?

    (a) A sample of 5 items is chosen from a shipment of 35 items.
    (b) A sample of 20 items is chosen from a shipment of 2000 items.

    What is the general rule for using the binomial distribution in problems where sampling is done without replacement?

30. Given the binomial distribution $f(x) = \binom{5}{x}(.05)^x(.95)^{5-x}$:

    (a) Is the distribution skewed?
    (b) Plot the distribution on a graph. Does the graph support your conclusion in part (a)? Explain.
    (c) In general, under what circumstance(s) is a binomial distribution symmetric? What happens to the variance under these circumstances? Explain.

31. The following two values correspond to Problem 18:

    (a) $P(x = 3) = 0.01$
    (b) $\mu = 0.5$

    Interpret these values specifically in terms of the problem.

32. The following two values correspond to Problem 20:

    (a) $P(x = 4) = 0.03$
    (b) $\mu = 8$

    Interpret these values specifically in terms of the problem.

33. Given a well-balanced coin, which is more likely to be observed: (a) 2 heads in 4 flips, or (b) 10 heads in 20 flips? Explain.

34. For a binomial distribution, is it true that the mean is the most likely value? Explain.

35. A large manufacturer has determined that 5 percent of the orders received come from new customers. This figure has been established over a long period of time and has remained stable. On a particular day, 40 orders were received.

    (a) Do you think the binomial probability distribution would be appropriate for determining the probability that x of these 40 orders would be from new customers? Why?
    (b) Assuming the binomial is appropriate, what is the expected number of these 40 orders that will be from new customers?
    (c) Do you think that precisely the number calculated in (b) will be from new customers? Why or why not?

CHAPTER 8

THE NORMAL CURVE

It was without a compeer among. . . . It was perfect, it was rounded, symmetrical, complete. . . .

Mark Twain

There are many probability distributions currently in use besides the binomial. One distribution that is used more than the others is called the *normal probability distribution*. It also is referred to as the normal distribution, the normal curve, and "the normal." Unlike the binomial, which applies to problems with one set of assumptions, the normal can be applied to many different situations.

An easy way to visualize the kinds of things described by the normal curve is to think of the smoothed histograms of Chapter 1 that were symmetric with a single peak. Many phenomena such as people's heights or weights, IQ's, dimensions of manufactured products, and sometimes department store sales for a given day have histograms of this kind. Under certain circumstances, the normal curve can be used to approximate the binomial distribution. In later chapters we shall use the normal distribution a great deal since it is useful for describing the way arithmetic means of samples are distributed.

An understanding of the normal and an appreciation of situations described by this distribution will be developed gradually in the next few chapters. This chapter has two objectives: (1) to introduce the normal curve and its properties, and (2) to develop the procedure for computing probabilities using the distribution.

The normal distribution is characterized by the following formula:

$$f(x) = \frac{1}{\sqrt{2\pi}\,\sigma} e^{-\frac{1}{2}\left(\frac{x-\mu}{\sigma}\right)^2}, \infty < x < \infty$$

Normal curve.

The formula represents the probability distribution of x, which stands for some measured value that varies at random according to the normal curve. Therefore, we can refer to x as a normally distributed random variable. The formula for f(x) is the basis for generating probabilities of variables that are normally distributed.

The formula for the normal curve appears complicated. Although we do not have to work with it directly in order to find probabilities, it is instructive to describe each of the components of the formula. The symbols π (read as Pi) and "e" are natural constants arising in mathematics. Numerically, they are given as

$$\pi = 3.14159 \ldots$$

and

$$e = 2.71828 \ldots$$

The dots trailing after each value signify unending decimals, where additional terms in each series provide a closer approximation to the true value.

Once it is established that a distribution is normal, a particular normal distribution is identified by specific values of μ and σ (read as Mu and Sigma). Different values of μ and σ represent different normal distributions and are referred to as the *parameters* of the normal. These parameters are the *mean* and the *standard deviation**.

When plotted in the form of a graph, the normal appears as a smooth, unbroken curve, as shown in Exhibit 1. This smooth appearance of the normal when graphed represents a key difference between the normal and the binomial distributions. The binomial is associated with x-values representing integer values or counts, whereas the normal is associated with *all* numerical values along the horizontal scale. Therefore, the normal is referred to as a *continuous* probability distribution rather than as a discrete one.

EXHIBIT 1
GRAPH OF A NORMAL CURVE

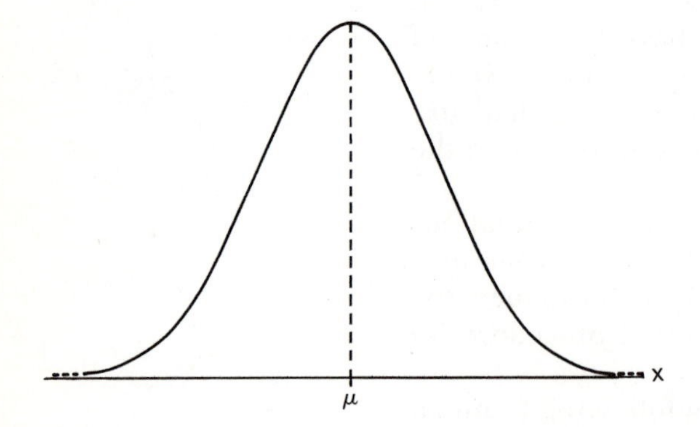

*An inconsistency exists regarding the way the symbols μ and σ are used, based upon certain conventions in the field of statistics. In general, they are used to denote the mean and standard deviation of any probability distribution, but also have been retained as the parameters of the normal curve. Mathematically, it can be shown that the two parameters are the mean and the standard deviation. Later, the same symbols will be used to denote the mean and standard deviation of populations or universes.

The normal is "bell-shaped" and is defined for values of x assuming any real number. The distribution has a single peak in the center, and drops smoothly in either direction of the peak, assuming lower and lower values as it approaches the limits of plus-and-minus infinity. Another way of describing this property is to say that it is *asymptotic* to the x-axis, or it continually approaches the x-axis but "never" touches it. As a result, it is not possible to draw the entire curve. The normal distribution is *symmetric* about the mean, μ.

The entire *area* under any normal curve is equal to one. This is essential for finding probabilities. Instead of substituting into the formula for the normal to find probabilities, probabilities are found as *areas* under the curve that are assigned to *intervals* rather than individual values of x. Since the total area under the curve is unity, the area under the curve corresponding to an interval of x-values is less than one and represents a probability.

Symbols used in this chapter are summarized as follows:

x = a value of a variable or measured quantity that varies at random
μ = the mean
σ = the standard deviation
z = the "standardized variate," needed when using Table 2 in the Appendix to find normal probabilities
$P(a \leq x \leq b)$ = the probability that a random variable x assumes a value in an interval between two specific numbers, "a" and "b"
$P(\mu \leq x \leq a)$ = the probability that a random variable x assumes a value in an interval between the mean and a specific number "a"

PROCEDURE

Due to the nature of the normal distribution, probability computations have been simplified and have been reduced to an elementary calculation that is used with a set of tabulated values given in the Appendix. The procedure for finding normal probabilities basically involves the following three steps:

Step 1. For given values of μ and σ and a specific numerical value of x, compute the standardized variate, z, by substituting into the following formula:

$$z = \frac{x - \mu}{\sigma}$$

Formula for the standardized variate.

In words, z is found by subtracting the mean from a value of x and dividing by the standard deviation.

Step 2. Refer to the table of "Areas of the Normal Curve" in the Appendix **(Table 2)**; the table provides areas corresponding to positive values of z. The use of the table is explained in the Appendix.

Step 3. From the table, determine the area corresponding to the value of "z" computed in Step 1.

> *Note.* Due to the way Table 2 is constructed, the three steps may have to be repeated a second time in order to determine a required probability.

The following points should be kept in mind:

(a) The entire area under the normal curve is equal to one.
(b) Because the normal distribution is symmetric, the area under the curve on either side of the mean, μ is equal to *0.5.*
(c) The table of Areas of the Normal Curve is tablulated only for the *right half* of the normal curve.
(d) All probability computations involving the normal are reduced to finding the probability.

$$P(\mu \leq x \leq a)$$

In other words, areas in the table correspond to probabilities for intervals between the mean, μ, and a number, "a," which is greater than the mean. The area tabulated corresponds to the shaded region in the diagram appearing in Exhibit 2. *The value of "z" computed in Step 1 pro-*

EXHIBIT 2
AREA TABULATED IN THE TABLE OF "AREAS OF THE NORMAL CURVE"

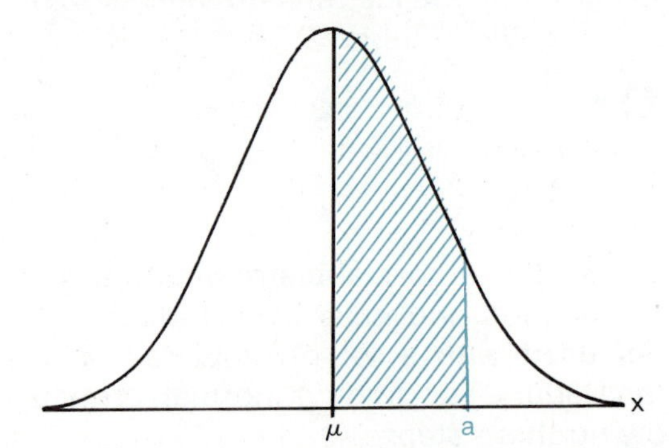

vides the area in the shaded region from the table directly. When "a" is to the left of the mean, the value of z will be negative; by ignoring the negative sign the value of z also will provide the required probability directly. In all other cases, the three steps listed above must be performed twice. Two values of z are obtained, and two probabilities that must be added or subtracted according to the specific probability required. This becomes clear by examining the examples that follow.

NUMERICAL EXAMPLES

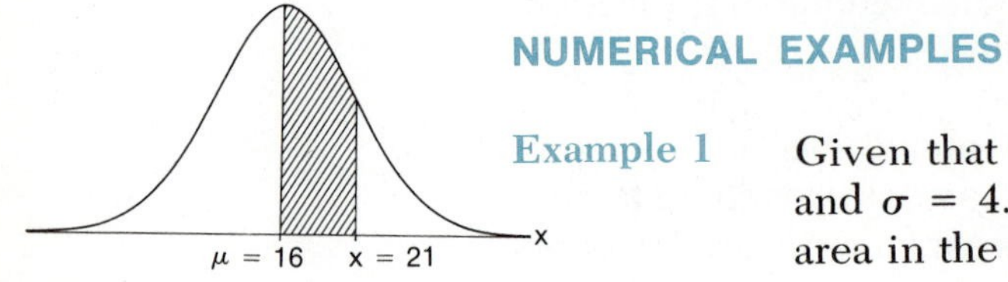

Example 1 Given that x is normally distributed with $\mu = 16$ and $\sigma = 4$. Find $P(16 \leq x \leq 21)$ or the shaded area in the diagram to the left.

Step 1. Compute, $z = \frac{x - \mu}{\sigma} = \frac{21 - 16}{4}$

$$= \frac{5}{4}$$

$$= 1.25$$

Step 2. From Table 2, z = 1.25 gives an area of 0.3944.

Step 3. Therefore, from the table

$$P(16 \leq x \leq 21) = 0.3944$$

Example 2 Given that f(x) is normal with $\mu = 3$ and $\sigma = 0.5$. Find $P(2 \leq x \leq 3.5)$.

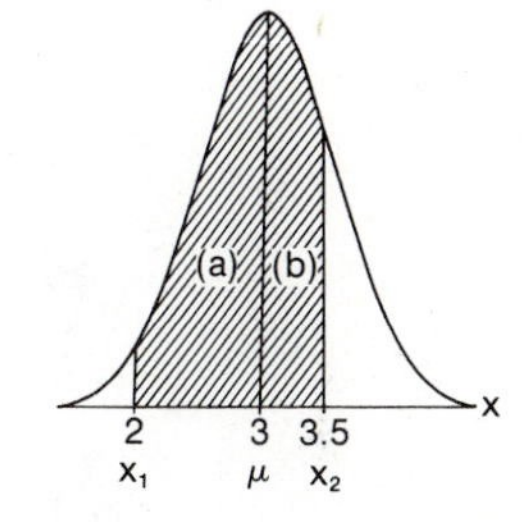

Due to the way the table of Areas of the Normal Curve is constructed, this problem must be handled in *two parts*, each corresponding to different sides of the mean labeled as (a) and (b) in the diagram to the right.

(a) $z_1 = \frac{x_1 - \mu}{\sigma} = \frac{2 - 3}{0.5}$

$= -2.00$ (left side of mean)

(b) $z_2 = \frac{x_2 - \mu}{\sigma} = \frac{3.5 - 3}{0.5}$

$= 1.00$ (right side of mean)

From Table 2, $z_2 = -2.00$ gives an area of 0.4772 (minus sign ignored when using table), and $z_2 = 1.00$ gives an area of 0.3413. These two areas must be *added* to obtain the desired probability. That is,

$$P(2 \leq x \leq 3.5) = 0.4772 + 0.3413$$

$$= 0.8185$$

Example 3 Given that a random variable is normally distributed with $\mu = 400$ and $\sigma = 75$. What is the probability that x falls between 425 and 475? This problem also must be done in two parts, but for a different reason. In this case, the required probability is associated with an interval whose values do not include the mean. This is shown in the diagram to the right.

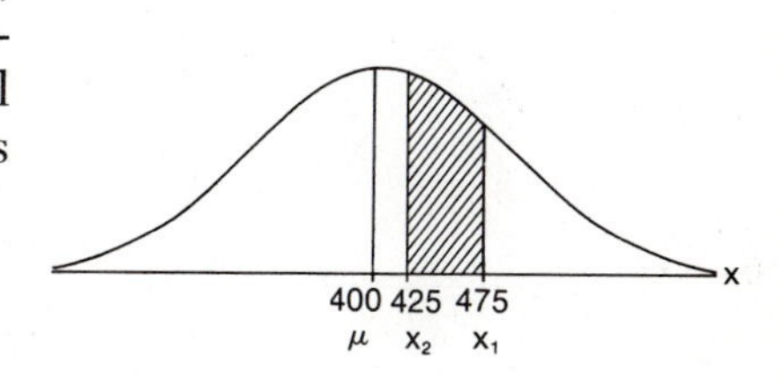

(a) $z_1 = \frac{x_1 - \mu}{\sigma} = \frac{475 - 400}{75}$

$= 1.00$ (corresponds to area between the mean and 475)

(b) $z_2 = \frac{x_2 - \mu}{\sigma} = \frac{425 - 400}{75}$

$= 0.33$ (corresponds to area between the mean and 425)

Using the table of Areas of the Normal Curve, $z_1 = 1.00$ yields an area of 0.3413, and $z_2 = 0.33$ yields an area of 0.1293. The two areas must be *subtracted* in order to find the required probability. Therefore,

$$P(425 \leq x \leq 475) = 0.3413 - 0.1293$$
$$= 0.2120$$

The reason the two areas are subtracted is that the first area equal to 0.3413 represents the area between 400 and 475 and includes the area between 400 and 425. By subtracting the area between 400 and 425 from 0.3413, we are left with the area between 425 and 475.

Example 4 Given that f(x) is normal with $\mu = 33$ and $\sigma = 8$. Find the probability that x falls between 28 and 31.

This problem is similar to Example 3, except we are dealing with an area to the *left* of the mean, which appears in the diagram below.

(a) $z_1 = \frac{x_1 - \mu}{\sigma} = \frac{28 - 33}{8}$

$= -0.63$ (corresponds to area between 28 and the mean)

(b) $z_2 = \frac{x_2 - \mu}{\sigma} = \frac{31 - 33}{8}$

$= -0.25$ (corresponds to area between 31 and the mean)

From Table 2, $z_1 = -0.63$ gives an area of 0.2357, and $z_2 = -0.25$ yields an area of 0.0987. The minus sign of both z values is ignored when using the table. *Subtracting* the two probabilities gives the desired result. That is,

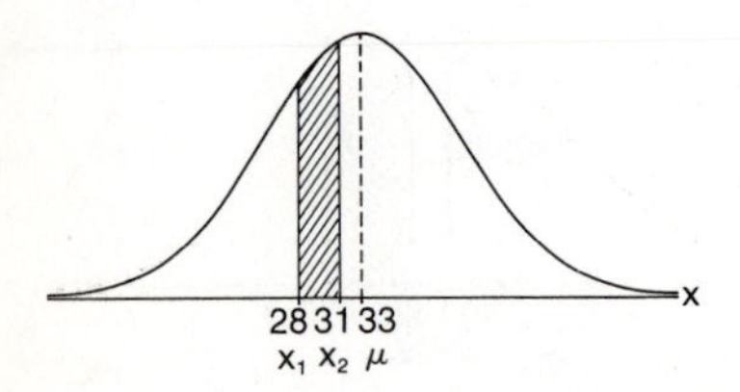

$$P(28 \leq x \leq 31) = 0.2357 - 0.0987$$
$$= 0.1370$$

Example 5 Given f(x) which is normal with $\mu = 33$ and $\sigma = 8$. Find the probability that x is *greater* than 43, or $P(x > 43)$. The required area is depicted in the diagram to the right. For this problem only one z-value must be computed; however, the corresponding probability must be subtracted from 0.5 in order to obtain the required probability, which lies in the "upper tail" of the distribution.

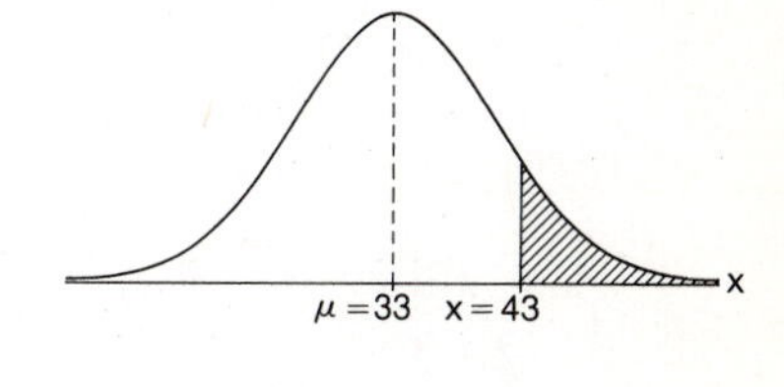

$$z = \frac{x - \mu}{\sigma} = \frac{43 - 33}{8}$$

$= 1.25$ (corresponds to area between mean and 43)

From Table 2, $z = 1.25$ yields an area of 0.3944. *Subtracting* this from 0.5, we get

$$P(x > 43) = 0.5 - 0.3944$$
$$= 0.1056$$

Example 6 Given that f(x) is normal with $\mu = 33$ and $\sigma = 8$. Find the probability that x is *less* than 21.

This problem is the same as Example 5 except the required probability lies in the "lower tail" of the distribution, which is shown in the diagram to the right.

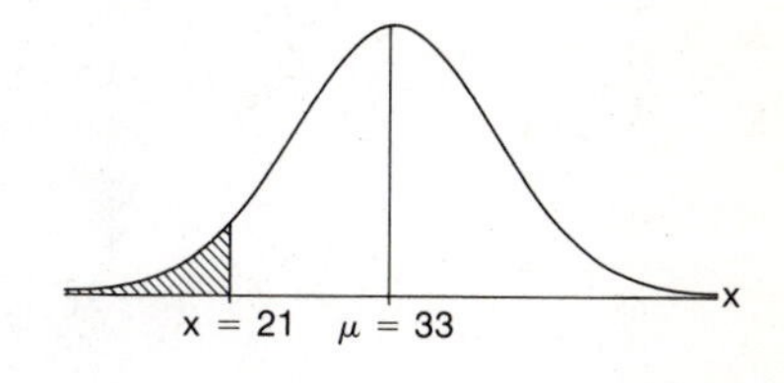

$$z = \frac{x - \mu}{\sigma} = \frac{21 - 33}{8}$$
$$= -1.50$$

From Table 2, $z = -1.50$ (ignoring sign) yields an area of 0.4332. *Subtracting* this from 0.5, we get

$$P(x < 21) = 0.5 - .4332$$
$$= 0.0668$$

Example 7 Suppose the distribution of weights (x-values) of items produced by a manufacturing process is "closely approximated" by a normal curve with mean weight equal to 13 pounds and standard deviation equal to 3 pounds. What *proportion* of the items produced have weights between 7 and 19 pounds?

Based upon the definition of probability given in Chapter 4, the proportion required in the problem is the same as the probability that an item has a weight between 7 and 19 pounds.

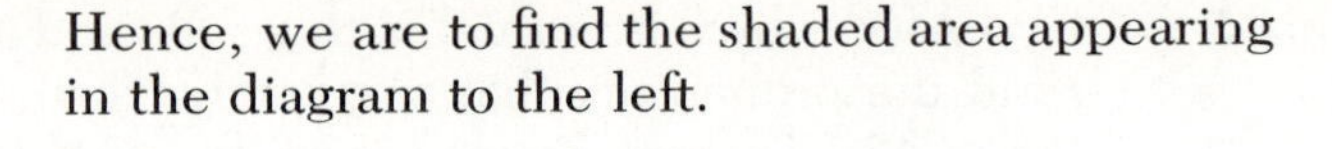

Hence, we are to find the shaded area appearing in the diagram to the left.

$$z_1 = \frac{7 - 13}{3} = -2.00$$

$$z_2 = \frac{19 - 13}{3} = 2.00$$

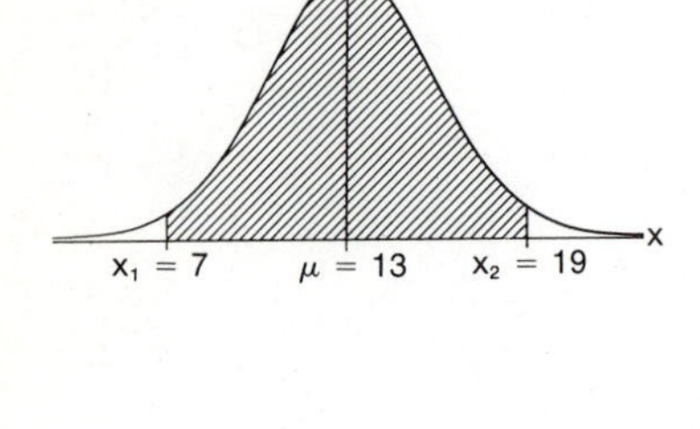

Both $z_1 = -2.00$ and $z_2 = 2.00$ yield an area of 0.4772 from Table 2. Therefore,

$$\begin{aligned} P(7 \leq x \leq 19) &= 0.4772 + 0.4772 \\ &= 2(0.4772) \\ &= 0.9544 \end{aligned}$$

Hence, the proportion of items with weights between 7 and 19 pounds equals 0.9544.

Note. Since the absolute difference between 7 and 13, and 19 and 13 is the same, equal to 6, the problem could be restated as finding the probability that an item has a weight *within* 6 pounds of the mean weight. Hence, only one z-value is necessary; the answer is obtained by multiplying the corresponding probability by 2.

Example 8 ***A Related Problem.*** Suppose we consider the normal curve with $\mu = 400$ and $\sigma = 75$ again. This time, we want to find the *value* of x below which the *lowest* 33 percent of x-values fall.

This is an exercise in the use of Table 2 that does not require a probability computation, but is useful in later work. In this case the area is given and you must *find the value of x*. Consider the diagram below.

You are told that the area to the *left* of the unknown x-value is 0.33. Since the table of Areas of the Normal Curve deals only with the cross-hatched region, or 0.17 (i.e., 0.5 − 0.33 = 0.17), it is this value that is found in the body of Table 2. Referring to the table, an area of 0.17 yields a value of z equal to 0.44. Since x is below the mean, z must be changed to −0.44. If we substitute all the known values into the expression of z, we can solve for x. That is,

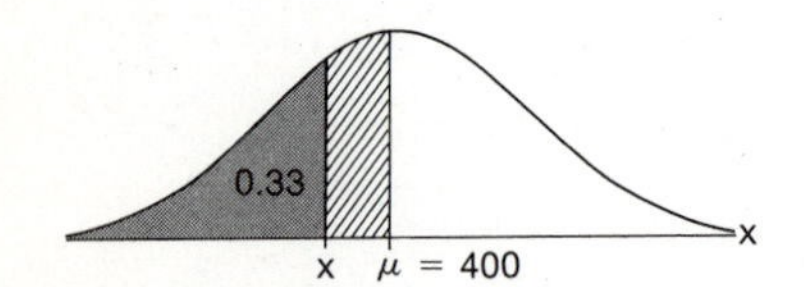

$$z = \frac{x - \mu}{\sigma}$$

$$-0.44 = \frac{x - 400}{75}$$

$$x = 400 - 0.44\,(75)$$

$$= 400 - 33$$

$$x = 367$$

Hence, 33 percent of the x-values lie below 367.

Note. Recognize that substituting and solving for x is equivalent to subtracting the standard deviation multiplied by z from the mean. The formula for x *less* than the mean takes the form

$$x = \mu - z\sigma$$

Formula for finding x below the mean when the normal probability is given.

When x is *above* the mean, the formula to find x for a given probability is given as

$$x = \mu + z\sigma$$

Formula for finding x above the mean when the normal probability is given.

Here, the standard deviation multiplied by z is added to the mean.

DISCUSSION

The use of the word "normal" to describe a probability distribution seems to imply that most random phenomena are distributed according to this distribution. This belief is supported by the frequency with which the normal curve appears in introductory textbooks as well as in a large number of applied problems. Although its usefulness is widespread, it should be emphasized that there are many phenomena that are described by *other* probability distributions. With this caution in mind, we shall explore some additional points about the normal.

Characteristics of the Normal. We stated that the parameters of the normal distribution are the mean and standard deviation. It is instructive at this point to examine these characteristics in more detail. If we assume for the moment that the standard deviation is held constant, a change in the value of μ, the mean, induces a shift in the peak of the distribution. The overall appearance of the curve remains the same, however. For example, consider the diagrams appearing in Exhibit 3. The three curves are normal with the same value of σ, but with different mean values of μ_1, μ_2, and μ_3. The different values of the mean correspond to different locations of the peak of the three distributions, which

A particular normal distribution is identified by its parameters, μ and σ, which are the mean and the standard deviation.

EXHIBIT 3
NORMAL CURVES WITH DIFFERENT MEANS

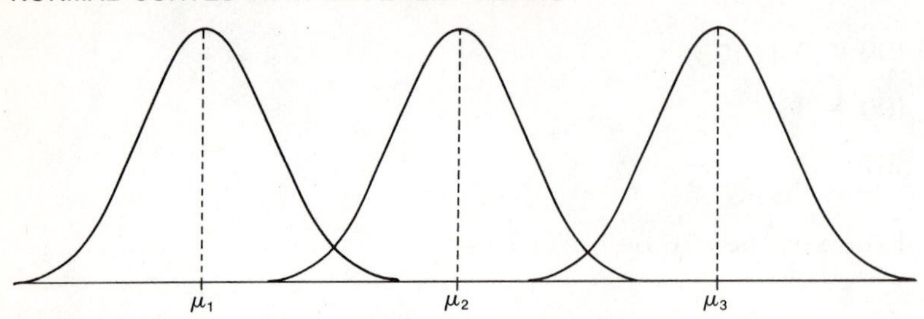

means the "bulk" of the area of the three curves also has a different location on the axis. This interpretation of the mean is similar to the one for frequency distributions presented in Chapter 2.

Now consider the case where the mean is held constant and we vary σ. This is depicted in Exhibit 4. Of the two distributions shown in the exhibit, the "narrower" curve has the smaller standard deviation, or σ_1 is less than σ_2. Also, we can say that the probability of an x-value falling within any set of fixed limits centered on the mean μ is larger for the distribution with the smaller standard deviation. In other words, it is more likely that a value falls within the vicinity of the mean when the standard deviation is smaller.

The z-value. In order to obtain "normal probabilities," it was necessary to compute a value of "z" corresponding to a particular value

EXHIBIT 4
NORMAL CURVES WITH DIFFERENT STANDARD DEVIATIONS, $\sigma_1 < \sigma_2$

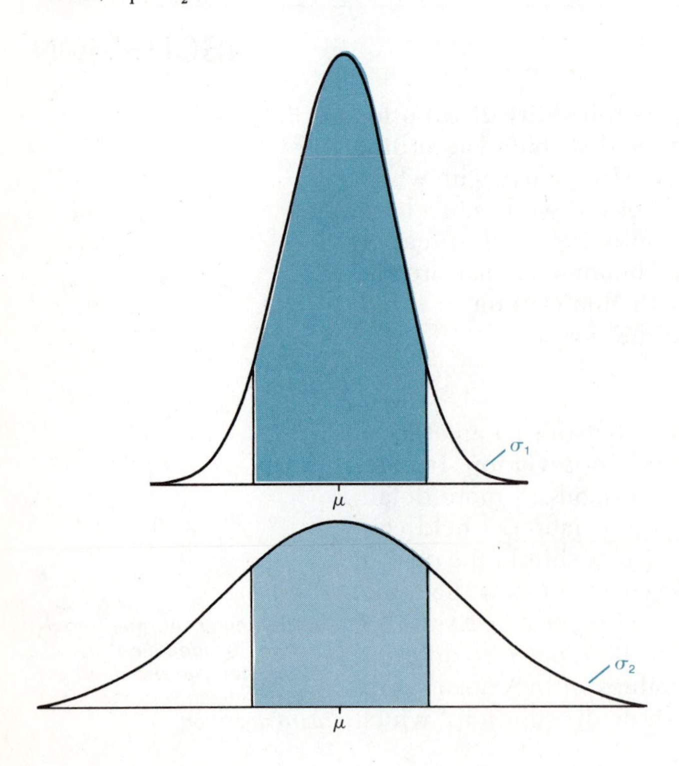

EXHIBIT 5
COMPARISON OF VALUES OF X AND Z

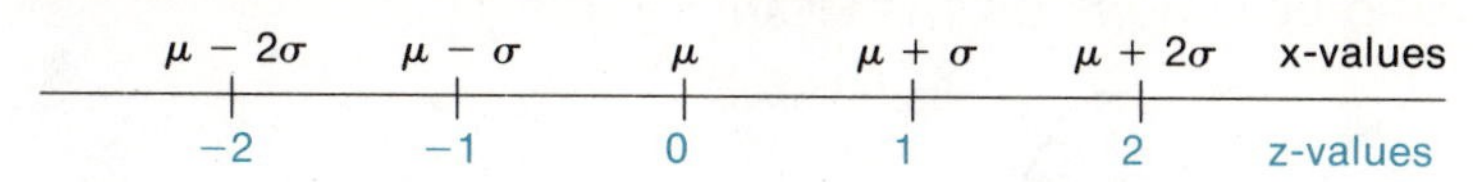

of x. The process of "changing from x into z" is called a *transformation.* In general, transformations allow us to deal with problems more simply. In the case of the normal, we are able to translate any normal curve with any set of μ and σ values into one particular normal curve that is tabulated in the Appendix.

The normal curve that is tabulated is called the **standard normal distribution.** It is a normal distribution with mean equal to *zero* and standard deviation equal to *one.* Any normal curve can be translated into the standardized curve. The unit of measurement of the standardized normal is the standard deviation of the original variable, x. In other words, if we begin with a variable x with a certain unit of measurement, we can express the scale of x-values in terms of multiples of its standard deviation from the mean. For example, consider the selected values appearing in Exhibit 5.

The standardized variate z represents the number of standard deviations that an x-value lies away from the mean of any normal distribution.

Notice when x assumes a value of $\mu + 2\sigma$ in the exhibit, the value of z equals 2. Hence, this value of x, whatever its number may be, is two standard deviations above the mean. If z equals −2, the corresponding value of x is 2 standard deviations below the mean. Any other value of x can be expressed in terms of z in a similar manner. In general, z represents the number of standard deviations that an x-value lies away from the mean of a normal distribution.

The Normal Curve and Reality. A few words are needed to clarify the notion of a "normal phenomenon." We have seen that the normal curve is defined for all the real numbers, or between plus-and-minus infinity. Can any "real life" quantity vary between these limits? Regardless of any set of measurements taken, no matter how large or how small, the range of values for which the normal is defined is unrealistic. Then how can we say that a variable is normally distributed and apply it to a realistic or practical problem?

If you think about the diagrams of the normal distribution used in this chapter, you will recall the "tails" of the distribution are very "close" to the x-axis. In terms of probability, this means that a "very small" proportion of the x-values occurs in the tails. In order to emphasize this point, consider the area or probability value in the table of Areas of the Normal Curve corresponding to a value of z equal to 3. The area given is 0.4987, which tells us that 49.87 percent of the x-values lies between the mean and some value that is 3 standard deviations from the mean. Furthermore, if we double the probability found in Table 2, we find that 99.74 percent of the x-values fall *within* 3 standard deviations of the mean. Diagrammatically, this is shown in Exhibit 6. Virtually all of the observations or x-values in the diagram lie within the "3-sigma" limits. Less than 0.3 percent lies outside these limits, which is the case for *any* normal distribution.

EXHIBIT 6
AREA UNDER A NORMAL CURVE WITHIN 3 STANDARD DEVIATIONS OF THE MEAN

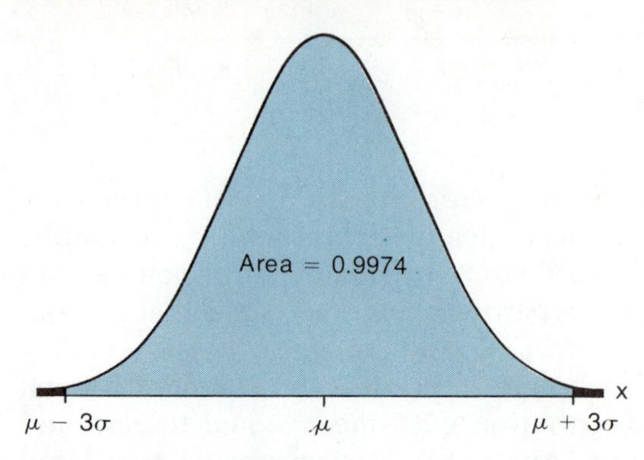

Although the normal is defined between "plus-and-minus infinity," such a small percentage lies in the "tails" that it can be applied to limits actually encountered in practical problems.

In practical terms, this result suggests that such a small percentage of observations lies in the tails of the normal distribution that, realistically speaking, we can ignore the tails. In other words, *within limits generally encountered* in practice, there are phenomena that can be approximated satisfactorily by a normal distribution well enough that it is useful.

CALCULATORS AND PROBABILITIES

This and the last chapter have exposed you to two special probability distributions, the normal and the binomial. Probabilities for the binomial are obtained by hand by directly substituting into the formula, and, for selected values of n and P, probabilities can be obtained from Table 1. As n becomes larger, hand computation of binomial probabilities becomes more difficult. Inexpensive scientific calculators with a "factorial key" and a "y^x key" can be used as an alternative to calculate binomial probabilities quite easily for samples as large as 69.

Probabilities for the normal, however, must be obtained by computing z and using Table 2 because these probabilities are not found by substituting into the formula for the normal. Fortunately, Table 2 is easy to use and does not pose any real problems. Later we shall use the normal distribution to approximate binomial probabilities that are too difficult to compute by hand.

It should be noted that more and more calculators, both desk-top and hand-held models with either stored programs or continuous memory, are equipped to deliver binomial probabilities for any sample size and normal probabilities at the press of a button. In addition, these calculators provide probabilities for other complicated distributions that we shall encounter in later chapters. We shall continue to use tables in the Appendix in order to illustrate various concepts in statistics. It should be kept in mind, however, that alternative ways of finding probabilities for the various distributions do exist.

QUESTIONS AND PROBLEMS

DEFINITIONAL QUESTIONS

1. What is a normal curve? What is meant by a normally distributed random variable?

2. What are the parameters of the normal distribution? What information is provided by these parameters?

3. Describe the general properties of all normal curves.

4. It is stated in this chapter that the normal is a continuous probability distribution. What is meant by this? If necessary, compare the normal with the binomial distribution in order to answer this question.

5. How are probabilities found using the normal curve? What information about a problem is necessary in order to compute "normal probabilities?"

6. What is a transformation? Define the term *standardized normal variate.* What are its properties?

7. In words, what is the meaning of the following two expressions?

   (a) $P(\mu \leq x \leq a)$
   (b) $P(a \leq x \leq b)$

COMPUTATIONAL PROBLEMS

8. Given a normal distribution with $\mu = 10$ and $\sigma = 5$. Find the following probabilities.

   (a) $P(10 \leq x \leq 16)$
   (b) $P(4 \leq x \leq 12)$
   (c) $P(13 \leq x \leq 14)$
   (d) $P(3 \leq x \leq 8)$
   (e) $P(x \geq 20)$
   (f) $P(x \leq 3)$

9. Given a normal distribution with $\mu = 50$ and $\sigma = 20$. Find the following probabilities.

   (a) $P(50 \leq x \leq 75)$
   (b) $P(25 \leq x \leq 60)$
   (c) $P(60 \leq x \leq 70)$
   (d) $P(20 \leq x \leq 35)$
   (e) $P(x \geq 95)$
   (f) $P(x \leq 25)$

10. Suppose a distribution of x-values is normal with a mean equal to 25 and standard deviation equal to 10.
    - (a) Find the probability that a value of x falls between the mean and 30.
    - (b) Find the probability that x falls between 20 and the mean.
    - (c) What is the probability that a value of x falls within 5 units of the mean?
    - (d) What is the probability that a value of x differs from the mean by at least 5 units?
    - (e) Below what value does the lowest 10 percent of the x-values fall?
    - (f) Above what value does the highest 25 percent of the x-values fall?
    - (g) Within what values do the middle 50 percent of the x-values fall? The middle 95 percent?

11. Find the value of x below which the lowest 30 percent of the x's fall for a normal distribution with $\mu = 35$ and $\sigma = 5$.

12. For a normal distribution with $\mu = 12$ and $\sigma = 3$, find the value above which the highest 10 percent of the x-values fall.

13. (a) For the distribution given in Problem 11, find the values within which the middle 90 percent of the x-values fall.
    - (b) For the distribution given in Problem 12, find the values within which the middle 80 percent of the x-values fall.

14. Assume that the height of adult males is normally distributed with a mean of 68 inches and a standard deviation of 2.5 inches. What proportion of adult males is over 6 feet tall?

15. The length of steel rods produced by a continuous manufacturing process are normally distributed with a mean length of 12 inches and a standard deviation of 0.2 inches. Rods less than 11.6 inches in length are too short and rods greater than 12.4 inches are too long.
    - (a) If one rod is drawn at random from the process, what is the probability that its length is between 11.8 and 12.3 inches?
    - (b) What proportion of the rods have a length less than 11.7 inches?
    - (c) If rods that are too short or too long are rejected, what proportion of the rods are rejected?
    - (d) If rods that are too short are scrapped and rods that are too long are reworked, what proportion of the rods are scrapped and what proportion are reworked?

16. Suppose the weekly demand for gasoline at a particular service station can be approximated by a normal distribution with a mean equal to 1500 gallons and standard deviation equal to 125 gallons.

(a) What is the probability that the weekly demand falls within 200 gallons of the mean?
(b) Find the probability that weekly demand is less than 1200 gallons.
(c) If the supply of gasoline received from the refinery is 1800 gallons, what percentage of the demand is not met?
(d) Below what number of gallons does the lowest 10 percent of demand fall? Above what number does the highest 5 percent of demand fall?

17. Assume the monthly accident rate per thousand employees at a certain industrial plant is normally distributed with a mean equal to 2 and standard deviation of 0.5 accidents per month.

(a) What is the probability that the accident rate is between 1 and 3 per month?
(b) Find the probability that the monthly accident rate exceeds 3 accidents.
(c) Above what value are the highest 10 percent of the monthly accident rates?

CONCEPTUAL QUESTIONS AND PROBLEMS

18. The normal distribution is symmetric with a single peak. Does this mean that all symmetric distributions are normal? Explain.

19. Criticize or explain the following statement. Two normal distributions can have the same mean and different standard deviations or they can have different standard deviations with equal means.

20. Probabilities using the binomial distribution are found by substituting directly into the formula for the binomial, whereas this is not the case for the normal. Why is there a difference in the way probabilities are found for the two distributions? Why must we compute a "z-value" in order to find a probability for a normal distribution?

21. When finding probabilities with a normal curve, we always deal with intervals; the probability of a single value of x is defined equal to zero? Why is this so?

22. When finding a normal probability, is there a difference between the values of $P(a < x < b)$ and $P(a \leq x \leq b)$, where a and b represent two numbers? Why or why not?

23. (a) When finding a normal probability of the form $P(a \leq x \leq b)$ for two numbers a and b, a z-value for a and b is computed, two probabilities are found from Table 2, and the prob-

abilities are either added or subtracted. Whether we add or subtract depends upon the values of a and b in relation to the value of the mean, μ. Why is it necessary to add or subtract probabilities in such a case?

(b) When finding probabilities like the one cited in part (a), some students are tempted to add or subtract the z-values first and use the result to find one probability from Table 2. What is wrong with this procedure?

24. Suppose you computed a z-value and obtained a value of 4. Refer to Table 2 in the Appendix and "look-up" the probability corresponding to z equal to 4. What did you find? Based upon your finding, what can you conclude about the probability associated with a z-value of 4? Of 5?

25. Refer to Table 2 in the Appendix and find the area corresponding to z equal to 1.00, 2.00, and 3.00. How do your answers relate to the probability that a value falls within one, two, or three standard deviations of the mean of *any* normal distribution? Based upon the areas found above, what general conclusions can be made about values in the "tails" of any normal curve?

26. Refer to Problem 17 where you are asked to assume that the monthly accident rate at an industrial plant is normally distributed. Compute the value of the accident rate that is 5 standard deviations below the mean. Is it possible for this value to occur? Based upon your answer to this question, is it realistic to use the normal distribution to compute probabilities? Explain.

27. Refer to Problem 17. The probability that the monthly accident rate is between 1 and 3 per month equals 0.95, approximately. Interpret this value.

28. Suppose the monthly accounts receivables for a hardware store chain is normally distributed with a mean value of $20,000 and a standard deviation equal to $1000. Find the probability that the accounts receivables for a particular month exceeds $21,500 and interpret the result.

29. Refer to Problem 15. Assume rods that are too short are scrapped and rods that are too long are reworked, and that the cost of rework is considerably less than the cost of scrapping a rod.

(a) If a setting change in the process results in a shift in the mean to 11.5 inches, what is the effect on the proportion of steel rods that are scrapped and reworked? What effect does this have on cost?

(b) Suppose the process mean shifts to 12.5 inches. What effect does this have on the proportion of rods that are scrapped and reworked? What is the effect on cost?

(c) Suppose it is desirable to scrap only 1 percent of the rods without regard to the number reworked. If the process mean can be shifted by a setting change, at what mean level should it be set?

(d) Suppose the mean length of the process remains the same, but the standard deviation changes to 0.3 inches. What is the effect on the proportion of rods that are scrapped and reworked?

30. Suppose letter grades are assigned to examination scores that can vary between 0 and 100 according to the following scale.

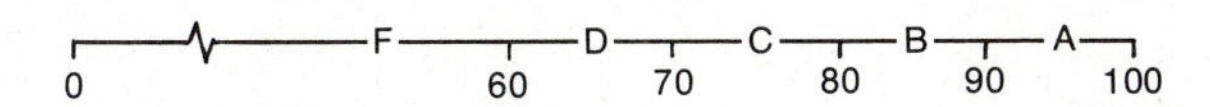

(a) Assuming exam scores are normally distributed with a mean equal to 70 and a standard deviation equal to 10, find the proportion of A's, B's, C's, D's, and F's.

(b) Suppose 100 students take the examination. Using the values found in part (a) rounded to 2 decimal places, find the number of students receiving each of the above letter grades.

(c) Using the scale provided above, a failure or grade of F is assigned to exam scores less than 60. If the criterion for failure is to be changed so that the percentage of failures is the same as the percentage of A's, what exam scores would lead to a failing grade?

(d) Use of the normal distribution to find answers to the above questions is based upon the assumption that examination scores are continuous. Since exam scores are recorded with no decimal places, what effect does this have on your results?

31. Refer to Problem 16. Is it reasonable to assume that the mean and standard deviation remain constant from week to week? Discuss.

STATISTICAL INFERENCE

CHAPTER 9

STATISTICAL SAMPLING

By a small sample we may judge of the whole piece.

Cervantes

I just examined about one dozen of asses and am perhaps too pessimistic.

Jerzy Neyman

You should never spoil a good story by checking your facts.

Mark Twain

At the very beginning of the book we defined a statistical **universe** as a group of elements about which a conclusion is made; recall, the term **population** is used interchangeably with universe. Whenever we treat a problem statistically, we reformulate it in terms of a universe of elements from which data about the problem can be collected. These data, when analyzed, provide a "solution" to the original problem.

Frequently, data are in the form of a **sample,** which we defined as a portion, or subset, of a population. Sample information is used in many instances because it is more practical than examining an entire population. Often, it is the only kind available. Furthermore, samples that are chosen properly generally provide much of the information obtained by taking a census, or a complete enumeration of the population. Consequently, time and money can be saved by sampling without losing much accuracy in the final result.

Data in the form of a sample frequently are used instead of a census.

It is interesting to note that the census of the population of the United States that is conducted every ten years employs sampling to obtain additional data about particular aspects of the population. On the one hand, a census of this kind is such an enormous job that it is not practical to conduct it frequently. On the other hand, when it is conducted it is difficult, time consuming, and costly to obtain great amounts of detail. Although most problems do not involve universes as large and spread-out as the population of the United States, timeliness, cost, and data availability are important considerations that contribute to the frequent use of sampling. In order to understand how sample data are used meaningfully, it is necessary to consider material from the areas

Special methods are required in order to make inferences using sample data.

of descriptive statistics and probability presented in the first two parts of the text. This material must be combined with some new ideas not covered in earlier chapters.

A sample of items can be drawn from a universe *randomly* or based upon *judgment*, or personal discretion. A **random sample**, or probability sample, is one where each element in the universe has a *known* probability of being selected. This means that the elements appearing in a sample are selected by chance. There is no way of knowing which elements from the universe actually are included in a random sample; however, it is possible to measure the frequency with which different elements will appear in repeated samples since there is a known probability of selection. When personal judgment is used to select samples, it is not possible to know this much about the sampling process.

Random samples are important since they make it possible to assess the reliability or measure sampling error associated with sample results.

The use of random samples is important in statistical work since it is the element of randomness that allows us to measure the uncertainty, or error, associated with a particular result or conclusion. Uncertainty is associated with sampled results because they are incomplete; a sample is only part of a population. Consequently, probabilities are attached to possible outcomes in order to assess the *reliability*, or amount of sampling error, associated with a conclusion about a population based upon a sample. The body of methods for selecting and analyzing random samples in order to make generalizations about populations is called **statistical inference** and is developed in this part of the text. This chapter serves as an introduction providing background for the chapters that follow.

Simple Random Samples. Although there are many types of random samples, each of these is a variation of one particular type called a **simple random sample.** This is a sample of items drawn from a population in such a way that each combination of elements of a given size has an *equal* probability of being the sample selected. When the term *random sample* is used without qualification, it usually refers to a simple random sample rather than to some other form of probability sample.

In order to select a simple random sample, a sampling frame must be available. A sampling **frame** is a complete listing of all elements in the population such that each element can be identified by a distinct number. Selecting a random sample is analogous to using a gambling device to generate numbers from this list. **Table 9** in the Appendix, entitled Random Digits, is an alternative that can be employed to select random samples. The description accompanying Table 9 provides an explanation regarding the way it can be used.

An entire area within the field of statistics is devoted to methods that can be used to select random samples under different circumstances. These methods are used to reduce cost, increase reliability of results, for convenience, or because it is impossible or impractical to select a simple random sample. All of these other methods, however, are based upon the simple random sample. Consequently, an understanding of statistical methods that are based upon the simple random sample represents the starting point for all work in statistical inference. Therefore, the material presented in this part of the text is based upon the

Simple random samples are assumed throughout the remainder of the text.

assumption that samples are selected in the form of a simple random sample.

Sampling With and Without Replacement. Typically when a sample is drawn from a population, items are successively removed from the population and are not replaced prior to the next draw. In other words, samples generally are drawn *without replacement* in practice. However, much of basic statistical theory is based upon sampling from infinite universes, which is equivalent to sampling *with replacement* from a finite population; in either case, draws can be made indefinitely without changing the composition of the universe.

When sampling without replacement where the sample size is small relative to the size of the universe, the composition of the universe changes so little with each draw that "with replacement methods" may be applied. Most of the material presented in the following chapters is based upon the assumption that sampling is done either with replacement or from an infinite universe.

Population Parameters and Sample Statistics. Characteristics of populations are referred to as **parameters.** For example, such quantities as the mean, variance, median, or mode computed from a universe are considered parameters that identify specific properties of a population. Since measurements are associated with the items in a population, we can construct a universe frequency distribution as well. By associating a probability distribution with the universe distribution we are able to describe the probability of universe values when sampled. It is therefore customary to consider identifying characteristics of probability distributions as parameters also. In many cases we shall deal with parameters in the form of symbols since the corresponding numerical value for the population will not be known. It is conventional to denote universe parameters with letters of the Greek alphabet; in some instances capital letters of the English alphabet are used.

Problems that are solved with statistical methods must be quantified in terms of one or more universe parameters.

When solving problems statistically they must be quantified in terms of one or more universe parameters. For example, if two cities are compared in order to determine which has the higher income level, we may choose to compare the median or the mean incomes. It may be of further interest to compare measures of variability since averages alone do not describe the way individual incomes are distributed.

Inferences about universe parameters are made in terms of sample statistics.

When a sample is used to make an inference about a population, a sample characteristic corresponding to a universe parameter is used. This sample characteristic is referred to as a sample statistic, or briefly, as a **statistic.** Sample statistics are frequently, but not always, defined similar to the corresponding universe parameter. For example, if a problem is expressed in terms of the universe mean, the sample arithmetic mean generally is the statistic used. Sample statistics usually are designated with lower case letters of the English alphabet.

Any sample statistic in the form of a single number used to estimate an unknown universe parameter is termed a **point estimate.** For example, suppose we want to know the average number of women employees per firm in the chemical industry. Instead of collecting data from every firm in the industry, we decide to estimate the average number by selecting a sample of forty chemical firms and computing the arithmetic mean number of women for these forty firms. The value of the sample arithmetic mean is a point estimate of the average number of women employees for the universe of chemical firms.

In the above example, the choice of the sample arithmetic mean appears obvious. However, considerable attention in the field of statistics has been devoted to the problem of determining which point estimate is best when estimating various parameters. This stems from the fact that different statistics have different sampling properties. We shall not consider the problem of choosing the best statistic to use in a given situation. We shall deal with ways in which particular statistics are used to draw inferences about specific population parameters.

Problem Definition, Research, and Statistics. The field of statistics is comprised of a specific set of principles that are useful in research, problem-solving, and decision-making. As stated earlier, statistical methods must be applied to other disciplines or fields of study in order for the methods to be of benefit. The solution of any problem with statistical methods involves more than just applying these individual methods. Much can be said about the process involved when undertaking a complete study for research or decision-making purposes. A discussion of the entire process is beyond the scope of this text. In order to provide some perspective to the remaining text material, the components of a complete statistical investigation are listed below.

1. Begin with a general problem statement that requires a conclusion to be drawn or a decision to be made.
2. Identify the relevant characteristics of the problem.
3. Formulate the problem in terms of the relevant characteristics.
4. Define the appropriate statistical universe.
5. Quantify the problem in terms of relevant universe parameters.
6. Decide what data will be needed to solve the problem.
7. Determine the method of data collection; if sample data are to be used, the sampling method and sample statistics must be determined.
8. Establish the procedure for analyzing the data and guidelines for basing conclusions upon the method of analysis.
9. Collect the data.
10. Analyze the data.
11. Draw conclusions based upon the analysis of the data.

Whenever statistical methods are applied, the problem must be carefully defined. Although this may seem obvious, it is not always as easy as it sounds. Frequently, elaborate studies are conducted but directed toward the solution of the wrong problem because of improper problem definition. Moreover, we strive to achieve as much control as possible so that conclusions can be safely attributed to the variables under study. In other words, it is desirable to achieve a level of control experienced under laboratory conditions. Unfortunately, when dealing with problems outside the laboratory this is not possible; in many cases we attempt to isolate the important parts of the problem using statistical procedures instead of applying direct controls.

Reference already has been made to the fact that error is associated with a sample result; *sampling error* is present because a sample is used rather than a census. Another kind of error is referred to as *systematic error*. This is a *non-sampling error* since it can arise whether a sample or a census is used. A systematic error is one that biases results persistently in one direction, and can be associated with either individual measurements or with the method of data collection used. For example, when individuals in high income brackets are questioned about their income, they have a tendency to understate the amount. In other cases where a questionaire is used that contains "leading questions," respondents are provided with suggestions regarding the way they should respond. Since systematic errors can occur in any investigation, they must be considered when developing the overall research design and eliminated whenever possible.

More than one kind of error is possible. The main concern in this part of the text is sampling error.

Although studies can be designed to measure the effect of systematic errors, these are not considered explicitly in the methods presented in this part of the text. Furthermore, other types of non-sampling errors such as improperly recorded measurements that are not systematic may be present in any investigation. Errors of this kind also are not considered in the methods presented.

In the last few paragraphs we touched on some of the considerations involved in designing a study or investigation where statistical methods are used. Although the field of statistics has a contribution to make in the design of an overall research method, we shall concentrate our attention upon specific methods of analyzing data and the way conclusions are made based upon these methods of analysis. Some insight about the other components of the above list are provided when appropriate.

What is to Come. There are very few entirely new topics to be considered in this part of the text. The material falls into four main categories: (1) random sampling distributions, (2) confidence interval estimation, (3) hypothesis testing, and (4) relationship.

The concept of a random sampling distribution, or probability distribution of a sample statistic, which is presented in the next chapter, is fundamental to statistical inference. With it we are able to account for the error in sample estimates. One way of doing this is to construct an interval estimate instead of a point estimate referred to earlier; the procedure for constructing interval estimates is called confidence interval estimation and is covered in Chapter 11. Related to confidence intervals is the area of hypothesis testing. Here we hypothesize the nature of the universe in terms of one or more parameters and then determine whether the hypothesis is tenable on the basis of sample evidence. The fundamentals of hypothesis testing are considered in

Chapter 12. Chapters that follow deal with extensions to a variety of problems. Estimation and hypothesis testing can be applied to relationships among variables for the purpose of making predictions or understanding the way in which a set of variables are related. This is considered in the chapter on regression and correlation analysis at the end of Part Three.

Different types of problems involve differences in procedure; however, the four main topic areas cited form the basis for the entire area of inference presented in this part of the book. Due to the prevalence of arithmetic means in statistical work and in the classroom, the main concepts are presented first in terms of means and then generalized to other areas in later chapters.

QUESTIONS AND PROBLEMS

DEFINITIONAL QUESTIONS

1. What is meant by the terms *universe, population,* and *sample?* What is the role of each in statistics?

2. Define each of the following terms.
   (a) Judgment sample
   (b) Random sample
   (c) Simple random sample

3. Distinguish between sampling with replacement and sampling without replacement.

4. Define the terms *reliability, sampling error,* and *non-sampling error.* What is the relationship between a non-sampling error and a systematic error?

5. What is meant by the term *sampling* as opposed to the word *sample?* How are sampling and statistical inference related?

6. Distinguish between a population parameter and a sample statistic.

7. What is meant by the phrase "point estimate of a universe parameter"?

8. Read over the list of steps involved in a statistical investigation appearing on page 166. Which are considered in the remainder of the text?

9. What type of random sampling is assumed throughout this part of the text?

NUMERICAL PROBLEMS

10. Using Table 9 in the Appendix, generate a sequence of ten digits between 0 and 9 at random. Repeated values may occur.

11. Using Table 9 in the Appendix, generate a sequence of five numbers at random between 50 and 75. Repeated values may occur.

12. Using Table 9 in the Appendix, select a simple random sample of five numbers with replacement between 0 and 99. Repeat the process for sampling without replacement.

13. The annual incomes of a universe of 50 families appear below:

Family	*Income*	*Family*	*Income*
1	$8850	26	$19790
2	$13409	27	$10123
3	$11417	28	$16936
4	$12932	29	$13199
5	$12772	30	$12141
6	$13902	31	$13262
7	$12588	32	$12950
8	$11622	33	$13380
9	$14091	34	$15470
10	$9540	35	$13725
11	$13683	36	$8900
12	$10977	37	$9976
13	$8335	38	$14174
14	$14819	39	$9271
15	$15785	40	$15262
16	$18550	41	$10790
17	$11370	42	$13878
18	$10494	43	$11865
19	$16119	44	$10500
20	$8743	45	$11799
21	$12253	46	$10460
22	$12764	47	$14174
23	$11240	48	$16392
24	$16423	49	$15585
25	$22136	50	$12841

Using Table 9 in the Appendix select a simple random sample of five families and their incomes.

14. Refer to the family income data presented in Problem 13.

    (a) Select a simple random sample of 10 from the universe of 50 families using Table 9 in the Appendix.
    (b) Record the annual incomes for the families in the sample.
    (c) Using formulas from Chapters 2 and 3, based upon the sample selected, calculate point estimates of the universe mean, variance, and standard deviation.

CONCEPTUAL QUESTIONS AND PROBLEMS

15. What is the essential difference between sample data and census data? Explain.

16. Why is sampling important? When sampling, as opposed to taking a census, what additional considerations must be taken into account?

17. Of the problems listed below, which reasonably could be solved using a census and which would be more reasonable to consider as sampling problems. Account for factors such as time, cost, necessity, and difficulty of data collection and processing in your response.
    (a) Television viewing habits of viewers throughout the entire country are to be investigated.
    (b) Sales volume of an industry of 10 firms listed on the New York Stock Exchange is to be acquired.
    (c) An estimate of the sales volume of all retail hardware stores across the country is to be determined.
    (d) Hour by hour predictions of election results are to be broadcast on a nationwide TV network.
    (e) The quality of a shipment of ballpoint pen refills is to be determined in terms of writing clarity and length of use.

18. Consider a shipment of 1000 items produced by a continuous manufacturing process. Under what circumstances can measurements taken from the items of the shipment be considered a census, and under what circumstances can they be considered a sample? Based upon your answer, what general conclusion can you draw about the definition of a sample and a census?

19. Why is it important to consider random samples?

20. What is meant by an infinite universe? Realistically speaking, does such a thing as an infinite universe exist? Explain.

21. Explain in terms of probability why sampling with replacement is equivalent to sampling from an infinite universe. Under what circumstances is it reasonable to assume that sampling is done with replacement even though a sample actually is selected from a finite universe without replacement? Where in the text have you encountered this problem before?

22. Read the explanation accompanying Table 9 in the Appendix regarding the nature of the random digits presented in the table. What are the similarities and differences between the numbers generated in the table and numbers generated by repeatedly rolling a die and recording the face appearing on each roll?

23. Suppose you are to construct a table consisting of a large sequence of integers between 1 and 10 such that each integer has

an equal probability of occurring at each position in the table. Can this be accomplished by repeatedly drawing a card from a deck of standard playing cards and recording the number appearing if it falls between 1 and 10 and ignoring the card if it does not lie between these values? Assume a card is replaced and the deck is reshuffled before each draw. Does the fact that a deck of cards has four cards of each face make a difference? Explain. What is the relationship between a table constructed in the above manner and the table of Random Digits, or Table 9, in the Appendix?

24. Of what use is a table of random numbers? What is the relationship between gambling devices, random number tables, and the sampling process?

25. What kind of error is considered in statistical inference based upon random samples? Is non-sampling error considered by the fact that "random" samples are used? Explain. When sampling, what kind of error can be reduced by increasing the sample size? Explain.

26. Refer to the family income data presented in Problem 13.

    (a) Select a simple random sample of five families from the universe of fifty given.
    (b) Calculate the mean, variance, and standard deviation of the sample selected in part (a). Use the formulas given in Chapters 2 and 3.
    (c) What information is provided by the values computed in part (b)? In what sense can these values be considered as descriptive measures? Can they be used to make inferences? Can they be considered as point estimates? Explain.
    (d) The mean and standard deviation of the entire universe of 50 families equal $13,033 and $2843, respectively. Why is there a difference between these values and the mean and standard deviation computed in part (b)?
    (e) Select another random sample of five families and compute the mean and standard deviation of the incomes of this sample. Compare the results with the mean and standard deviation calculated in part (b); why do they differ?
    (f) Based upon your observations in parts (d) and (e), what general conclusions can you make about the value of a sample statistic and point estimates? What is the relationship between your conclusions and sampling error and the problem of sampling?

CHAPTER 10

A BASIC TOOL
The Random Sampling Distribution

Nine times out of ten, in the arts as in life, there is actually no truth to be discovered; there is only error to be exposed.

H. L. Mencken

Basic to statistical inference is the random sampling distribution. Recall, a statistic such as the sample arithmetic mean is defined as a characteristic of a sample. Since different samples from the same population have *different values* of a particular statistic, we can describe the variation of these values in the form of a distribution. When random sampling is used, we can determine the probability of occurrence of the values of a sample statistic. Hence, under random sampling the value of a statistic varies according to chance and is a random variable. The probability distribution of *all possible* values of a sample statistic is defined as the **random sampling distribution,** or simply the sampling distribution, of that statistic.

The concept of a sampling distribution is fundamental but somewhat difficult to understand. In order to understand the nature of a sampling distribution, imagine that you have a list of all working residents of a large city together with the annual income of each worker. Based upon the material in Chapter 1 we can construct a frequency distribution of the annual incomes and graph the result as a histogram; we shall refer to this as the population, or universe, distribution. We also can compute the arithmetic mean of all the workers' incomes; we shall call this the universe mean.

Suppose we select a random sample of 100 workers from the population of working residents and record their incomes. We could, of course, construct a frequency distribution for this group of workers' incomes. The frequency distribution of this sample should reflect the nature of the universe distribution, but obviously it would not be exactly the same. An arithmetic mean of this sample also can be computed.

Envision a process where samples of the same size *repeatedly* are drawn from the universe until *all* possible combinations of workers are selected. Recognize that there are many unique samples of workers and that an arithmetic mean income can be computed for each sample. Since the values of the sample mean will differ, we can construct *another* frequency distribution that is associated with these sample arithmetic means. Hence, from the original distribution of universe measurements we can derive a *distribution of a sample statistic.* In the situation described it is the sample arithmetic mean; however, a similar distribution could be developed for any other sample statistic. The frequency distribution of a sample statistic developed in this way is defined as a *sampling distribution.*

Probability distributions that describe the relative frequency of occurrence of numerous sample statistics are available. These probability distributions provide the frequency of occurrence of values of various statistics used to estimate universe parameters. Consequently, it is *not necessary* to undertake the repeated sampling process outlined above in order to describe the random behavior of many sample statistics used in practice. With the aid of theoretical probability distributions we can determine the frequency with which values of a sample statistic occur and describe the variation, or extent of error, associated with various statistics used as estimates of universe parameters.

A sampling distribution describes the way the values of a statistic vary from sample to sample. Regardless of the statistic, virtually every sampling distribution has a mean, a variance, and a standard deviation. Because of its importance, the standard deviation of a sampling distribution is distinguished from the population standard deviation by a special name; it is referred to as a **standard error.** The population standard deviation describes the variation among elements of the universe, whereas the standard error measures the variability in a statistic due to sampling; it is a *measure of sampling error.*

In the case of the distribution of a sample mean, the arithmetic mean represents the average of *all possible* sample means, or the "mean of the means" so to speak; the corresponding variance and standard deviation measure the variability among all possible values of the sample mean. In this case, the standard deviation is called the standard error of the mean. It is instructive to examine the table of symbols appearing in Exhibit 1 in order to understand the meaning of these characteristics.

EXHIBIT 1

CHARACTERISTICS OF DISTRIBUTIONS ASSOCIATED WITH THE DISTRIBUTION OF THE SAMPLE ARITHMETIC MEAN

	Universe	*Any Sample of the Many Possible*	*Sampling Distribution*
Mean	μ	$\bar{x}$	$\mu_{\bar{x}}$
Variance	σ^2	s^2	$\sigma^2_{\bar{x}}$
Standard Deviation	σ	s	$\sigma_{\bar{x}}$

The symbols in the universe column of the exhibit are familiar from previous chapters, and those in the second column are symbols of descriptive measures used in Chapters 1 and 2; lower-case letters denote sample characteristics The universe characteristics, μ, σ^2, and σ, correspond to single numbers since they represent parameters, whereas different values of $\bar{x}$, s^2, and s exist that correspond to each of the many possible samples that can be selected from a universe.

The symbols appearing in the last column are new and are introduced here for the first time. The subscript attached to each symbol denotes the variable described by each measure and is used to distinguish these measures from the universe parameters. Hence, the symbol $\mu_{\bar{x}}$ (read Mu-sub-x-bar) represents the mean of the *distribution of all possible sample arithmetic means* where the sample mean is the variable described. Similarly, $\sigma^2_{\bar{x}}$ and $\sigma_{\bar{x}}$ are the variance and standard deviation of the distribution of the sample mean. Each of these quantities represents a single number and is a parameter of the *sampling distribution* of the sample arithmetic mean. $\sigma_{\bar{x}}$ is called the standard error of the mean and measures the error due to sampling the mean. The diagram appearing in Exhibit 2 summarizes the material presented above. Histograms indicate that we are dealing with an empirical or observable phenomenon; frequency curves drawn through the universe distribution and the sampling distribution signify that these frequency

EXHIBIT 2

DEVELOPMENT OF THE RANDOM SAMPLING DISTRIBUTION OF THE SAMPLE MEAN

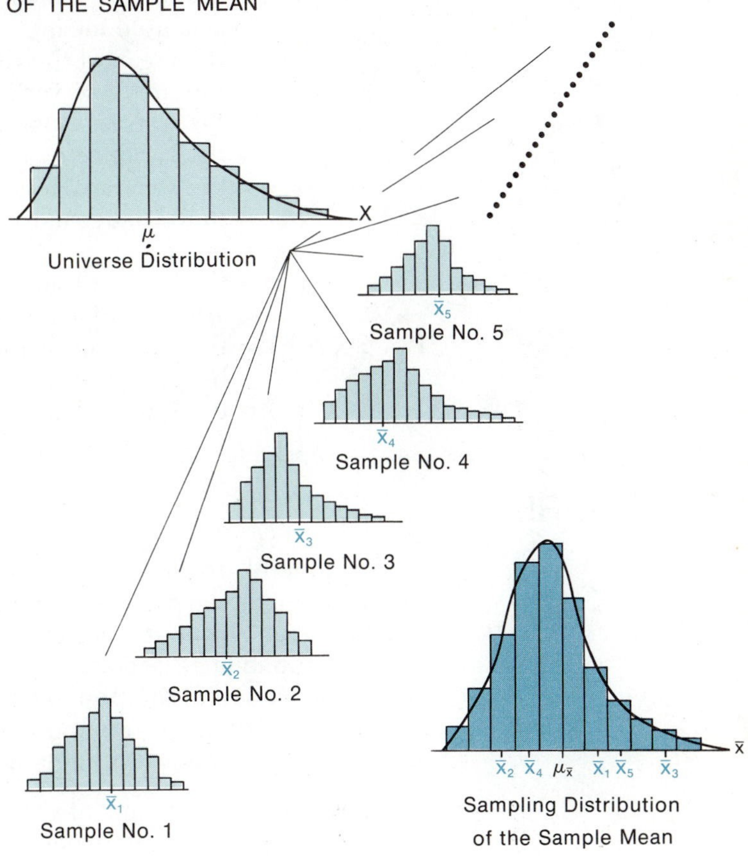

distributions can be described by probability distributions. The diagrams are associated with the development of the distribution of the sample mean; however, the general concept applies to *any* statistic.

Based upon the above discussion and a careful examination of the diagram, you should be aware of five important points: (1) one universe distribution exists with fixed parameters, (2) for any universe many possible samples exist, (3) the value of any sample statistic varies from sample to sample (4) one sampling distribution with a particular set of parameters corresponds to a given statistic, and (5) the sampling distribution describes the way the values of a sample statistic vary from sample to sample.

This chapter has two main goals: (1) to introduce the concept of a sampling distribution of a statistic, and (2) to demonstrate how to compute probabilities associated with values of the sample arithmetic mean when the values of the parameters are known. The second goal is accomplished in the procedure and example sections. Emphasis is placed upon means at this point because they are widely used and easy to understand.

A summary of symbols used in this chapter is as follows:

X = a value of a measurement in a population or universe
x = a value of a measurement in a sample
N = the number of elements in the universe
n = the sample size
μ = the universe mean
$\bar{x}$ = a sample mean
$\mu_{\bar{x}}$ = the mean of the distribution of the sample mean
σ^2 = the universe variance
σ = the universe standard deviation
s^2 = the variance of a sample
s = the standard deviation of a sample
$\sigma^2_{\bar{x}}$ = the variance of the distribution of the sample arithmetic mean
$\sigma_{\bar{x}}$ = the standard deviation of the distribution of the sample arithmetic mean, or the standard error of the mean
$P(a \leq \bar{x} \leq b)$ = the probability that the sample arithmetic mean falls between the specific numerical values, a and b
z = the standardized normal variate

PROCEDURE

In general, to find the probability of occurrence of a value of a sample statistic we must know the form of the probability distribution of that statistic; probabilities then are found in the standard way. When finding probabilities associated with the sample arithmetic mean, the numerical values of the population mean and standard deviation must be *known;* probability computations for the sample mean then are based upon the following fundamental facts:

(1) The arithmetic mean of the sampling distribution of the sample mean, $\mu_{\bar{x}}$, equals the mean of the population, μ, regardless of the

form of the population distribution. That is,

$$\mu_{\bar{x}} = \mu$$

Mean of the distribution of sample means.

(2) The variance of the sampling distribution of the sample mean, $\sigma^2_{\bar{x}}$, equals the population variance, σ^2, *divided* by the sample size, n. That is,

$$\sigma^2_{\bar{x}} = \frac{\sigma^2}{n}$$

Variance of the distribution of sample means.

Hence, the standard deviation of the distribution of sample means equals the square root of the variance.

$$\sigma_{\bar{x}} = \sqrt{\sigma^2_{\bar{x}}}$$

or

$$\frac{\sigma}{\sqrt{n}}$$

Standard deviation of the distribution of sample means or the standard error of the mean.

(3) Sample means from normal populations are normally distributed, regardless of the size of the sample.

(4) Sample means from populations that are *not* normal are not normally distributed, *but* the distribution approaches the normal as the sample size approaches infinity. This is known as the *Central Limit Theorem.* Consequently, the normal distribution *approximates* the distribution of the sample mean for large samples regardless of the shape of the universe distribution.

These four facts* enable us to find the probability of occurrence of values of the sample mean for all sample sizes when the population is normal, and for large samples when the population is not normal; simply use the appropriate parameter values and proceed according to the steps provided in Chapter 8 for the normal curve. The formula for the standardized variate remains the same except different symbols are used to reflect the difference in the variable involved. That is,

$$z = \frac{\bar{x} - \mu_{\bar{x}}}{\sigma_{\bar{x}}}$$

or

$$\frac{\bar{x} - \mu}{\sigma_{\bar{x}}}$$

Standardized normal variate for distribution of sample means.

In words, z is found by subtracting the mean of the sampling distribution from the value of the sample mean and dividing by the standard

*The four facts about the sampling distribution of the mean represent results from the fields of probability theory and mathematical statistics. By actually drawing repeated samples from specific populations, it is possible to show empirically that sample means behave according to the results presented.

error. Since the mean of the sampling distribution of sample means equals the mean of the universe, the second of the two expressions is the one actually used to compute z.

NUMERICAL EXAMPLES

Example 1 A random sample of 9 items is drawn from a normal population with mean, μ, equal to 20 and standard deviation, σ, equal to 15. Find the probability that the sample arithmetic mean falls between 10 and 25, $P(10 \leq \bar{x} \leq 25)$.

The following information is provided in the problem statement:

$$n = 9$$

$$\mu = 20$$

$$\sigma = 15$$

Population is normally distributed

Since the population is normal, the distribution of the sample mean is normal with mean and standard deviation equal to

$$\mu_{\bar{x}} = \mu = 20$$

$$\sigma_{\bar{x}} = \frac{\sigma}{\sqrt{n}} = \frac{15}{\sqrt{9}} = \frac{15}{3} = 5$$

The required probability, $P(10 \leq \bar{x} \leq 25)$, is represented by the shaded region in the following diagram of the normal curve.

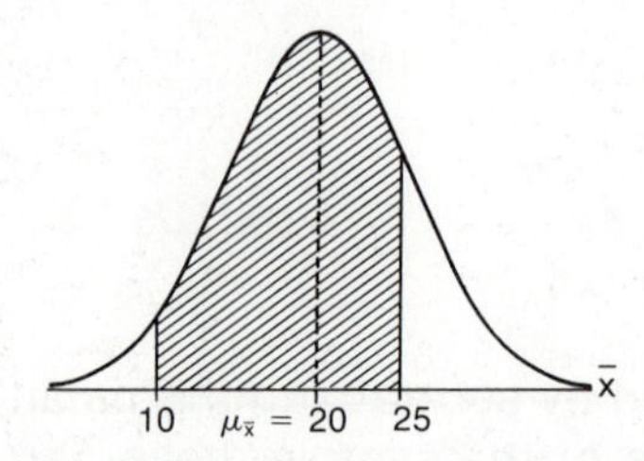

Hence, we compute z-values for the normal curve in the usual way, except we use the z-value in terms of the mean and standard deviation of the sampling distribution of the mean:

$$z = \frac{\bar{x} - \mu_{\bar{x}}}{\sigma_{\bar{x}}}$$

$$z_1 = \frac{10 - 20}{5}$$

$$= -2$$

$$z_2 = \frac{25 - 20}{5}$$

$$= 1$$

Based upon Table 2 of the Appendix a z-value of 2 yields a probability of 0.4772 and a z-value of 1 yields 0.3413. The probability is found by *adding* these two probabilities. Hence,

$$P(10 \leq \bar{x} \leq 25) = 0.4772 + 0.3413$$
$$= 0.8185$$

The probability that the mean of a sample of 9 items falls between 10 and 25 is 0.8185.

Example 2 A continuous manufacturing process produces items whose weights are normally distributed with a mean weight of 8 pounds and a standard deviation of 3 pounds. A random sample of 16 items is to be drawn from the process. What is the probability that the *arithmetic mean of the sample* exceeds 9 pounds? Interpret the result.

Given: Population normally distributed

$$\mu = 8$$
$$\sigma = 3$$
$$n = 16$$

Find: $P(\bar{x} > 9)$

Compute: $\mu_{\bar{x}} = \mu = 8$

$$\sigma_{\bar{x}} = \frac{\sigma}{\sqrt{n}}$$

$$= \frac{3}{\sqrt{16}}$$

$$= \frac{3}{4} \text{ or } 0.75$$

By way of contrast, the population distribution

and the sampling distribution of the mean are shown below.

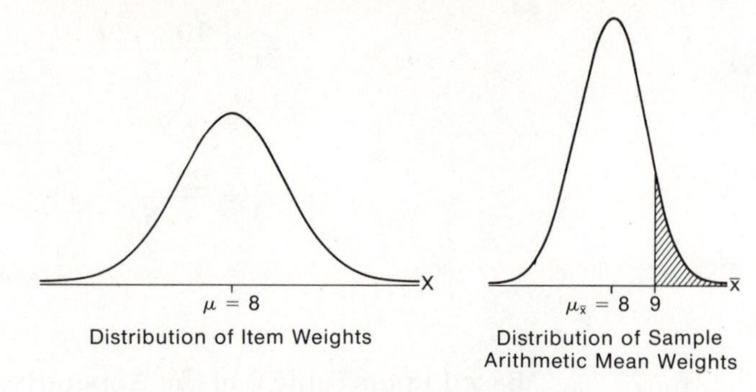

Distribution of Item Weights

Distribution of Sample Arithmetic Mean Weights

The required probability is represented as the shaded region in the *second* diagram. Hence,

$$z = \frac{\bar{x} - \mu_{\bar{x}}}{\sigma_{\bar{x}}}$$

$$= \frac{9 - 8}{0.75}$$

$$= \frac{1}{0.75} = 1.33$$

A z-value of 1.33 from the table of Areas of the Normal Curve yields 0.4082. The required probability is obtained by *subtracting* this number from 0.5, which yields

$$P(\bar{x} > 9) = 0.5 - 0.4082$$

$$= 0.0918 \text{ or } 0.09 \text{ (rounded)}$$

Using the rounded result, this probability can be *interpreted* to mean that 0.09 or 9 percent *of all possible samples* of 16 items drawn from the population will possess a sample mean value greater than 9 pounds.

Example 3 A population of items has an unknown distribution but a known mean and standard deviation of 50 and 100, respectively. Based upon a randomly drawn sample of 81 items drawn from the population, what is the probability that the sample arithmetic mean does not exceed 40?

In this problem we are given the following information.

$$\mu = 50$$
$$\sigma = 100$$
$$n = 81$$

We are asked to find the probability

$$P(\bar{x} \leq 40)$$

Based upon the information provided, we can determine the parameters of the sampling distribution as

$$\mu_{\bar{x}} = \mu = 50$$

$$\sigma_{\bar{x}} = \frac{\sigma}{\sqrt{n}}$$

$$= \frac{100}{\sqrt{81}}$$

$$= \frac{100}{9} = 11.\dot{1}$$

Although the form of the population distribution is unknown, we can assume that a sample size of 81 is large enough to apply the *Central Limit Theorem.* Hence, the normal distribution can be used to find the required probability that appears as the shaded region in the following diagram:

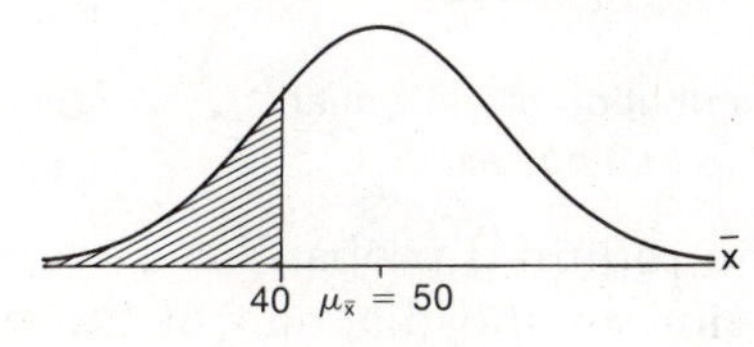

Hence,

$$z = \frac{\bar{x} - \mu_{\bar{x}}}{\sigma_{\bar{x}}}$$

$$= \frac{40 - 50}{11.1}$$

$$= -0.90$$

Based upon the table of Areas of the Normal Curve, a z-value of 0.90 yields a probability of 0.3159. The required probability is found by *subtracting* this from 0.5. Hence,

$$P(\bar{x} \leq 40) = 0.5 - 0.3159$$

$$= 0.1841$$

> *Note.* Since the Central Limit Theorem was employed, we can interpret the result as *approximately* 0.1841 or 18.41 percent of all possible samples of 81 items possess a mean value not exceeding 40.

Example 4 A random sample of 100 is to be selected from a large group of wage earners whose mean annual income is $9500, with a standard deviation equal to $800. Find and interpret the probability that the sample arithmetic income falls between $9404 and $9612.

The result is based upon the *Central Limit Theorem* since we are not told how individual incomes are distributed; as indicated earlier, income distributions generally are skewed to the right and therefore are not normal. Once again, by way of contrast, consider the following diagrams of the universe and the sampling distribution.

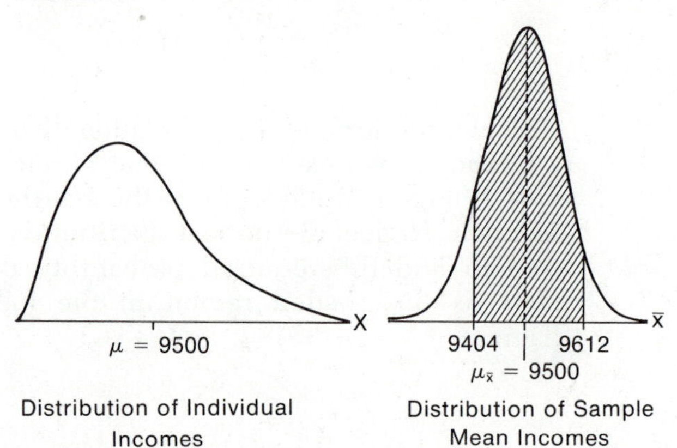

Distribution of Individual Incomes

Distribution of Sample Mean Incomes

The required probability is given as the shaded region in the diagram of the normal curve appearing on the right. Hence,

$$\mu_{\bar{x}} = \mu = 9500$$

$$\sigma_{\bar{x}} = \frac{\sigma}{\sqrt{n}}$$

$$= \frac{800}{\sqrt{100}} = 80$$

and

$$z = \frac{\bar{x} - \mu_{\bar{x}}}{\sigma_{\bar{x}}}$$

$$z_1 = \frac{9404 - 9500}{80}$$

$$= -1.2$$

$$z_2 = \frac{9612 - 9500}{80}$$

$$= 1.4$$

Based upon the table of Areas of the Normal Curve, a z-value of 1.2 yields 0.3849 and a z-value of 1.4 yields 0.4192. The answer is obtained by *adding* these two probabilities. Hence,

$$P(9404 \le \bar{x} \le 9612) = 0.3849 + 0.4192$$
$$= 0.8041 \text{ or } 0.8 \text{ (rounded)}$$

Using the rounded figure, we can interpret this result as approximately 0.8 or 80 percent of all possible samples of 100 wage earners have mean incomes between \$9404 and \$9612.

> *Note.* The probability computed in this example reveals nothing about the proportion of individuals possessing various incomes; this only can be determined directly from the universe distribution, provided its form is known.

Example 5 ***A Related Problem.*** Using the information provided in Example 2, find the values of the sample arithmetic mean within which the middle 95 percent of all sample means will fall.

In this problem we are given a probability and must *determine values of the sample mean*, indicated as $\bar{x}_1$ and $\bar{x}_2$ in the diagram below.

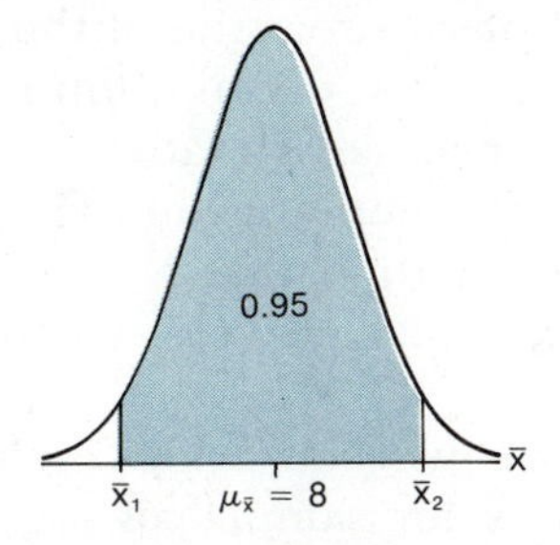

Since we are to find the values of the sample mean corresponding to an interval centered on the mean of the distribution, we can use the same value of z for both values of $\bar{x}$. This z-value is obtained by using half the specified probability, or 0.4750 (i.e., .95/2), and locating z in the table of Areas of the Normal Curve. Hence, z = 1.96. Based upon the formula for z, we must solve for the values of $\bar{x}$ in terms of the known values which yields:

$$\bar{x}_1 = \mu_{\bar{x}} - z\sigma_{\bar{x}}$$
$$= 8 - 1.96(0.75)$$
$$= 8 - 1.47$$
$$= 6.53$$

$$\bar{x}_2 = \mu_{\bar{x}} + z\sigma_{\bar{x}}$$
$$= 8 + 1.96(0.75)$$
$$= 8 + 1.47$$
$$= 9.47$$

Note. This problem is solved in the same way as Example 8 in Chapter 8 on the Normal Curve, except we are dealing with the sample mean instead of an individual value of x.

DISCUSSION

After reading the last three sections you should understand the concept of a sampling distribution and know how to determine probabilities for the sample arithmetic mean when its distribution is assumed to be normal. Before closing this chapter, we shall discuss some additional points about the material covered.

The Central Limit Theorem is important because it enables us to describe the shape of the distribution of the mean for large samples without knowledge of the shape of the universe.

The Central Limit Theorem. The Central Limit Theorem is a powerful concept since it allows us to describe the random behavior of the sample arithmetic mean for large samples without knowledge of the shape of the universe. Previously, we encountered the problem of deciding when a sample is large. In general, we cannot say whether a sample size is sufficiently large to apply the Central Limit Theorem until the required decimal place precision of the solution is specified. Moreover, the closer the universe distribution is to normal, the smaller the sample size necessary to achieve a satisfactory approximation of the sampling distribution of the sample mean using the normal curve. In other words, a universe distribution that is skewed more than another will require a larger sample size to achieve the same degree of approximation.

We shall not consider the sampling distribution of the mean for small samples when the universe is not normal.

For classroom work, we shall adopt the **general rule** that the Central Limit Theorem applies for sample sizes *greater than 30.* That is, in cases where the form of the universe distribution is unknown, we shall assume the sampling distribution of the sample mean is normal if the sample size exceeds 30. The shape of the distribution of the mean for small samples depends upon the exact form of the universe distribution and is different in each case; we shall not consider such cases in this text.

The Standard Error of the Mean. The standard deviation of the sampling distribution, $\sigma_{\bar{x}}$, is referred to as the standard error of the mean in order to distinguish it from the universe standard deviation, σ. The universe standard deviation measures variability among the original measurements, whereas the standard error measures variability among values of a sample statistic.

A standard error measures variability among values of a sample statistic and is a measure of sampling error.

The formula for the standard error of the mean (i.e., $\sigma_{\bar{x}} = \sigma/\sqrt{n}$) reveals that larger variability in the universe will result in greater variability due to sampling the sample mean. Furthermore, the standard error becomes smaller as the sample size increases, which implies that the variability due to sampling diminishes as more information is collected.

Ideally, it is desirable to have no sampling error, but this would occur only if all possible sample results were identical. Since this is not possible when sampling, we strive to achieve the smallest amount of sampling error based upon available resources. The standard error is a measure of sampling error since it measures how close the results from each of the samples are to one another; in the case of the sample mean, the standard error is a measure of the closeness of individual means among samples.

Sampling Without Replacement From Finite Universes. The formula for the standard error of the mean assumes sampling is from an infinite universe or, equivalently, with replacement from a finite universe. When sampling without replacement from a finite universe the standard error can be modified to account for the continued change in the composition of the universe from draw to draw.

The formula for the standard error of the mean when sampling without replacement is

$$\sigma_{\bar{x}} = \sqrt{\frac{N - n}{N - 1}}\left(\frac{\sigma}{\sqrt{n}}\right)$$

Standard error of the mean for sampling without replacement.

The term under the square-root sign is called the **finite correction factor** and adjusts the standard error for the relationship between the size of the sample, n, and the size of the universe, N.

It should be clear from the above formula that the finite correction factor equals zero when the sample size and the universe size are equal; obviously, the standard error of the mean then will be zero. Hence, when the sample and universe sizes are the same, there will be no sampling error since the situation is equivalent to a census rather than a sample. In general, using the formula for an infinite universe, $\sigma/\sqrt{n}$, is conservative since the result is larger than the one assuming no replacement. When the sample size is small relative to the size of the universe, the two formulas provide similar results. We shall assume that sampling is from an infinite universe or its equivalent throughout the remaining chapters. Consequently, we shall ignore the finite correction factor.

Sampling Distributions in General. Although we outlined the general concept of a sampling distribution of a statistic, we centered most of our attention on the sample mean. The concept for other statistics is similar; hence, the sampling distribution describes the way any statistic varies from sample to sample. The *shape* of the distribution varies, however, depending upon the statistic. Other sampling distributions are introduced in later chapters where they actually are used to make inferences. In many cases the shape of the sampling distribution is based upon an *assumption* about the shape of the universe distribution. Most of the methods in the text are based upon assumptions about the universe. Therefore, it is important to be aware of the underlying assumptions as they are introduced.

The concept of a sampling distribution has been introduced so that it can be applied to problems of inference where universe parameters are unknown.

Known vs. Unknown Universe Parameters. The probability calculations illustrated in this chapter are based upon known values of universe parameters. This was necessary in order to introduce the concept of a sampling distribution. Recognize, however, that sampling typically is undertaken because the value of a population parameter is unknown. Consequently, in the chapters that follow, we shall apply the concept of a sampling distribution to problems where inferences can be made about *unknown* universe parameters.

QUESTIONS AND PROBLEMS

DEFINITIONAL QUESTIONS

1. Distinguish between a population, or universe, and a sample.
2. Distinguish between a parameter and a statistic.
3. Briefly, but clearly, define the following terms.
   (a) Universe distribution
   (b) Universe mean
   (c) Universe variance
   (d) Universe standard deviation
4. Briefly, but clearly, define each of the following terms.
   (a) Distribution of a sample
   (b) Sample arithmetic mean
   (c) Sample variance
   (d) Sample standard deviation
5. What is a sampling distribution or a random sampling distribution?
6. What is meant by the following terms?
   (a) Mean of a sampling distribution
   (b) Variance of a sampling distribution
   (c) Standard deviation of a sampling distribution
7. What is the meaning of the terms *sampling error* and *standard error?*
8. What information is provided by the Central Limit Theorem?
9. What is the general rule we shall use to determine if a sample size is large enough to apply the Central Limit Theorem?
10. Distinguish between sampling with replacement and sampling without replacement. Which is to be assumed in the remaining chapters in this part of the text?

11. What is the finite correction factor? When should it be used? Is it to be used in the remaining chapters in this part of the text?

12. For any two numbers "a" and "b," what is the meaning of the following two expressions?

    (a) $P(a \leq X \leq b)$
    (b) $P(a \leq \bar{x} \leq b)$

13. What are the five important points associated with the introduction to this chapter and Exhibit 2?

14. What are the four facts that form the basis for the computations illustrated in the examples of this chapter?

COMPUTATIONAL PROBLEMS

15. Given a universe of measurements with mean equal to 35 and variance equal to 15. Based upon a sample of 9 items drawn at random from the population, what are the values of $\mu_{\bar{x}}$, $\sigma^2_{\bar{x}}$, and $\sigma_{\bar{x}}$?

16. The weight of metal carried by railroad freight cars varies about a mean weight of 45 tons with a standard deviation of 3 tons. What is the mean, variance, and standard error of the distribution of the sample mean weight based upon a sample size of 10 freight cars?

17. Based upon a universe assumed to be normally distributed with mean equal to 50 and standard deviation equal to 15, find the following probabilities associated with the mean of a sample of 9 randomly drawn items.

    (a) $P(50 \leq \bar{x} \leq 60)$
    (b) $P(40 \leq \bar{x} \leq 50)$
    (c) $P(45 \leq \bar{x} \leq 55)$
    (d) $P(55 \leq \bar{x} \leq 65)$
    (e) $P(35 \leq \bar{x} \leq 40)$

18. Refer to Problem 17. What is the probability that the sample mean will fall within 7 units of the mean of the universe? What is the probability that the difference between the mean of the sample and the universe mean will be at least 8?

19. Refer to Problem 17. Above what value of the sample arithmetic mean will the highest 20 percent of the sample means fall? Below what value will the lowest 10 percent fall?

20. Given a universe with an unknown probability distribution that has a mean equal to 150 and a standard deviation of 60. Find the following probabilities associated with the sample arithmetic mean based upon a random sample of 100 items.

(a) $P(150 \leq \bar{x} \leq 175)$
(b) $P(140 \leq \bar{x} \leq 150)$
(c) $P(135 \leq \bar{x} \leq 160)$
(d) $P(155 \leq \bar{x} \leq 165)$
(e) $P(138 \leq \bar{x} \leq 144)$

21. Refer to Problem 20. Find the probability that the mean of the sample will fall within 10 units of the mean of the universe. What is the probability that the difference between the sample mean and universe mean will be at least 15?

22. Refer to Problem 20. Within what values will the middle 95 percent of the sample means fall? The middle 80 percent?

23. The distribution of the lengths of steel rods produced by a continuous manufacturing process is approximated by a normal curve with a mean of 16 inches and a standard deviation of 0.5 inch. A random sample of 9 rods is selected from the process.

(a) Find the probability that the sample arithmetic mean falls between 16.15 and 16.30 inches.
(b) What is the probability that the mean of the sample falls between 15.7 and 16.2 inches?
(c) Find the probability that the mean of the sample differs from the process mean by more than 0.3 inch.
(d) What proportion of samples have an arithmetic mean exceeding 16.4 inches?
(e) If 20 percent of samples of 9 rods have a mean length below a certain value, what is this value?

24. A random sample of 81 invoices is to be selected from the sales invoices of a large merchandising outlet in order to estimate the mean amount per sale. Suppose the actual mean amount equals \$3800 and the standard deviation is \$900.

(a) Find the probability that the mean of the sample exceeds \$4000.
(b) What is the probability that the sample arithmetic mean amount of sales lies between \$3900 and \$4100?
(c) Find the probability that the sample mean falls within \$150 of the actual mean sales.
(d) What proportion of the samples has a mean that is less than \$3650?
(e) If 30 percent of the samples have mean sales greater than a particular dollar value, what is this value?

CONCEPTUAL QUESTIONS AND PROBLEMS

25. What is the relationship between a universe distribution, the distribution of a sample, and a random sampling distribution. Why is it important to distinguish among the three types of distributions? Why is the concept of a sampling distribution of

special importance?

26. What is the relationship between the universe mean, the mean of a sample, and the mean of the distribution of the sample mean?

27. What is the relationship between the population variance, the variance of a sample, and the variance of the sampling distribution of the sample mean? Basically, would your answer be the same if you were asked about the standard deviation of the three types of distributions instead of the variance? Explain. How is the standard error of the mean related to the measures cited in this question?

28. How does the standard error of the mean measure sampling error? Is the amount of sampling error in the sample mean affected by the amount of variability in the universe? Explain. Why is it important to consider sampling error?

29. If only one sample is selected in a sampling problem, how is it possible to have an entire distribution of the sample mean?

30. If a census of an entire population of items is undertaken, is it meaningful to consider sampling error? What are the values of the universe standard deviation and the standard error of the mean when a census is taken? What do your answers suggest about the concept of a sampling distribution, the standard error of the mean, and sampling error?

31. Based upon the formulas for the variance of the sampling distribution of the mean for sampling with and without replacement, explain what the effect of taking a census has on sampling error. What is the effect of the sample size and the universe size on sampling error in general? Under what circumstances is sampling error affected most by the sample size?

32. It has been said that sampling can do as good a job as a census. Explain the meaning of this statement in terms of the standard error of the mean and the probability of occurrence of the sample mean.

33. Why is the Central Limit Theorem an important concept?

34. The concept of a random sampling distribution of the sample mean is emphasized in this chapter. Is it possible to develop a sampling distribution for other statistics besides the sample mean? Explain.

35. Based upon the material presented in this chapter, it is possible to compute probabilities associated with the occurrence of values of the sample mean under certain circumstances. What are these circumstances? Under what circumstances is it im-

possible for you to compute these probabilities? Why? In general, what is the role of assumptions in performing probability calculations for values of a sample statistic?

36. When it can be assumed that a universe is normally distributed, it is not necessary to consider the sample size in order to establish the shape of the distribution of the sample mean. When must the size of the sample be considered?

37. When computing probabilities in this chapter, the value of z that is used equals $(\bar{x} - \mu_{\bar{x}})/\sigma_{\bar{x}}$ whereas the formula $(x - \mu)/\sigma$ was used earlier. What are the similarities and differences between the two formulas? Under what circumstances should each of the formulas be used?

38. A random sample of the monthly charge account balances of a large department store is to be selected. Assume the mean balance of all accounts equals $450 and the standard deviation equals $125.

    (a) Find the probability that the mean monthly balance of a sample of 100 accounts falls between $435 and $475.
    (b) In terms of the definition of probability used throughout the text, interpret the value found in part (a) in terms of the problem.

39. Refer to Problem 38.

    (a) For a sample of 49 accounts, find the probability that the sample mean will fall within $35 of the mean balance of all accounts.
    (b) In order to find the probability in part (a), it is not necessary to know the value of the universe mean even though the question is posed in terms of this value. Why is this true?
    (c) Recompute the probability required in part (a) for a sample of 81 accounts.
    (d) Compare your answers to part (a) and part (c). Should there be a difference? Explain. In general, what conclusion can you make about the probability of occurrence of sample values based upon the comparison of the two probabilities? Relate your answer to the concept of sampling error.

40. Assume that the annual salaries of the 5000 salesmen of a large retail chain can be approximated by a normal curve with a mean equal to $16,000 and a standard deviation equal to $1500.

    (a) What is the probability that the income of a randomly drawn salesman falls between $15,000 and $17,000? Interpret the result.
    (b) What is the probability that the mean of a random sample of 25 salesmen falls between $15,800 and $16,200? Interpret the result.
    (c) What distribution did you use to compute the probabilities in parts (a) and (b) above? Should these distributions be different? Explain.

CHAPTER 11

ESTIMATION AND CONFIDENCE INTERVALS

Lest men suspect your tale untrue, keep probability in view.

John Gay

When in doubt use a bigger hammer.

Thomas L. Martin, Jr.

Whenever we use sample data to determine the magnitude of an unknown universe parameter, we are faced with the possibility of being wrong. The value of a sample statistic used as a point estimate rarely equals the parameter estimated. Recall from Chapter 9 that a point estimate is defined as any sample statistic in the form of a single number that is used to estimate an unknown universe parameter. A point estimate rarely equals the universe parameter because of sampling error, or variability of the value of a sample statistic from sample to sample. We learned how to describe sampling error in the last chapter in terms of the random sampling distribution and the standard error of a statistic.

Based upon the random sampling distribution it is possible to construct an estimate that explicitly accounts for sampling variability. In this respect, parameter estimation is similar to shooting at a target. The probability of getting a "bulls-eye" is very small, whereas the chance of hitting the target is greater since the target is larger. If the size of the target is increased, the chance of hitting it becomes even greater. Hence, if instead of using a single value to estimate a universe parameter we use a set of values, or an *interval estimate,* our chances of correctly estimating the parameter increase. The larger the interval, the greater the probability that it will contain the universe parameter. For example, suppose you are told that the average cost of hospital care for a sample of patients is $1500. It is unlikely that this value which is in the form of a point estimate is a correct estimate of the mean cost of all patients. If, however, we state that the mean cost of all patients lies between $900 and $1800 there is a greater chance that the estimate is correct. In other words, it is more likely that the true mean falls within the interval used as an estimate than it is to equal $1500 exactly.

The statistical method by which interval estimates of parameters are determined is referred to as **confidence interval estimation.** This method possesses a special feature that enables us to establish a *known,* or measurable, probability that the estimating procedure is correct. An estimate based upon this procedure is called a *confidence interval* estimate. We use a confidence interval estimate in order to increase our chances of being correct and to establish the worth of the estimate in terms of probability. The probability assessment is based upon the *random sampling distribution* of the sample statistic used.

A general approach exists for constructing confidence intervals for any parameter. Since different sampling distributions apply in particular cases, the specific method of computation may vary. In this chapter confidence interval estimation is presented in terms of means, since this represents the simplest case computationally and commonly is used in practice.

In the last chapter we learned about the sampling distribution of the sample mean under two circumstances when the value of the universe standard deviation is known. There we found that the distribution is normal for all sample sizes when the universe is normal and that it is approximately normal for large samples when the universe is not normal. These two facts are helpful to us in this chapter, but they do not satisfy our needs completely.

Whenever we use the sample mean to construct a confidence interval estimate of the universe mean, we need to know something about the universe standard deviation. If it is known, there is no problem. However, if the universe standard deviation is unknown, which is the usual case, we also must estimate it from the sample. In this case, the sampling distribution of the mean is not normal but can be approximated by a normal curve for large samples.

The procedure and example sections that follow present the way to construct a confidence interval estimate of the universe mean when the distribution of the sample mean is assumed to be normal. For our work in this chapter, we can *assume* that the sampling distribution of the mean is normal: (1) for any sample size when the population is normal and the universe standard deviation is known, and (2) for large samples for any universe when the standard deviation is either known or unknown. In the second case, the normal is an approximation. After working with these cases in detail, we shall consider the problem of constructing a confidence interval for means of normal populations based upon unknown standard deviations and a small sample. This is presented in the discussion section.

The symbols used in this chapter are:

x = the value of a measured quantity, or a variable
n = the sample size
μ = the universe mean

$\bar{x}$ = the sample arithmetic mean
σ = the universe standard deviation
s = the standard deviation of a sample
$\sigma_{\bar{x}}$ = the standard error of the sample mean
$s_{\bar{x}}$ = an estimate of the standard error of the sample mean
z = the standardized normal variate needed when using the table of Areas of the Normal Curve, Table 2
t = the standard variate needed when using the table of Values of Student's t-Distribution, Table 4
DF = degrees of freedom
E = the maximum tolerable error of estimating μ
$\hat{S}$ = an alternative estimate of σ similar to s but based upon a divisor of n − 1

PROCEDURE

Based upon a random sample of n observations, a confidence interval estimate of an unknown universe mean, μ, is found according to the four steps that follow. The procedure applies when the distribution of the sample mean is assumed to be normal.

Step 1. Choose a *level of confidence* representing the probability of being correct. The only guidance provided by statistical theory is that the probability should be "high." Typical values used are 0.90, 0.95, 0.98, and 0.99.

Step 2. Find the *confidence coefficient,* z, corresponding to the chosen level of confidence, from the table of Areas of the Normal Curve.

Step 3. Compute the standard error of the mean, $\sigma_{\bar{x}}$, using the formula

$$\sigma_{\bar{x}} = \frac{\sigma}{\sqrt{n}}$$

The standard error of the sample mean.

in cases where the population standard deviation, σ, assumes a known value. When σ is unknown, compute the sample estimate $s_{\bar{x}}$, of the standard error using the formula

$$s_{\bar{x}} = \frac{s}{\sqrt{n-1}}$$

Sample estimate of the standard error of the mean.

where s is the standard deviation of the sample selected.* The reason for dividing by "n − 1" instead of "n" to obtain the sample estimate of the standard error is considered in the discussion section.

*The formula for the sample standard deviation, s, is the same as the one presented in Chapter 3, and can be used when necessary to compute the standard deviation of a set of sample observations.

Step 4. Construct a confidence interval on the basis of the values obtained in Step 2 and Step 3 by substituting into the following expressions

Known σ		*Unknown σ*
Normal Universe—Any n Any Universe—Large n		Any Universe—Large n
$\bar{x} \pm z\sigma_{\bar{x}}$	*Or*	$\bar{x} \pm zs_{\bar{x}}$

Confidence interval for μ assuming a normal sampling distribution.

where $\bar{x}$ is the mean of a specific random sample.* When a normal sampling distribution is assumed, a confidence interval estimate of the population mean is obtained by multiplying the standard error by the confidence coefficient for a selected level of confidence and adding the result to and subtracting it from the sample mean. This results in two values or limits that represent the actual confidence interval estimate.

NUMERICAL EXAMPLES

Example 1 A population of items is assumed to be normally distributed with an unknown mean and known standard deviation equal to 36. Based upon a random sample of 9 items drawn from the population, a sample arithmetic mean of 62 is computed. Construct a confidence interval estimate of μ.

Step 1. Suppose we choose a confidence level of 0.95 as high enough for our purposes.

Step 2. Using the table of Areas of the Normal Curve, or Table 2, look up half of 0.95, which yields a confidence coefficient of z = 1.96.

Step 3. Compute $\sigma_{\bar{x}} = \frac{\sigma}{\sqrt{n}}$

$$= \frac{36}{\sqrt{9}}$$

$$= 12$$

Step 4. Substitute and compute

$$\bar{x} \pm z\sigma_{\bar{x}}$$

$$62 \pm 1.96(12)$$

$$\pm 23.5$$

$$38.5 - 85.5$$

*If the sample mean, $\bar{x}$, is not given, it can be calculated from the sample observations by using the formula for the arithmetic mean presented in Chapter 2.

The mean, μ, is estimated to lie between 38.5 and 85.5 with 95 percent confidence.

Example 2 Given a population with unknown mean and standard deviation. Construct a 98 percent confidence interval estimate of the mean, based upon a sample mean and standard deviation of 22 and 11 computed from a random sample of 100 items.

This problem is similar to the one above except the level of confidence is different, and we must use the estimated standard error, $s_{\bar{x}}$, since σ is not known.

The following information is given:

$$\bar{x} = 22$$

$$s = 11$$

$$\text{Confidence level} = 98 \text{ percent}$$

Based upon the above information, the confidence interval estimate is found as follows.

$z = 2.33$ (corresponds to 98 percent confidence obtained from Table 2 for half .98, or .49)

$$s_{\bar{x}} = \frac{s}{\sqrt{n-1}}$$

$$= \frac{11}{\sqrt{100-1}} = \frac{11}{\sqrt{99}} = \frac{11}{9.95}$$

$$= 1.11$$

Substituting and computing,

$$\bar{x} \pm z s_{\bar{x}}$$

$$22 \pm 2.33(1.11)$$

$$\pm 2.59$$

$$19.41 - 24.59$$

Example 3 You are interested in estimating the mean family income of a small town with 90 percent confidence. Based upon personal interviews of a random sample of 900 families, you find that the sample mean is \$8536 and the sample standard deviation is \$423. The confidence interval is obtained as

$$\bar{x} \pm zs_{\bar{x}}$$

$$8536 \pm 1.65\left(\frac{423}{\sqrt{900 - 1}}\right)$$

$$\pm 1.65(14.1)$$

$$\pm 23.27$$

$$8512.73 - 8559.27$$

The mean family income is estimated to lie between \$8512.73 and \$8559.27. In this example all of the steps are combined within the formula for the interval estimate.

DISCUSSION

The preceding sections dealt with the construction of confidence interval estimates. Little was said about the confidence level or the probability required to construct the interval, and the way the interval can be interpreted was omitted. These are important considerations that will now be discussed.

The meaning of a confidence interval.

Consider, as an example, a 95 percent confidence interval used to estimate a universe mean, μ, when the distribution of the sample mean is normal. This kind of interval estimate leads to the following probability statement.

$$P(\bar{x} - z\sigma_{\bar{x}} \leq \mu \leq \bar{x} + z\sigma_{\bar{x}}) = 0.95$$

The confidence level provides the proportion of all possible intervals based upon repeated samplings that will contain the universe parameter.

This statement is interpreted to mean that the probability that an interval of the form $\bar{x} \pm z\sigma_{\bar{x}}$ contains the universe mean is 0.95. Put another way, 95 percent of the intervals constructed for all possible sample means (based on a fixed sample size) drawn from a given population will contain the mean, μ. If we sample repeatedly from the same population, 95 percent of all intervals will contain the population mean, or 95 percent of the time we will correctly estimate the population mean. Five percent of the time the interval will lead to an incorrect estimate.

The above interpretation treats the expression $\bar{x} \pm z\sigma_{\bar{x}}$ as a random quantity similar to the sample mean, $\bar{x}$, since both vary from sample to sample. It should be stressed, however, that *once* numbers based upon a *particular* sample are substituted into the expression for the interval estimate, the probability statement does not hold. Since μ is a fixed number, although unknown, it either is or is not contained within the particular interval. We never know which of the two cases is true. The probability statement applies to the entire estimation process considered in terms of repeated samplings, while an interval estimate is based upon one sample result. Once numbers actually are substituted to obtain a specific interval estimate we associate the word confidence instead of probability with the interval. We always act as if this interval contains the universe mean.

The reason that we can act as if the universe mean is contained within a particular interval is *based upon our choice of the confidence level,* which is subject to our control. We preset the confidence at a level sufficiently high or close to one so that we are not concerned about committing an error. Hence, for practical purposes, the chance of error is small enough and can be ignored.

We always act as if a particular interval estimate contains the universe parameter due to the way we choose the confidence level.

The reasoning suggested is not very different from the kind we use in everyday decisions. As an example, consider an extremely ordinary case. Hardly anyone doesn't cross a street. When you actually cross, do you think that the probability of being hit by a car is zero? Obviously, this is not the case; people are injured occasionally. If we know there is a chance of being injured and we actually decide to cross, it might be that we feel this chance is small enough to ignore. This is not to suggest that each time we make an ordinary decision we should refer to our probability tables and pocket calculators. This kind of decision is made rather quickly on the basis of a somewhat elusive but satisfactory thinking process. However, the same kind of reasoning can be used explicitly in dealing with problems of a "less ordinary" nature, and forms the basis for confidence interval estimation and other types of inference.

In order to illustrate the meaning of a confidence interval, a confidence level of 0.95 was suggested. In practical work, such a value is frequently used. Sometimes, however, other levels are considered due to the nature of a particular problem. The choice of a confidence level depends upon an individual's decision about the probability of being correct or the risk of being wrong. By increasing the confidence level, the chances of being correct in the long-run are increased and the risk of error is reduced.

While affording greater protection against error, increased confidence imposes a loss in another sense. The interval must widen, which means that a particular estimate is not as precise. Consequently, confidence and precision must be balanced in order to obtain a useful and informative estimate. Only through an understanding of a particular problem can the required amount of precision be established. The field of statistics can determine how to achieve a given amount of precision, but not how much precision is necessary.

The width of the interval varies with the level of confidence.

Sample Size Determination. Related to the width of a confidence interval is the sample size. The size of the sample affects the size of the standard error, which in turn influences the width of the interval. Therefore, the interval size can be decreased by increasing the size of the sample. This fact can be helpful before an actual sample is selected.

Suppose we are interested in estimating the universe mean, μ, subject to the requirement that the error of estimation must not exceed a fixed value, E, with a certain probability. In other words, we require that the sample mean falls within the range $\mu \pm E$ with the specified probability. If we assume that the distribution of the sample mean is normal, the problem is to find the sample size, n, such that the sample mean falls within the limits $\mu - E$ and $\mu + E$ shown in the diagram opposite this paragraph.

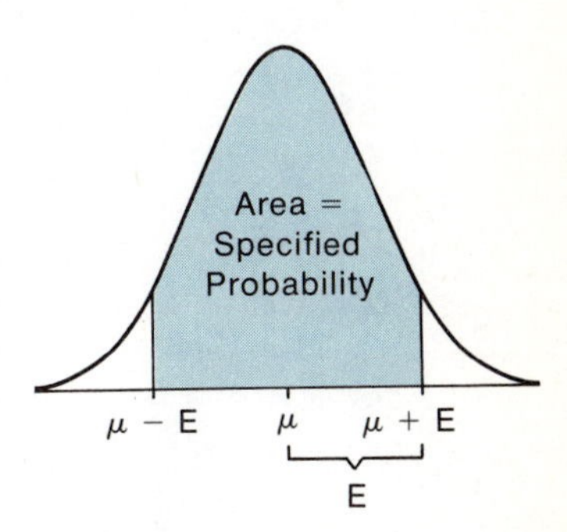

Since the distance between a value of a normal variable and its mean corresponding to any probability can be expressed in terms of a multiple of the standard deviation, the error E can be written as

$$E = z\sigma_{\bar{x}}$$

where z corresponds to the specified probability and $\sigma_{\bar{x}}$ is the standard deviation. By substituting the formula for $\sigma_{\bar{x}}$ into the above expression, we obtain another expression relating E to the sample size n.

$$E = z\sigma_{\bar{x}}$$
$$= z\left(\frac{\sigma}{\sqrt{n}}\right)$$

Using simple algebra, we can solve for the square root of n as

$$\sqrt{n} = \frac{z\sigma}{E}$$

If we square both sides of this expression, we isolate n in terms of the other quantities and obtain a formula for determining the size of the sample which takes the form

$$n = \frac{z^2\sigma^2}{E^2}$$

Formula for determining sample size for estimating μ.

By specifying a level of confidence, z is determined. σ^2 is the *population* variance, and E represents the error or maximum tolerable difference between the unknown population mean and the sample estimate at the prescribed level of confidence.

In order to illustrate the use of the above formula, suppose we are interested in estimating the mean family income of a certain town. We specify that the sample estimate will not be useful if it differs from the universe mean by more than \$25 at a 98 percent level of confidence. Our best guess regarding the population standard deviation, based upon other sources of information, is \$423. Using the formula for n and substituting, we have

$$n = \frac{z^2\sigma^2}{E^2}$$
$$= \frac{(2.33)^2(423)^2}{(25)^2}$$
$$= 1554$$

A random sample of 1554 families is required in order to estimate the mean family income within \$25 with 0.98 confidence. The value of z equal to 2.33 corresponds to the confidence level of 0.98 and is obtained from Table 2.

Notice the sample size calculation includes a value of the population standard deviation. Typically, although not always, when the mean is unknown, the standard deviation also is unknown; yet information about σ is needed to determine the sample size necessary to meet the given requirements. Pilot or preliminary studies may be conducted in order to estimate σ and thereby determine the proper sample size, n.

Unknown σ and Student's t-Distribution. Earlier, it was indicated that the formulas given in the procedure section for constructing a confidence interval estimate of μ do not apply when σ is unknown and n is small for normal and non-normal populations. Of these two, cases of non-normal populations will not be considered since the results depend upon the specific shape of the universe distribution and vary with each distribution.

Using another sampling distribution, we can establish a confidence interval estimate for μ when σ is unknown and n is small. This is based upon the fact that *the sample mean is distributed as Student's t-Distribution for random samples drawn from normal populations with unknown σ.** Although the t-distribution appears similar to the normal when graphed, the formula for this probability distribution is complicated and the distribution has different characteristics. Since the t-distribution is available in tabular form, it is not necessary to deal with the formula directly.

Note. The t-distribution also is a continuous bell-shaped curve that is symmetric about a single peak and asymptotic to the horizontal axis. The t-distribution has one parameter called the *degrees of freedom,* which is given by the following formula when used to find a confidence interval for the mean:

$$DF = n - 1$$

Degrees of freedom for confidence intervals of μ using the t-distribution.

where the symbol "DF" represents the degrees of freedom. The number of degrees of freedom in this case is obtained by subtracting "one" from the sample size, n. This quantity identifies a particular t-distribution just as values of n and P identify a particular binomial distribution and μ and σ identify a particular normal curve. Different t-distributions can be identified by differing numbers of degrees of freedom.

The t-distribution is tabulated in the Appendix as **Table 4,** which is entitled "Values of Student's t-Distribution." The table is not constructed in the same way as the table of the normal curve, but provides values of "t" for selected *two-tailed probabilities* for various degrees of freedom. Probabilities provided in the table are shown in Exhibit 1. The values of t correspond to given probability levels such that *half* of the probability appears in each tail of the distribution. This is shown as the shaded regions in the exhibit. Specific instructions for using the table appear in the Appendix.

In order to construct an interval estimate using the t-distribution,

*Strictly speaking, the following quantity, called Student's t-statistic,

$$t = \frac{\bar{x} - \mu}{s_{\bar{x}}}$$

is distributed according to a t-distribution. The difference between "t" and "z," used for the normal distribution, lies in the denominator of the statistic where the estimated standard error always is used.

EXHIBIT 1
PROBABILITIES ASSOCIATED WITH THE TABLE OF "VALUES OF STUDENT'S t-DISTRIBUTION"

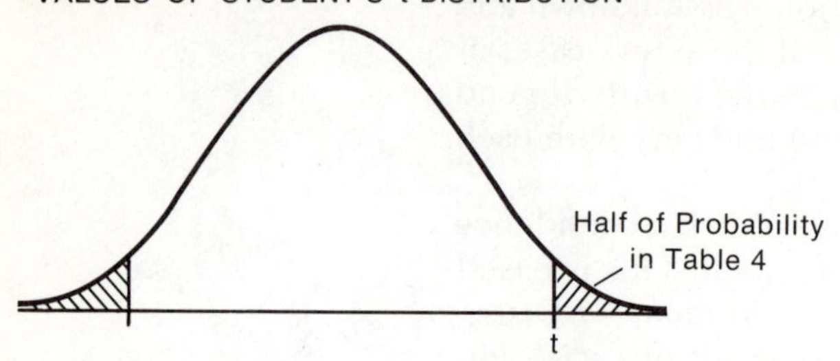

we follow a format similar to the one outlined in the procedure section; the difference lies in the fact that the estimated standard error always is used and a value of "t" is obtained from Table 4 rather than "z" from the normal table. The confidence interval estimate takes the form

$$\bar{x} \pm ts_{\bar{x}}$$

Confidence interval for μ (normal population and unknown σ).

where $\bar{x}$ is the sample mean, $s_{\bar{x}}$ is the estimated standard error, and "t" is a value from Table 4 corresponding to a chosen level of confidence. By substituting these values into the above expression an interval estimate is obtained.

Consider an example illustrating the use of the t-distribution. Suppose we select a random sample of 10 women from the customers of a department store in order to estimate the average age of women shoppers with 95 percent confidence. We find the sample mean equals 36.1 years and the sample standard deviation to be 11 years. Based upon the problem statement we are given the following information:

$$\text{Confidence Level} = 95 \text{ percent, or } 0.95$$
$$\text{Sample Mean, } \bar{x} = 36.1$$
$$\text{Sample Standard Deviation, } s = 11$$
$$\text{Sample Size, } n = 10$$

Corresponding to this information, we compute the degrees of freedom as

$$\begin{aligned} DF &= n - 1 \\ &= 10 - 1 \\ &= 9 \end{aligned}$$

The value of t is found in Table 4 by locating 9 degrees of freedom and the complement of the confidence level of 0.95, or 0.05. This is based on the fact that the table is in terms of the area in the tails whereas the confidence level corresponds to the center of the distribution. For 9 degrees of freedom and a two-tailed probability level of 0.05, the value of t from Table 4 is

$$t = 2.262$$

The estimated standard error of estimate is computed using the formula given earlier as

$$s_{\bar{x}} = \frac{s}{\sqrt{n - 1}}$$

$$= \frac{11}{\sqrt{10 - 1}}$$

$$= \frac{11}{\sqrt{9}} = \frac{11}{3}$$

$$= 3.67$$

Substituting into the expression for the confidence interval estimate and computing, we have

$$\bar{x} \pm ts_{\bar{x}}$$

$$36.1 \pm 2.262(3.67)$$

$$\pm 8.30$$

$$27.8\text{–}44.4$$

Hence, our estimate of the average age of women shoppers lies between 27.8 and 44.4 years with 95 percent confidence.

The procedure illustrated above can be applied to any other confidence level beside 0.95 in cases where the t-distribution is the appropriate sampling distribution. The value of t will differ depending upon the sample size and the chosen level of confidence. Although the limits of the interval estimate using t are different than ones obtained by using the normal curve, once the interval estimate is constructed it always is used and interpreted in the same way.

If you look back at Table 4 in the Appendix, the values of "t" on the last row corresponding to an infinite (∞) degrees of freedom are the same as the z-values obtained from the normal table at each given probability level. This is due to the fact that the t-distribution *approaches* the normal in the limit as the sample size becomes larger. Although the t-distribution is the exact distribution to use for all sample sizes when the standard error is estimated, the real benefit of "t" is associated with

The t-distribution can be approximated with the normal for large samples.

EXHIBIT 2
COMPARISON OF THE NORMAL CURVE AND STUDENT'S t-DISTRIBUTION

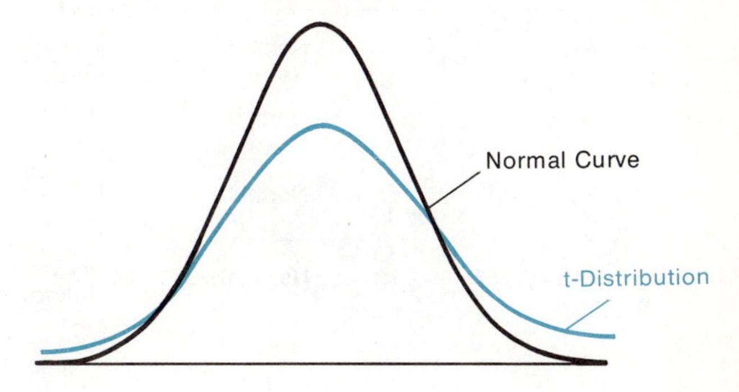

small samples. The normal is a satisfactory approximation for large samples.

By graphically comparing the t-distribution with the normal we get a better understanding of its properties. Consider the diagram presented in Exhibit 2. Although the t-distribution looks like a normal curve, it is "flatter" and "wider" than the normal. Consequently, values of t are larger than values of z and lead to wider confidence intervals. This occurs because the t-distribution accounts for the added uncertainty associated with estimating σ, whereas the normal treats this as a known value. As the sample size increases, there is less uncertainty in the sample estimate of σ, which explains why the t-distribution becomes narrower and closer to the normal.

It is interesting to conclude this discussion of the t-distribution with a historical note. The t-distribution was developed by a brewer named William Sealy Gosset in 1907. Gosset was a "company man" who doubled as a statistician and published under the name of Student. The brewery had a continuing need to test product quality with small samples, yet no accurate method currently was available. Gosset went to work in the interest of the firm and developed a small sample theory which resulted in Student's t-Distribution as his major contribution to the field of statistics. The distribution is useful in other applications beside confidence intervals for means and is used again in later chapters.

The t-distribution is important in many other problems.

Method Summary. Throughout this chapter three formulas for constructing a confidence interval estimate of the mean are presented. These formulas are associated with five situations based upon varying assumptions about the universe, σ, and n. The table presented in Exhibit 3 provides a summary of these formulas together with the underlying assumptions. This table can be used to select the appropriate interval to use in a particular problem. Recognize that the t-distribution provides an exact confidence interval when the population is normal and σ is unknown, or estimated; when the sample size is large, an approximate answer to the same problem can be obtained by using the normal.

For our purposes a sample can be considered large if it is greater than 30.

EXHIBIT 3
SUMMARY OF CONFIDENCE INTERVALS TO ESTIMATE THE UNIVERSE MEAN, μ

Confidence Interval	*Standard Error*	*Universe*	σ	*n*	*Applicability*
$\bar{x} \pm z\sigma_{\bar{x}}$	$\frac{\sigma}{\sqrt{n}}$	Normal Non-normal	Known Known	Any Large	Exact Approximate
$\bar{x} \pm zs_{\bar{x}}$	$\frac{s}{\sqrt{n-1}}$	Normal Non-normal	Unknown Unknown	Large Large	Approximate Approximate
$\bar{x} \pm ts_{\bar{x}}$	$\frac{s}{\sqrt{n-1}}$	Normal	Unknown	Any	Exact
NOT DISCUSSED		Non-normal	Known & Unknown	Small	

It is satisfactory to use the large sample result when the sample size exceeds 30.

The last row of the table indicates that confidence intervals for small samples from non-normal universes have not been considered. In order to deal with these situations we must know the form of the sampling distribution that varies with each universe distribution or use other methods that do not depend on assumptions about the universe. The cases presented are sufficient to illustrate the concept of confidence interval estimation and include many applications encountered in practice. The table helps to emphasize the point that assumptions are important in statistical work; assumptions should match the realities of a particular problem in order to obtain a valid result. Moreover, exact results are not always available and approximations frequently are employed.

Estimating Standard Deviations. When estimating the population mean with a confidence interval it is necessary to include the standard deviation. In other problems knowledge of the population standard deviation is essential in its own right. When only sample information is available we must estimate σ. A point estimate that can be used is the sample standard deviation.

$$s = \sqrt{\frac{\Sigma(x - \bar{x})^2}{n}}$$

Sample standard deviation.

This is the formula for the standard deviation presented in Chapter 3. Lower case letters are used instead to indicate that it is based upon a sample.

Frequently, a closely related alternative to the sample standard deviation is used as an estimate of the universe value which takes the form

$$\hat{S} = \sqrt{\frac{\Sigma(x - \bar{x})^2}{n - 1}}$$

Alternative point estimate of σ.

This alternative is basically the same as the sample standard deviation except the sum of the squared deviations from the mean is divided by "$n - 1$" instead of n. Division by $n - 1$ can be considered as an adjustment since the unadjusted sample standard deviation tends to understate, or bias, the true value of σ. Generally, the adjusted version of the standard deviation is the one provided by calculators that are preprogrammed to calculate the standard deviation.

When estimating the standard error of the mean, $\sigma_{\bar{x}}$, we used the following formula.

$$s_{\bar{x}} = \frac{s}{\sqrt{n - 1}}$$

Estimated standard error of the mean.

This formula for the estimated standard error of the mean is based upon the unadjusted sample standard deviation, s, divided by $\sqrt{n - 1}$. The same result can be obtained by dividing the adjusted value, $\hat{S}$, by $\sqrt{n}$.

Hence, the formula for the estimated standard error of the mean contains the adjustment for the bias in estimating $\sigma_{\bar{x}}$ directly in the formula. When n is *large*, it is not necessary to consider the "minus one" in either formula since its effect is negligible.

The interpretation of the confidence interval given earlier applies to all cases; the method of computation varies with the parameter estimated.

Other Considerations. Confidence intervals can be constructed for many other parameters beside means. Regardless of the parameter estimated, the *interpretation* provided earlier applies to *any* confidence interval. What differs is the way an interval estimate is computed, since the sampling distribution will vary depending upon the sample statistic used as an estimate.

The material in this chapter is concerned with fundamental concepts regarding the estimation of parameters using sample information. Means are used to illustrate the basic ideas. Keep in mind that the problems considered deal only with one type of error, that due to sampling. There are, however, additional considerations when undertaking an actual estimation problem. For instance, the last numerical example referred to a personal interview of families in order to obtain information about their income. It is well known that individuals are reluctant to divulge such information. As a consequence, the possibility exists that incomes are incorrectly reported. This could happen in a census as well as a sample, and consequently this *non-sampling error* is not considered in any of the formulas or calculations presented.

Many types of non-sampling errors can be made when using data to solve a problem. Realistically, the nature of non-sampling errors should be discussed in terms of specific problems, and generally are treated in courses on research methods rather than statistics. The point made here is that considerations beside knowledge of a specific statistical method are necessary when using data to draw inferences; our primary goal is to understand the nature of statistical methods that are, in turn, used in problems of larger scope.

QUESTIONS AND PROBLEMS

DEFINITIONAL QUESTIONS

1. Define the following terms:

   (a) Estimation
   (b) Point estimate
   (c) Interval estimate
   (d) Confidence interval estimate

2. Distinguish between the terms *confidence interval*, *confidence level*, and *confidence coefficient*.

3. Distinguish between the standard error of the mean and the estimated standard error.

4. Under what circumstances can the normal distribution be used to construct a confidence interval estimate of the universe mean?

5. What is Student's t-distribution? Under what circumstances can it be used to construct a confidence interval estimate of the universe mean? When should it be used?

6. What is the meaning of the term *degrees of freedom* as it is used in this chapter?

7. In words, describe the general procedure for constructing a confidence interval estimate of the universe mean. Specify the formulas used to construct interval estimates of the mean and the circumstances under which they apply.

8. What is the general rule for selecting a level of confidence? What are the commonly used levels?

9. What is the difference between an "adjusted" and "unadjusted" standard deviation?

10. What is the general rule for deciding when a sample size is large? Under what circumstances when constructing confidence intervals should the rule be applied?

COMPUTATIONAL PROBLEMS

11. A sample of 25 items is randomly drawn from a normal universe with unknown mean and a standard deviation equal to 15. The mean of the sample equals 12. Construct a 95 percent confidence interval estimate of the universe mean.

12. Suppose that a random sample of 50 items drawn from a particular universe has a mean equal to 30 and a standard deviation equal to 28. Construct a 98 percent confidence interval estimate of the universe mean.

13. A random sample of 10 items is selected in order to estimate the mean of a normal population. The mean and standard deviation of the sample equal 50 and 24, respectively. Construct a 99 percent confidence interval estimate of the population mean.

14. Appearing below are 10 values corresponding to a random sample drawn from a population with an unknown mean and standard deviation.

13	9	19	20	7
8	15	17	11	14

(a) Using the information from the sample, find point estimates of the population mean and standard deviation.
(b) Construct a 90 percent confidence interval estimate of

the population mean. Assume the population is normally distributed.

15. The average monthly electricity consumption for a sample of 100 families equals 1250 units. Assuming the standard deviation of electric consumption of all families is 150 units, construct a 95 percent confidence interval estimate of the actual mean electric consumption.

16. A random sample of 122 apartment dwellers was selected in order to estimate the mean rental paid by residents of a large apartment complex. The mean of the sample is $325 with a standard deviation of $50. Estimate the mean rental paid by the residents of the complex with a 99 percent confidence interval.

17. In order to establish a mileage rating for a particular brand of automobile, a sample of 10 cars is tested resulting in a mean of 19.7 miles per gallon (mpg) and a standard deviation of 2.4. Assuming car mileage is normally distributed, construct a 90 percent confidence interval estimate of the mean mpg for this brand of car.

18. As part of a general survey of hospitals in a large city, the daily room rate for a sample of five hospitals was observed. These appear below.

$135 $216 $165 $143 $197

(a) Based upon the sample data, find point estimates of the mean and standard deviation of hospital room rates throughout the city.
(b) Assuming hospital room rates are normally distributed, estimate the mean daily room rate of all hospitals with a 95 percent confidence interval.

19. The weight of scrap metal hauled in freight cars varies among loads. As part of a study of railroad freight carriers, it is necessary to estimate the mean weight of scrap metal hauled per freight car. How many freight cars should be sampled in order to be 95.5 percent confident that the sample mean will fall within 1000 pounds of the true mean weight? Assume the standard deviation, σ, equals 6000 pounds and a normal sampling distribution.

20. The length of life of brake linings used in automobiles exhibits variability like all other manufactured products. In order to study the durability of a particular brand of brake lining, a sample of linings is to be tested, and the sample arithmetic mean length of life is to be used as an estimate of the true mean. How many linings should be sampled for the sample mean to fall within 300 miles of the true mean with a probability of 0.99? Assume the standard deviation, σ, equals 800 miles and the sampling distribution of the mean is normal.

21. Why is estimation important? What are the advantages of using interval estimates instead of point estimates? What are the disadvantages?

22. Why is estimation a problem of statistical inference? Is there any relationship between estimation and descriptive statistics? Explain.

23. What is the relationship between a point estimate, a sample statistic, a random sampling distribution, and a confidence interval? Of what importance is a random sampling distribution in confidence interval estimation? A random sample?

24. In your own words, explain how a confidence interval accounts for sampling variability.

25. (a) In general terms, explain how the probability statement associated with a confidence interval is interpreted.
    (b) Given the following probability statement expressed in the form of symbols:

$$P(\bar{x} - z\sigma_{\bar{x}} \leq \mu \leq \bar{x} + z\sigma_{\bar{x}}) = 0.9$$

    What is the meaning of this probability statement? Can you give the same interpretation to the expression $P(10 \leq \mu \leq 20) = 0.9$? Explain.

26. Explain the role played by assumptions when constructing confidence interval estimates of universe parameters.

27. Criticize or explain the following statements:
    (a) Confidence interval estimates can only be made for means since these are the only ones presented in this chapter.
    (b) Since we always act as if the parameter to be estimated falls within confidence limits based upon a particular sample, only one kind of sampling error can be committed. The error that can be committed is that we incorrectly act as if the parameter falls within the confidence limits when it really does not.
    (c) When constructing a confidence interval estimate, we have complete control over the probability of committing an error since we can make this as small as possible.
    (d) The confidence level represents the probability of making a correct decision.
    (e) Since the confidence level has meaning as a probability only in terms of repeated trials, it has no meaning in terms of a specific estimate based upon a particular sample.
    (f) Since the width of a confidence interval is unimportant,

it is necessary to be concerned about the confidence level rather than the size of the sample.

(g) Although the standard deviation is needed in order to construct a confidence interval for the mean, it makes no difference whether the population standard deviation is known or not.

28. Of the 14,714 families in West Falls, a random sample of 220 families was taken in order to determine the mean family income in this depressed area. A 95 percent confidence interval ($3812 to $4116) was established on the basis of the sample results.

Using only the above information which of the following statements are valid? For statements that are not valid, explain why they are not.

(a) Of all possible samples of size 220 drawn from this population, 95 percent of the sample means will fall in the interval ($3812 to $4116).
(b) Of all possible samples of size 220 drawn from this population, 95 percent of the universe means will fall in the interval ($3812 to $4116).
(c) Of all possible samples of size 220 drawn from this population, 95 percent of the confidence intervals established by the above method will contain the universe mean.
(d) 95 percent of families in West Falls have incomes between $3812 and $4116.
(e) The distribution of family incomes in West Falls is practically normal.
(f) Using the above method, exactly 95 percent of the intervals so established will contain the sample mean $\bar{x}$.
(g) West Falls needs Federal aid.
(h) We do not know whether the universe mean is in the interval ($3812 to $4116) or not.

29. Refer to Problem 16. The 99 percent confidence interval estimate of the mean rental paid by apartment dwellers is $313–$337. In what way can this interval be interpreted?

30. Refer to Problem 17. The 90 percent confidence interval estimate of the mean miles per gallon is 18.2–21.2 miles per gallon. Specifically in terms of the problem, how can this interval be interpreted?

31. The time taken by a smoke detector to sound an alarm after a fire has started depends upon many factors in addition to the basic amount of variability resulting from the fact that it is a manufactured item. Assume it is of interest to determine the mean time that it takes a particular brand of detector to react to fire considering only the basic variability among different detectors. With all other factors subject to control, a random sample of 40 smoke detectors is tested resulting in a mean of 75 seconds and a standard deviation of 15 seconds.

(a) Construct a 90 percent confidence interval estimate of the mean reaction time, based upon the above information.
(b) Interpret the interval found in part (a) specifically in terms of the problem.

32. An owner of a delicatessen located in a large shopping center has observed that his facilities accommodate weekday customers without difficulty; however, conditions of overcrowding exist on Saturdays. Before deciding what to do about the problem, the owner decides to estimate the average number of customers based upon a sample of the number of Saturday customers. A sample of 12 Saturdays yields a mean number of customers equal to 850 with a standard deviation of 75.

(a) Assuming the number of Saturday customers is normally distributed, construct a 95 percent confidence interval estimate of the average number of Saturday customers.
(b) In terms of the problem, interpret the estimate found in part (a).

33. Refer to Problem 31. If the confidence level were changed to 0.95, what would be the effect on the estimate found in part (a)? What would be the effect on the estimate if the sample size were larger? Give reasons for your answers.

34. Refer to Problem 32. If the confidence level were reduced to 0.90, what would be the effect on the estimate found in part (a)? If the sample size were 25 instead of 12, would this have an effect on the estimate? What is the benefit of using a larger sample size?

35. A random sample of 50 invoices is selected in order to estimate the mean dollar amount of orders placed with a distributor of home appliances. The mean of the sample equals $15,000 with a standard deviation equal to $1000.

(a) Construct a 95 percent confidence interval estimate of the mean number of orders.
(b) Construct a 90 percent confidence interval estimate of the mean number of orders and compare the answer with the one found in part (a).
(c) Construct a 95 percent confidence interval of the mean number of orders assuming the sample size equals 100 and compare the result with the one in part (a).
(d) What overall conclusions can you make about the effect of different confidence levels and sample sizes on confidence interval estimates? Discuss the effect of these differences on the usefulness of an interval estimate.

36. Why is it important to consider the size of the sample when making estimates? Relate your answer to point and interval estimates. What factors must be considered when determining sample size before sampling actually is performed?

37. Refer to Problem 19. Suppose it had been decided that no more than $2000 would be spent in order to estimate the mean weight of scrap metal hauled by freight cars. Assuming it costs $50 to weigh the contents of one freight car, what effect would this have on the solution to the problem? In general, what is the effect of the cost of sampling upon an estimate based on sample evidence?

38. Suppose you are considering the purchase of a health insurance policy. Your reason for obtaining health insurance is for protection against health costs that you cannot afford rather than receiving payment for each and every medical care cost. For the foreseeable future assume you anticipate that you cannot afford more than $2000 at one time for medical problems. An insurance agent presents you with a lot of figures among which is a statement that the typical cost incurred beyond that paid through insurance to persons covered by the policy you are considering is estimated to lie between $1000 and $12,000 with 95 percent confidence!

    (a) Is the estimate provided by the insurance agent useful? Explain. Would a point estimate be more useful?
    (b) By decreasing the level of confidence, the width of the interval estimate would decrease. Is it possible that a narrower interval at a lower level of confidence would be useful? Explain.
    (c) Suppose you ask the agent for the size of the sample underlying the interval estimate cited and you are told that it is based upon a sample of records of 5 policyholders. Would this information change the way you view the estimate in terms of making a decision? Explain.

39. Suppose you are interested in purchasing a new car that you must order from the factory because you want special features not available in showroom models. One thing that concerns you, however, is the length of time you must wait for delivery. When asked about delivery time, the salesman replies in terms of the following set of interval estimates.

Delivery Time	*Level of Confidence*
8–14 months	0.999
4–6 months	0.900
2–6 weeks	0.450
1–2 weeks	0.300
1–2 days	0.001

    (a) What relationship between confidence and precision is implied by the above response?
    (b) Is the information provided by the salesman useful? Explain. Could the information be useful to anyone interested in ordering a car from the factory? Discuss.
    (c) If you actually have purchased a new car ordered from the factory, did the salesman give you a point or an interval estimate of the time for delivery? Was the estimate close

to the actual delivery time? What kind of estimate, point or interval, would you prefer if you were faced with the above decision problem? Why?

40. In what way does the t-distribution account for the uncertainty resulting from estimating the standard deviation? Why do values of t for a given level of confidence vary with the size of the sample when this is not the case for values of z based upon the normal curve?

41. (a) Under what circumstances are the following measures of variability used in constructing a confidence interval for the mean: s, $s_{\bar{x}}$, σ, $\sigma_{\bar{x}}$? Be sure to identify specifically each measure in your answer.
    (b) Why is a divisor of $n - 1$ instead of n used when computing a standard deviation? Is it ever necessary to use $n - 1$ when calculating a universe standard deviation? Explain. When can the divisor of $n - 1$ be ignored and n used instead? Why?

42. It has been said that sampling is as good as a census. In what sense is this statement correct? Explain in terms of the concept of estimation of a universe parameter.

43. In your own words discuss the importance of descriptive statistics and probability in the field of statistical inference.

44. (a) In this and the last chapter, the random sampling distribution of the mean is used but in different ways. Explain the meaning of this statement.
    (b) Generally, when a probability statement is made, it is in terms of a random variable assuming one or more possible values. A statement of confidence that is related to a probability, however, is associated with a specific numerical value of a universe parameter. In terms of the concept of the probability of a random variable, explain how it is possible to talk in terms of the probability associated with a specific or constant value.

45. Virtually every set of measurements is subject to variability. Another way of saying the same thing is that universes dealt with in practice exhibit variability. What is the relationship between this "inherent" variability and the problem of sampling?

46. Statistical theory provides us with the general rule that a confidence level should be a "high" probability, but does not tell us the specific value to choose. This choice is made by an individual involved with a particular problem requiring a confidence interval estimate. Since we always act as if the true universe value falls within a specific set of confidence limits based upon a particular sample, why not choose a confidence level of

"one" that would lead to a correct conclusion 100 percent of the time and with no probability of error?

47. Refer to Problem 17. In that problem you are asked to construct a confidence interval estimate of the average miles per gallon delivered by a particular brand of car in order to establish a mileage rating. The problem dealt purely with the procedure for constructing a confidence interval estimate based upon given numerical results.

    (a) In order to present a "true" picture of the mileage performance of the particular brand of automobile, what other factors should be taken into consideration?
    (b) Do you think it would make a difference if the tests were performed by an independent consumer organization rather than the manufacturer of the particular brand of automobile? Explain.
    (c) In order to obtain an accurate estimate of the average miles per gallon, does an independent consumer organization face problems that are not present for the manufacturer? Explain.
    (d) What general conclusions might you make about reported mileage ratings for automobiles based upon your responses to the above questions? How do these conclusions relate to the problem of sampling as it is presented in this chapter?

CHAPTER 12

HYPOTHESIS TESTING

It is no act of common passage, but a strain of rareness.

William Shakespeare

If there is an opportunity to make a mistake, sooner or later the mistake will be made.

Edmond C. Berekely

Another way to make an inference in addition to a confidence interval is to test a hypothesis. A **hypothesis** is a statement about the nature of a universe. A **test of a hypothesis** is a procedure that is used to determine whether the statement about the universe is tenable based upon sample data. Hypothesis testing usually is employed when it is of interest to reach a conclusion or make a choice between specific courses of action. The procedure is related to interval estimation, but it is performed differently owing to the difference in emphasis between the two approaches.

Consider a problem involving the manufacture of light bulbs. A shipment is received by a distributor who must determine if it is of acceptable quality before it is sold. A defective shipment is returned to the manufacturer. Since it is impossible to produce bulbs that have identical lengths of life, quality must be considered in terms of one or more parameters of the shipment; a high mean life with low variance is desirable.

Assuming the variance exhibits a satisfactory amount of stability, let us concentrate on the mean life of the shipment. Assume that the distributor requires that the mean life should be at least 1000 hours. The problem can be expressed in terms of two *quantified* statements or hypotheses about the shipment quality, or about the universe of bulbs:

(1) The mean length of life is at least 1000 hours ($\mu \geq 1000$)
(2) The mean length of life is less than 1000 hours ($\mu < 1000$)

Corresponding to each hypothesis is a *course of action* that depends upon the actual state of the shipment:

(1) Accept the shipment; hold for sale
(2) Do not accept the shipment; return to manufacturer

As the problem is stated, emphasis is placed upon making a decision regarding the acceptability of the shipment rather than knowing the magnitude of the mean life of the bulbs.

Since tested light bulbs no longer are useful, sampling is the only way to assess shipment quality. Whenever sampling is employed to make a decision, we need a procedure to account for errors arising because the entire universe is not observed. Hypothesis testing is one such procedure.

In order to perform a test we must distinguish between the two hypotheses or statements about the universe. The one "tested" is referred to as the **null hypothesis,** or the hypothesis tested. The other represents the negation of the null hypothesis and is called the **alternate hypothesis.** The alternate hypothesis must be defined in such a way that rejection of the null hypothesis automatically leads to the acceptance of the alternate. No overlap in the values of the two hypotheses can exist.

Special attention must be given to the way a null hypothesis is chosen since the entire testing procedure is based upon this choice. The null hypothesis *always* is chosen as the hypothesis that contains the *equality.** For example, in the light bulb problem presented above, the null hypothesis is "the mean length of life is at least 1000 hours," or $\mu \geq 1000$. Of the two possible alternatives this one contains the "equal-sign" in addition to the "greater-than" inequality. The value specified in the null hypothesis is referred as the **hypothesized value** or the hypothesized parameter. Hence, "1000" is the hypothesized value of the mean length of life in our example.

Special symbols are used to distinguish between the null and alternate hypotheses. The null hypothesis is designated by the symbol H_0 (read as "H-sub-zero") whereas the symbol H_1 (read as "H-sub-one") is used for the alternate hypothesis. Therefore, the two hypotheses in the light bulb example can be written in terms of these symbols as

$$H_0: \mu \geq 1000 \text{ (Null Hypothesis)}$$
$$H_1: \mu < 1000 \text{ (Alternate Hypothesis)}$$

The manner in which hypotheses are formulated *depends* upon the statement of the original problem to be solved. The problem statement in the above problem requires that the mean life of an acceptable shipment must be "at least 1000 hours," which is translated in the form of symbols as $\mu \geq 1000$. On the other hand, suppose the distributor specifies that a shipment is acceptable only if the mean life is "greater than

*There are problems where no inequalities appear in either hypothesis, so that each is expressed in terms of a "strict equality." For example, the light bulb problem might be simplified in terms of the two hypotheses, $\mu = 1000$ and $\mu = 800$. In other words, it is assumed that the mean possesses one of two possible values. In such a case, it does not matter which of the two hypotheses is the null hypothesis. When dealing with means, however, situations like this are not very realistic and are not commonly considered.

1000 hours." In this case, the hypothesis leading to an acceptable shipment is written as $\mu > 1000$, which differs from the original example. The other possibility is that the mean life is "at most 1000 hours" and is written as $\mu \leq 1000$. Since the equality appears in the second of the two hypotheses, the second hypothesis becomes the null hypothesis and is written as $H_0\colon \mu \leq 1000$. The alternate hypothesis in this case automatically becomes $H_1\colon \mu > 1000$.

The last paragraph introduces the idea that *it is the problem statement that determines the possible hypotheses* that eventually lead to the null and alternate hypothesis used in the testing procedure. Two different problem statements were considered that led to two different sets of hypotheses. A third possibility exists. Instead of being concerned with accepting or rejecting a particular shipment, suppose we are interested in reaching a conclusion about the process that actually produces the bulbs. The problem is to determine whether the quality of the output of the process remains at the "same" level, or that the mean "equals 1000 hours." This problem suggests the null hypothesis $H_0\colon \mu = 1000$ against the alternative $H_1\colon \mu \neq 1000$. In this case, the alternate hypothesis specifies that the mean life is "not equal" to 1000 hours, which is an abbreviated way of specifying that the mean is either less than 1000 (i.e., $\mu < 1000$) *or* greater than 1000 hours (i.e., $\mu > 1000$); the null hypothesis appears as a "strict equality" with no inequality signs.

Exhibit 1 summarizes the results of the above discussion about the relationship between the problem statement and the formulation of hypotheses. This is done in terms of the light bulb problem. Notice, in each case the null hypothesis represents the hypothesis with the equality. Furthermore, the actual problem statement when translated into symbols is not necessarily the alternative that appears in the null hypothesis. This can be seen by examining the second case where the alternative hypothesis corresponds to the statement of the problem. Finally, although each case presented in the exhibit is different, the "hypothesized value" equal to 1000 remains the same since the hypothesized value corresponds to the equality.

Once a problem is formulated in terms of hypotheses about universe parameters, a **decision rule** must be developed that tells us which values of a sample statistic lead to acceptance and rejection of the null hypothesis. The statistic that is used to perform a test of a hypothesis is referred to as the **test statistic.** Since the value of a sample statistic exhibits variability from sample to sample, a test of a hypothesis must account for this variability. For example, suppose we hypothesize the mean life

EXHIBIT 1

HYPOTHESES RELATING TO POSSIBLE PROBLEM STATEMENTS IN LIGHT BULB EXAMPLE

Problem Statement	*Null Hypothesis*	*Alternate Hypothesis*
Shipment mean is at least 1000 hours	$H_0\colon \mu \geq 1000$	$H_1\colon \mu < 1000$
Shipment mean is greater than 1000 hours	$H_0\colon \mu \leq 1000$	$H_1\colon \mu > 1000$
Process mean equals 1000 hours	$H_0\colon \mu = 1000$	$H_1\colon \mu \neq 1000$

of the light bulbs is 1000 hours and decide to test whether this is true on the basis of the mean of a sample of items drawn from the shipment. Even if this hypothesis is correct, it is unlikely that the mean of the sample will be equal to the shipment mean. This results from the fact that the sample mean may differ from the universe mean owing to chance or sampling variability. Hence, it is necessary to decide how much of a difference between the sample result and the hypothesized value is tolerable before being able to conclude that the shipment mean does not equal 1000. In other words, it is necessary to develop a way of deciding how much variability between the sample result and the hypothesized value can be attributed to chance and how much cannot. If we conclude that the variability is due to chance, we accept the null hypothesis to be true and if not we reject the null hypothesis and conclude that it is false.

Aided by the sampling distribution, we are able to establish the maximum difference between the value hypothesized and the sample statistic that is consistent with the null hypothesis; the set of values of the statistic corresponding to this difference and leading to the acceptance of the null hypothesis is called the **region of acceptance.** Conversely, the set of values of the statistic leading to rejection of the null hypothesis is referred to as the **region of rejection,** or critical region. The value of the sample statistic that defines the regions of acceptance and rejection is referred to as the **critical value.**

In order to understand these points, consider the null hypothesis in the light bulb example, $H_0: \mu \geq 1000$, against the alternate hypothesis, $H_1: \mu < 1000$. Suppose we intend to select a sample from the shipment, compute the arithmetic mean of this particular sample, and use the result to decide which of the two hypotheses is correct. Hence, we are to test a hypothesis where the sample mean is the test statistic.

In order to perform the test we first must decide on the values of the sample arithmetic mean that will lead to the conclusion that the null hypothesis is true and those values that will lead us to conclude that it is false. If we conclude the null hypothesis is false we reject it and automatically conclude that the alternate hypothesis is true. Obviously, if the value of the sample mean happens to be greater than or equal to 1000, we would accept the null hypothesis since the sample result would be consistent with this hypothesis; the sample mean has a value that falls in the same range as the universe mean described in the null hypothesis.

For the example we are considering, the real problem is to decide what values of the sample mean *below* the hypothesized value of 1000 would lead to the rejection of H_0. This results from the fact that the alternate hypothesis specifies values of the universe mean that are "less than" 1000, or $\mu < 1000$. In other words, values of the sample mean that are below the hypothesized mean are not consistent with the null hypothesis and become "candidates" that may lead to the rejection of H_0.

Since the sample arithmetic mean is subject to variability due to chance, values of the sample mean close to the hypothesized mean are more likely to occur than others when the null hypothesis is true. Consequently, it is reasonable to reject the null hypothesis when the sample mean is "much less" than the hypothesized mean since these values of the sample mean are unlikely to occur when the null hypothesis is true. Hence, "low values" of the sample mean below 1000 lead to rejection of H_0 and acceptance of H_1. For example, consider the following diagram:

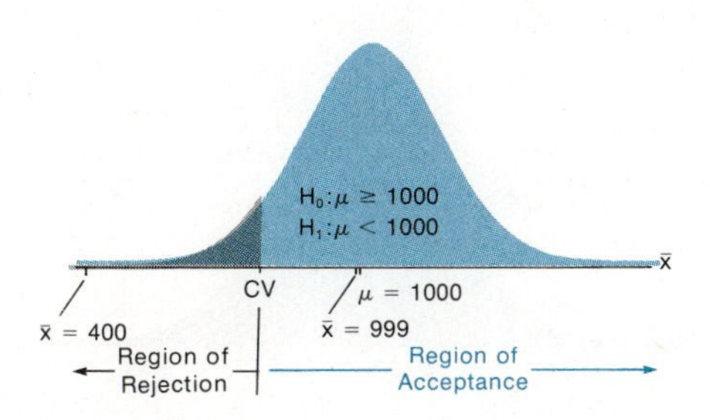

Suppose the sample arithmetic mean equals 999. Although this value is less than the hypothesized value, it is so close that it would not be reasonable to reject the null hypothesis. On the other hand, suppose the sample mean equals 400. This value is so much less than the hypothesized value that it would seem to be unlikely for it to occur if the null hypothesis is true. Consequently, such a result would lead to the rejection of the null hypothesis.

The two values of the sample mean just cited represent extremes. In such cases, there really is no difficulty in deciding whether to accept or reject the null hypothesis. The real problem is to decide on the point within the extremes beyond which we feel that it is not likely that the sample result comes from a universe possessing the hypothesized value. This point is the critical value, designated as CV in the above diagram, which provides the decision rule used to choose between the two hypotheses in terms of a region of rejection and a region of acceptance. Values of the sample mean falling in the region of rejection lead to the rejection of the null hypothesis, and values of the sample mean falling in the region of acceptance lead to the acceptance of the null hypothesis.

Probabilities can be attached to values of the sample mean using the sampling distribution of the mean. In the example we are considering, we can use the sampling distribution of the mean to identify values of the sample mean that are "much lower" than 1000 and are unlikely to occur if the universe mean actually is 1000. Hence, we use the sampling distribution of the mean to identify values of the sample mean that are "much lower" than 1000 and are unlikely to occur if the universe mean actually is 1000. Hence, we use the sampling distribution to find critical values that define the region of rejection. The way a critical value is found is presented in the procedure section.

In general, there are two types of tests of hypotheses. These are referred to as *one-tailed* and *two-tailed* tests. The type of test used *depends* upon the way the hypotheses are formulated. For the example just considered, a one-tailed test would be used. In that example, the region of rejection corresponds to values of the sample mean that are "much lower" than the hypothesized mean. Hence, the region of rejection appears in the lower or "left tail" of the sampling distribution.

If we were to test the null hypothesis $H_0: \mu \leq 1000$ against the alternate $H_1: \mu > 1000$, a one-tailed test also would be used. In this case, however, the region of rejection would correspond to values of the sample mean that are above the hypothesized value of 1000, or in the "right tail" of the sampling distribution of the mean since the alternate hypothesis specifies that the mean is "greater than 1000." This is evident by examining the following diagram:

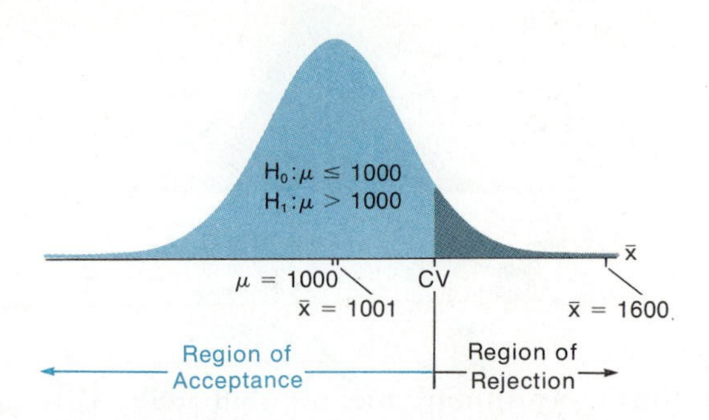

A sample mean equal to 1001 is greater than 1000 but extremely close; it would not be reasonable to reject the null hypothesis on the basis of such evidence. A sample mean equal to 1600, however, is extremely greater than the hypothesized value and would lead to rejection of the null hypothesis. The critical value lies between these extremes and defines a region of rejection in the upper tail corresponding to values of the sample mean that are "much higher" than 1000 and unlikely to occur if the null hypothesis is true.

When testing the null hypothesis $H_0: \mu = 1000$ against the alternate $H_1: \mu \neq 1000$, a two-tailed test is used. This results from the fact that "very low" *and* "very high" values of the sample mean lead to rejection of the null hypothesis. The region of rejection appears in "both tails" of the sampling distribution. This can be seen by examining the following diagram.

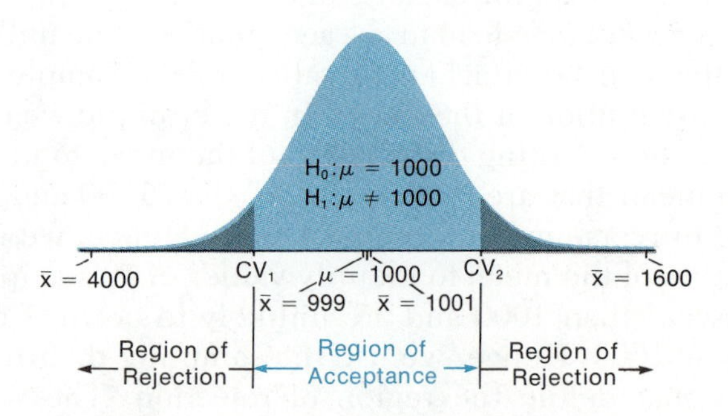

The same reasoning used in the other two cases applies here also, except that low *and* high values of the sample mean that are unlikely to occur if the null hypothesis is true lead to rejection of the null hypothesis. Consequently, two critical values, CV_1 and CV_2, are necessary to define the decision rule needed to perform the test.

In general, the type of test is based upon the tail of the sampling distribution corresponding to the region of rejection, which always appears in one or both of the tails. Appearing in Exhibit 2 is a summary of the types of tests corresponding to the different sets of hypotheses cited earlier. From the exhibit, we can see that the type of test is determined by the alternate hypothesis since the tail where the region of rejection

EXHIBIT 2
TYPES OF TESTS RELATING TO POSSIBLE HYPOTHESES IN THE LIGHT BULB EXAMPLES

Null Hypothesis	Alternate Hypothesis	Type of Test	Region of Rejection: Values of $\bar{x}$	Region of Rejection: Portion of Sampling Distribution
$H_0: \mu \geq 1000$	$H_1: \mu < 1000$	One-tailed	Low	Left tail
$H_0: \mu \leq 1000$	$H_1: \mu > 1000$	One-tailed	High	Right tail
$H_0: \mu = 1000$	$H_1: \mu \neq 1000$	Two-tailed	High & Low	Both tails

appears corresponds to the direction of the inequality shown in the alternate hypothesis.*

There are two types of errors associated with any test. These *always* are related to the null hypothesis and are defined as follows:

Type I error: Rejection of a true null hypothesis
Type II error: Acceptance of a false null hypothesis

Whenever a hypothesis is formulated, we do not know whether it is true or false. However, we can distinguish between the types of errors that can be made conditional upon the possible alternatives. For example, consider the first light bulb illustration where the null hypothesis is that the mean life is at least 1000 hours, or $\mu \geq 1000$. If this represents the actual state of the shipment and we conclude the mean life is less than 1000 hours, we have committed a Type I error; in practical terms, an acceptable shipment is rejected. A Type II error would be committed if an unacceptable shipment were accepted by concluding the mean life is at least 1000 hours when it really is less.

The types of errors that can be made are summarized in Exhibit 3 in terms of the first light bulb example. We can see from the exhibit that no error is committed if the hypothesis chosen as the conclusion is the same as the true state of the shipment. If, however, we reject the null hypothesis and conclude that the alternate is true when it is not, a Type I error is committed. On the other hand, if we conclude that the null hypothesis is true and accept it when the alternate is true, a Type II error is committed.

By employing the sampling distribution of a statistic, we can measure in advance the probabilities of committing the two types of errors. The probability of committing a Type I error is essential when constructing a test of a hypothesis, and is considered in this chapter. The probability of a Type I error is given a special name; it is called the **level of significance.** The probability of a Type II error is useful in evaluating a test and is considered at the end of the discussion section.

---

*Since it is the alternate hypothesis that determines the type of test, there is no difference from a testing point of view between the following sets of hypotheses:

(1) $H_0: \mu = 1000$, $H_1: \mu < 1000$ and $H_0: \mu \geq 1000$, $H_1: \mu < 1000$

(2) $H_0: \mu = 1000$, $H_1: \mu > 1000$ and $H_0: \mu \leq 1000$, $H_1: \mu > 1000$

In other words, as long as an equality appears in the null hypothesis, whether it is accompanied by an inequality or not, the type of test remains the same.

EXHIBIT 3
SUMMARY OF TYPES OF ERRORS IN LIGHT BULB EXAMPLE

True State of Shipment	***Conclusion*** $H_0: \mu \geq 1000$	$H_1: \mu < 1000$
$\mu \geq 1000$	No Error	Type I Error
$\mu < 1000$	Type II Error	No Error

The preceding exposition providing background material is rather long. Actually, the entire chapter is somewhat longer than others in the text. This, unfortunately, cannot be avoided since many new definitions and ideas are needed in order to fully understand hypothesis testing. This chapter has two main goals: (1) to introduce the general concepts underlying all tests of hypotheses, and (2) to provide specific methods for testing hypotheses about means. The general ideas presented in this chapter form the basis for other types of tests considered in the four remaining chapters in this part of the text. Consequently, this chapter is designed to provide more than just an understanding of one specific set of procedures.

Like interval estimation, different computational procedures are required in particular cases, although there is a standard format for conducting a test. We shall employ the approach of the last chapter and assume a normal sampling distribution for means in the procedure and example sections. Extensions to the t-distribution and other cases appear in the discussion section after additional points about the concept of a test of a hypothesis are presented.

The symbols used in this chapter are defined as follows:

μ = the universe mean
σ = the universe standard deviation
n = the sample size
$\bar{x}$ = the arithmetic mean of a sample
s = the standard deviation of a sample
$\sigma_{\bar{x}}^2$ = the variance of the distribution of the sample mean
$\sigma_{\bar{x}}$ = the standard error of the sample mean
$s_{\bar{x}}$ = an estimate of the standard error of the sample mean
H_0 = the null hypothesis, or the hypothesis tested
H_1 = the alternate hypothesis
μ_0 = a "number," representing a hypothesized value of the universe mean (read as "mu-sub-zero")
μ_1 = *a* value of the universe mean specified in the alternate hypothesis (read as "mu-sub-one")
α = the level of significance, or the probability of committing a Type I error (read as "alpha")
CV = the critical value, which defines the regions of acceptance and rejection
β = the probability of a Type II error
P_w = the power of the test

z	= the standardized normal variate used to find probabilities from Table 2
	= the z-statistic (this represents an alternate use of z that appears in the discussion)
z_α	= the standardized normal variate from the table of Areas of the Normal Curve used in a one-tailed test
$z_{\alpha/2}$	= the standardized normal variate from the table of Areas of the Normal Curve used in a two-tailed test
t_α	= the standardized variate needed when using the table of Student's t-Distribution in a one-tailed test
$t_{\alpha/2}$	= the standardized variate needed when using the table of Student's t-Distribution in a two-tailed test
DF	= degrees of freedom
t	= the t-statistic
$\mu_{\bar{x}_1 - \bar{x}_2}$	= the mean of the sampling distribution of the difference between sample means from two independent samples
$(\mu_1 - \mu_2)_0$	= a "number" representing a hypothesized value of the difference between population means
$\sigma_{\bar{x}_1 - \bar{x}_2}$	= the standard error of the difference between sample means
$s_{\bar{x}_1 - \bar{x}_2}$	= an estimate of the standard error of the difference between sample means based upon sample standard deviations

PROCEDURE

The following general steps apply when performing a test of any hypothesis:

Step 1. State the problem in terms of two hypotheses about the universe.

Step 2. Quantify the two hypotheses in terms of a relevant universe parameter. Identify the null and the alternate hypotheses, such that the null hypothesis always contains an "equality."

Step 3. When appropriate, specify courses of action corresponding to the null and alternate hypotheses.

Step 4. Identify the test statistic.

Step 5. Determine the sampling distribution of the test statistic under the assumption that the "equality" in the null hypothesis is true.

Step 6. Select the value of the level of significance, α, representing the probability you personally are willing to risk of committing a Type I error. The guide provided by statistical theory is that this probability must be "small." Conventional values used are 0.01, 0.02, 0.025, 0.05, and 0.10.

Step 7. Based upon the chosen level of significance, compute the critical value(s) of the sample statistic that provides a decision rule leading to the acceptance or rejection of the null hypothesis. The region of rejection always appears in one or both tails of the sampling distribution of the test statistic.

Step 8. Select a random sample and compute the value(s) of the relevant statistic(s).

Step 9. Based upon the calculated value of the test statistic, apply the decision rule established in Step 7.

Step 10. Draw a conclusion or make a decision in terms of the original problem.

Of the ten steps listed above, we shall concentrate on Step 7 in the remainder of this procedure section. Here we present the way the critical value and the decision rule are determined in order to test a hypothesis about a *universe mean* when the distribution of the sample mean, used as a test statistic, is assumed to be *normal.* All of the steps are considered in the example section. The formula for the critical value(s) and the decision rule vary with the type of test and the hypothesis tested. Each case is presented separately below.

One-Tailed Test. REGION OF REJECTION IN LEFT TAIL

Null Hypothesis, $H_0: \mu \geq \mu_0$
Alternate Hypothesis, $H_1: \mu < \mu_0$

The critical value is found by substituting into the formula for CV below.

Known σ		**Unknown σ**
Normal Universe—Any n *Any Universe—Large n*	*or*	*Any Universe—Large n*
$CV = \mu_0 - z_\alpha \sigma_{\bar{x}}$		$CV = \mu_0 - z_\alpha s_{\bar{x}}$

Critical values when testing against alternate hypothesis, $\mu < \mu_0$.

The symbol μ_0 is the value of the hypothesized mean. z_α represents the standardized normal variate corresponding to an area equal to α, the chosen level of significance, in the *left* tail of the normal curve. The critical value is obtained by *subtracting* z_α multiplied by the standard error of the mean from the hypothesized value. If the universe standard deviation, σ, is known, the true standard error, $\sigma_{\bar{x}}$, is used. Otherwise, it is estimated by $s_{\bar{x}}$.

The corresponding decision rule, based upon the mean of a random sample, is

Accept H_0 when $\bar{x} > CV$
Reject H_0 when $\bar{x} \leq CV$

Decision rule when testing against alternative hypothesis, $\mu < \mu_0$.

This rule tells us to accept the null hypothesis when the sample mean,

$\bar{x}$, is greater than the calculated critical value and reject the null hypothesis when the sample mean is less than or equal to the critical value.

One-Tailed Test. REGION OF REJECTION IN RIGHT TAIL

Null hypothesis, $H_0: \mu \leq \mu_0$
Alternate hypothesis, $H_1: \mu > \mu_0$

The critical value is found by solving for CV using the following formula:

Known σ		Unknown σ
Normal Universe—Any n *Any Universe—Large n*	or	*Any Universe—Large n*
$CV = \mu_0 + z_\alpha \sigma_{\bar{x}}$		$CV = \mu_0 + z_\alpha s_{\bar{x}}$

Critical value when testing against alternate hypothesis, $\mu > \mu_0$.

The critical value in this case is found by *adding* z_α multiplied by the standard error of the mean to the hypothesized value, μ_0. z_α is the standardized normal variate corresponding to an area equal to the level of significance, α, in the *right* tail of the normal curve. The choice between the use of $\sigma_{\bar{x}}$ or $s_{\bar{x}}$ depends on whether σ is known or unknown. The corresponding decision rule, based upon the mean of a randomly drawn sample, is

Accept H_0 when $\bar{x} < CV$
Reject H_0 when $\bar{x} \geq CV$

Decision rule when testing against alternate hypothesis, $\mu > \mu_0$.

The decision rule in this case tells us to accept the null hypothesis when the sample mean, $\bar{x}$, is less than the calculated critical value and to reject the null hypothesis when the sample mean is greater than or equal to the critical value.

Two-Tailed Test. REGION OF REJECTION IN BOTH TAILS

Null Hypothesis, $H_0: \mu = \mu_0$
Alternate Hypothesis, $H_1: \mu \neq \mu_0$

In this case there are two critical values, CV_1 and CV_2, one for each tail of the sampling distribution, which are found by substituting into the expressions

Known σ		Unknown σ
Normal Universe—Any n *Any Universe—Large n*	or	*Any Universe—Large n*
$CV_1 = \mu_0 - z_{\alpha/2}\sigma_{\bar{x}}$ $CV_2 = \mu_0 + z_{\alpha/2}\sigma_{\bar{x}}$		$CV_1 = \mu_0 - z_{\alpha/2}s_{\bar{x}}$ $CV_2 = \mu_0 + z_{\alpha/2}s_{\bar{x}}$

Critical values when testing against alternate hypothesis, $\mu \neq \mu_0$.

The two critical values in the case of a two-tailed test are determined so that *half* the level of significance, $\alpha/2$, appears in *each* tail of the distribution of the mean. Hence, $z_{\alpha/2}$ represents the standardized normal variate corresponding to $\alpha/2$ in the tail of the normal curve. The critical values are determined by multiplying this value of z from the table by the standard error and *adding* the result to the hypothesized mean, μ_0, and also *subtracting* it from μ_0. The decision rule based upon the sample mean for this test takes the form

$$\begin{aligned} &\text{Accept } H_0 \text{ when } CV_1 < \bar{x} < CV_2 \\ &\text{Reject } H_0 \text{ when } \quad \bar{x} \leq CV_1 \\ &\qquad\qquad\qquad \text{or} \\ &\qquad\qquad\qquad \bar{x} \geq CV_2 \end{aligned}$$

Decision rule when testing against alternate hypothesis, $\mu \neq \mu_0$.

The decision rule for the two-tailed test tells us to accept the null hypothesis when the sample mean falls within the critical values and reject it if the sample mean is either less than or equal to the lower critical value, CV_1, or if the sample mean is greater than or equal to the upper critical value, CV_2.

Note. The formulas for the standard error of the mean to find a critical value are the same as the ones used in the last chapter on estimation. When the universe standard deviation, σ, is *known*, the formula for the standard error of the mean is

$$\sigma_{\bar{x}} = \frac{\sigma}{\sqrt{n}}$$

In the case of an *unknown* universe standard deviation, the standard error of the mean that is used is the estimated value

$$s_{\bar{x}} = \frac{s}{\sqrt{n-1}}$$

where s is the standard deviation of a sample.

A diagram summarizing the tests of hypotheses presented in this section appears in Exhibit 4. It can be seen that the region of rejection found by calculating the critical value(s) represents the extreme values of the sample mean associated with the tails of the *random sampling distribution* of the mean. Critical values are computed in the same way as any other values of a normally distributed variable corresponding to a given probability level; the probability level is specially chosen so that there is small probability of a Type I error. The exhibit provides all the information necessary to perform tests for means based upon a normal sampling distribution.

NUMERICAL EXAMPLES

The examples in this section are divided into three parts corresponding to the two types of one-tailed tests and the two-tailed test.

EXHIBIT 4

SUMMARY OF TESTS OF HYPOTHESES FOR THE UNIVERSE MEAN USING A NORMAL SAMPLING DISTRIBUTION

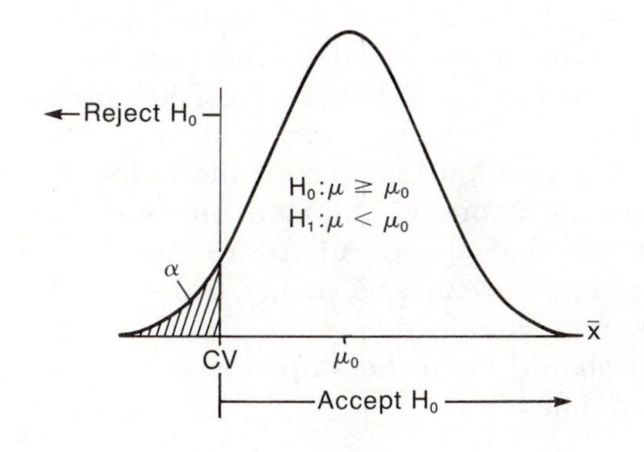

One-Tailed Test—
Left Tail

$CV = \mu_0 - z_\alpha \sigma_{\bar{x}}$

Or

$\mu_0 - z_\alpha s_{\bar{x}}$

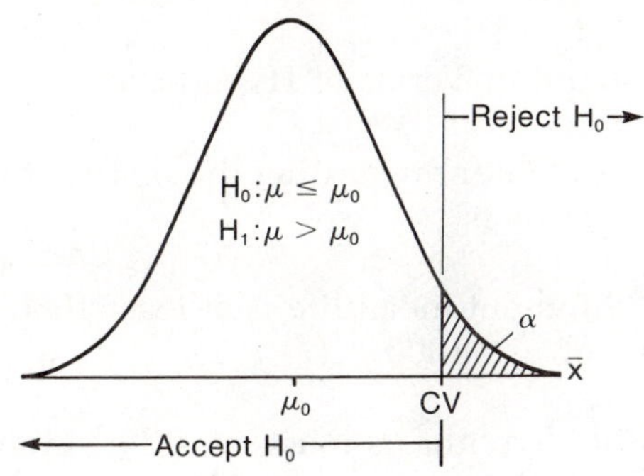

One-Tailed Test—
Right Tail

$CV = \mu_0 + z_\alpha \sigma_{\bar{x}}$

Or

$\mu_0 + z_\alpha s_{\bar{x}}$

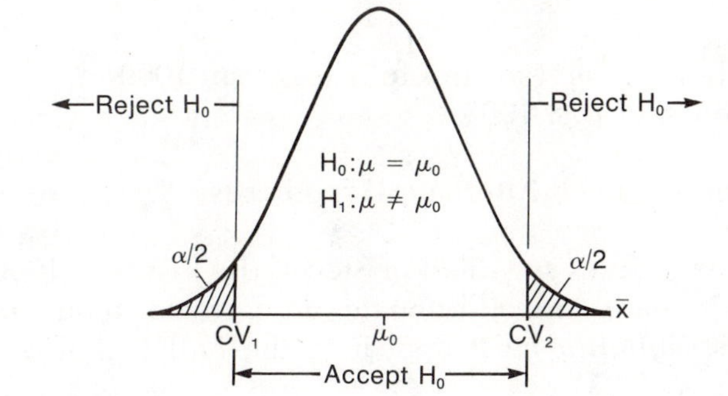

Two-Tailed Test—
Both Tails

$CV_1 = \mu_0 - z_{\alpha/2} \sigma_{\bar{x}}$

$CV_2 = \mu_0 + z_{\alpha/2} \sigma_{\bar{x}}$

Or

$CV_1 = \mu_0 - z_{\alpha/2} s_{\bar{x}}$

$CV_2 = \mu_0 + z_{\alpha/2} s_{\bar{x}}$

Example 1, appearing in the first section, considers each of the steps listed at the beginning of the procedure section. This is done in order to bring all the ideas about hypothesis testing together at one point in the text. The remaining illustrations consider selected aspects of the procedure.

One-Tailed Test. REGION OF REJECTION IN LEFT TAIL

Example 1 Shipments of light bulbs meet distributor specifications provided the mean length of life is at least 1000 hours. If a shipment does not meet this requirement, it is returned to the manufacturer. Shipments received in the past have consistently had a standard deviation of 50 hours, and length of life can be assumed to be normally distributed. A random sample of 49 bulbs is drawn from an incoming shipment in order to determine whether specifications are met. The sample arithmetic mean is 955 hours. Based upon the information given, determine whether the distributor should keep the shipment or return it to the manufacturer.

Step 1. Problem Goal Stated in Terms of Hypotheses

Based upon the statement regarding the distributor's specification, one hypothesis is

The shipment mean life is at least 1000 hours, or $\mu \geq 1000$

The other possible hypothesis must consider shipment states not consistent with the stated specification, resulting in the hypothesis

The shipment mean life is less than 1000 hours, or $\mu < 1000$

Step 2. Choosing The Null and Alternate Hypotheses

Of the two hypotheses specified in Step 1, the first ($\mu \geq 1000$) contains an equality in addition to the "greater-than" inequality. The *equality* must appear in the null hypothesis. Hence,

$$H_0: \mu \geq 1000$$
$$H_1: \mu < 1000$$

Step 3. Specifying Courses of Action

This follows directly from the problem statement:

H_0: Shipment meets specifications – Keep the shipment
H_1: Shipment does not meet specifications – Return shipment to manufacturer

Step 4. Test Statistic

Since the hypotheses are stated in terms of the universe mean, the test statistic is the sample arithmetic mean, $\bar{x}$.

Step 5. Identify the Sampling Distribution of the Test Statistic

Since shipment standard deviations exhibit stability over time, we shall treat the universe standard deviation as known, equal to 50 hours (i.e., $\sigma = 50$). Based upon this assumption and the fact that we are told that the distribution of light bulb life is normal, the distribution of the sample mean can be assumed to be normal.

Step 6. The Level of Significance

Since a level of significance is not provided in the problem, we must select one. Assume we are willing to take a small risk of returning an acceptable shipment and choose a 1 percent level, or $\alpha = 0.01$.

Step 7. Critical Value and Decision Rule

Based upon the alternate hypothesis ($H_0: \mu < 1000$), "low values" of the sample mean lead to rejection of the null hypothesis. Consequently, the region of rejection is in the left tail of the sampling distribution. This is defined in terms of the critical value:

$$
\begin{aligned}
CV &= \mu_0 - z_\alpha \sigma_{\bar{x}} \\
&= \mu_0 - z_{.01}\left(\frac{\sigma}{\sqrt{n}}\right) \\
&= 1000 - 2.33\left(\frac{50}{\sqrt{49}}\right) \\
&= 1000 - 2.33\left(\frac{50}{7}\right) \\
&= 1000 - 2.33\,(7.14) \\
&= 1000 - 16.64 \\
&= 983.36
\end{aligned}
$$

The value of z substituted above corresponds to an area in the tail of a normal curve equal to 0.01, the value of α. The value of the standard error is obtained by substituting into the formula for $\sigma_{\bar{x}}$, which is included as part of the calculations.

The value of μ_0, the hypothesized mean, is found from H_0 in Step 2. Based upon the critical value, the decision rule is

$$\text{Decision Rule: Accept } H_0 \text{ if } \bar{x} > 983.36$$
$$\text{Reject } H_0 \text{ if } \bar{x} \leq 983.36$$

Step 8. Select Sample and Compute Value of Test Statistic

The mean of the random sample is 995 hours. If this were not given, it would have to be computed from individual sample observations.

Step 9. Applying the Decision Rule

$$\bar{x} = 995 > CV = 983.36$$

The sample arithmetic mean equal to 995 is greater than the critical value of 983.36. Therefore, based upon the decision rule found in Step 7 we *accept* the null hypothesis.

Step 10. Conclusion in Terms of Original Problem

Based upon the courses of action specified in Step 3, we *conclude* that the shipment meets specifications since the null hypothesis was accepted; hence, the shipment should be kept for sale.

The results of the testing procedure are illustrated in the following diagram:

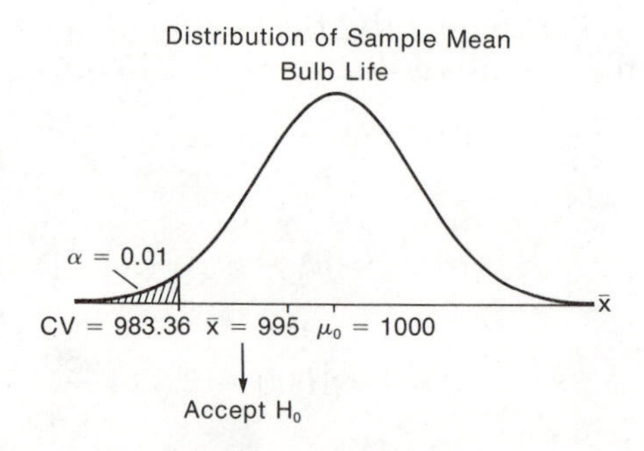

Example 2 A random sample of 65 compact cars of the same make and style is selected to determine whether the gas mileage of this model meets manufacturer claims. Mileage claims are specified in terms of the average miles per gallon (mpg) per car, which is to be *at least* 35 mpg under highway driving conditions. The sample of cars tested under required conditions had a sample arithmetic mean and standard deviation equal to 30 mpg and 5 mpg, respectively. In order to test whether the claim is valid, assume a risk of 0.01

of concluding that the claim is not met when, in fact, it actually is.

In terms of procedure, this problem is similar to the first example except the universe standard is not provided; the standard error must be estimated with $s_{\bar{x}}$. We can assume the sample is large enough that a normal sampling distribution is applicable.

The null and alternate hypotheses with the corresponding conclusions are

$H_0: \mu \geq 35$ (Mileage requirement met)
$H_1: \mu < 35$ (Mileage requirement not met)

The following information is given:

$$n = 65$$
$$\bar{x} = 30$$
$$s = 5$$
$$\alpha = 0.01$$

The critical value is computed as

$$\begin{aligned} CV &= \mu_0 - z_\alpha s_{\bar{x}} \\ &= \mu_0 - z_{.01}\left(\frac{s}{\sqrt{n-1}}\right) \\ &= 35 - 2.33\left(\frac{5}{\sqrt{65-1}}\right) \\ &= 35 - 2.33\left(\frac{5}{\sqrt{64}}\right) \\ &= 35 - 2.33(.625) \\ &= 35 - 1.5 \\ &= 33.5 \end{aligned}$$

The decision rule is

Accept H_0 if $\bar{x} > 33.5$
Reject H_0 if $\bar{x} \leq 33.5$

Applying the rule, we have

$$\bar{x} = 30 < CV = 33.5$$

We *reject* the null hypothesis and *conclude* the mileage claim for this brand of car is not met. Diagrammatically, we have

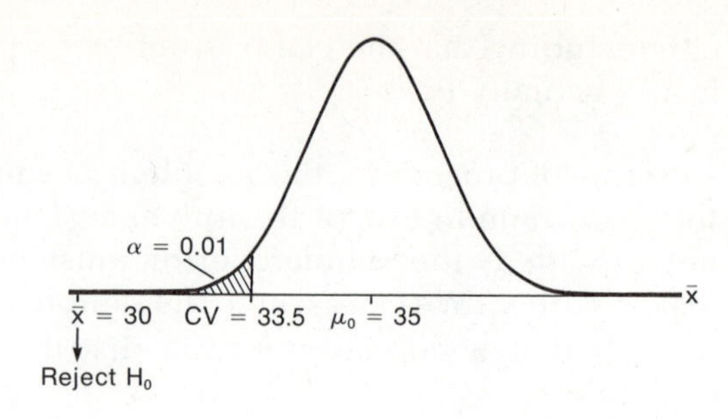

One-Tailed Test. REGION OF REJECTION IN THE RIGHT TAIL

Example 3 In order to determine whether students at a particular school scored higher than the national average on a standardized aptitude test, a sample of 50 students was selected at random. The mean and standard deviation of the sample were computed as 505 and 35. The national average is 500. Assume a level of significance of 0.025 in order to test the following hypothesis:

$H_0: \mu \leq 500$ (School average not greater than national average)

$H_1: \mu > 500$ (School average greater than national average)

Given: $n = 50$
$\bar{x} = 505$
$s = 35$
$\alpha = 0.025$

Since the alternate hypothesis indicates that "high values" of the sample mean lead to rejection of the null hypothesis (i.e., $H_1: \mu > 500$), the region of rejection is in the right tail of the sampling distribution of the mean. The universe standard deviation is not given. Therefore, the critical value is based upon the estimated standard error and is found using the formula for a one-tailed test with the region of rejection in the upper tail.

Compute: $CV = \mu_0 + z_\alpha s_{\bar{x}}$

$$= \mu_0 + z_{.025}\left(\frac{s}{\sqrt{n-1}}\right)$$

$$= 500 + 1.96\left(\frac{35}{\sqrt{50-1}}\right)$$

$$= 500 + 1.96\left(\frac{35}{\sqrt{49}}\right)$$

$$= 500 + 1.96\left(\frac{35}{7}\right)$$

$$= 500 + 1.96(5)$$

$$= 500 + 9.8$$

$$= 510 \text{ (rounded)}$$

Decision Rule: Accept H_0 if $\bar{x} < 510$
Reject H_0 if $\bar{x} \geq 510$

Since $\bar{x} = 505$ is less than CV $= 510$ we *accept* the null hypothesis and *conclude* that the school average is *not* higher than the national average.

Diagrammatically, we have

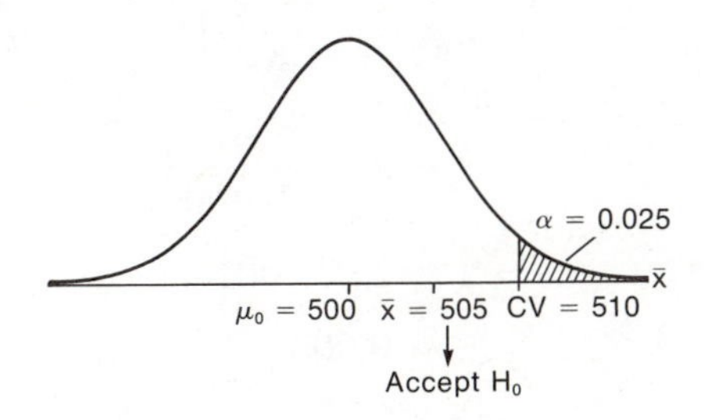

Two-Tailed Test. Region of Rejection in Both Tails

Example 4 A continuous manufacturing process of steel rods is said to be in a "state of control" and produces acceptable rods if the mean diameter of all rods produced is 2 inches. Although the process standard deviation exhibits stability over time ($\sigma = 0.01''$), the process mean may vary due to operator error or problems of process adjustment. Periodically, random samples of 100 rods are selected to determine whether the process is producing acceptable rods. A two-tailed test (with a 0.01 level of significance) is used since rods that are either too narrow or too wide are unacceptable. If the result of a test indicates the process is out of control, it is stopped and the source of trouble is sought; otherwise, it is allowed to continue operating.

A random sample of 100 rods is selected resulting in a mean of 2.1 inches. Test a hypothesis to determine whether the process should be continued.

To Test: $H_0: \mu = 2$ (Continue process)
$H_1: \mu \neq 2$ (Stop the process)

Given: $\sigma = 0.01$
$n = 100$
$\bar{x} = 2.1$
$\alpha = 0.01$
$\alpha/2 = 0.005$

A two-tailed test is required in this example since the alternate hypothesis $H_1 : \mu \neq 2$ specifies that rods that are too narrow as well as too wide on the average are not acceptable. Consequently, "low values" and "high values" of the sample mean lead to rejection of the null hypothesis resulting in a region of rejection in both tails of the sampling distribution of the mean.

Compute:
$$
\begin{aligned}
CV_1 &= \mu_0 - z_{\alpha/2}\sigma_{\bar{x}} \\
&= \mu_0 - z_{.005}\left(\frac{\sigma}{\sqrt{n}}\right) \\
&= 2 - 2.58\left(\frac{.01}{\sqrt{100}}\right) \\
&= 2 - 2.58(.001) \\
&= 2 - 0.003 \\
&= 1.997
\end{aligned}
$$

$$
\begin{aligned}
CV_2 &= \mu_0 + z_{\alpha/2}\sigma_{\bar{x}} \\
&= \mu_0 + z_{.005}\left(\frac{\sigma}{\sqrt{n}}\right) \\
&= 2 + 2.58\left(\frac{.01}{\sqrt{100}}\right) \\
&= 2 + 0.003 \\
&= 2.003
\end{aligned}
$$

Decision Rule: Accept H_0 if $1.997 < \bar{x} < 2.003$
Reject H_0 if $\bar{x} \leq 1.997$
or
$\bar{x} \geq 2.003$

Since $\bar{x} = 2.1$ exceeds the upper critical value of 2.003, the null hypothesis is *rejected* and the *decision* is to stop the process in order to determine the source of trouble. We can conclude further that the process produces rods that are *too wide* since the sample mean fell in the region of rejection in the *upper* tail of the sampling distribution.

The example is summarized in the following diagram:

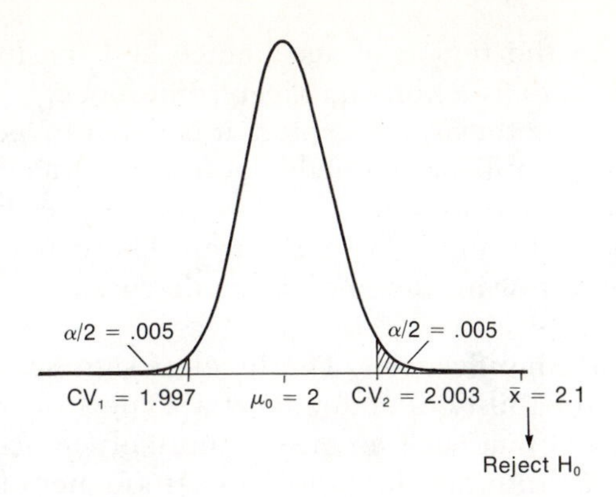

Note. In the case of a two-tailed test as illustrated in this example, the level of significance, α, is split such that each half appears in each tail of the sampling distribution. Hence, the value of z is designated as $z_{\alpha/2}$ and is found in the table of Areas of the Normal Curve corresponding to half the specified level of significance (i.e., $\alpha/2 = .005$).

DISCUSSION

The basis for testing hypotheses lies in the fact that the value of a sample statistic varies from sample to sample. Consequently, the value from a single sample will, in all likelihood, be different from the parameter of the universe sampled. By employing the procedures for testing a hypothesis, it is possible to determine whether a sample result differs from a hypothesized value as a result of *chance* or whether the difference reflects an *actual* difference between the true universe value (which is unknown) and the value hypothesized.

Tests of hypotheses are used to determine whether or not a sample result differs from a hypothesized value due to chance.

When a value of a sample statistic falls in the region of rejection, the difference between the sample statistic and the hypothesized parameter is referred to as a **significant difference:** the difference is greater than can be attributed to chance alone. Hence, we reason that if chance did not produce the difference, it can be attributed to a "non-chance" factor and results from the fact that the sample came from a universe with a parameter value specified in the alternate hypothesis.

A significant difference leads to the rejection of the null hypothesis, since such a difference is considered to be unlikely to occur if the null hypothesis were true. A significant difference is one that is greater than one that is attributed to chance.

By specifying the level of significance, α, as "small," we can act as if it is unlikely to reject a true null hypothesis. Therefore, when a value of a statistic does fall in the region of rejection, we act as if the result was not due to chance. Since α is subject to our control, the risk of rejecting a true null hypothesis can be made as small as desired. This approach is similar to the kind we applied in confidence interval estimation, except emphasis is placed on the probability of error rather than on the probability of a correct decision.

When a sample statistic falls in the region of acceptance, although its value differs from the hypothesized value, the difference is attributed to chance. This is due to the fact that we consider such a difference likely to occur if the null hypothesis is true. A difference between a sample

A non-significant difference leads to the acceptance of the null hypothesis. Such a difference is considered to be attributable to chance. We do not prove that the null hypothesis is true when we accept it, but accept it to be true because we do not have sufficient evidence to reject it.

statistic falling in the region of acceptance and the hypothesized parameter is referred to as a **non-significant difference.**

Whenever a hypothesis is accepted, it is not "proved" to be true but is held only to be tenable or reasonably defensible. In other words, there is insufficient evidence to reject it. Of course, an accepted hypothesis can be false, constituting a Type II error. The concept of a Type II error is considered toward the end of this discussion.

The Level of Significance. The level of significance, α, has been defined as the probability of committing a Type I error, and is chosen to be small enough that such an error is unlikely to occur. Since α is a probability, we can interpret it in terms of a frequency of occurrence in a large number of repeated trials. Hence, α provides the proportion of the time that a true null hypothesis will be rejected if we *repeated* the sampling and testing procedure a large number of times under similar conditions. For example, consider the gas mileage illustration in which we used a level of significance of 0.01 to test the hypothesis

$$H_0 : \mu \geq 35 \text{ mpg (Mileage requirement met)}$$

against the alternative

$$H_1 : \mu < 35 \text{ mpg (Mileage requirement not met)}$$

We can interpret the value 0.01 to mean that 0.01 or 1 percent of a large number of tests of the same hypothesis (based upon different random samples of the same size) would lead us to conclude incorrectly that the average miles per gallon is less than 35, and that the mileage claim is not met when in fact it really is. Consequently, 1 percent of the time we would incorrectly reject cars as unacceptable if we continually applied the same testing procedure.

The level of significance, α, can be interpreted as the proportion of the time that repeated sample results will lead to the incorrect rejection of the null hypothesis.

Situations where the consequences of committing a Type I error are serious require a smaller level of significance than others where the consequences are less severe. The size of α, however, should be based upon a personal choice as it relates to a particular problem. Consequently, the value of α may be different for different problems.

There is one very important point about α to keep in mind. α should be chosen independently and should not be based upon knowledge of the value of a test statistic from a particular sample. Otherwise, the value of the test statistic could influence the final conclusion regarding acceptance or rejection of the null hypothesis; this would invalidate the test.

The level of significance that is chosen should not be influenced by the sample results used to draw a conclusion.

Testing Means with the z-Statistic. Alternative procedures for performing a test of a hypothesis exist. In general, these are used either for convenience or because of the nature of a particular test and the available tables. Whenever a *normal* sampling distribution applies, as in the case of means presented earlier, a test can be performed in terms of the standardized variate, z. Briefly stated, a comparison is made between the z-value computed on the basis of the observed sample mean and the hypothesized mean and a z-value corresponding to a chosen level of significance.

The test statistic is calculated using the formula

$$z = \frac{\bar{x} - \mu_0}{\sigma_{\bar{x}}}$$

or

$$\frac{\bar{x} - \mu_0}{s_{\bar{x}}}$$

z-Statistic for testing means.

This value of z is referred to as the "z-statistic" where $\bar{x}$ is the sample mean of a particular sample, μ_0 is the value hypothesized in the null hypothesis, and $\sigma_{\bar{x}}$ is the standard error of the mean. The estimated standard error, $s_{\bar{x}}$, is used in place of $\sigma_{\bar{x}}$ when this is unknown.

By transforming tests for means in terms of the z-statistic, an alternative way of performing these tests is available. Conclusions are made in the same way except values of z are compared instead of values of $\bar{x}$.

In order to perform a test of a hypothesis, the value of z calculated on the basis of the above formula is compared with a critical value of z based upon α obtained from the table of Areas of the Normal Curve. Critical values and decision rules for testing means in terms of the z-statistic are presented in Exhibit 5. The decision rules are similar to the ones presented earlier except that values of z are used instead of values of the sample mean directly.

In order to understand how to perform a test using the z-statistic, consider Example 2 where we tested the null hypothesis $H_0: \mu \geq 35$ against the alternative $H_1: \mu < 35$, where $n = 65$, $\alpha = 0.01$, $\bar{x} = 20$, and $s = 5$. The value of z corresponding to $\bar{x}$ under the null hypothesis is computed using the formula for the z-statistic as

$$\begin{aligned} z &= \frac{\bar{x} - \mu_0}{s_{\bar{x}}} \\ &= \frac{30 - 35}{5/\sqrt{65 - 1}} \\ &= \frac{-5}{.625} \\ &= -8 \end{aligned}$$

EXHIBIT 5

CRITICAL VALUES AND DECISION RULES FOR TESTING MEANS USING THE z-STATISTIC

Null Hypothesis	*Alternate Hypothesis*	*Critical Value(s)*	*Decision Rule*
$H_0: \mu \geq \mu_0$	$H_1: \mu < \mu_0$	$CV = -z_\alpha$	Accept H_0 if $z > -z_\alpha$ Reject H_0 if $z \leq -z_\alpha$
$H_0: \mu \leq \mu_0$	$H_1: \mu > \mu_0$	$CV = z_\alpha$	Accept H_0 if $z < z_\alpha$ Reject H_0 if $z \geq z_\alpha$
$H_0: \mu = \mu_0$	$H_1: \mu \neq \mu_0$	$CV_1 = -z_{\alpha/2}$ $CV_2 = z_{\alpha/2}$	Accept H_0 if $-z_{\alpha/2} < z < z_{\alpha/2}$ Reject H_0 if $z \leq -z_{\alpha/2}$ *or* $z \geq z_{\alpha/2}$

EXHIBIT 6
AN EXAMPLE USING THE z-STATISTIC TO PERFORM A TEST OF A HYPOTHESIS

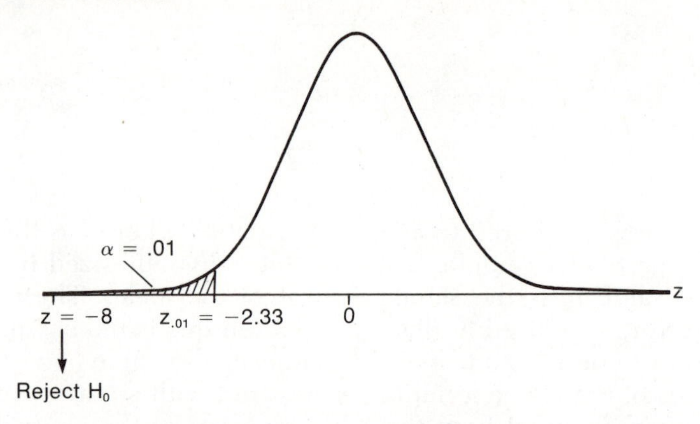

The formula with $s_{\bar{x}}$ is used since the known value of the standard deviation is not available. The value of z corresponding to a level of significance, α, equal to 0.01 is obtained from Table 2 and equals 2.33. Since the region of rejection is in the left tail, the critical value of z is negative and equals

$$\begin{aligned} CV &= -z_{\alpha} \\ &= -2.33 \end{aligned}$$

Based upon the decision rule for this type of test specified in Exhibit 5, we find that the calculated value of z equal to −8 is *less* than the critical value equal to −2.33, which leads to the *rejection* of the null hypothesis. The results of this example are summarized in Exhibit 6.

The example using the z-statistic is in terms of a one-tailed test with a region of rejection in the left tail. The overall procedure applies to the other cases also. In general, the same conclusion is drawn whether $\bar{x}$ is used directly or if the tests are performed using z.

By comparing the results shown in the exhibit with those of Example 2, you will find that the observed and critical values of z are located in the same positions on the z-scale as the observed and critical values of the sample mean on the $\bar{x}$ scale. In other words, the same conclusion is drawn, regardless of the statistic used to perform the test. When a value of the z-statistic falls in the region of rejection, so does the value of the sample mean. Consequently, the decision rules presented in Exhibit 5 in terms of z are equivalent to the rules summarized in Exhibit 4 in terms of $\bar{x}$ for each type of test. Both lead to the same conclusions. The form of the test for means that is chosen is a matter of personal preference.

When sampling from normal populations with an unknown standard deviation, a t-test for means using Student's t-distribution should be used.

The t-Test. When testing hypotheses about the mean of a normal universe when the standard deviation, σ, is *unknown,* the t-distribution should be used instead of the normal. As in the case of confidence intervals, the real advantage of the t-distribution is associated with *small* samples since the normal provides a satisfactory approximation when n is large. The procedures for testing hypotheses using the t-distribution are similar to the ones using the normal; the difference lies in the use of the t-table, which is constructed differently.

As in the case of the normal, tests for means using the t-distribution can be performed directly in terms of the sample mean, $\bar{x}$, or in terms of a t-statistic similar to the z-statistic used in the last section of this discussion. In either case, the formulas are the same except t is used instead of z and the estimated standard error, $s_{\bar{x}}$, *always* is used. In order to illustrate the t-test, we shall use the t-statistic since this is similar to the format used in later chapters.

A t-test can be performed directly in terms of the sample mean or in terms of a t-statistic, which is similar to the use of the normal. The t-statistic is used here and throughout the remainder of the text.

The test statistic in the form of the t-statistic for testing hypotheses about means is calculated using the following formula:

$$t = \frac{\bar{x} - \mu_0}{s_{\bar{x}}}$$

Student's t-statistic for testing means.

The t-statistic is computed by taking the difference between the mean of a particular sample, $\bar{x}$, and the hypothesized mean, μ_0, and dividing the difference by the estimated standard error of the mean, $s_{\bar{x}}$.

In order to perform the test, the calculated value of t using the above formula is compared with a critical value of t from Table 4 corresponding to a chosen level of significance, α. Critical values and decision rules for performing the t-test are presented in Exhibit 7. By examining the exhibit we can see that the null and alternate hypotheses are the same as the ones presented earlier. The test is performed by comparing the calculated value of the t-statistic with the critical value of t. The regions of acceptance and rejection depend upon the type of test and are determined similarly to the normal case. This easily is seen from the exhibit.

Note. Recall from Chapter 11 where the t-distribution is introduced that it is dependent upon an identifying characteristic or parameter called the degrees of freedom, which is given by the formula

$$DF = n - 1$$

Degrees of freedom for the t-test for the mean, μ,

In other words, the same t-distribution applies here and is identified by the number of degrees of freedom equal to the sample size minus one. For a given number of degrees of freedom, values of t are obtained from Table 4 in terms of *two-tailed probabilities.* When testing hypotheses, this means that a value of $t_{\alpha/2}$ directly corresponds to a two-tailed probability in Table 4 equal to the level of significance, α, in a two-tailed test. For a one-tailed test, the value of t_α corresponds to *twice* α in Table 4.

EXHIBIT 7

CRITICAL VALUES AND DECISION RULES FOR TESTING MEANS USING THE t-STATISTIC

Null Hypothesis	*Alternate Hypothesis*	*Critical Value(s)*	*Decision Rule*
$H_0: \mu \geq \mu_0$	$H_1: \mu < \mu_0$	$CV = -t_\alpha$	Accept H_0 if $t > -t_\alpha$ Reject H_0 if $t \leq -t_\alpha$
$H_0: \mu \leq \mu_0$	$H_1: \mu > \mu_0$	$CV = t_\alpha$	Accept H_0 if $t < t_\alpha$ Reject H_0 if $t \geq t_\alpha$
$H_0: \mu = \mu_0$	$H_1: \mu \neq \mu_0$	$CV_1 = -t_{\alpha/2}$ $CV_2 = t_{\alpha/2}$	Accept H_0 if $-t_{\alpha/2} < t < t_{\alpha/2}$ Reject H_0 if $t \leq -t_{\alpha/2}$ *or* $t \geq t_{\alpha/2}$

In order to illustrate the test for means using the t-statistic, consider a one-tailed test based upon the following hypotheses:

$$H_0: \mu \leq 20$$
$$H_1: \mu > 20$$

where

$$\begin{aligned} n &= 5 \\ DF &= n - 1 \\ &= 5 - 1 = 4 \\ \bar{x} &= 28.9 \\ s &= 6 \\ \alpha &= 0.05 \end{aligned}$$

and the universe distribution is assumed normal.

From Table 4 in the Appendix, the critical value of t corresponding to a one-tailed 0.05 level of significance with 4 degrees of freedom is

$$\begin{aligned} CV &= t_\alpha \\ &= t_{.05} \\ &= 2.132 \end{aligned}$$

In order to find this value from the table, we must locate 4 degrees of freedom and the two-tailed probability of 0.10 (i.e., twice 0.05). The t-statistic under the null hypothesis is computed as

$$\begin{aligned} t &= \frac{\bar{x} - \mu_0}{s_{\bar{x}}} \\ &= \frac{28.9 - 20}{6/\sqrt{5 - 1}} \\ &= 2.97 \end{aligned}$$

The format of the t-test is the same as one using the normal except the table that must be used is different.

Since the computed t-statistic of 2.97 exceeds the critical value of 2.132, we *reject* the null hypothesis. Diagrammatically, the example is summarized in Exhibit 8.

Testing Differences Between Means. There are many problems where a comparison is made between two groups or populations. For example, it may be of interest to know whether one product is better than another, whether incomes of a particular group of workers are greater than those of another, or whether the consumption of some commodity differs between two groups of consumers. Like all problems handled statistically, these must be quantified in terms of one or more universe parameters before they can be solved. In this section we shall consider one particular case where a problem can be translated into a comparison between the means of two populations.

Suppose a savings and loan association intends to open a new branch in one of two possible localities, a northern or a southern suburb

EXHIBIT 8
AN EXAMPLE TESTING A HYPOTHESIS WITH THE t-STATISTIC

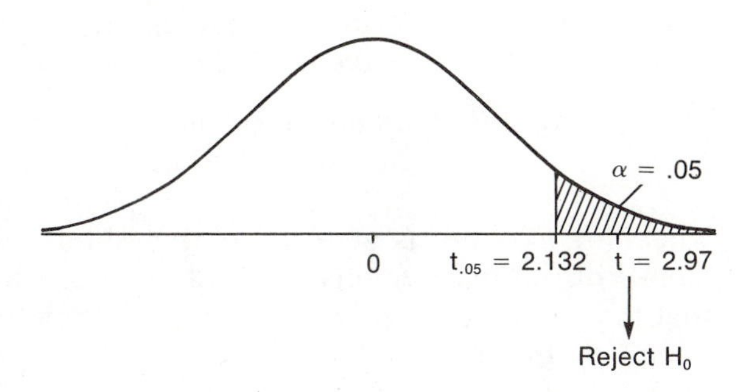

adjoining the city where the main office is located. The most important consideration in making the choice is the income level of the two regions. It has been decided that if the average income levels are different, the region with the higher average income will be the chosen location; otherwise, additional criteria will be considered in order to make the choice. Since census figures cannot be isolated for these specific suburbs, comparison of incomes is to be done on a sample basis. Independently drawn random samples of 50 families from each suburb reveal average incomes of $13,124 and $12,974 for the northern and southern suburbs, respectively. For simplicity, assume the standard deviations of incomes are known and equal $250 and $200.

The problem is to determine whether the mean incomes of the two suburbs are equal, given the information in the following table.

INFORMATION GIVEN IN SAVINGS AND LOAN ASSOCIATION PROBLEM

	Northern Suburb (N)	*Southern Suburb (S)*
Standard deviation	$\sigma_N = \$250$	$\sigma_S = \$200$
Sample size	$n_N = 50$	$n_S = 50$
Sample mean	$\bar{x}_N = \$13,124$	$\bar{x}_S = \$12,974$

In this problem we are interested in a comparison between the means of two populations, the values of which are unknown to us. We do know the values of the means of particular samples drawn from the individual populations. Direct comparison of these values without a test is not sufficient since the results are subject to chance variability because they are based upon a sample.

As the problem is stated, there are two possibilities to consider: the mean incomes of the two suburbs are equal (i.e., $\mu_N = \mu_S$) or the means are not equal (i.e., $\mu_N \neq \mu_S$). In order to solve the problem in terms of a test, we need to formulate it in terms of a null and an alternate hypothesis. Because of the nature of the testing procedure, it is more meaningful to express the two alternatives in terms of the *difference* between the universe means. In this form, the problem is translated in terms of a null and alternate hypothesis as follows:

$H_0: \mu_N - \mu_S = 0$ (The mean incomes are equal; consider other criteria to make choice between suburbs)

$H_1: \mu_N - \mu_S \neq 0$ (The mean incomes are *not* equal; open branch in region with the larger mean)

The null hypothesis states that the difference between the mean incomes of the two regions is equal to zero, which is the same as stating that the means are equal. The alternate hypothesis states that the difference is not equal to zero, or that the mean incomes are not equal. As in the case of tests involving one sample, the alternative containing the "equality" is the null hypothesis, or hypothesis tested.

In order to perform a test of the hypothesis we need a test statistic in the form of a single value that is based upon the sample data. A reasonable statistic to use is the *difference between the sample means* of the two samples. In general, the statistic used to test the difference between two universe means is given as

Statistic used to test difference between means.

$$\bar{x}_1 - \bar{x}_2$$

where the subscripts apply to any two populations. When the difference between the sample means is "small" or close to zero, it is reasonable to assume that the corresponding universe means are equal. On the other hand, when the difference between the sample means is "large," there is possible evidence that the universe means are not equal.

By testing a hypothesis we can distinguish between "large and small" differences between the sample results and reach a conclusion about the universe means. Recall from the material presented earlier in this chapter that the sampling distribution of the test statistic is used to determine large and small differences that lead to rejection and acceptance of the null hypothesis. Hence, in order to solve the savings and loan association problem we need to know the properties of the sampling distribution of the difference between sample means of two independently drawn samples.

Just as we spoke of a distribution of the mean based upon a single sample, we can think in terms of the difference between means varying among samples when sampling repeatedly from two populations. Under certain assumptions, the distribution of the difference between means based upon all possible samples has known properties. Here, we shall consider one set of assumptions that can be used to solve the savings and loan association problem.

In general, when two independent random samples are drawn from any two normal populations with known but not necessarily equal variances, the distribution of the difference between sample means also is *normal* with the following properties:

Mean of sampling distribution of the difference between sample means.

$$\mu_{\bar{x}_1 - \bar{x}_2} = \mu_1 - \mu_2$$

$$\sigma_{\bar{x}_1 - \bar{x}_2} = \sqrt{\frac{\sigma_1^2}{n_1} + \frac{\sigma_2^2}{n_2}}$$

Standard error of sampling distribution of the difference between sample means.

In other words, the sampling distribution of the difference between sample means has a mean equal to the difference between the universe means and a standard deviation equal to the square root of the sum of the variances of the sampling distributions of the means considered from each population separately. The formulas hold for any two sample sizes, n_1 and n_2, which do not have to be equal.

Based upon these properties, we can test the hypothesis that there is no difference between the mean incomes of the two suburbs.* Since the alternate hypothesis contains a "not equal" sign, we must perform a *two-tailed* test. The format of the test is identical to the test for a two-tailed test for the mean based upon a single sample considered earlier. In this case, however, the formulas to compute the standard error and the test statistic are different.

Assuming a 0.05 level of significance, α, the critical values for the test are computed as

$$\begin{aligned}
CV_1 &= (\mu_N - \mu_S)_0 - z_{\alpha/2}\sigma_{\bar{x}_N - \bar{x}_S} \\
&= 0 - z_{.025}\sqrt{\frac{\sigma_N^2}{n_N} + \frac{\sigma_S^2}{n_S}} \\
&= 0 - 1.96\sqrt{\frac{(250)^2}{50} + \frac{(200)^2}{50}} \\
&= -1.96\sqrt{\frac{62500}{50} + \frac{40000}{50}} \\
&= -1.96\sqrt{1250 + 800} \\
&= -1.96\sqrt{2050} \\
&= -1.96(45.28) \\
&= -88.75
\end{aligned}$$

Notice that the lower critical value, CV_1, is computed like the one sample case where the value of $z_{\alpha/2}$ from Table 2 is multiplied by the standard error (of the difference) and the result is subtracted from the hypothesized value, or "hypothesized difference" between universe means, $(\mu_N - \mu_S)_0$. Since the hypothesized value equals zero, the upper critical value, CV_2, has the same numerical value as the one computed above, but it is positive. That is,

$$\begin{aligned}
CV_2 &= (\mu_N - \mu_S)_0 + z_{\alpha/2}\sigma_{\bar{x}_N - \bar{x}_S} \\
&= 88.75
\end{aligned}$$

*Implicit in the application of the results to this example is the assumption that incomes in the two suburbs are normally distributed, whereas previous comments indicate that income distributions generally are skewed. The assumption made here is not necessarily unrealistic, since suburbs may be somewhat homogeneous socio-economically, and may not exhibit extremes encountered by the overall population.

EXHIBIT 9

AN EXAMPLE TESTING A HYPOTHESIS FOR THE DIFFERENCE BETWEEN MEANS

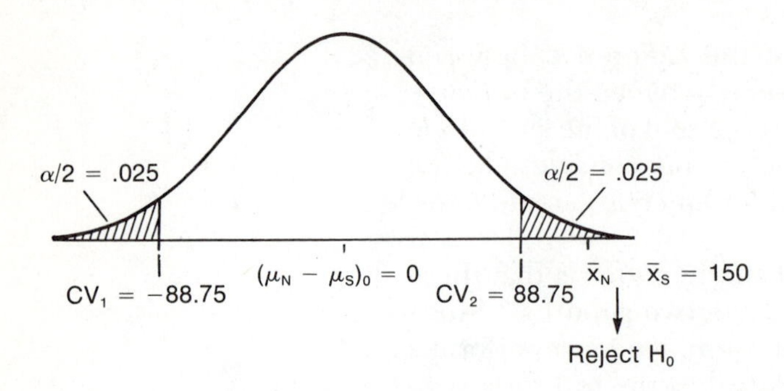

The corresponding decision rule takes the form

$$\begin{aligned}
&\text{Accept } H_0 \text{ if } -88.75 < \bar{x}_N - \bar{x}_S < 88.75\\
&\text{Reject } H_0 \text{ if } \quad \bar{x}_N - \bar{x}_S \leq -88.75\\
&\qquad\qquad\qquad\qquad \textit{or}\\
&\qquad\qquad\qquad \bar{x}_N - \bar{x}_S \geq 88.75
\end{aligned}$$

The value of the test statistic is computed as the difference between the sample means observed.

$$\begin{aligned}
\bar{x}_N - \bar{x}_S &= 13124 - 12974\\
&= 150
\end{aligned}$$

Since the observed difference equal to 150 exceeds the upper critical value CV_2 equal to 88.75, we *reject* the null hypothesis and *conclude* that the new branch should be opened in the northern suburb. The northern suburb is chosen since the difference $\bar{x}_N - \bar{x}_S$ is positive, which implies that the mean income of the northern suburb, μ_N, exceeds the mean income of the southern suburb, μ_S. The results of the test are summarized in Exhibit 9. We can see from the exhibit that the format of the test is similar to ones discussed earlier.

The procedure for testing the difference between means on the basis of two samples is similar to the one sample case except that the test statistic and the corresponding standard error are computed differently.

The problem of testing the difference between universe means was introduced in terms of a specific example. Recognize that the formulas apply to other problems meeting the same assumptions. Furthermore, the example considered the hypothesis that the difference between universe means equals zero. The procedure can be used for other hypothesized differences beside zero. Also, by extending the ideas for the one sample case, one-tailed tests for differences can be performed. Recall that the type of test is determined on the basis of the initial statement of the problem to be solved. Lastly, the test statistic used to perform the test for the difference between means was directly in terms of the difference between sample means. The z-statistic can be used as an alternative, except the terms in the formula correspond to differences rather than individual means.

Note. The test for differences between means presented in this section is based upon the assumption that the samples are drawn from normal populations with known variances. Based upon this assumption, the sampling distribution of the difference between sample means is assumed to be normal.

The normal also can be used to approximate the distribution of differences between means when "large" samples are drawn from non-normal populations, whether or not the universe variances are known. When the universe variances are unknown and a normal sampling distribution is assumed, sample standard deviations can be substituted into the formula for the standard error of the difference between means.

When *small* samples are drawn from normal populations, the distribution of the difference between sample means is not normal but a t-distribution. The formula for the standard error of the difference is not the same and is not presented here.

Other Considerations. The computational procedures presented in this chapter apply to tests of hypotheses for means. The overall approach, however, applies to tests for any parameter. What differs is the shape of the sampling distribution of the test statistic and the way the critical values are computed. The tests for means are unique to the extent that the normal and t-distribution used as sampling distributions are symmetric. This is not true for all tests of hypotheses. Regions of rejection always are associated with the tails of a sampling distribution; however, they are not necessarily found in terms of the mean of the distribution as in the case of the normal and t. This point becomes clear after reading later chapters relating to other types of tests.

Tests of hypotheses are important to consider when using sample data to make a decision. The general approach for performing a test is the same regardless of the parameter tested. Specific differences in the procedure exist due to the need to use different test statistics that have different sampling distributions.

The concept of hypothesis testing is important when working with decision problems involving sample data. When census data are available there is no reason to test hypotheses with these procedures since these specifically are designed to cope with errors due to sampling. Of course, non-sampling errors may be associated with the measurement of universe parameters. Non-sampling errors, however, are not considered directly in the procedures for testing hypotheses.

There is one final consideration worth noting. When the null hypothesis is rejected, we refer to the difference between the value of the test statistic and the hypothesized parameter as significant. Frequently, it is automatically but incorrectly assumed that such a difference also is important. A significant difference implies that sufficient evidence is available to reject the null hypothesis and conclude that the universe parameter is not equal to the value hypothesized. This being the case, it does not immediately follow that the observed difference is useful or important.

A significant difference is not necessarily an important difference.

As an example, consider the savings and loan association problem where we conclude that there is a difference between the mean incomes of the two suburban areas. Assuming the difference between the sample means of $150 is the actual difference between the mean incomes of the two suburbs, one could question whether this amount will have an impact upon company profits. If some portion or all of the difference, on the average, were deposited with the association from year to year, the suburb with the higher income should receive the new branch.

If, on the other hand, it is assumed that the difference of $150 is absorbed in consumption or is not reflected in higher mortgage loans, then the conclusion made really is of no importance when reaching a decision regarding the location of the new branch. In general it is important to consider the magnitude of the difference that is significant to determine whether it actually is useful. In practical work, a significant difference may not be sufficient evidence upon which to base a decision.

Type II Errors and the Power of the Test *(Optional)*. The hypothesis tests of this chapter are constructed on the basis of α, the level of significance. Once the null and alternate hypotheses are specified, the critical values, regions of acceptance and rejection, and decision rule can be determined using α. If the null hypothesis is true, we automatically know the probabilities of accepting and rejecting the null hypothesis. Moreover, we have control over these probabilities. By focusing on the probabilities of accepting and rejecting *false* null hypotheses, we can fully appreciate how a test works and understand why a particular test can be considered as a "good" test.

When a null hypothesis is accepted, it does not necessarily mean that it is true. The ability of a test to distinguish between true universe values and false null hypotheses is determined in terms of the probability of a Type II error or the power of a test.

The null hypothesis is false whenever a universe possesses a parameter value consistent with the alternate hypothesis. It is important to know how well this is detected by a specific test. In other words, it is important to know the extent to which a test can discriminate between true and false null hypotheses. This is accomplished either in terms of the *probability* of a Type II error or in terms of the power of the test. Recall, a **Type II error** is the acceptance of a false null hypothesis. The **power of a test** is defined as the probability of rejecting a false null hypothesis. Power, therefore, is the probability of drawing a correct conclusion *when* the null hypothesis is false.

The probability of committing a Type II error and the power of the test assume different values depending upon the value of the universe parameter consistent with the alternate hypothesis. In other words, many different values are related to an alternate hypothesis when it is specified in terms of an inequality; a different probability of a Type II error and a different power correspond to each value.

In this section we shall learn how to calculate the probability of a Type II error and the power of a test and then discuss the meaning of these quantities. We shall focus on tests concerning universe means when the distribution of the sample mean is normal in order to convey the main ideas.

In general, the probability of a Type II error is found as the area or probability associated with values of the test statistic in the region of acceptance, which is based upon the sampling distribution corresponding to a universe value consistent with the alternate hypothesis. The power of the test can be found as the area or probability associated with values of the statistic in the region of rejection, or by subtracting β from one.

In general, the *probability of a Type II error,* β, is determined by finding the probability associated with values of the test statistic in the region of acceptance, *conditional* upon a parameter value specified in the alternate hypothesis. The sampling distribution of the test statistic is used to find the required probability.

To determine β for a test of the universe mean when the distribution of the sample mean is assumed to be *normal,* find the area in the region of acceptance under the normal curve with mean, μ_1, specified in the alternate hypothesis. Three points should be noted: (1) the region of acceptance is found according to the steps outlined in the procedure section of this chapter, (2) a different β exists for different values of μ_1, specified in the alternate hypothesis, and (3) areas under the normal curve are found in the usual way according to the procedures introduced earlier.

The power of the test, P_w, or the probability of rejecting a false null hypothesis can be found as the complement of β by the formula

$$P_w = 1 - \beta$$

Power of the test.

In other words, P_w can be found by subtracting the probability of a Type II error from one. Power also can be found directly as the area under the sampling distribution in the region of rejection. For a given value of μ_1, the sum of β and power *must* equal one.

In order to understand these ideas, consider the light bulb example again. Recall, shipments of bulbs are acceptable to the distributor if the mean life is at least 1000 hours. The problem was translated into a test of the hypothesis $H_0: \mu \geq 1000$ (Accept the shipment) against the alternative $H_1: \mu < 1000$ (Return the shipment). Based upon a random sample of 49 bulbs and a level of significance, α, equal to 0.01, the critical value for the test is

$$CV = 983.36$$

where σ equals 50 and $\sigma_{\bar{x}}$ equals 7.14.

Assume the shipment mean equals 980 hours instead of the hypothesized value of 1000. Since the null hypothesis would be false under this assumption, we can compute the probability of committing a Type II error. The results are described in Exhibit 10. The first distribution

EXHIBIT 10
ILLUSTRATION OF THE PROBABILITY OF A TYPE II ERROR IN THE LIGHT BULB EXAMPLE

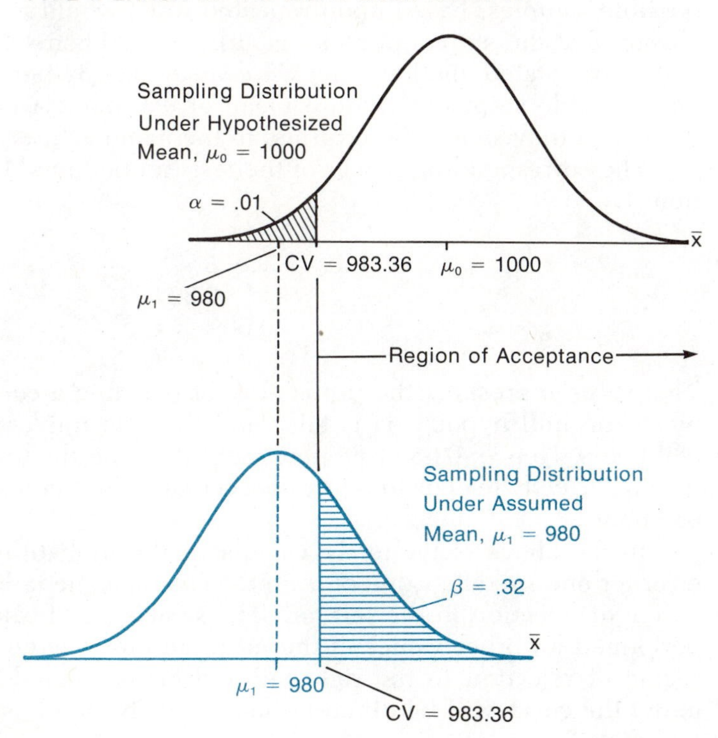

shown in the exhibit is the sampling distribution of the mean under the null hypothesis. This is the distribution used to compute the critical value that forms the basis for the test.

The second distribution shown in the exhibit (color) is the sampling distribution of the mean under the assumption that the universe mean equals 980. The probability of a Type II error equal to 0.32 is the area under this distribution associated with the region of acceptance. Assuming normality, this value is found by computing z corresponding to the region between the assumed mean, μ_1, and the critical value, finding the probability from Table 2, and subtracting this from 0.5.

The value of z is computed as

$$\begin{aligned} z &= \frac{CV - \mu_1}{\sigma_{\bar{x}}} \\ &= \frac{983.36 - 980}{7.14} \\ &= \frac{3.36}{7.14} \\ &= 0.47 \end{aligned}$$

Based upon Table 2, this value of z yields a probability of 0.1808. The probability of a Type II error in this case is

$$\begin{aligned} \beta &= 0.5 - 0.1808 \\ &= 0.3192 \text{ or } 0.32 \text{ (rounded)} \end{aligned}$$

β can be interpreted as the proportion of all possible samples that leads to the incorrect acceptance of the null hypothesis when a specific parameter value consistent with the alternate hypothesis is the true value; P_w can be interpreted as the proportion of all possible samples that leads to the correct rejection of the null hypothesis.

This value can be interpreted to mean that 0.32 or 32 percent of all possible samples, based upon repeated tests, would result in the conclusion that the shipment mean is at least 1000 hours when it really is 980 hours. Stated another way, we can say that 32 percent of the time unacceptable shipments with a mean of 980 hours would be kept for sale when they should be returned to the manufacturer.

The corresponding power of the test can be found by subtracting β from 1.

$$\begin{aligned} P_w &= 1 - \beta \\ &= 1 - 0.32 \\ &= 0.68 \end{aligned}$$

This value represents the probability of reaching a *correct* conclusion when the null hypothesis is false and the true universe mean equals 980 hours. Hence, 0.68 or 68 percent of the time the testing procedure would correctly lead us to reject unacceptable shipments with a mean of 980 hours.

In the above example we calculated the probability of a Type II error for one specific value of μ_1 in the case of a one-tailed test with the region of rejection in the left tail. The same type of calculation can be performed for other values of the mean and for one-tailed tests with a region of rejection in the right tail and for two-tailed tests. The area under the curve will be different but always is found as the area under the alternative distribution in the region of acceptance. The power of the test always can be found as the complement of β.

The interpretations of β and P_w in the above example are *conditional* upon a specific mean of 980 hours. In practice, we do not know the magnitude of the universe mean; therefore, β and P_w are related to the value of a universe mean that we *assume* to be true. By doing this for many universe values consistent with the alternate hypothesis, we are able to determine the ability of a test to discriminate between true universe values and a false null hypothesis.

Exhibit 11 presents the probability of committing a Type II error and the power of the test for selected values of the universe mean in the light bulb example. β and P_w are given for values of the universe mean that are consistent with the alternate hypothesis: all values of the universe mean are less than 1000. The larger the difference between the value of the universe mean shown in the table and the hypothesized value of 1000, the smaller the probability of committing a Type II error. Conversely, the power of the test increases as the difference becomes greater.

EXHIBIT 11
β AND P_w FOR SELECT VALUES OF μ CORRESPONDING TO THE LIGHT BULB EXAMPLE

Universe Mean (μ)	*Probability of a Type II Error (β)*	*Power of The Test (P_w)*
970	0.03	0.97
975	0.12	0.88
980	0.32	0.68
985	0.59	0.41
990	0.82	0.18
995	0.95	0.05

The probability of committing a Type II error generally is "high" when the difference between the universe mean and the hypothesized value is small. In other words, small differences between a true value and the one hypothesized are not detected very often. The corresponding power is low so that a correct conclusion using sample data is infrequent when the true value is "close" to the value hypothesized. As the difference between the actual value and the one hypothesized increases, β becomes smaller and the power increases. In other words, larger differences will lead to correct decisions more frequently when using sample information. The ability of a test to detect false null hypotheses becomes greater when the difference between the hypothesized value and the true value is larger.

The power of the test increases and the probability of a Type II error decreases as the difference between the hypothesized parameter and the actual value increases.

By graphing the probability of a Type I and a Type II error on a single graph, we can visualize how well a test draws correct and incorrect conclusions in terms of the possible values that can be assumed by the universe mean. This is accomplished with a **power curve,** which is a graph that depicts the probability of *rejecting* the null hypothesis for all possible values of the universe parameter. In order to illustrate the concept, consider Exhibit 12, which presents the power curve for the test used in the light bulb example.

The power curve is a graph depicting the probability of rejecting a null hypothesis when it is true or false. It provides the power of the test and the probability of a Type I error on one graph.

The first point to note about the exhibit is that all values of μ to the left of 1000 correspond to the alternate hypothesis and the remaining

EXHIBIT 12
POWER CURVE CORRESPONDING TO LIGHT BULB EXAMPLE $H_0: \mu \geq 1000$ $H_1: \mu < 1000$

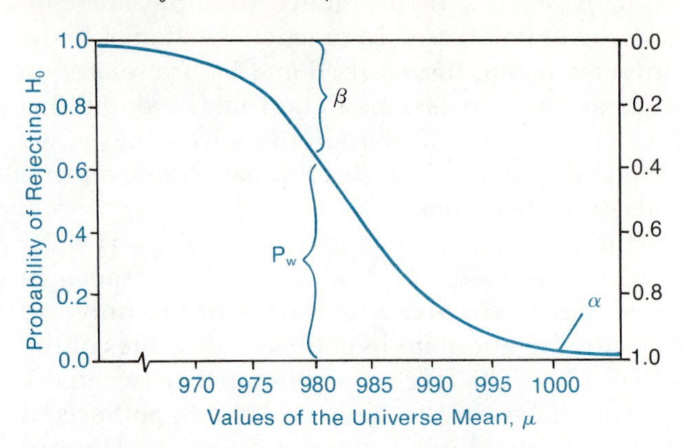

values correspond to the null hypothesis. Hence, values of μ equal to or greater than 1000 correspond to the true null hypothesis, whereas values to the left of 1000 correspond to a false null hypothesis. Although the power of the test is associated with false null hypotheses, the power *curve* considers the probability of rejecting H_0 in cases where it can be true *or* false.

The height of the curve corresponding to values of μ consistent with the alternate hypothesis, or a false null hypothesis, represents the power of the test, P_w. This can be read on the left-hand scale. The distance from the top axis to the curve gives the probability of a Type II error, β, which is read on the right-hand scale. P_w increases and β decreases as the difference between the actual and hypothesized mean increases.

Special attention should be given to values of μ that correspond to the null hypothesis. When the universe mean is 1000, which represents the "equality" in the null hypothesis, the height of the curve is α, the probability of committing a Type I error. At this point, the distance from the top axis to the curve is the complement of α but does *not* equal β since a Type II error cannot occur when the null hypothesis is true. The complement of α is the probability of drawing a correct conclusion when the null hypothesis is true.

The values of μ greater than 1000 correspond to the "inequality" associated with the null hypothesis. Since these values also correspond to a true null hypothesis, the height of the curve in this region represents the probability of committing a Type I error. Since these probabilities are less than α, they can be ignored. By constructing a one-tailed test on the basis of the "equality" of the null hypothesis, α represents the *maximum* probability of committing a Type I error. Hence, better protection against a Type I error is achieved if the actual parameter corresponds to the "inequality" of the null hypothesis.

The power curve illustrated corresponds to a one-tailed test with a region of rejection in the left tail of the sampling distribution. Power curves also can be constructed for one-tailed tests with the rejection region in the right tail and for two-tailed tests. Power curves for the three possible cases are shown in Exhibit 13. The quantity μ_0 appearing in the

EXHIBIT 13
POSSIBLE TYPES OF POWER CURVES

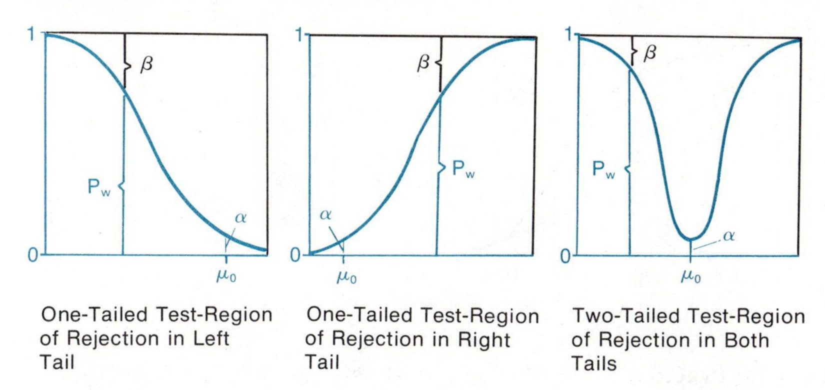

diagrams represents the hypothesized mean specified by the equality of the null hypothesis. The power curve for a one-tailed test with the region of rejection in the upper tail basically is the same as the one in the case already discussed except the pattern is reversed. For the two-tailed test, the power curve assumes a different shape. The probability of committing a Type I error exists only at the value of the hypothesized mean, μ_0. Interpretations of β and P_w remain the same, but apply to values above and below the hypothesized mean.

Power curves can be constructed for all three types of tests. For one-tailed tests, the value on the curve corresponding to the hypothesized value represents the maximum probability of a Type I error, whereas for a two-tailed test, the height of the curve is the probability of a Type I error.

The power curve can be used to evaluate how well a test performs. Every test of a hypothesis has a power curve that is based upon the value of α and the sample size, n, and is independent of a specific sample result. In other words, the power curve is determined by the test, not the results of a sample; the sample results are used to perform the test and reach a conclusion. In order to understand how to relate the power curve to a particular test, it is useful to introduce the concept of a power curve under ideal conditions. This is the curve that corresponds to a test of a hypothesis where it is not possible to commit an error of either kind: false null hypotheses are always rejected ($P_w = 1$), and true null hypotheses are never rejected ($\alpha = 0$). When sampling, the ideal cannot be realized; however, a test whose power curve is closest to the ideal when compared with other tests is the best test.

An ideal test is one where the probability of rejecting false null hypotheses always is one and the probability of rejecting true null hypotheses is zero. When sampling, the ideal can not be achieved, but it is a useful concept used to evaluate a particular test.

The power of a test can be increased in two ways: (1) by increasing the sample size, n, or (2) by increasing the level of significance, α. By holding α fixed and increasing n, the power of the test increases, which means that the power curve is closer to the ideal. Similarly, when n is held constant and α is increased, the power curve is closer to the ideal. In the first case, protection against a Type I error remains the same and increased power is obtained at a higher cost of sampling. In the latter case, the cost of sampling is held constant, but the risk of committing a Type I error increases.

The power of a test can be increased by increasing α or by increasing n.

Consider the light bulb problem where, on the one hand, we increase α from 0.01 to 0.05 and, on the other, we increase n from 49 to 100. The resulting power curves are shown in Exhibit 14. In both cases we can see that power is increased and the curve is closer to the ideal. Obviously, as the power increases the probability of a Type II error diminishes.

EXHIBIT 14
INCREASING POWER IN THE LIGHT BULB EXAMPLE

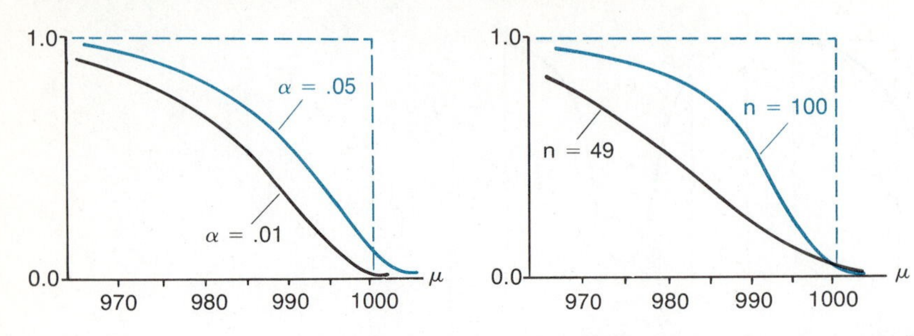

Theory vs. Practice. A test of a hypothesis can be performed on the basis of α without considering the risks of committing Type II errors. In practice, this is quite common. In some cases, the consequences of committing a Type II error are not of great importance so that β can be ignored. In others, the importance of Type II errors depends upon the goals of those doing the testing. For example, the light bulb manufacturer probably feels it is more costly to reject a good shipment (Type I error) than to accept a bad one (Type II error). Consequently, the manufacturer would be interested in controlling α, whereas the distributor is more interested in β.

Since β depends upon the actual value of the universe parameter, it is not possible to control β for every possibility. An acceptable level of β can be established, however, for an important difference between the hypothesized value and some universe value. A sample size can then be determined that meets this requirement of detecting a false hypothesis for an important difference while providing adequate protection against a Type I error.

The concepts of β and P_w are general, although they are illustrated in terms of means. Both are obtained from the sampling distribution of the test statistic corresponding to the alternate hypothesis, which may not always have the same form as the one under the null hypothesis.

The testing procedures discussed to this point dealt exclusively with means. Consequently, general concepts associated with tests of hypotheses were discussed in terms of means. Since the shape of the distribution of the sample mean is the same under the null and the alternate hypothesis, it is relatively simple to compute the power of the test and β. Although these concepts are general, the mechanics for other parameters vary with the form of the sampling distribution of the test statistic used. In addition, in some cases the shape of the sampling distribution is not the same for the null as it is for the alternate hypothesis. When this occurs it sometimes is difficult to determine power and the probability of a Type II error. These may still be important to consider, but because of the difficulties they may be ignored. It is important to remember that the concepts of power and a Type II error do apply to other types of tests beside ones for means.

CALCULATORS AND COMPUTERS

Since this chapter serves as a general introduction to hypothesis testing in addition to considering specific tests for means, it is of interest to mention the role of computing devices in hypothesis testing at this point. When performing individual tests like the ones presented in this chapter, there is no real problem doing the calculations by hand or with a simple "4-function" calculator, especially for small samples. In situations where sample statistics must be computed directly from large samples or when elaborate procedures involve many kinds of calculations, and in situations where more than one test is performed on the same data, there can be savings in time and effort by using a more sophisticated computing device such as a computer or a calculator with a stored program feature. Programs exist for performing many kinds of tests in addition to the ones presented here. These are available in the form of "canned" computer programs as well as programs that are part of optional libraries that can be purchased with various hand-held and desk-top calculators. Many of the programs represent subroutines that are part of larger integrated programs used to accomplish multiple tasks. Since the testing procedures used in this chapter are very simple and do not require a lot of computation, it is not necessary to provide an example of the output of a computer for these tests.

The output for a particular test of a hypothesis generally is the numerical value of the test statistic based upon the sample input data. The resulting test statistic then can be compared with a tabulated critical value to determine whether the result is significant or not. Many computer programs and some calculators with a stored program feature can provide what is called a "p-value" along with the value of the test statistic presented. A **p-value** is the probability of obtaining a difference equal to or greater than the one observed between the value of the test statistic and the hypothesized value under the assumption that the null hypothesis is true.

An alternative way to perform a test of a hypothesis is with a p-value. Although p-values can be obtained on a limited basis with some tables, the p-value is especially useful to use when it can be obtained automatically from a sophisticated calculating device.

When a p-value is available, obtained from a table or a calculating device, it can be compared directly to the level of significance in order to perform a test. In the case of a one-tailed test, the null hypothesis is rejected if the p-value is less than α. For a two-tailed test, the null hypothesis is rejected if the p-value is less than half the level of significance, or $\alpha/2$. Using this procedure the conclusion always will be the same as one based on the other procedures discussed.

QUESTIONS AND PROBLEMS

DEFINITIONAL QUESTIONS

1. What is meant by the terms *hypothesis* and *a test of a hypothesis?*

2. Distinguish between the terms *null hypothesis* and *alternate hypothesis.* What is a hypothesized value?

3. Define the following terms:

   (a) Type I Error (c) Probability of a Type I Error
   (b) Type II Error (d) Probability of a Type II Error

4. Define the terms *decision rule* and *critical value*. What is the relationship between the two concepts?

5. Explain the meaning of the following terms:
   (a) Acceptance of a Null Hypothesis
   (b) Rejection of a Null Hypothesis
   (c) Region of Acceptance
   (d) Region of Rejection

6. Define the term *level of significance*. How is it related to the probability of committing a Type I error?

7. What is a test statistic? How is it used in hypothesis testing?

8. Distinguish between a one-tailed test and a two-tailed test. Under what circumstances is each type of test used?

9. What is a significant difference? A non-significant difference? What kind of conclusion is based upon each?

10. What is meant by the term *normal sampling distribution?* Under what circumstances is it used in this chapter?

11. What is Student's t-distribution? Under what circumstances is it used in this chapter?

12. What is the meaning of the following terms?
    (a) z-statistic
    (b) t-statistic

    What are the similarities and differences between the two statistics?

13. What is meant by the term *degrees of freedom* as it is used in this chapter?

14. List the general steps needed to carry out a test of any hypothesis.

15. What is the general rule for choosing a level of significance? What values of the level of significance generally are used in practice?

16. In words, explain the general procedure for determining a critical value needed to perform a test of a hypothesis. What basic information is required in order to determine a critical value for any test?

17. What is meant by the term *sampling distribution of the difference between means?* How is it used in hypothesis testing?

18. Define the following terms:

    (a) Difference Between Universe Means
    (b) Difference Between Sample Means
    (c) Mean of the Distribution of Differences Between Means
    (d) Standard Error of the Distribution of Differences Between Means

    How are each of the concepts listed above used in hypothesis testing?

19. What is the power of a test? How is it related to the probability of a Type II error?

20. Distinguish between the symbols α, β, and P_w.

21. What is a power curve?

22. In general terms, describe the procedure used to find the probability of a Type II error and the power of the test.

COMPUTATIONAL PROBLEMS

23. Given a universe of items that is normally distributed with unknown mean, and standard deviation, σ, equal to 5. Based upon a random sample of 9 items with mean, $\bar{x}$, equal to 29, test the hypothesis $H_0: \mu \geq 35$ against the alternate $H_1: \mu < 35$ using a 0.05 level of significance.

24. Test the null hypothesis $H_0: \mu \leq 20$ against the alternate $H_1: \mu > 20$ based upon the following information.

    $\sigma = 10$
    $n = 36$
    $\bar{x} = 24$
    $\alpha = 0.05$

25. Test the null hypothesis $H_0: \mu = 165$ against the alternate $H_1: \mu \neq 165$. The following information is given.

    Universe distribution is normal
    $\sigma = 40$
    $n = 25$
    $\bar{x} = 158$
    $\alpha = 0.05$

26. A random sample of 10 items is drawn from a normal population. The mean and standard deviation of the sample are 73 and 19, respectively. Based upon a 0.01 level of significance, test the hypothesis $H_0: \mu \leq 68$ against the alternate $H_1: \mu > 68$.

27. A random sample of 5 items is selected from a normal universe

with the following results:

16 19 24 21 17

Based upon the above sample data, test the hypothesis $H_0: \mu \leq 14$ against the alternate $H_1: \mu > 14$. Use a 0.05 level of significance.

28. Test the null hypothesis that the means of two populations are equal, $H_0: \mu_1 - \mu_2 = 0$, against the alternative that they are not equal, $H_1: \mu_1 - \mu_2 \neq 0$. Assume the standard deviations of the two universes are known and are equal to the following values: $\sigma_1 = 17$ and $\sigma_2 = 19$. Independent random samples are drawn from the two populations with the following results:

1	2
$n_1 = 40$	$n_2 = 35$
$\bar{x}_1 = 32.4$	$\bar{x}_2 = 38.9$

29. In order to respond to a claim made by an independent consumer group that local postal deliveries take longer than in the past, a postal administrator randomly selects 150 local letters and traces them to their destination. The arithmetic mean delivery time for the sample is two-and-one-half days with a standard deviation of one-half day. Based upon an intensive study of mail delivery in the past, the established mean time for local delivery of a letter is one-and-three-quarter days. Using a 0.05 level of significance, test the claim made by the consumer group based upon the administrator's data.

30. A firm engaged in holding educational programs throughout the country has determined that the business remains profitable as long as the average number of participants is at least 17 per program. From the current records, a sample of 40 programs yields a mean number of participants equal to 13.6 with a standard deviation of 6.8. Based upon this evidence, is it reasonable to conclude that the average number of participants per program is at a profitable level?

31. A traffic safety director of a local police department is interested in determining whether there has been a change in the average number of accidents per week at a particular intersection after introduction of a "right turn on red" law. A sample of 12 weeks yields an average of 2.7 accidents per week. Prior to the new law, it had been established that the number of accidents is normally distributed with a mean of 2.2 and a standard deviation of 0.3 accidents. Has a change in the average number of accidents occurred since the new law was introduced?

32. An inspector for a municipal license bureau suspects that a certain franchised fast-food operation serves less beef in its adver-

tised "$\frac{1}{4}$-pound" hamburger. In order to check this suspicion, five members of the bureau are sent to the store. Each purchases one hamburger which is later inspected for beef content and weight. The average amount of beef found was 0.21 pound with a standard deviation of 0.06. Is this sufficient evidence to justify the inspector's suspicion? Assume that hamburger weights naturally vary according to a normal distribution.

33. An owner of various types of vending machines periodically checks his soft drink machines to determine whether they dispense the proper amount. In general, if a cup is underfilled customers are dissatisfied whereas overfilled cups lead to wasted soft drink. Assume the vendor has established that the average amount dispensed should be 6 ounces in order to maintain a proper balance between underfill and overfill. A sample of 5 cupfuls is discharged from a particular machine with the following results:

5.3 oz   7.2 oz   8.7 oz   5.1 oz   6.9 oz

Assuming the amount dispensed per cup is normally distributed, what should the vendor conclude about this particular machine?

34. In order to determine whether two differently priced brands of flashlight batteries are equally effective, a consumer testing bureau tests 45 batteries of each brand for length of life. The results are presented in the following table.

Brand 1	Brand 2
$\bar{x}_1 = 165$ hours	$\bar{x}_2 = 177$ hours
$s_1 = 15$ hours	$s_2 = 19$ hours

Using a 0.01 level of significance, determine whether there is a difference in the effectiveness of the two brands of batteries.

35. The following information is related to the test of the hypothesis $H_0: \mu \leq 20$ against the alternative $H_1: \mu > 20$.

$$\sigma = 10$$
$$n = 36$$
$$\alpha = 0.05$$
$$CV = 22.8$$

Find β and P_w for the test based upon the critical value, CV, when the universe mean, μ_1, equals 24.

36. The information presented below is associated with the test of the hypothesis $H_0: \mu = 35$ against the alternative $H_1: \mu \neq 35$.

$$\sigma = 5$$
$$n = 9$$
$$\alpha = 0.05$$
$$CV_1 = 32$$
$$CV_2 = 38$$

Find β and P_w for the test under the assumption that the universe mean μ_1, equals 36.

37. The owner of a chain of jewelry stores is considering a new location for one of the stores in the chain. Unless the mean family income in the new locality is at least $16,000, it is not worthwhile relocating the store. In order to determine whether a particular location is worthwhile, a survey of 100 families is to be undertaken. On the basis of the survey data, a hypothesis is to be tested and a decision is to be made. Assume the standard deviation of family incomes, σ, equals $1200 and that a 0.02 level of significance is used to make the test.

    (a) Find the critical value and the decision rule that can be used to perform the test.
    (b) Based upon the results of part (a), find the probability of a Type II error and the power of the test corresponding to a mean family income of $15,800.

38. Refer to Exhibit 11. Based upon the information given regarding the test in the light bulb example, verify the results in the exhibit.

39. Refer to Problem 35. Calculate P_w for values of the universe mean equal to 21, 22.8, 23, and 25. Plot P_w on a graph against the values of the universe mean. Trace a power curve through the plotted points.

40. Refer to Problem 36. Calculate P_w for values of the universe mean equal to 30, 32, 34, 36, 38, and 40. Plot P_w on a graph against values of the universe mean. Trace a power curve through the plotted points.

CONCEPTUAL QUESTIONS AND PROBLEMS

41. In general terms, explain why hypothesis testing is important. Why is hypothesis testing considered a part of statistical inference? Explain.

42. Before a problem is solved using hypothesis testing, generally it is expressed in terms of a universe parameter(s). What is meant by this statement? Why is it necessary to translate a problem in this way before a test is performed?

43. In general terms, discuss the importance of the sampling dis-

tribution of a statistic when testing a hypothesis. What is the importance of a random sample?

44. In your own words, explain the reasoning underlying conclusions made when a significant and a non-significant difference results from a test of a hypothesis.

45. Criticize or explain the following statements: Hypotheses are generated from the general statement of a problem rather than from the field of statistics, whereas the field of statistics does provide a rule for choosing the null and alternate hypotheses and a way of testing the null hypothesis. Furthermore, it is the alternate hypothesis that determines the type of test.

46. Criticize or explain each of the following statements:

    (a) When a confidence interval estimate is used, there is only one kind of error that can occur: concluding that the universe parameter falls within the interval estimate when it really does not. The probability of error is subject to the control of the investigator and equals one minus the level of confidence. The same is true of hypothesis testing since the level of significance, α, is subject to control.
    (b) Since two types of errors are associated with tests of hypotheses, it is not possible to perform a test without considering the probabilities of both types of errors.
    (c) The level of significance, α, is not affected by the size of the sample, n, however, the sample size does have an effect on the critical value and the decision rule used to perform a test of a hypothesis.
    (d) When testing a hypothesis, the decision rule is used to accept or reject a null hypothesis and has nothing to do with choosing a course of action relating to the original problem.
    (e) Random samples are unimportant in hypothesis testing since a test is based upon the probability of committing a Type I error that is fixed in advance.
    (f) Hypothesis tests can be performed only for means since these are the ones considered in this chapter.
    (g) Since the level of significance has meaning as a probability in terms of repeated trials, it has no meaning in terms of a specific test of a hypothesis based upon one sample.
    (h) When performing a test of a hypothesis, we do not know whether we have reached a correct conclusion. Consequently, a test really is of no use.

47. A firm has decided that a certain amount of bad debts on receivables can be ignored without changing its collection policy. If, on the average, bad debts do not exceed $45 per customer, no corrective measures are to be instituted. In order to determine whether the collection policy should be changed, a sample of 100 customer accounts is selected at random from the firm's

files. The mean and standard deviation amount of bad debts are calculated as $57 and $28, respectively.

(a) Based upon the information given above, test a hypothesis in order to determine whether the collection policy should be changed. What course of action is suggested by the result of the test?
(b) In your own words, explain why you were led to the conclusion in part (a).
(c) In order to perform the test in part (a), you had to choose a level of significance. Specifically in terms of the problem, interpret the value of the level of significance you chose.

48. Recent attention has been given to statements that there exists a decline in the average aptitude test score of college applicants and that the decline can be attributed to poorer quality instruction in the secondary schools. A school administrator of a small municipality is concerned about the standing of students in the town since she feels that these students should be above the national average of 470. In order to test this assumption, a random sample of 25 college applicants from the town's high schools is selected. The mean aptitude test score for the sample equals 475 with a standard deviation of 30.

(a) Based upon the information provided, does the evidence support the administrator's assumption about the town's students?
(b) In your own words, explain why you were led to the conclusion in part (a).
(c) In order to answer the question in part (a), you should test a hypothesis. Since a level of significance is not provided, you must supply your own. Interpret the value of the level of significance you chose specifically in terms of the problem.

49. Refer to Problem 47. If you have not already done so, solve the problem in terms of the "z-statistic."

50. Refer to Problem 48. If you have not already done so, solve the problem in terms of the "t-statistic."

51. As part of a study of agencies providing services for the mentally retarded, a governmental bureau wants to determine whether agencies receiving its funds received more than an average of $75,000 of its funds during the preceding funding period. Although the bureau should know how its money is spent, this is not possible due to many reasons relating to the way funds actually reach an intended source through our political system. In order to determine whether funds received exceeded the stipulated average amount, a random sample of 100 agencies is selected. The mean amount of funds received in the sample equals $80,000 with a standard deviation of $20,000.

(a) Assuming the sample standard deviation equals the true value, calculate the probability of observing a sample mean in excess of $80,000, given that the universe mean equals $75,000. Is it likely that the sample result came from a universe with a mean equal to $75,000? Explain.
(b) Based upon your answer to part (a), is it possible to conclude that the average amount of funds received per agency exceeds $75,000? Explain.
(c) What is the relationship between the calculation performed in part (a) and a test of a hypothesis? Does this suggest an alternative way of performing a test? Why?

52. An environmental law relating to a certain atomic power plant specifies that the average temperature of water cycled back into a nearby river cannot exceed 82° F. Fifty water samples from the river yield an average of 79° F with a standard deviation of 41° F.

(a) Based upon the sample evidence, what conclusion should be reached regarding compliance with the environmental law?
(b) In general, should a hypothesis be tested in order to solve problems like the one above? Explain.
(c) For the specific sample information given above, is it necessary to perform a test of a hypothesis? Why or why not?

53. In order to determine whether automobile maintenance costs have increased, a government agency interviewed a random sample of 100 car owners regarding the amount of money spent on car maintenance during the past year. The average cost of the sample was reported as $375 with a standard deviation of $85. The established average maintenance cost based upon past studies is $325.

(a) Should the agency conclude that maintenance costs have increased based upon the data collected? Support your answer with a statistical test.
(b) Based upon the conclusion reached in part (a), is it reasonable to conclude that maintenance costs have increased because prices charged by repair shops have increased?
(c) Based upon the conclusion in part (a), is it reasonable to conclude that maintenance costs increased because cars manufactured today are of poorer quality? Without examining prices charged by repair shops, is it possible to separate the effect of price changes and manufacturing quality in order to draw the type of conclusions suggested in parts (b) and (c)?
(d) What is suggested by the questions in parts (b) and (c) about drawing general conclusions on the basis of a test of a hypothesis?
(e) The statement of the problem presented above is very general. It speaks of maintenance costs but does not specify

whether these costs are for repairs only or whether they include fuel and insurance costs as well. Furthermore, such factors as age of the cars represented in the sample, make and model, initial purchase price, etc. were not considered in the problem statement. What effect would these factors have on the overall conclusion reached as a result of the test performed in part (a)? What is suggested by this question about conclusions drawn from tests of hypotheses and their relationship to the way a problem initially is defined?

54. Suppose that you take a census of the incomes of all lawyers in Chicago and New York and compute the average yearly income as follows:

	Mean Income
New York	\$63,000
Chicago	\$57,000

Can you conclude "statistically" that the average income of New York lawyers exceeds that of Chicago lawyers? Would you test a hypothesis? If yes, what hypothesis? If no, why not?

55. A statewide commission on human relations claims that wages paid to minority workers in the steel industry are less than wages paid to other workers. In order to support this claim, independent random samples of 100 minority and 100 non-minority workers are drawn with the following results:

WEEKLY WAGES

	Mean	*Standard Deviation*
Minority	$\bar{x}_1 = \$375$	$s_1 = \$65$
Non-Minority	$\bar{x}_2 = \$425$	$s_2 = \$50$

Based upon this evidence, is the claim made by the commission valid?

56. A basic assumption of our legal system is that a person is innocent until proven guilty. Suppose we translate this as a hypothesis to be tested where the alternate hypothesis is that a person is guilty.
    (a) In terms of the problem, explain the meaning of a Type I and a Type II error.
    (b) Which of the two types of errors specified in part (a) is more serious?
    (c) Which of the two types of errors is more serious for society in general? For the defendant? Explain. What does your answer suggest about the importance of the two types of errors relating to a particular problem?

57. Assume specifications for testing shipments of batteries are expressed in terms of the mean length of life.

    (a) If you are the manufacturer, what would be the null and alternate hypotheses that you would test? Which is more meaningful, a one- or a two-tailed test? Explain.
    (b) Based upon the hypotheses specified in part (a), would a Type I or a Type II error be more serious? Explain.
    (c) Reconsider your answers to parts (a) and (b) separately in terms of batteries used in ordinary four-function pocket calculators and in pacemakers for heart patients.

58. Refer to Exhibit 1 at the beginning of this chapter. Appearing in the exhibit are three different problem statements relating to light bulbs together with the corresponding sets of hypotheses. Compare each set of hypotheses in terms of the way the critical values are determined, the courses of action associated with acceptance and rejection of the null hypothesis, and the meaning of a Type I and a Type II error. In general, what conclusions can be made about the way a problem is formulated in terms of hypotheses and the results of a test of a hypothesis?

59. The rule for determining the level of significance, α, is not very specific since it tells us that α must be a small probability but does not tell us the exact value to choose. Why shouldn't α always be set at zero, since this would mean that a Type I error will never occur?

60. At the end of the chapter in the section on Calculators and Computers, the use of a "p-value" is cited as an alternative way of testing a hypothesis.

    (a) Use the p-value approach cited to solve Problem 24.
    (b) By computing a p-value, we are able to obtain a measure of how unlikely it is for an observed result to have come from a hypothesized universe. There is a problem, however, with the approach since the p-value generally is computed before α is selected. Explain why this can be a problem.

61. What information is provided by the probability of a Type II error and the power of a test? Of what use are these values?

62. In words, explain the relationship that exists between α, β, and P_w. How are these considered on a power curve?

63. When a test of a hypothesis is performed, the value of the test statistic based upon a specific sample is required. When computing the probability of a Type II error or the power of a test, however, the specific value of the test statistic is not required. Explain the meaning of these statements.

64. Why is it impossible to compute β and P_w when the true universe parameter equals the hypothesized value?

65. (a) What effect do changes in the sample size have on β and P_w?
(b) What effect do changes in α have on β and P_w?

66. Refer to Problem 37. If the true mean family income equals \$15,900, β equals 0.89 and P_w equals 0.11. In terms of the problem, interpret the values of β and P_w given above. Which of the two types of errors, I or II, is more important to consider by the owner of the jewelry store chain? Discuss.

CHAPTER 13

INFERENCES ABOUT PROPORTIONS AND PERCENTAGES

There is no excellent beauty that hath not some strangeness in the proportion.

Francis Bacon

Many problems involving parameters other than means are solved with the approach presented in the preceding chapters. In this chapter we concentrate on problems formulated in terms of the proportion, or equivalent percentage, of a population that possesses a particular characteristic. A **population proportion** is defined as the ratio of the number of elements possessing a characteristic divided by the total number of elements in the population. The percentage is obtained by multiplying the proportion by 100; the two measures impart the same information and can be used interchangeably. The symbol P is used to represent the population proportion.

There are many situations where it is natural to consider the population proportion to solve a problem. For example, students may be for or against a university policy, and the administration is interested in the proportion of the student body opposing the policy. Assembly-line items can be classed either as acceptable or as defective, and a plant manager must know the proportion that are defective. It would be of interest to a state department of motor vehicles to know the proportion of cars driven that are unsafe.

In each of the above examples the basic data associated with an element of the population naturally assumes one of two possibilities. This is the simplest form of information that is termed *qualitative* or *categorical*, since an element either falls or does not fall into a category where no unit of measurement such as dollars or pounds are used. In other cases, it is convenient or useful to assign elements to a particular category based upon recorded measurements. For example, a firm's uncollectable receivables specified in dollar amounts may be divided into two groups according to whether or not they exceed $1000; interest may focus on the proportion of uncollected receivables that exceed $1000. Similarly, family incomes may be characterized as above or below the subsistence level, where there is an interest in the proportion of the population that does not earn a subsistence income. In these two cases, a measured characteristic has been translated into two qualitative categories, whereas in the earlier cases the elements fall into categories naturally and are not based upon some underlying measurement.

This chapter is concerned with problems where inferences are made about the universe proportion, P, regardless of the way elements are assigned to one of two possible categories. Since P also can be considered as a probability, the material presented provides an introduction to problems of inference about probabilities.

When a sample is used to make an inference about a population proportion, the basic data is in the form of *counts*. That is, the *number* of elements in a sample possessing a characteristic of interest is observed. The sample statistic generally used is based upon this number and is referred to as the sample proportion. The **sample proportion** is defined as the ratio of the number of elements in a sample possessing a characteristic of interest divided by the sample size. The formula for the sample proportion is

$$p = \frac{x}{n}$$

The sample proportion or point estimate of P.

where x equals the number of sample elements possessing a characteristic and n is the sample size. The sample proportion, p, represents a point estimate of P, the universe value.

For random samples, the sample proportion is a random variable that varies from sample to sample according to a known probability distribution, or random sampling distribution. When sampling is done with replacement or its equivalent, the distribution of the sample proportion is like the binomial distribution presented in Chapter 7 but differs slightly in terms of its characteristics.

The mean and standard deviation, or standard error, of the sampling distribution of the sample proportion are given by the following formulas.

$$\mu_p = P$$

Mean of the distribution of the sample proportion.

$$\sigma_p = \sqrt{\frac{P(1 - P)}{n}}$$

Standard error of the sample proportion.

The mean of the sampling distribution of p, designated by the symbol μ_p, equals the population proportion, P. The standard error of p equals the square root of P times 1 − P divided by the sample size. Although the mean of p is like the mean of the distribution of sample means presented in earlier chapters since it equals the corresponding universe parameter, the standard deviation is not. Notice, the standard deviation of the sample estimate, p, is a function of the universe proportion, P. In other words, σ_p depends upon the value of P and varies for different values of the population proportion.

Since the standard error depends upon P, σ_p always is unknown when P is unknown. When it is necessary to use an estimate of σ_p, the following sample estimate can be used.

$$s_p = \sqrt{\frac{p(1-p)}{n}}$$

Sample estimate of the standard error of p.

In other words, the estimated standard error, s_p, is obtained by using the sample proportion, p, in place of P in the formula for σ_p.

In this chapter, we will apply these concepts to confidence intervals and tests of hypotheses about universe proportions in much the same way as we did for means earlier. Fortunately, for large samples, the distribution of the sample proportion can be *approximated* by a normal curve with a mean and a standard deviation equal to the values specified above. Consequently, the procedures for large samples are the same as the ones for means except the standard error is computed differently. Since the large sample procedures using the normal curve are more common, these are considered in detail in the next two sections. Small sample problems involving proportions are more difficult since the binomial distribution must be used directly. These are considered briefly in the discussion section along with a number of related topics.

The symbols used in this chapter are summarized as follows:

P = the population proportion
= the proportion of elements in a population possessing a characteristic of interest
n = the sample size
x = the number of elements in a sample possessing the characteristic of interest
p = the sample proportion, or point estimate of P
= the proportion of elements in a sample possessing the characteristic of interest
μ_p = the mean of the sampling distribution of the sample proportion
= P
σ_p = the standard error of the sampling distribution of the sample proportion
s_p = the sample estimate of the standard error of the sample proportion
H_0 = the null hypothesis
H_1 = the alternate hypothesis
P_0 = a number representing a hypothesized value of the population proportion
α = the level of significance, or the probability of a Type I error

CV	= the critical value
z	= the value of a standardized normal variate corresponding to a given probability or confidence level
	= the z-statistic (this represents an alternate use of z which appears in the discussion)
z_α	= the standardized normal variate used in a one-tailed test
$z_{\alpha/2}$	= the standardized normal variate used in a two-tailed test
E	= the maximum tolerable difference or error between P and p used to determine sample size.
$\mu_{p_1 - p_2}$	= the mean of the sampling distribution of the difference between sample proportions based upon two independent samples
$(P_1 - P_2)_0$	= a number representing a hypothesized value of the difference between population proportions
$\sigma_{p_1 - p_2}$	= the standard error of the difference between sample proportions
$s_{p_1 - p_2}$	= an estimate of the standard error of the difference between sample proportions

PROCEDURE

Confidence Interval Estimation. In order to construct a confidence interval estimate of the population proportion, P, based upon *large* samples, use the normal approximation by substituting into the following expression.

$$p \pm z s_p$$

Confidence interval for P for large samples.

p is the sample proportion based upon a specific set of sample results, s_p is the estimated standard error of the sample proportion, and z equals the standardized normal value from Table 2 corresponding to a *chosen* level of confidence. Hence, a confidence interval estimate of P is obtained by adding and subtracting the estimated standard error times z to the estimate of the sample proportion.

Hypothesis Testing. For *large* samples, tests of hypotheses for proportions can be performed with a normal sampling distribution. Overall, the procedures are the same as the ones presented earlier for means. A problem stated in terms of a universe proportion must be formulated in terms of a null and an alternate hypothesis. Based upon a chosen level of significance, α, a critical value(s) leading to regions of acceptance and rejection or a decision rule is established and a test is performed on the basis of the value of the test statistic computed from a specific sample. The test statistic in this case is the sample proportion, p, which has a sampling distribution that is approximately normal for

Critical Values and Decision Rules for Testing Proportions With Large Samples

Null Hypothesis	Alternate Hypothesis	Type of Test	Region of Rejection	Critical Value(s)	Decision Rule
$H_0: P \geq P_0$	$H_1: P < P_0$	One-tailed	Left tail	$CV = P_0 - z_\alpha \sigma_p$	Accept H_0 if $p > CV$ Reject H_0 if $p \leq CV$
$H_0: P \leq P_0$	$H_1: P > P_0$	One-tailed	Right tail	$CV = P_0 + z_\alpha \sigma_p$	Accept H_0 if $p < CV$ Reject H_0 if $p \geq CV$
$H_0: P = P_0$	$H_1: P \neq P_0$	Two-tailed	Both tails	$CV_1 = P_0 - z_{\alpha/2}\sigma_p$ $CV_2 = P_0 + z_{\alpha/2}\sigma_p$	Accept H_0 if $CV_1 < p < CV_2$ Reject H_0 if $p \leq CV_1$ *or* $p \geq CV_2$

large samples. The regions of acceptance and rejection are based upon the type of test that is dictated by the alternate hypothesis. Critical values for each type of test can be computed using the formulas provided in the above table; the corresponding decision rule is given in each case.

In all three cases presented in the table, the hypotheses have the same form as the ones for means except they are expressed in terms of the proportion P and a *hypothesized* value P_0. Critical values for each case are computed using the formulas in the next to the last column of the table; z_α represents the standardized normal value corresponding to the level of significance, α, in one tail and $z_{\alpha/2}$ represents the standardized value corresponding to half the level of significance in both tails of the sampling distribution. The decision rules appearing in the last column of the table are expressed in terms of the sample proportion, p. The procedures for testing hypotheses about proportions for large samples are summarized diagrammatically in Exhibit 1 which appears on the following page.

Note. The standard error, σ_p, in the formulas for the critical values are based upon the sampling distribution of p *under* the null hypothesis. Since σ_p is dependent upon the value of P, the hypothesized value is used. Hence, the standard error needed to find the critical value is computed using the formula

$$\sigma_p = \sqrt{\frac{P_0(1 - P_0)}{n}}$$

The standard error of p used to find the critical value in tests for proportions.

Hence, the standard error used in a test for proportions is obtained by substituting the hypothesized value, P_0, into the formula for σ_p. It is not necessary to consider the sample estimate of the standard error when performing a test for P based upon a single sample.

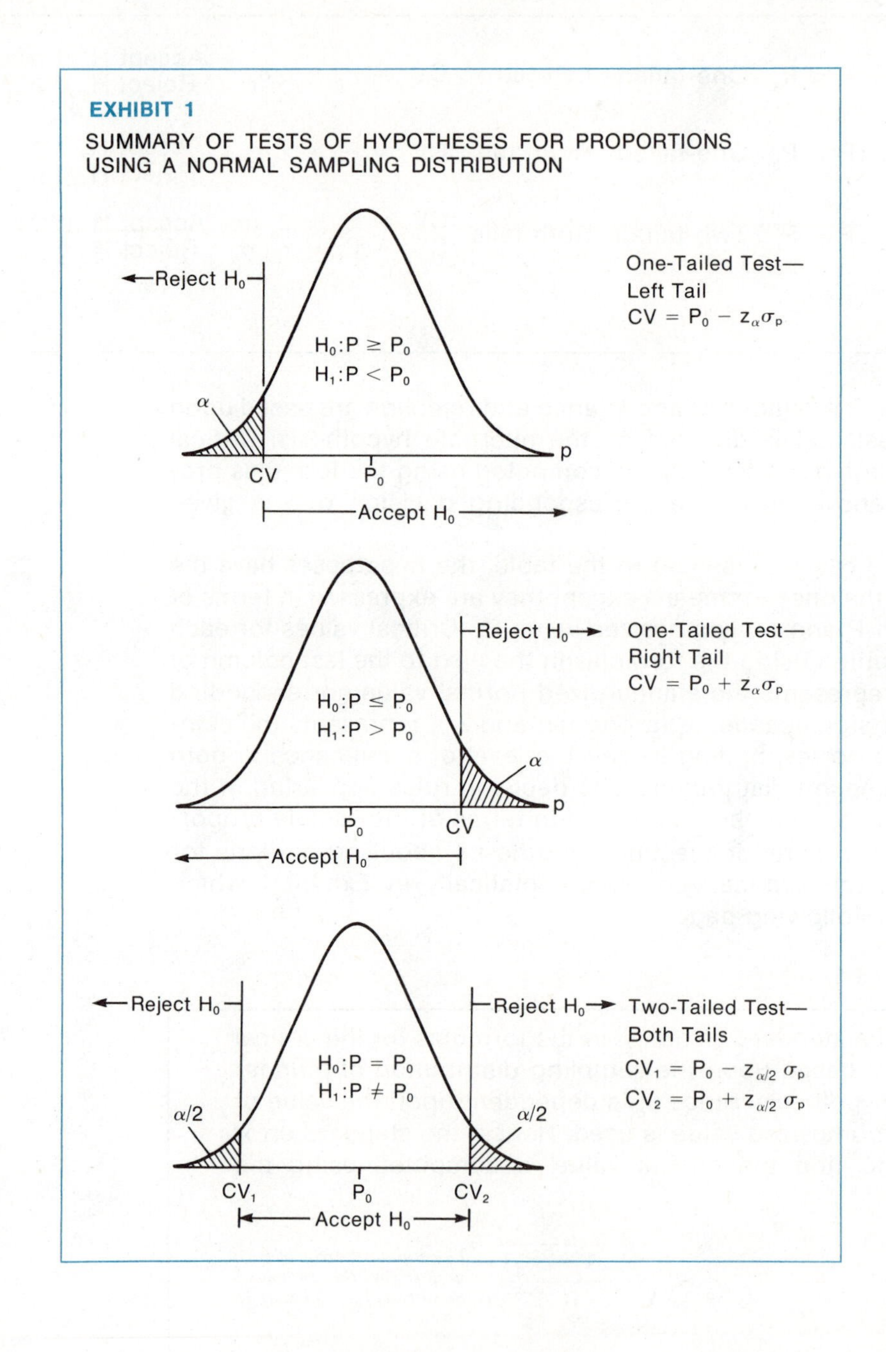
EXHIBIT 1
SUMMARY OF TESTS OF HYPOTHESES FOR PROPORTIONS USING A NORMAL SAMPLING DISTRIBUTION
Reject H_0
One-Tailed Test—
Left Tail
$CV = P_0 - z_\alpha \sigma_p$
$H_0: P \geq P_0$
$H_1: P < P_0$
α
p
CV
P_0
Accept H_0
Reject H_0
One-Tailed Test—
Right Tail
$CV = P_0 + z_\alpha \sigma_p$
$H_0: P \leq P_0$
$H_1: P > P_0$
α
p
P_0
CV
Accept H_0
Reject H_0
Reject H_0
Two-Tailed Test—
Both Tails
$CV_1 = P_0 - z_{\alpha/2}\, \sigma_p$
$CV_2 = P_0 + z_{\alpha/2}\, \sigma_p$
$H_0: P = P_0$
$H_1: P \neq P_0$
$\alpha/2$
$\alpha/2$
CV_1
P_0
CV_2
Accept H_0

NUMERICAL EXAMPLES

Example 1 ***Confidence Interval.*** Suppose we are interested in estimating the proportion of registered voters favoring a particular candidate in a pending mayoral election that is to be done with a 95 percent confidence interval estimate. Based upon a survey of 200 randomly selected registered voters, 90 stated a preference for the candidate.

To Find: 95 percent confidence interval estimate of P, the proportion of registered voters favoring the mayoral candidate

Given: $n = 200$
$x = 90$
$z = 1.96$ (corresponds to .95 confidence, based upon Table 2)

Compute: $p = \frac{x}{n}$

$$= \frac{90}{200} = 0.45$$

$$s_p = \sqrt{\frac{p(1 - p)}{n}}$$

$$= \sqrt{\frac{.45(1 - .45)}{200}}$$

$$= \sqrt{\frac{.45(.55)}{200}}$$

$$= \sqrt{.001238} = 0.035 \text{ (rounded)}$$

The confidence interval is found by substituting into the following expression and computing.

$$p \pm zs_p$$

$$.45 \pm 1.96(.035)$$
$$.45 \pm 0.07$$
$$0.38 - 0.52$$

Our estimate of the proportion favoring the candidate lies between 0.38 and 0.52 of the registered voters with 95 percent confidence.

Example 2 ***One-Tailed Test.*** Region of Rejection in Left Tail. The administration of a university has decided that unless at least half of the undergraduate student body favors the required physical education program, it will be discontinued. Since printing and processing costs are too high to permit administering the questionnaire to the entire student body of 15,000, the administration decides to sample 150 students at random and base its decision on the sampled result. From the sample selected, 66 students favor continuation of the program. Use a 1 percent level of significance to perform a test of a hypothesis to determine whether at least half of the students favor the program or not.

To Test: $H_0: P \geq 0.5$ (At least half the students favor continuation and the program should be continued)

$H_1: P < 0.5$ (Less than half the students favor continuation and the program should be discontinued)

Given: $n = 150$

$\alpha = 0.01$ (1 percent level of significance)

$x = 66$

Compute:
$$
\begin{aligned}
CV &= P_0 - z_\alpha \sigma_p \\
&= P_0 - z_{.01}\sqrt{\frac{P_0(1 - P_0)}{n}} \\
&= 0.5 - 2.33\sqrt{\frac{.5\,(1 - .5)}{150}} \\
&= 0.5 - 2.33\sqrt{.001667} \\
&= 0.5 - 2.33\,(0.041) \\
&= 0.5 - 0.096 \\
&= 0.40 \text{ (rounded)}
\end{aligned}
$$

$P_0 = 0.5$ is obtained from the null hypothesis and is used to compute σ_p. $z_{.01} = 2.33$ corresponds to a normal curve with 0.01 of the area in the left tail. The value of z is obtained from Table 2 in the Appendix.

Decision Rule: Accept H_0 if $p > 0.40$
Reject H_0 if $p \leq 0.40$

Since $p = 0.44 \left(\text{i.e., } p = \frac{x}{n} = \frac{66}{150}\right)$ is greater than the critical value of 0.40, we *accept* the null hypothesis and *conclude* that at least half the undergraduate student body favors continuation of the required physical education program. Therefore, the program should be continued. This example is summarized in the following diagram:

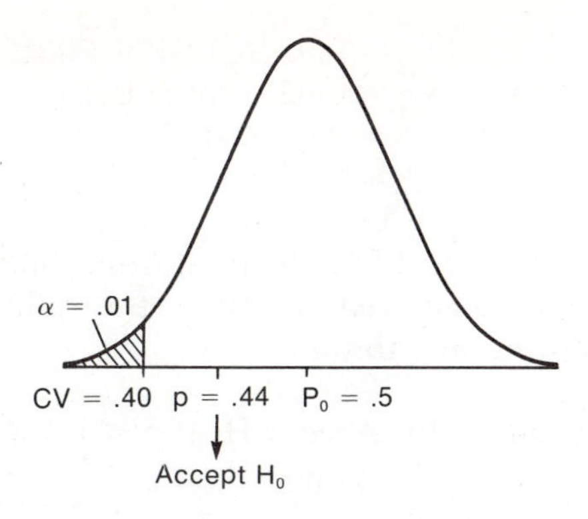

Example 3 ***One-Tailed Test.*** REGION OF REJECTION IN RIGHT TAIL. A large equipment manufacturer has established a policy of monitoring its uncollectable receivables or bad debts based on periodic samples of records of its current accounts. It is felt that at most 2 percent of bad debts in excess of \$1000 is acceptable. When it is determined that more than 2 percent of its accounts receivable in excess of \$1000 are uncollectable, an investigation of the problem is undertaken that usually leads to a tightening of the firm's collection and credit policies. A 5 percent level of significance is used to make the determination. Based upon a random sample of 100 uncollected accounts, 7 are in excess of \$1000.

Recognize the criterion of "2 percent" for choosing among alternatives provides the same information as a proportion of 0.02. Hence,

To Test: $H_0: P \leq 0.02$ (Maintain current policy regarding bad debts)

$H_1: P > 0.02$ (Undertake investigation of bad debts problem)

Given: $n = 100$

$\alpha = 0.05$ (5 percent level of significance)

$x = 7$

Compute:
$$CV = P_0 + z_\alpha \sigma_p$$
$$= P_0 + z_{.05}\sqrt{\frac{P_0(1 - P_0)}{n}}$$
$$= 0.02 + 1.65\sqrt{\frac{.02(1 - .02)}{100}}$$
$$= 0.02 + 1.65\sqrt{\frac{.02(.98)}{100}}$$

$$= 0.02 + 1.65\sqrt{.000196}$$
$$= 0.02 + 1.65(.014)$$
$$= 0.02 + 0.023$$
$$= 0.043$$

where $z_{.05} = 1.65$, obtained from Table 2, corresponds to an area of 0.05 in the right tail of the sampling distribution.

Decision rule: Accept H_0 if $p < 0.043$
Reject H_0 if $p \geq 0.043$

Since $p = 0.07 \left(\text{i.e., } \frac{x}{n} = \frac{7}{100}\right)$, we *reject* the null hypothesis because it exceeds 0.043. Hence, we *conclude* that the proportion of bad debts in excess of \$1000 is greater than 0.02 and the bad debt problem should be investigated. A summary of this example appears in the following diagram:

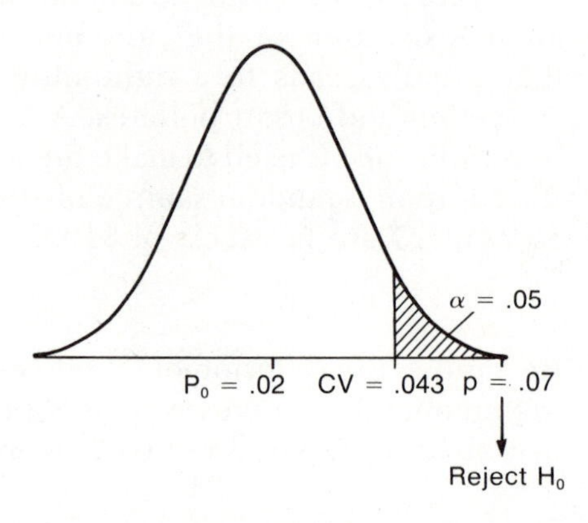

Example 4 ***Two-Tailed Test.*** REGION OF REJECTION IN BOTH TAILS. One of the ways a merchandising firm determines the effectiveness of its TV advertising is to periodically poll a sample of viewers. It has been decided that an advertisement performs acceptably if 40 percent of all viewers remember the ad; less than 40 percent is unacceptable, whereas more than 40 percent indicates superior performance. Therefore, in order to determine how an ad performs, a two-tailed test is required. A 5 percent level of significance is used. Associated with a particular advertisement, 96 viewers from a random sample of 200 remember seeing the firm's ad.

To Test: $H_0: P = 0.4$ (40 percent of viewers remember the ad)

$H_1: P \neq 0.4$ (The percentage of viewers remembering the ad is not 40 percent)

Given: $n = 200$
$\alpha = 0.05$
$x = 96$

Compute:

$$CV_1 = P_0 - z_{\alpha/2}\sigma_p$$
$$= P_0 - z_{.05/2}\sqrt{\frac{P_0(1 - P_0)}{n}}$$
$$= P_0 - z_{.025}\sqrt{\frac{.4(1 - .4)}{200}}$$
$$= .4 - 1.96\sqrt{\frac{.4(.6)}{200}}$$
$$= .4 - 1.96\sqrt{.0012}$$
$$= .4 - 1.96(0.035)$$
$$= 0.331$$

$$CV_2 = P_0 + z_{\alpha/2}\sigma_p$$
$$= P_0 + z_{.025}\sigma_p$$
$$= .4 + 1.96(.035)$$
$$= 0.469$$

where $z_{.025} = 1.96$ from Table 2 corresponds to half the level of significance or an area of 0.025 in each tail of the sampling distribution.

Decision rule: Accept H_0 if $0.331 < p < 0.469$
Reject H_0 if $p \leq 0.331$
or
$p \geq 0.469$

The sample proportion, computed as

$$p = \frac{x}{n}$$
$$= \frac{96}{200} = 0.48$$

exceeds the upper critical value of 0.469. Therefore we *reject* the null hypothesis and conclude that the proportion of viewers remembering the firm's advertisement exceeds 0.4 or 40 percent. Since the sample value falls in the upper region of rejection, we conclude that the performance of the ad was superior. The test is summarized graphically in the following diagram.

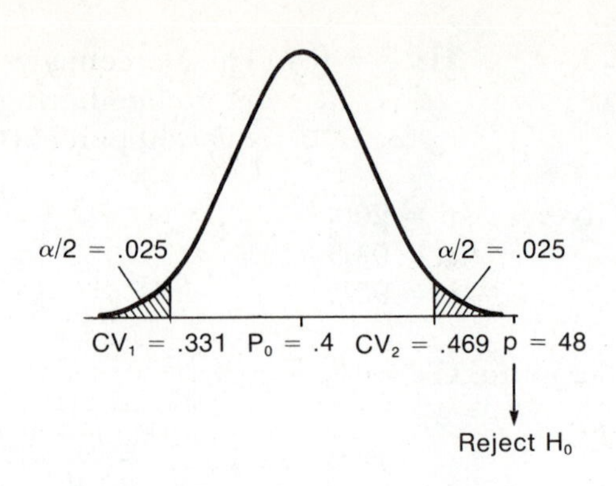

DISCUSSION

In this chapter large sample methods for constructing confidence intervals and testing hypotheses about proportions have been presented. These large sample procedures basically are the same as those used for means in Chapters 11 and 12 since they are based upon the assumption of a normal sampling distribution. The underlying reasoning and interpretations are the same.

The essential difference between the procedures is the formula for the standard error. For means, the standard error is independent of the universe mean, whereas the standard error of the sample proportion, σ_p, is related to the population proportion and changes with the value of P.

Proportions vs. Percentages. Since percentages are obtained from a proportion by moving the decimal point two places to the right, they impart the same information. Consequently, the large sample formulas presented apply directly to percentages also. Whether a proportion or a percentage is used in a particular problem is a matter of personal preference, although it is sometimes simpler to talk in terms of percentages. Each of the formulas presented earlier can be used directly by substituting a percentage instead of a proportion wherever p or P is encountered.

Sample Size Determination. As in the case of means, the size of the sample affects the width of a confidence interval estimate of P. Consequently, we may want to determine the sample size needed to provide a required amount of precision at a given level of confidence prior to selecting a sample. This can be done for *large* samples by substituting into the formula

$$n = \frac{z^2P(1 - P)}{E^2}$$

Formula for determining sample size when estimating P.

By specifying a level of confidence, z is determined. P represents the actual proportion in the population possessing a particular characteristic. Since this value is unknown before a sample is drawn, prior knowledge or a reasonable guess can be used to establish the value of P needed to calculate n. If no prior information is available and we are unwilling to guess, we can substitute a value of *0.5* for P since this provides the *maximum* possible standard error.* This will lead to a conservatively large value of n.

The quantity E represents the maximum difference or error between the population proportion, P, and the sample estimate, p, that we are willing to tolerate at the chosen level of confidence. This difference must be specified and substituted into the formula.

In order to illustrate how the formula is used for determining sample size, suppose we want to estimate a population proportion with a sample estimate that differs by no more than 0.04 with a probability of 0.95. Assume the population value cannot exceed 0.3. Substituting into the formula, we have

$$n = \frac{z^2P(1 - P)}{E^2}$$

$$= \frac{(1.96)^2(.3)(.7)}{(.04)^2}$$

$$= 504 \text{ (rounded)}$$

where $z = 1.96$ corresponds to a normal probability of 0.95 and is found in Table 2. The result indicates that in order to meet the requirements specified above it is necessary to sample 504 elements from the universe. If we do not make an assumption about the size of P, we could substitute 0.5 instead of 0.3; the value of n that results is 600 which is larger.

Using the z-Statistic. The z-statistic can be used as an alternative when performing tests of hypotheses about proportions with a normal sampling distribution. The formula for computing the z-statistic is

$$z = \frac{p - P_0}{\sigma_p}$$

$$= \frac{p - P_0}{\sqrt{\dfrac{P_0(1 - P_0)}{n}}}$$

z-Statistic for testing the population proportion.

In other words, z is found by subtracting the hypothesized proportion, P_0, from the sample proportion, p, and dividing the result by the standard error of p *under* the null hypothesis.

In order to perform a test, the computed value of z is compared with a value of z corresponding to a chosen level of significance, and the null hypothesis is accepted or rejected according to a decision rule similar to the ones presented for means in Chapter 12. For one-tailed tests z_α is used, whereas $z_{\alpha/2}$ is used in a two-tailed test.

In order to illustrate this alternative, consider the two-tailed test

*The fact that the standard error is largest when P equals 0.5 can be seen by examining the formula for σ_p, or $\frac{P(1 - P)}{n}$. Since P and 1 − P sum to one, one of these quantities decreases when the other increases. When both approach 0.5, their product is larger than the product of any other numbers that sum to one. Hence, σ_p is largest when P and 1 − P equal 0.5.

performed in Example 4. The null and alternate hypotheses are $H_0:P = 0.4$ and $H_1:P \neq 0.4$, where $P_0 = 0.4$, $n = 200$, $p = 0.48$, and $\alpha = 0.05$. The z-statistic is calculated as

$$z = \frac{p - P_0}{\sqrt{\dfrac{P_0(1 - P_0)}{n}}}$$

$$= \frac{.48 - .4}{\sqrt{\dfrac{.4(1 - .4)}{200}}}$$

$$= \frac{.48 - .4}{\dfrac{.4\,(.6)}{200}}$$

$$= \frac{.08}{.035} = 2.29$$

This value must be compared with the value corresponding to half the level of significance, $z_{.025}$, since we are performing a two-tailed test. From Table 2, $z_{.025}$ equals 1.96. Since the value of the z-statistic equal to 2.29 exceeds 1.96, the null hypothesis is rejected. Since the value of the z-statistic is positive, we conclude that P is greater than 0.4.

If the z-statistic were negative and less than −1.96, the null hypothesis would be rejected; however, the conclusion would be that P is less than 0.4. Values of the z-statistic between −1.96 and 1.96 would lead to the acceptance of H_0 and the conclusion that P equals 0.4.

Testing Differences Between Proportions. There are problems when it is of interest to determine if two population proportions are equal, based upon separate samples from each of the two populations. The null and alternate hypotheses in such cases can be written in terms of the difference between proportions as

$H_0:P_1 - P_2 = 0$ (Population proportions are equal)

$H_1:P_1 - P_2 \neq 0$ (Population proportions are not equal)

where the subscripts denote the respective populations.

When large samples are drawn independently from both populations, we can use the difference between sample proportions, $p_1 - p_2$, as the *test statistic*. The sampling distribution of this statistic is approximately normal. Consequently, we can apply the same format as we did in the case of the difference between means. In other words, the sampling distribution of the difference between sample proportions, $p_1 - p_2$, based upon large, independently drawn samples of sizes n_1 and n_2 is approximately normal with mean and standard deviation given as

Mean of sampling distribution of differences between sample proportions.

$$\mu_{p_1 - p_2} = P_1 - P_2$$

$$\sigma_{p_1 - p_2} = \sqrt{\frac{P_1(1 - P_1)}{n_1} + \frac{P_2(1 - P_2)}{n_2}}$$

Standard Error of differences between sample proportions for large samples.

In words, the mean of the distribution of differences between sample proportions equals the difference between the population proportions. The standard error equals the square root of the sum of the variances of the distributions of the individual sample proportions.

In order to test the hypothesis specified above, we can use the same format for determining critical values and decision rules that we have for all tests involving the normal curve. Since specific values of P_1 and P_2 are not specified in the null hypothesis and $\sigma_{p_1 - p_2}$ is dependent upon these values, we must use a sample estimate. A satisfactory estimate is given by the formula

$$s_{p_1 - p_2} = \sqrt{\frac{p_1(1 - p_1)}{n_1} + \frac{p_2(1 - p_2)}{n_2}}$$

Estimated standard error of the difference between sample proportions for large samples.

This formula has the same form as the one given earlier except the true proportions are replaced by their sample estimates, p_1 and p_2.*

Consider the following example. An entrance examination is given to candidates for a particular job. Those who pass the examination are allowed into a training program that, upon successful completion, leads to employment with the firm. A question has been raised regarding the possibility that there is a disparity between the proportion of men and women allowed into the training program. In order to determine whether a disparity exists, independent samples of men and women who have taken the examination are selected. Results of the two samples are summarized in the following table.

	Men	*Women*
Sample Size	$n_1 = 100$	$n_2 = 150$
Proportion Passing	$p_1 = 0.65$	$p_2 = 0.58$

We can formulate the problem in terms of a test of a hypothesis as

$H_0: P_1 - P_2 = 0$ (The proportion of men and women passing is equal and no disparity exists)

$H_1: P_1 - P_2 \neq 0$ (The proportion of men and women passing is not equal and a disparity exists)

*Under the assumption that the null hypothesis is true, the population proportions are specified as equal. In such a case, some prefer to use an estimate which takes this into consideration. This is accomplished with the formula

$$\sqrt{\bar{p}(1 - \bar{p})\left(\frac{1}{n_1} + \frac{1}{n_2}\right)}$$

where $\bar{p}$ is the "pooled" estimate of the common population proportion, or the weighted average of p_1 and p_2, given as

$$\bar{p} = \frac{n_1p_1 + n_2p_2}{n_1 + n_2}$$

Based upon a 5 percent level of significance, the critical values are determined as

$$
\begin{aligned}
CV_1 &= (P_1 - P_2)_0 - z_{\alpha/2} s_{p_1 - p_2} \\
&= (P_1 - P_2)_0 - z_{.025} \sqrt{\frac{p_1(1-p_1)}{n_1} + \frac{p_2(1-p_2)}{n_2}} \\
&= 0 - 1.96 \sqrt{\frac{.65(.35)}{100} + \frac{.58(.42)}{150}} \\
&= 0 - 1.96\sqrt{.002275 + .001624} \\
&= 0 - 1.96\sqrt{.003899} \\
&= 0 - 1.96(.062) \\
&= -0.122
\end{aligned}
$$

$$
\begin{aligned}
CV_2 &= (P_1 - P_2)_0 + z_{\alpha/2} s_{p_1 - p_2} \\
&= 0 + 1.96(.062) \\
&= +0.122
\end{aligned}
$$

The term $(P_1 - P_2)_0$ represents the hypothesized value of the difference between the population proportions, which in this case is zero. The value of $z_{.025}$ equal to 1.96 corresponds to half the level of significance of 0.05 in each tail, since a two-tailed test is required. The estimated standard error, $s_{p_1 - p_2}$ must be used since P_1 and P_2 are not specified in the null hypothesis. The decision rule for the test based upon the calculated critical values takes the form

$$
\begin{aligned}
&\text{Accept } H_0 \text{ if } -0.122 < p_1 - p_2 < 0.122 \\
&\text{Reject } H_0 \text{ if } \quad p_1 - p_2 \leq -0.122 \\
&\qquad\qquad\qquad \textit{or} \\
&\qquad\qquad\qquad p_1 - p_2 \geq 0.122
\end{aligned}
$$

The value of the test statistic in terms of the difference between the sample proportions is computed as

$$
\begin{aligned}
p_1 - p_2 &= 0.65 - 0.58 \\
&= 0.07
\end{aligned}
$$

Since this value falls between the critical values of −.122 and .122, we *accept* the null hypothesis and *conclude* that no disparity exists between the proportion of men and women passing the test. Diagrammatically, the test is described in Exhibit 2.

Tests for differences between proportions based upon large samples can be performed for any value beside zero using the same format as the one illustrated above; the value of the hypothesized difference is substituted for $(P_1 - P_2)_0$ in the formulas for the critical values. The procedure also can be applied to one-tailed tests by using z_α instead of $z_{\alpha/2}$. Furthermore, the alternative testing procedure using the z-statistic applies to differences between proportions for large samples.

EXHIBIT 2

AN EXAMPLE OF A TEST OF A HYPOTHESIS ABOUT THE DIFFERENCE BETWEEN PROPORTIONS FOR LARGE SAMPLES

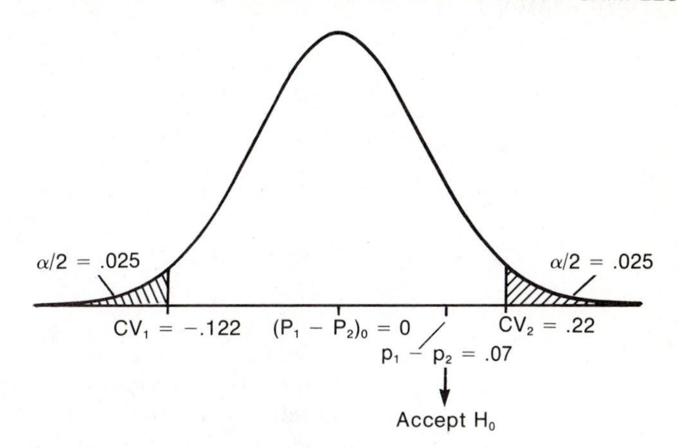

Small Samples. In this section of the discussion we shall see how the binomial distribution is used directly to test hypotheses about proportions. The binomial is the exact sampling distribution to use for all sample sizes when sampling with replacement, but it is of real benefit when the sample size is *small*. In principle, all of the previous ideas about a test of a hypothesis apply; however, modifications are necessary to account for the fact that the binomial is discrete and usually is not symmetric for small samples.

The simplest way to perform a test of a hypothesis involving the binomial is in terms of a probability or "p-value" associated with the number of times, x, that an event occurs in a random sample of n elements. x is used instead of the sample proportion, p, since the formula for the binomial distribution usually is expressed in terms of x. Furthermore, tables of the binomial such as Table 1 are presented in terms of x.

In order to understand the small sample case, consider the problem of determining whether a shipment of items is of an acceptable quality based upon sample information. Suppose the shipment is acceptable if no more than 25 percent of the items are defective. The problem can be expressed in terms of hypotheses about the shipment proportion defective, P, as

$H_0: P \leq 0.25$ (Shipment is acceptable)
$H_1: P > 0.25$ (Shipment is not acceptable)

By examining the alternate hypothesis we can see that it corresponds to a one-tailed test where the region of rejection lies in the "upper" or right tail. In other words, a "very large" number of defectives in a sample may indicate that the null hypothesis should be rejected.

Suppose we select a sample of seven items from the shipment (i.e., $n = 7$) and find four that are defective (i.e., $x = 4$). If the probability of observing this many defectives or more given that P equals 0.25 is "high," we can conclude that such a sample result is likely to occur under the null hypothesis. Consequently, there would not be sufficient evidence to reject H_0. On the other hand, if the probability of observing 4

EXHIBIT 3
BINOMIAL PROBABILITY DISTRIBUTION FOR n = 7 AND P = 0.25 USED TO FIND P-VALUE IN SMALL SAMPLE EXAMPLE

x	f(x)
0	.1335
1	.3115
2	.3115
3	.1730
4	.0577
5	.0115
6	.0013
7	.0001

or more defectives in a sample taken from a shipment with 25 percent defectives is "small," we can conclude that the sample result is unlikely to occur under the null hypothesis. Such evidence would lead to the rejection of H_0.

For this particular example, our problem is to determine the probability of observing "4 or more" defectives in a sample of 7 items drawn from a shipment with the hypothesized proportion defective of 0.25. This probability is referred to as the *p-value,* which is compared with a chosen level of significance, α. If the p-value is greater than α, we accept the null hypothesis and if it is less we reject the null hypothesis. Assume the level of significance is 5 percent (i.e., $\alpha = 0.05$).

Reproduced in Exhibit 3 is the entire probability distribution of x for n = 7 and P = 0.25. The probabilities are taken from the table of Binomial Probabilities, Table 1, in the Appendix. The probability that the number of defectives, or x, equals "4 or more" is obtained by adding the last four probabilities in the table. That is,

$$\begin{aligned} \text{p-value} &= P(x \geq 4 | n = 7, P = 0.25) \\ &= .0577 + .0115 + .0013 + .0001 \\ &= 0.0706 \end{aligned}$$

Since this p-value is greater than the level of significance, $\alpha = 0.05$, we *accept* the null hypothesis and *conclude* the shipment is acceptable, or that the proportion defective does not exceed 0.25.

In general, when performing a test such as this we find the probability of obtaining a discrepancy equal to or greater than the one observed conditional upon the hypothesized value. If the probability found is greater than α we accept, and if less we reject, the null hypothesis. Hence, if there were a "less-than" inequality in the alternate hypothesis, we would find the probability associated with an observed x-value or less and compare. For a two-tailed test, we must find the probability of at least *or* at most the observed value of x, depending upon which side of the average number the x-value falls. The average number equals nP, where P assumes the value specified in the null hypothesis. The computed probability then is compared with α *divided* by 2, or half the level of significance.

QUESTIONS AND PROBLEMS

DEFINITIONAL QUESTIONS

1. What is meant by the term *population proportion,* or *percentage?* Cite three problems where such a measure might be useful.

2. Define the following terms.
   (a) Sample Proportion
   (b) Distribution of the Sample Proportion
   (c) Mean of the Sample Proportion
   (d) Standard Error of the Sample Proportion
   (e) Estimated Standard Error of the Sample Proportion

3. Under what circumstances is the sample proportion rather than the population proportion used?

4. In terms of proportions, define the following terms.
   (a) Estimation
   (b) Point Estimate
   (c) Interval Estimate
   (d) Hypothesis
   (e) Null Hypothesis
   (f) Alternate Hypothesis
   (g) Hypothesized Value
   (h) Test of a Hypothesis

5. Define the following terms.
   (a) Difference Between Population Proportions
   (b) Difference Between Sample Proportions
   (c) Distribution of the Difference Between Sample Proportions
   (d) Mean of the Difference Between Sample Proportions
   (e) Standard Error of the Difference Between Sample Proportions
   (f) Estimated Standard Error of the Difference Between Sample Proportions
   (g) Test of a Hypothesis of the Difference Between Proportions

6. Under what circumstances can the normal distribution be used when drawing inferences about proportions?

7. Under what circumstances is the binomial distribution used in this chapter to draw inferences about proportions?

8. What test statistics can be used to test hypotheses about proportions? Explain how they are used in this chapter.

9. How can the "z-statistic" be used to test hypotheses about proportions?

10. In words, describe the procedure for constructing a confidence interval estimate of a universe proportion based upon large samples.

11. In words, describe the procedure for testing hypotheses about proportions.

12. Why is it important to consider the size of the sample when estimating a proportion?

COMPUTATIONAL PROBLEMS

13. A random sample of 100 items is drawn from a population in order to estimate the proportion, P. Construct a 95 percent confidence interval estimate based upon the sample result, x = 45.

14. Test the null hypothesis H_0:P ≤ 0.35 against the alternate H_1:P > 0.35 based upon the following information?

$$n = 80$$
$$x = 34$$
$$\alpha = 0.05$$

15. Test the null hypothesis H_0:P = 0.5 against the alternate H_1:P ≠ 0.5 on the basis of the following information

$$n = 120$$
$$p = 0.44$$
$$\alpha = 0.02$$

16. As part of a study of the effectiveness of a firm's maintenance service, the firm is interested in determining the proportion of small pieces of equipment awaiting repair. A random sample of records associated with 150 pieces of equipment uncovered 12 that are scheduled to be repaired. Based on the information given, construct a 98 percent confidence interval estimate of the proportion of all pieces of equipment waiting to be repaired.

17. A consumer group opposing telephone charges for directory assistance challenges the phone company's claim that the proportion of calls made for directory assistance that can be obtained from the phone directory is in excess of 0.6. In order to test the claim a random sample of 200 calls for directory assistance is selected. It is determined that 128 of these calls could have been obtained from available phone directories. Using a 5 percent level of significance, determine whether the above claim is valid.

18. A firm had established that in the past 25 percent of all orders for merchandise came from new customers. A random sample of 100 current orders indicates that 14 are from new customers. Has there been a change in the percentage of orders placed by new customers between the two periods? Justify statistically.

19. A particular assembly-line operation operates effectively as

long as the proportion of items that are defective is *less* than 0.1. A periodic check on the operation based upon a sample of 75 items indicates that 6 are defective. Using a 2 percent level of significance, determine whether the assembly line is operating effectively.

20. Calculate the maximum sample size that is necessary for a sample proportion used as an estimate to fall within 0.1 of the universe proportion with a probability of 0.98. Repeat the calculation under the assumption that P equals 0.2.

21. A petition for the recall of the mayor of a certain city must be validated before the recall question is considered. In other words, it is necessary to determine the percentage of signatures on the petition that actually represents registered voters. Determine the number of signatures that must be investigated in order to estimate the proportion of registered voters signing the recall petition within 0.05 of the true value with a probability of 0.99.

22. Using the information in the table below, test the hypothesis $H_0: P_1 = P_2$ against the alternate $H_1: P_1 \neq P_2$ at a 5 percent level of significance.

$n_1 = 100$	$n_2 = 150$
$x_1 = 20$	$x_2 = 35$

23. Periodically, a seaside resort conducts a survey by mail questionnaire in order to establish current attitudes about the resort's image. Although mail questionnaires represent a relatively inexpensive way of obtaining information, they are not as effective as other methods since there is a high degree of non-response using mail questionnaires.

    In order to determine whether the response rate can be improved, the resort decides to mail questionnaires to two separately drawn random samples, each 250 in size, where a small gift is attached to the questionnaire sent to individuals in one of the samples. After conducting the survey, it was found that 85 responses resulted from the sample with no gift and 110 responses from the sample where the small gift was sent along with the questionnaire. Based upon the evidence, can one conclude that a different response results by giving a small gift?

CONCEPTUAL QUESTIONS AND PROBLEMS

24. Why is it important to consider problems in terms of proportions or percentages?

25. When is it necessary to perform a test of a hypothesis about a proportion?

26. Why is it important to consider the distribution of the sample proportion when making inferences about universe proportions?

27. What is measured by the standard error of the sample proportion, σ_p? Does s_p measure the same thing? Explain.

28. Criticize or explain the following statements. In general, the standard error of a sample statistic is a function of the size of the sample. This is true for the sample mean and the sample proportion. There is a major difference between the standard errors of these measures, since the standard error of the sample mean is independent of the universe mean whereas this is not the case for the sample and universe proportion.

29. What is the relationship between the material presented in this chapter and the material in Chapter 7 regarding the binomial distribution?

30. An examination of the records of a state facility for car inspection indicates that 58 of the last 125 cars that were inspected passed inspection the first time they were brought to the facility.

    (a) Construct a 95 percent confidence interval estimate of the percentage of all cars inspected by the facility that pass the first time.
    (b) Interpret the estimate you found in part (a) specifically in terms of the problem.
    (c) If the sample size used to construct the estimate in part (a) equaled 250, what effect would this have on the estimate? Would the interpretation in part (b) be different? Explain.

31. What are the similarities and differences between tests of hypotheses for proportions presented in this chapter and tests for means presented in Chapter 12?

32. Suppose you suspect that a particular coin is biased and you want to determine whether the suspicion is true.

    (a) Should the problem be treated as one in estimation or hypothesis testing? Explain.
    (b) As a problem in hypothesis testing, how would you determine whether a coin is biased or not?
    (c) What is the relationship between this problem and the problems presented in the discussion section of Chapter 7 on the binomial distribution?
    (d) Suppose you have a coin with an unknown probability of heads and you want to estimate this probability. How many times must you flip the coin so that your estimate is within 0.01 of the true probability of a head with 0.99 confidence? How is this question related to the material presented in Chapter 4 on Basic Probability.

CHAPTER 14

THE ANALYSIS OF VARIANCE

Experimental observations are only experience carefully planned in advance, and designed to form a secure basis of new knowledge.

R. A. Fisher

Hypothesis testing was introduced in Chapter 12 in terms of the arithmetic mean. Emphasis was placed upon tests to determine whether a mean from a single sample is drawn from a universe with a mean assuming some specified or hypothesized value. In the discussion section of the same chapter, the concept was extended to a test to determine if two universe means are equal based upon two independent random samples. There are many problems where it is necessary to determine whether the means of *more* than two populations are equal. For example, six different promotional devices may be compared to determine if they are equally effective with respect to sales. A distributor of appliances may want to know whether the quality of batteries produced by four manufacturers is the same based upon the mean life of a sample from each manufacturer. It may be of interest to a school administrator to determine whether five distinct methods of teaching are equally effective, where average grade performance is used as a measure of effectiveness. In cases such as these, the methods of Chapter 12 are not appropriate.

In this chapter we focus on a method for testing whether the means of any number of populations are equal. The method is based upon an unusual result in statistics that the *equality of several universe means can be tested by comparing sample variances.* The general term for methods employing variances to test means is the **analysis of variance.** Here, we shall examine the procedure for performing an analysis of variance in a very simple case. Extensions of this procedure have far reaching implications for experimental work involving sample data and will be discussed briefly.

Before presenting the detailed procedure for performing an analysis of variance, let us explore the concept with the following example. When a worker is considered for internal promotion to an administrative position in a particular firm, he or she must take a standard qualifying examination that is the same regardless of the applicant's specialty or department within the firm. The examination has been challenged on the grounds that certain groups of workers are unable to do as well as others, owing to the nature of the examination. In order to determine whether there is any merit to the claim, it is decided to randomly select samples within different worker groups and administer the examination to the chosen workers. If it is determined that the mean examination score for the groups is the same, the conclusion is that all worker groups do equally well on the examination.

In order to keep the example simple, assume samples of 5 workers are selected from 3 groups (e.g., machine operators, expediters, and dispatchers), and the same standard examination is given to those selected. Examination scores can vary on a scale between 0 and 100. Exhibit 1 presents results for the three groups designated as (1), (2), and (3). The scores for each group are provided in separate columns. Below each column is the arithmetic mean examination score for the three groups.

EXHIBIT 1

EXAMPLE OF DATA USED TO ILLUSTRATE THE PROBLEM OF ANALYSIS OF VARIANCE

	Examination Scores		
Worker		*Group*	
	(1)	*(2)*	*(3)*
1	73	95	79
2	83	89	82
3	74	93	81
4	80	86	74
5	77	94	84
Mean Score	77.4	91.4	80.0

From the exhibit we can see there are differences in the scores *within* each column or group. This variability can be expected to occur by chance since it reflects the natural differences that occur within each of the universes from which the samples were drawn. By examining the arithmetic means at the bottom of the table, we can see that another kind of variability is present. Since these means are different, there is variability *among* the group scores.

The *problem* we face is to determine whether the differences among group means of the three samples also is due to chance, or whether the differences result because the mean scores for the three groups of workers from which the samples were drawn actually are different. Hence, we are faced with the problem of testing the hypothesis that the means of the three groups of workers are the same. Rejection of this hypothesis leads to the conclusion that at least one of the universe means is different from the others.

Testing the above hypothesis using the analysis of variance is based upon the *assumption* that the variability among the sample means of the groups is the same as the variability within each group *if* the population means are identical. In other words, if the means of the populations are the same, the difference between the two types of variability would result only from chance. On the other hand, if the population means are not the same, the variability among the groups would differ from the variability within the groups by more than what can be attributed to chance. The test for means using the analysis of variance is based upon a comparison of measures of the two types of variability.

The measure of variability used in the analysis of variance is called a *mean square.* A mean square is similar to a variance since it represents the sum of squared deviations about a mean divided by the number of degrees of freedom. Recall, when performing the t-test for a single mean, the estimated variance is defined as the sum of the squared deviations from the mean divided by the "sample size minus one." The divisor in that case was referred to as the degrees of freedom, which identified a particular t-distribution that is used to solve a problem based upon a given sample size. In general, we can think of degrees of freedom as it is used in this chapter as a quantity that is related to the number of terms in the sum of the squared deviations from the mean.

The mean squares appearing in this chapter are similar, in a sense, to the variance cited above except they relate to different types of variability and are based upon different degrees of freedom. One mean square is used to measure the variation *within* the samples. This is based upon the sum of the squared deviations of the elements within each group or column computed about the individual group means. The *within group sum of squares* divided by the appropriate number of degrees of freedom provides the *within group mean square* and represents a measure of variability due to chance, or experimental error.

The other mean square is based upon the sum of the squared deviations of the individual sample or group means computed about the grand mean of *all* the observations considered as one sample. This sum of squared deviations is referred to as the *between group sum of squares.* When this is divided by the appropriate number of degrees of freedom, the result is the *between group mean square.* This measures the group effect or differences existing among the various groups. If the means of the populations are equal, there is no group effect, and the between group mean square also will represent variability due to chance. Consequently, when the group means in the universe are equal, the within group mean square and the between group mean square should not be much different and their ratio should be close to one. Unusually large ratios would indicate that the group means are not equal.

The procedure section that follows provides a method for comparing the two mean squares in terms of their ratio in order to determine if there is a group effect or whether differences observed among the group means are the result of chance. The procedure applies to any number of groups, but is restricted here to samples that are of the same size. Unequal samples are considered briefly in the discussion.

The symbols used in this chapter are summarized as follows:

k = the number of independent samples or groups compared
= the number of separate columns of data

n = the sample size, which is the same for all groups

x_j = an observation in the j-th column, or of the j-th sample or group

μ_j = the universe mean corresponding to the j-th group
T = a column total, or the sum of the observations in any column
G = the grand total, or the sum of *all* observations in *all* columns
GS = the grand total of the squares, or the sum of the squares of *all* sample observations in *all* columns
SV = source of variation
SS = sum of squares, or the sum of the squared *deviations* about a mean
SS_{bg} = between group sum of squares
SS_{wg} = within group sum of squares
DF = degrees of freedom
DF_{bg} = between group degrees of freedom
DF_{wg} = within group degrees of freedom
MS = mean square, or a sum of squares divided by a corresponding number of degrees of freedom
MS_{bg} = between group mean square
MS_{wg} = within group mean square
F = F-statistic, or a sample statistic distributed according to an F-distribution
α = the level of significance, or the probability of committing a Type I error
F_α = the value of F tabulated in Table 6 of the Appendix corresponding to a level of significance α
H_0 = the null hypothesis, or the hypothesis tested

Note. Most of the symbols in this chapter conventionally are used in analysis of variance problems. An effort has been made to eliminate more complicated symbols involving double subscripts that typically are used in the formulas. The symbol x_j appears in the above list, but is not used in the formulas that follow. It is used exclusively as a column heading to indicate the group to which the observations in a particular column belong. For example, the symbol x_2(i.e., j = 2) represents the second column of observations. Since the symbol T is used instead of Σx, the total of the observations in the second column would be labeled as T_2. The usual notation for the sum of squared observations is used; hence, the sum of the squares in the second column would be labeled as Σx_2^2. It should be noted that the term "sum of squares" (SS) refers to the sum of squared *deviations* from a mean, *not* the sum of squares of individual observations.

PROCEDURE

The procedure presented in this section can be used to test whether the arithmetic means of any number of groups are equal, based upon independently drawn samples of the same size from each of the groups. The null hypothesis can be written generally as

$$H_0 : \mu_1 = \mu_2 = \ldots = \mu_k$$

The null hypothesis underlying the analysis of variance.

where k is the number of groups considered. The null hypothesis states that the means of groups 1 through k are equal. The alternative hypothesis is that at least one of the means is not equal to the others. In other words, the null hypothesis can be false in many ways since any combination of the k means can be unequal for the null hypothesis to be false.

The null hypothesis is tested by performing the following steps:

Step 1. In order to simplify the work, perform the following *preliminary computations:*

(a) Calculate the totals of *each* column, Σx, and label these as T.
(b) Add the column totals, ΣT, which results in the grand total of all the observations, G.
(c) Calculate the square of *all* observations and compute the grand total of these squared observations, GS. A convenient way to do this is to find the sum of the squares for each column, Σx^2, and add these to find GS.

Step 2. Based upon the preliminary calculations, find the between group and within group sum of squares using the formulas

$$SS_{bg} = \frac{\Sigma T^2}{n} - \frac{G^2}{nk}$$

Between group sum of squares.

$$SS_{wg} = GS - \frac{\Sigma T^2}{n}$$

Within group sum of squares.

where the expression ΣT^2 represents the sum of each of the column totals squared, G^2 is the square of the grand total of all the observations, and n is the sample size common to all groups.

Step 3. Calculate the between group and within group degrees of freedom using the formulas

$$DF_{bg} = k - 1$$

Between group degrees of freedom.

$$DF_{wg} = nk - k$$

Within group degrees of freedom.

The degrees of freedom between groups is found by subtracting one from the number of columns, k. The degrees of freedom within groups is obtained by subtracting the number of columns, k, from the total number of observations. The total number of observations equals the common sample size of each group multiplied by the number of columns.

Step 4. Based upon Step 2 and Step 3, calculate the between group and within group mean square using the formulas

$$MS_{bg} = \frac{SS_{bg}}{DF_{bg}}$$

Between group mean square.

$$MS_{wg} = \frac{SS_{wg}}{DF_{wg}}$$

Within group mean square.

Each of the mean squares is obtained by dividing the sum of squares by the corresponding number of degrees of freedom.

Step 5. Find the value of the test statistic, F, which is the ratio of the above mean squares:

$$F = \frac{MS_{bg}}{MS_{wg}}$$

F-statistic for the analysis of variance for testing the equality of means.

This statistic, called the F-statistic or F-ratio, is obtained by dividing the between group mean square by the within group sum of squares.

Note. The ratio of mean squares, or F-ratio, given in Step 5 has a known sampling distribution that can be used to perform the test of the hypothesis that the group means are equal. Assuming this is the case defined under the null hypothesis, the ratio of mean squares based upon k independent random samples of size n has a sampling distribution called the F-distribution. This is a continuous probability distribution that is defined for all non-negative numbers and is skewed to the right.

The F-distribution that applies to a particular analysis of a variance problem depends upon the degrees of freedom corresponding to the between group sum of squares *and* to the within group sum of squares. Since the between group mean square always appears in the numerator of the F-ratio, the corresponding degrees of freedom, DF_{bg}, is referred to as the *numerator degrees of freedom.* The degrees of freedom corresponding to the within group mean square, DF_{wg}, is referred to as the *denominator degrees of freedom.* The F-distribution that is tabulated in the Appendix as **Table 6** is based upon the numerator and denominator degrees of freedom directly.

Step 6. Based upon a chosen level of significance, α, and the degrees of freedom calculated in Step 3, DF_{bg} and DF_{wg}, find the critical value from Table 6 in the Appendix. The critical value is given as

$$CV = F_{\alpha}$$

Critical value for the analysis of variance for means.

where F_{α} represents the value of F corresponding to a level of significance α.

> *Note.* Since the F-distribution is identified by the two degrees of freedom, separate tables are needed for different levels of significance. Table 6 is given in two parts, (a) and (b), which correspond to a 0.05 and 0.01 level of significance that can be used in our tests of the hypothesis in the analysis of variance.
>
> Values of F provided in Table 6 correspond to the two probabilities in the right tail of the distribution. Since the test is one-tailed, the value of F obtained from the table is the critical value F_{α}, which equals $F_{.01}$ or $F_{.05}$.

Step 7. In order to perform the test, apply the following decision rule:

$$\text{Accept } H_0 \text{ when } F < CV$$
$$\text{Reject } H_0 \text{ when } F \geq CV$$

Decision rule for the analysis of variance for means.

Therefore, we *accept* the null hypothesis if the computed value of F is less than the critical value equal to F_{α} and conclude that the means are equal. If the computed value of F is greater than or equal to the critical value, we *reject* the null hypothesis and conclude that all of the means are not equal.

> *Note.* The formulas for finding the F-ratio are summarized in the form of an analysis of variance table shown in Exhibit 2. After finding the sum of squares and the degrees of freedom by the formulas given, the mean squares are found by dividing the degrees of freedom into the sum of squares. The value of F is obtained by dividing the first mean square by the second. Frequently, the results of these computations are reported in the form of a summary table similar to the one shown in the exhibit. A summary of the F-test based upon these computations is summarized in Exhibit 3.

EXHIBIT 2
ANALYSIS OF VARIANCE TABLE

Source of Variation(SV)	*Sum of Squares(SS)*	*Degrees of Freedom(DF)*	*Mean Square(MS)*	*F-ratio*
Between Groups(bg)	$\frac{\Sigma T^2}{n} - \frac{G^2}{nk}$	$k - 1$	SS_{bg}/DF_{bg}	MS_{bg}/MS_{wg}
Within Groups(wg)	$GS - \frac{\Sigma T^2}{n}$	$nk - k$	SS_{wg}/DF_{wg}	

EXHIBIT 3

SUMMARY OF THE F-TEST FOR THE ANALYSIS OF VARIANCE

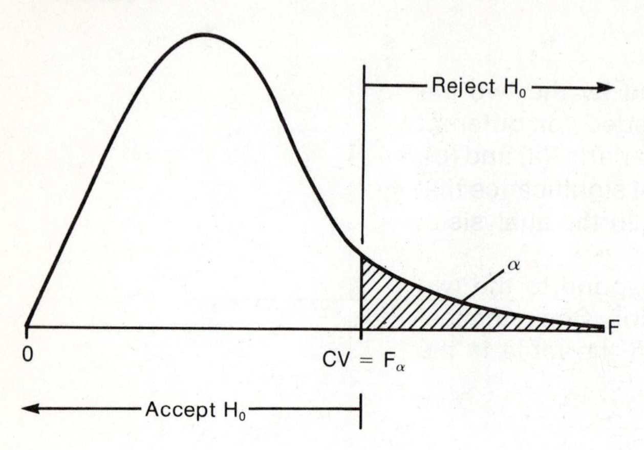

NUMERICAL EXAMPLE

In order to illustrate the procedure described in the last section, let us use the data presented in Exhibit 1. Recall, the basic problem is to determine whether or not the mean examination scores for the three groups of workers are the same, based upon random samples of five workers (n = 5) from the three groups (k = 3). The null hypothesis states that the mean scores for the worker groups are equal.

Step 1. Preliminary Calculations

CALCULATION OF TOTALS OF OBSERVATIONS AND SQUARES BY COLUMNS

Original Data and Totals Group (1) x_1	Group (2) x_2	Group (3) x_3	*Squares of Observations and Totals by Column* x_1^2	x_2^2	x_3^2
73	95	79	5329	9025	6241
83	89	82	6889	7921	6724
74	93	81	5476	8649	6561
80	86	74	6400	7396	5476
77	94	84	5929	8836	7056
387 (T_1)	457 (T_2)	400 (T_3)	30023 (Σx_1^2)	41827 (Σx_2^2)	32058 (Σx_3^2)

CALCULATION OF GRAND TOTAL OF OBSERVATIONS AND GRAND TOTAL OF THE SQUARES

Column Totals of Observations		*Column Totals of Squares*	
T_1	387	Σx_1^2	30023
T_2	457	Σx_2^2	41827
T_3	400	Σx_3^2	32058
G =	1244	GS =	103908

Step 2. Finding the Sum of Squares Based Upon the Preliminary Calculations

$$SS_{bg} = \frac{\Sigma T^2}{n} - \frac{G^2}{nk}$$

$$= \frac{387^2 + 457^2 + 400^2}{5} - \frac{1244^2}{5 \times 3}$$

$$= \frac{149769 + 208849 + 160000}{5} - \frac{1547536}{15}$$

$$= \frac{518618}{5} - \frac{1547536}{15}$$

$$= 103723.6 - 103169.1$$

$$= 554.5$$

$$SS_{wg} = GS - \frac{\Sigma T^2}{n}$$

$$= 103908 - 103723.6$$

$$= 184.4$$

> *Note.* Since the term $\Sigma T^2/n$ equal to 103723.6 appears in both formulas, the entire calculation is not repeated to find SS_{wg}. The result from the first calculation is used in the second. Also, if there are a large number of groups, it may be easier to square each of the T's, or totals of the observations, and find the sum, ΣT^2, in table form before substituting into the formula.

Step 3. Calculating the Degrees of Freedom

$$DF_{bg} = k - 1 = 3 - 1 = 2$$

$$DF_{wg} = nk - k = 5 \times 3 - 3 = 15 - 3 = 12$$

Step 4. Calculating the Mean Squares

$$MS_{bg} = \frac{SS_{bg}}{DF_{bg}} = \frac{554.5}{2} = 277.25$$

$$MS_{wg} = \frac{SS_{wg}}{DF_{wg}}$$
$$= \frac{184.4}{12}$$
$$= 15.37$$

Step 5. Finding the F-ratio

$$F = \frac{Ms_{bg}}{MS_{wg}}$$
$$= \frac{277.25}{15.37}$$
$$= 18.04$$

The results of Step 2 through Step 5 are summarized in the following table.

Analysis of Variance Table for Exam Score Problem

SV	SS	DF	MS	F
Between Group	554.5	2	277.25	18.038
Within Group	184.4	12	15.37	

Step 6. Finding the Critical Value, F_α.

Based upon a 5 percent level of significance, $\alpha = 0.05$, numerator degrees of freedom equal to 2, $DF_{bg} = 2$, and denominator degrees of freedom equal to 12, $DF_{wg} = 12$, the value of $F_{.05}$ from Table 6a in the Appendix is 3.89. Hence,

$$CV = F_\alpha$$
$$= F_{.05}$$
$$= 3.89$$

Step 7. Performing the Test

In order to perform the test, we compare the computed value of the F-ratio with the critical value of $F_{.05}$ according to the following decision rule:

$$\text{Accept } H_0 \text{ if } F < 3.89$$
$$\text{Reject } H_0 \text{ if } F \geq 3.89$$

Since $F = 18.04$ is greater than $F_{.05} = 3.89$, we *reject* the null hypothesis and *conclude* there is a difference in the mean examination scores among the three worker groups.

The results of the entire test illustrated in this example are summarized in the following diagram:

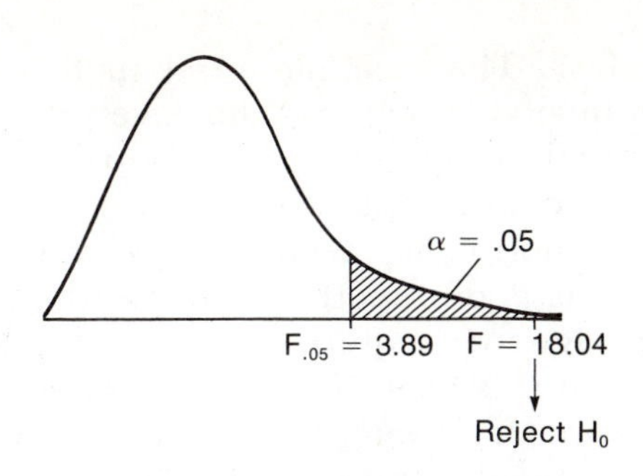

DISCUSSION

The procedure just illustrated applies to problems for determining whether any number of samples of the same size are drawn from populations having the same mean. Under the null hypothesis of equal means, the ratio of mean squares has sampling properties described by an F-distribution. Each of the mean squares appearing in the F-ratio is a variance. The reason the formulas do not look like variances we have seen earlier is that they represent *shortcuts* that are similar to the shortcut formula for the variance presented at the end of Chapter 3. The shortcut formulas are much easier to use to calculate mean squares; however, it is not as easy to understand what kind of variability is measured when examining these formulas.

The formulas for mean squares used in the procedure section actually are shortcuts for variances describing different types of variability.

The within group mean square appearing in the denominator of the F-ratio is a measure of the chance variability that is expected to occur naturally among observations. The between group mean square in the numerator of F is a measure of the variation among groups and is comparable to the variability occurring by chance when there are no differences among the actual group means. Hence, when the null hypothesis is true, the F-ratio should be close to 1, and departures from 1 occur owing to sampling variability only. The F-test is used to determine how much of a difference between the observed ratio and 1 is acceptable before we can conclude that the group means are different.

Assumptions. Any analysis of variance is based upon certain assumptions. Like all hypothesis testing problems, it is assumed that sampling is done at *random*. Hence, the elements assigned to each of the groups compared must constitute a random sample. Furthermore, the method presented in this chapter is based upon the assumption that each of the samples is *independently* drawn. The method can be modified when this assumption does not hold. A common problem where the independence assumption does not hold occurs when the same subjects are observed at different points in time under varying circumstances.

The procedure used in this chapter assumes that independently drawn random samples are selected from normal universes with the same variance.

The final assumption underlying the analysis of variance is that each of the samples are drawn from groups or populations that are normally distributed with the same variance. It has been shown that the F-test used here is not sensitive to departures from normality but is sensitive to differences among universe variances. Moreover, modifications to the technique can be applied to data that do not conform to all of the assumptions, so the procedure is applicable to a wide variety of problems.

Unequal Samples. The example used in this chapter is based upon a comparison involving 3 groups; however, the procedure is general and can be applied to any number of groups. The procedure given here is restricted to equal sample sizes corresponding to each of the groups. A minor variation of the technique is available when sample sizes are unequal. This is of importance for two basic reasons. On the one hand, a study design may require equal samples, but some data may be missing at the time it is analyzed. On the other hand, data may be acquired from a secondary source so that no control over the study design exists. Consequently, the data that is acquired may be based upon unequal samples.

Note. When sample sizes are unequal, the term $\Sigma T^2/n$ in each sum of squares is affected. Each of the T^2 values must be divided by the corresponding sample size before summing. Hence, the formulas for the sums of squares become

$$SS_{bg} = \sum\left(\frac{T_j^2}{n_j}\right) - \frac{G^2}{N}$$

$$SS_{wg} = \Sigma GS - \sum\left(\frac{T_j^2}{n_j}\right)$$

Sums of squares for unequal sample sizes.

where T_j and n_j represent the sum of the observations and the sample size of the j-th sample. The symbol N represents the total number of observations in all samples combined. The within group degrees of freedom, therefore, equals $N - k$ instead of $nk - k$. The mean squares and the F-ratio are found according to the formulas in the procedure section.

Factorial Experiments. In this chapter we focussed on testing differences among means when more than two groups are involved. The groups, however, were associated with one particular characteristic. For example, in the illustration used earlier we were concerned with worker classifications according to type of job. This is called a one-factor experiment. The concept of analysis of variance can be extended to problems with more than one characteristic or factor. In the same example, it might be of interest to determine whether workers in different job categories score differently on different examinations. Here, two factors are involved, worker category and type of examination. Another example of a two-factor problem would be to determine whether product quality is the same for various manufacturers and at different points of delivery.

In many cases, already collected data are broken down into the effects of different characteristics or factors using the analysis of variance. In other cases, a study is designed in advance in order to determine if different factors have an effect upon some variable of interest. In both cases, the effects of these factors are called *main effects*. These are analogous to the between group effect used in this chapter. When more than one factor is considered, a separate mean square is associated with each effect. In experimental work where different techniques or methods are tested, the main effects are referred to as *treatment effects*. Using an

F-test, these effects are compared with a within group mean square that also is referred to as a measure of *experimental error*.

In addition to the main effects, an additional effect can be isolated when more than one factor is considered. This effect is called an *interaction*. An interaction occurs when the combined effect of two or more factors is different than the sum of their separate effects. For example, suppose we want to determine whether different counter displays of a certain item produce different sales. The counter displays are tried in different store locations. If the combined sales based upon given combinations of counter displays and locations is different from the total sales associated with individual locations and counter displays, an interaction occurs between the two main effects, location and type of display.

The procedure presented in this chapter applies to the simplest analysis of variance problem. Extensions and modifications exist that are applicable to a variety of experimental problems.

The technique of analysis of variance can be modified to isolate a measure of variability or a mean square associated with the interaction of two or more factors. Further, interactions can be tested with an F-statistic similar to tests of main effects. If the interactions are significant, however, a problem arises with respect to the interpretation of main effects, since each main effect is dependent upon the particular levels of the treatments used to perform the tests. Consequently, conclusions about main effects cannot be generalized to combinations of levels not considered in the testing procedure.

The technique of analysis of variance can be applied to a variety of experimental problems that are very complex, where different forms of sampling and assignment of test items or subjects are employed. Associated with each type of problem or design is a specific set of formulas that are needed to measure the various effects. Because of the importance of the concept of analysis of variance in experimental work, many of the specific methods can be found in literature devoted to experimental design. It is not the purpose of this chapter to consider alternative designs and tests. The purpose of this chapter is to introduce the concept of the analysis of variance and its role in decision-making with the simplest case.

CALCULATORS AND COMPUTERS

The simplest problems in the analysis of variance involve more computational work than the basic methods presented earlier. This is obvious from the example in this chapter that involves only three groups and equal samples of five elements per group. As the number of groups or the sample size increases, the amount of computation required becomes much greater. This is especially true for more complicated designs involving unequal samples, dependent observations, and an increased number of factors.

Computer programs are available to solve many types of problems using the analysis of variance. These are of great help in cases where designs are more complicated than the one illustrated in this chapter. Output of a computer program is similar to the analysis of variance table presented earlier. One- and two-factor analysis of variance problems now can be solved easily with programs incorporated in libraries accompanying hand-held programable calculators. Within the next few years, calculators of this kind will be able to solve many different problems involving more complicated designs.

QUESTIONS AND PROBLEMS

DEFINITIONAL QUESTIONS

1. What is meant by the term *analysis of variance?* What kinds of problems are solved using the analysis of variance? Explain.

2. What is the difference between a "group mean" and the "grand mean"? In your answer, distinguish between sample and universe values.

3. Define the following terms.

   (a) Sum of Squares
   (b) Degrees of Freedom
   (c) Mean Square

4. What is the meaning of the following terms?

   (a) Within Group Sum of Squares
   (b) Within Group Degrees of Freedom
   (c) Denominator Degrees of Freedom
   (d) Within Group Mean Square
   (e) Between Group Sum of Squares
   (f) Between Group Degrees of Freedom
   (g) Numerator Degrees of Freedom
   (h) Between Group Mean Square

5. What is the null hypothesis underlying the analysis of variance?

6. What is meant by the critical value used in the analysis of variance? How is it found?

7. What is the F-statistic or F-ratio? How is it used in the analysis of variance?

8. What is the F-distribution? What characteristics uniquely identify a particular F-distribution?

9. What is an analysis of variance table?

10. Describe the procedure for performing a test of a hypothesis in the analysis of variance. What is the basic assumption underlying this test?

COMPUTATIONAL PROBLEMS

11. Given below are observations based upon samples of 5 items drawn randomly from three groups.

Observation Number	*Group* A	B	C
1	32	25	24
2	30	22	29
3	24	30	27
4	29	27	25
5	31	26	30

Perform an analysis of variance in order to determine whether or not the group means are the same.

12. Three brands of radial tires are tested in order to determine whether they are of equal quality. Based upon a sample of 9 tires of each brand the following results are obtained:

LENGTH OF TIRE LIFE (THOUSAND MILES)

Brand 1	2	3
35.9	32.8	33.0
35.4	34.5	30.9
31.4	29.5	29.0
33.3	28.9	32.7
31.6	32.2	31.6
34.5	29.8	29.6
36.8	30.9	31.3
36.1	27.0	31.5
33.1	31.5	35.5

Determine whether or not the three brands of tires are of equal quality in terms of mean length of life.

13. An independent consumer group is interested in evaluating 4 brands of scientific calculators. All of the calculators perform the same functions, but each operates differently owing to differences in design. Consequently, the time it takes to perform a given task may vary among the four calculators.

    In order to determine whether there is a difference in performance time, 24 people in groups of 6 are randomly assigned to the four brands of calculators. Each person performs an assigned task and the performance time is recorded. The results appear in the table below.

TIME TO PERFORM ASSIGNED TASK (MINUTES)

Brand 1	Brand 2	Brand 3	Brand 4
3.2	1.5	2.9	3.4
3.0	1.9	3.2	2.2
1.5	2.6	2.7	2.0
2.2	1.3	2.4	3.2
1.6	1.8	2.3	2.3
2.7	2.0	3.5	2.7

Based upon the data provided in the table, determine whether there is a difference in the mean time to perform the assigned task among the different brands.

CONCEPTUAL QUESTIONS AND PROBLEMS

14. In your own words, explain how it is possible to perform a test regarding the equality of means in terms of variances.

15. What kind of variability is measured by the within group mean square? By the between group mean square? Under what circumstances do both measure the same thing? Explain.

16. In your own words, explain why the within group mean square is a measure of chance variability.

17. How are "group means" and the "grand mean" related to analysis of variance problems?

18. What are the similarities and differences between the test of a hypothesis presented in this chapter and the ones for means presented in Chapter 12?

19. Refer to Problem 13 and *assume* the result of the F-test is significant.

    (a) What overall conclusion can be made about the performance time associated with the brands of calculators?
    (b) Compute the arithmetic mean performance time for each of the samples associated with the individual brands.
    (c) Based upon the results of part (b), can you conclude that there is a difference in performance time between Brand 1 and Brand 2? Between Brand 2 and Brand 3? Explain.
    (d) On the basis of your answers to part (c), what general conclusion can you draw about significant results of tests associated with the analysis of variance?
    (e) Assuming the differences observed among the arithmetic means computed in part (b) represent "real" differences, can we automatically conclude that these differences are of practical importance? Explain.

20. In general terms, what is meant by the following terms?

    (a) One-Factor Experiment
    (b) Two-Factor Experiment
    (c) Many-Factor Experiment
    (d) Main Effect or Treatment Effect
    (e) Interaction Effect

21. Refer to Problem 13. In that problem, 24 people are assigned in groups of 6 to four brands of calculators in order to determine whether a difference exists in the time to perform a given task.
    (a) In terms of the original problem, why is it important to consider the difference among performance times in terms of means?
    (b) Why is it important to observe more than one performance time for each brand of calculator?
    (c) Suppose it is not possible to find 24 separate people to perform the test and a fewer number are available. Given below are three methods of assigning fewer people to the different brands.

        i. Six people perform the task once on each of the four types of calculators.
        ii. Four people are assigned randomly to the four brands. Each performs the task 6 times on the same calculator.
        iii. One person performs the same task 6 times on each of the four types of calculators.

        Explain how each of the above methods of assignment differs from the original design, even though 24 observations are made in all cases. What kinds of difficulties can arise by using these three methods of assignment in terms of the original problem to be solved?
    (d) Suppose that in the original problem, the people involved in the test had varying amounts of experience with the use of scientific calculators. What effect would this have on the results of the test?
    (e) In the original problem, it is assumed that everyone is given the same task to perform so that the variability among the different brands can be measured. Suppose the task that is assigned is relatively simple. Assuming the result of the test is significant, is it reasonable to assume that the same conclusion holds for a more complicated or more difficult task? Why or why not? How can the overall design be modified so that a meaningful conclusion may be reached about simple and difficult tasks?

22. Refer to the numerical example presented in this chapter on page 292.

    (a) Based upon the original data given, compute the following quantities using the formulas given.

        i. $\sum^{3} n(\bar{x} - \bar{\bar{x}})^2$
        ii. $\sum^{3} \sum^{5} (x - \bar{x})^2$
        iii. $\sum^{15} (x - \bar{\bar{x}})^2$

Each of these quantities represents a different sum of squares, or sum of squared deviations from a mean, where x represents an individual worker's examination score, $\bar{x}$ is the mean of a particular group of workers, and $\bar{\bar{x}}$ represents the grand mean or mean of all workers in all groups. The expression in *ii.* represents a double summation where a sum is performed over observations within a sample of 5 observations and then summed over the 3 samples.

(b) Compare the result of the computation in part (a), *i.* and *ii.*, with the between group sum of squares and the within group sum of squares calculated in the example. Based upon this comparison and the formulas given above, do you get a better idea of the meaning of between and within group sum of squares? Explain.

(c) What variation is measured by the quantity found in part (a), *iii.*?

(d) Add the results obtained in part (a), *i.* and *ii.* and compare this to the result obtained in *iii.* How does the result relate to the concept of testing means with variances?

CHAPTER 15

CONTINGENCY TABLES The Chi-Square Test for Association

People can be divided into three groups: those who make things happen, those who watch things happen, and those who wonder what happened.

John W. Newbern

An approximate answer to the right problem is worth a good deal more than an exact answer to an approximate problem.

John Tukey

Early in our treatment of probability we dealt with relationships among characteristics with the concept of dependence among events. We accomplished this in terms of conditional probabilities. In order to illustrate the concept at that time, we considered a problem relating job qualification with workers' area of residence by means of a crossclassified, or contingency, table. Recall, a crossclassified table presents the frequency, or number of elements, falling into two or more categories corresponding to various characteristics simultaneously.

There are many problems where it is of interest to establish relationships among characteristics that naturally are described by various categories. For example, we constantly hear of the results of studies relating smoking to lung cancer and heart disease. Since the enactment of new discrimination laws, there is a great deal of interest in the relationship between hiring and promotion practices and race and sex. Market researchers frequently are concerned with the relationship between consumer attitudes and product success.

CONTINGENCY TABLES
The Chi-Square Test for Association

In this chapter we again are interested in the problem of dependence among characteristics. As indicated by the title, the material to be covered falls under the general heading of association. Hence, characteristics that are statistically dependent are said to be associated. Recognize that we will not employ the everyday usage of the term *association* here. It is commonly accepted that two characteristics are associated if they occur together often, whereas in a statistical sense association means that two characteristics occur together more often than they do in the absence of the other characteristic. For example, in order to say that smoking and respiratory diseases are associated, or related statistically, it is not sufficient to establish that the disease occurs frequently among smokers; we must further determine that it occurs less frequently among non-smokers.

When the concept of dependence was introduced with a crossclassified table in Chapter 4, we assumed that the frequencies in the table represent actual values existing in the universe. Consequently, dependence could be established by comparing conditional probabilities computed directly from the table. In this chapter we focus on problems where frequencies are based upon *sampled* data. In such cases, direct comparison of conditional probabilities will tell us they are not the same, but this does not disclose whether differences observed merely are due to chance or whether they reflect real differences existing in the universe. Hence, it is necessary to use a statistical test to determine whether a relationship that is observed results from sampling or whether it reflects a true relationship in the universe.

In order to understand the problem fully, consider an example where we are interested in the possible effectiveness of a drug. Suppose we have a sample of 200 people with a disease, where 100 are given the drug and the remaining 100 are not. The resulting data in terms of the number cured and the number not cured are summarized in Exhibit 1. The table presents the number of people, all with the disease, that fall within various categories corresponding to all possible combinations of column and row headings. For example, the number in the upper left corner of the table tells us that 65 of the 100 people given the drug were cured. This frequency and the others in the body of the table are referred to as *joint frequencies*, since they provide the number of elements, or people in this case, corresponding to more than one category simultaneously, or jointly.

The row and column totals are termed *marginal frequencies*, or marginals, because they appear in the margins of the table. The marginals provide the frequency in each category without reference to any of the other categories. For example, the 80 appearing at the end of the

EXHIBIT 1
AN EXAMPLE OF DATA IN THE FORM OF A CROSSCLASSIFIED TABLE

	Number of People		
	Drug	*No Drug*	*Total*
Cured	65	55	120
Not Cured	35	45	80
Total	100	100	200

second row represents the number of people not cured of the disease, whether they received the drug or not. Recognize that in this example the marginal totals corresponding to those given and not given the drug were specified as fixed in advance, whereas the totals representing the number cured and not cured were not determined before the study was conducted. The latter totals are assumed to vary freely and are known only after the data are collected. We shall deal with the way marginal totals can be generated after we develop the procedure to test for association.

Recall from Chapter 4 that if we divide the joint frequencies by the total number of elements, which in this example equals 200, we obtain *joint probabilities*. With respect to examples considered in this chapter, probabilities obtained in this way would represent *estimates* of joint probabilities since they are based upon sampled data. Likewise, marginal frequencies divided by the total number of elements yield estimates of marginal probabilities. Also, recall that joint frequencies divided by marginal totals yield *conditional probabilities*. Consequently, when these are calculated from sample data they represent estimates also.

A conditional probability is defined as the probability that an event will occur given the occurrence of, or conditional upon the occurrence of, another event. With respect to the above drug example, consider the conditional probability of being cured in cases where the drug was administered. This is obtained as the number given the drug who were cured (65) divided by the total number of persons given the drug (100), resulting in a value of 0.65. Hence, 65 percent of those given the drug were cured. The conditional probability of being cured when the drug is not given is 0.55, or 55/100, based upon the sampled data. Although these probabilities are not equal, we cannot automatically conclude that dependence exists and that the drug is effective. The difference between the two conditional probabilities may be the result of chance, or sampling variability, rather than the result of a true difference in the cure rates. If the true cure rates are unequal, we can conclude that dependence exists.

In cases where the sample results are extreme, showing evidence of either perfect independence or perfect dependence, there is little difficulty in making a decision. Perfect dependence, or association, is shown in Exhibit 2 and perfect independence, or non-association, is demonstrated in Exhibit 3. The tables in the exhibits provide the joint frequencies or number of people associated with each cell in the table

EXHIBIT 2
EXAMPLES OF PERFECT ASSOCIATION OR DEPENDENCE

Number of People

	Drug	*No Drug*	*Total*
Cured	100 (1.0)	0 (0.0)	100
Not Cured	0 (0.0)	100 (1.0)	100
Total	100	100	200

	Drug	*No Drug*	*Total*
Cured	0 (0.0)	100 (1.0)	100
Not Cured	100 (1.0)	0 (0.0)	100
Total	100	100	200

EXHIBIT 3
EXAMPLES OF PERFECT NON-ASSOCIATION OR INDEPENDENCE

Number of People

	Drug	No Drug	Total
Cured	20 (0.2)	20 (0.2)	40
Not Cured	80 (0.8)	80 (0.8)	160
Total	100	100	200

	Drug	No Drug	Total
Cured	50 (0.5)	50 (0.5)	100
Not Cured	50 (0.5)	50 (0.5)	100
Total	100	100	200

	Drug	No Drug	Total
Cured	100 (1.0)	100 (1.0)	200
Not Cured	0 (0.0)	0 (0.0)	0
Total	100	100	200

along with the conditional probabilities, in parentheses, based upon the column totals.

In the example of perfect association shown in Exhibit 2, the table on the left indicates that everyone who received the drug was cured and everyone who did not receive the drug was not cured. Similarly, in the second table everyone who received the drug was not cured whereas everyone who did not receive the drug was cured. Both cases represent extremes where a perfect relationship exists in the sample between the characteristics of cure and drug administration; knowing whether or not the drug was administered tells us exactly what happened to the patients. Recognize, in the second case the results are strange since all who received the drug were not cured and those who did not receive the drug were cured; the drug in question must have a completely adverse effect to obtain such a result. Nevertheless, the example serves to illustrate a situation of perfect association.

On the other hand, consider the examples in Exhibit 3. Although the joint frequencies are different among tables, they are exactly the same in the "drug" and "no drug" columns within each table. Regardless of the specific values, we can say that the same percentage will be cured whether or not the drug is administered. Hence, there is evidence in the sample of perfect non-association or independence since it is impossible to determine what will happen to a patient when we are given knowledge of the administration of the drug.

In this chapter, we are interested in cases similar to the one shown in Exhibit 1 where the frequencies in the body of the table yield conditional probabilities that neither are exactly equal nor are they at the other extreme where one is zero and the other equals 100 percent. In such cases, we need a *test* to determine whether the disparity in the conditional probabilities reflected in the observed frequencies is due to chance or to a real difference in the universe.

In order to perform a test for association we must compare joint frequencies observed in a sample to frequencies expected to occur by chance. A test statistic based upon calculated values of the expected frequencies is used to perform the test. The test statistic we shall use is referred to as "chi-square," which generally is applied to problems where expected and observed frequencies are compared. Details concerning chi-square and its sampling properties are reviewed in the procedure section.

When tests about contingency tables are introduced, the symbols typically used are more confusing than ones related to other tests. This is due to the fact that joint frequencies must be identified in terms of a corresponding row and column simultaneously. In order to aid our understanding of the procedure presented in the next section, let us examine a 2 × 2 table (i.e., 2 rows and 2 columns) using symbols rather than numbers. This should help us understand the important components of the formulas so that we can avoid using cumbersome symbols.

Shown in Exhibit 4 is an example of a contingency table in symbol form. The exhibit provides frequencies in the form of symbols corresponding to any two characteristics, A and B, each of which are broken down into two categories, 1 and 2. The joint frequencies in the body of the table are accompanied by 2 subscripts; the first identifies the *row* and the second identifies the *column* where each is located. For example, O_{11} represents the number of elements in the first row and first column, or category 1 of characteristic A and category 1 of characteristic B; O_{21} corresponds to the second row and first column, or category 2 of characteristic A and category 1 of characteristic B.

The marginal frequencies are represented by an upper case C or R, indicating either a column or a row total; a lower case c or r will be used later for the *number* of columns and rows. The single subscript accompanying a marginal total indicates the row total or column total corresponding to a particular category. Hence, R_1 represents the total number of elements in the first row of the table (i.e., $R_1 = O_{11} + O_{12}$), or the total in category 1 of characteristic A. C_2, for example, is the total number of elements in the second column (i.e., $C_2 = O_{12} + O_{22}$), or the total in category 2 of characteristic B.

The total number of elements in the entire sample is denoted, as usual, by the symbol "n." The sum of the row totals and the sum of the column totals both equal n (i.e., $R_1 + R_2 = C_1 + C_2 = n$). Furthermore, the sum of all the joint frequencies in the body of the table also equals n (i.e., $O_{11} + O_{12} + O_{21} + O_{22} = n$).

In order to perform the test described in the next section, an expected frequency corresponding to each observed joint frequency must be computed. Each expected frequency is based upon the row and column total associated with an observed frequency. Hence, we could denote expected frequencies with 2 subscripts also. For example, the

EXHIBIT 4
USE OF SYMBOLS TO DESCRIBE FREQUENCIES IN A CONTINGENCY TABLE

	Characteristic B		
Characteristic A	*Category 1*	*Category 2*	*Total*
Category 1	O_{11}	O_{12}	R_1
Category 2	O_{21}	O_{22}	R_2
Total	C_1	C_2	n

symbol E_{21} would represent the expected frequency corresponding to O_{21}, or the second row and first column. E_{21}, therefore, is based upon the row total R_2 and the column total C_1.

The sections that follow *omit* the subscripts from the formulas. Keep in mind, however, that each expected frequency used in the test corresponds to a particular joint frequency, row total, and column total as though the subscripts were included.

The symbols used in the remaining sections of the chapter are as follows:

- n = the sample size, or the total number of elements sampled
- O = an observed frequency, or a joint frequency in the body of a contingency table
- R = a row total
- C = a column total
- r = the number of rows in a contingency table
- c = the number of columns in a contingency table
- E = an expected frequency
- H_0 = the null hypothesis, or hypothesis tested
- H_1 = the alternate hypothesis
- α = the level of significance, or the probability of a Type I error
- χ^2 = the sample statistic used to test for association in a contingency table
  = the chi-square statistic
- χ^2_α = the value of chi-square corresponding to the level of significance, α, obtained from Table 5
- CV = the critical value
- DF = degrees of freedom

PROCEDURE

The procedure presented in this section is used to test for association between characteristics where the sample data is presented in the form of a contingency table with *any* number of rows, r, and columns, c. The null and alternate hypotheses for the test are

H_0: No association or relationship exists between the characteristics; they are independent

H_1: An association or relationship exists between the characteristics; they are dependent

Null and alternate hypotheses for the test of association.

In order to perform the test of hypothesis, the following steps are required:

Step 1. Corresponding to every joint frequency observed, O, *compute* an expected frequency, E, using the formula

$$E = \frac{R}{n} \times C$$

Formula for computing an expected frequency in a test for association.

According to the formula, an expected frequency corresponding to a particular cell in the table is found by dividing the row total, R, by the sample size, n, and multiplying by the column total, C.

Step 2. Based upon the observed values and corresponding expected frequencies, *compute* the test statistic or value of chi-square according to the formula*

$$\chi^2 = \sum \frac{(O - E)^2}{E}$$

Chi-square statistic for testing association.

The value of the chi-square statistic is found by finding the difference between each observed and expected frequency, squaring these differences and dividing by the expected frequency, and adding each of these computed values.

Note. When the null hypothesis is true, or when no association exists, the chi-square statistic given by the above formula is approximately distributed according to a chi-square sampling distribution. This is a continuous probability distribution that is defined for all non-negative numbers and is skewed to the right.

The chi-square distribution is tabulated in the Appendix as **Table 5.** The values of chi-square presented in the table correspond to probabilities in the *right tail* of the distribution and correspond directly to the level of significance used in the test for association.

When testing for association, different chi-square distributions exist and depend upon the number of rows and columns of a contingency table. More specifically, the chi-square distribution that applies is identified by the number of "degrees of freedom." With respect to tests for association, the number of degrees of freedom is given by the formula

$$DF = (r - 1)(c - 1)$$

Degrees of freedom for the chi-square test of association.

The number of degrees of freedom equals the number of rows minus one multiplied by the number of columns minus one. Chi-square values for given probabilities are provided in Table 5 for selected numbers of degrees of freedom.

Step 3. Find the critical value of chi-square, χ^2_α, corresponding to a chosen level of significance, α, from Table 5 in the Appendix.

*Strictly, the formula for the chi-square statistic should be written in terms of a "double-sum" as

$$\chi^2 = \sum^{c} \sum^{r} \frac{(O - E)^2}{E}$$

where the expression following the summation signs is first summed over all rows and these row sums are then summed over all columns. Therefore, it is implied in the formula given in the procedure that the double summation process is employed. This is equivalent to summing each value of $(O - E)^2/E$ for every cell in the contingency table.

EXHIBIT 5
SUMMARY OF THE CHI-SQUARE TEST FOR ASSOCIATION

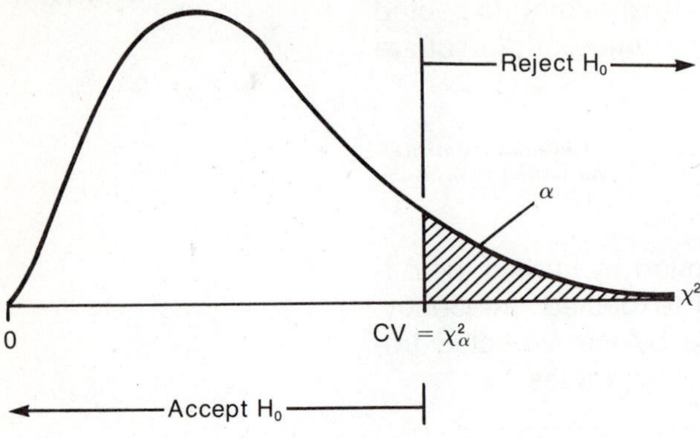

Hence,

$$CV = \chi^2_\alpha$$

Critical value for test of association.

where the tabulated value of chi-square corresponds to an area α in the *right* tail of the distribution. The value of chi-square is identified in terms of the number of degrees of freedom that is found by the formula for DF given in the above note.

Step 4. Compare the computed and tabulated values of chi-square using the decision rule

$$\text{Accept } H_0 \text{ when } \chi^2 < CV$$
$$\text{Reject } H_0 \text{ when } \chi^2 \geq CV$$

Decision rule for test of association.

In other words, accept the null hypothesis and conclude there is no association when the computed value of the chi-square statistic is less than the tabulated critical value equal to χ^2_α; otherwise reject the null hypothesis and conclude that association between the characteristics exists. A summary of the testing procedure based upon a diagram of the chi-square distribution is presented in Exhibit 5.

NUMERICAL EXAMPLES

Example 1 Consider an example similar to the drug illustration presented at the beginning of the chapter. There we were given a contingency table with 2 rows and 2 columns that in modified form appears below:

NUMBER OF PATIENTS

	Drug	*No Drug*	*Total*
Cured	65	80	145
Not cured	35	70	105
Total	100	150	250

Recall, the table provides us with the number of patients given and not given the drug, respectively, and the number in each group that are cured and not cured. The total number of people in the sample, n, equals 250.

The null and alternate hypotheses for the test are

H_0: No relationship exists between treatment and cure

H_1: A relationship does exist between treatment and cure

In this example, which is the first of three, each of the steps in the procedure section will be identified and explained in detail in order to provide a clear understanding of the procedure. In the examples that follow, the general procedure is the same and, consequently, the details are ignored.

Step 1. Compute Expected Frequencies

An expected frequency is computed corresponding to each observed frequency in the body of the contingency table. The table below lists each observed frequency, read *left to right* from the contingency table, and illustrates the computation of each expected frequency.

COMPUTING EXPECTED FREQUENCIES IN THE DRUG EXAMPLE

	Observed Frequency O	*Expected Frequency* $E\left(=\frac{R}{n} \times C\right)$
	65	58(=145/250 × 100)
	80	87(=145/250 × 150)
	35	42(=105/250 × 100)
	70	63(=105/250 × 150)
Total	250	250

Consider the first computation in the table corresponding to the observed frequency of 65 in the first row and first column. The corresponding row total (R) equals 145 and the column total (C) is 100. Hence, the expected frequency is found as

$$E = \frac{R}{n} \times C$$

$$= \frac{145}{250} \times 100$$

$$= .58 \times 100$$

$$= 58$$

Each of the remaining values of E are found in the same way except the values of R and C vary depending upon the cell of the contingency table considered.

> *Note.* The value of O was not needed in the computation of E, but is introduced in order to identify the corresponding row and column totals used in each computation.

Step 2. Compute Chi-Square

The value of chi-square is found by substituting into the formula

$$\chi^2 = \sum \frac{(O - E)^2}{E}$$

This is best performed in tabular form which appears in the table below.

Computing the Value of the Chi-Square Statistic

(1) O	*(2)* E	*(3)* O − E	*(4)* (O − E)²	*(5)* (O − E)²/E
65	58	7	49	0.8448
80	87	−7	49	0.5632
35	42	−7	49	1.1667
70	63	7	49	0.7778
Total				χ^2 = 3.3525

The expected frequency in column (2) is subtracted from the observed frequency in column (1) in order to obtain O − E in column (3). Each of the values in column (4) is obtained by squaring the differences in (3). Column (5) is obtained by dividing each squared value in (4) by the corresponding expected frequency in column (2). The value of χ^2 is obtained as the sum of the values in column (5) as

$$\chi^2 = \sum \frac{(O - E)^2}{E}$$

$$= .8448 + .5632 + 1.1667 + .7778$$

$$= 3.3525 \text{ or } 3.353 \text{ rounded}$$

Step 3. Find the Critical Value of Chi-Square

The critical value of chi-square is found from Table 5 in the Appendix. *Assuming* a 5 percent level of significance, the critical value is

$$CV = \chi^2_\alpha$$
$$= \chi^2_{.05}$$
$$= 3.841$$

which corresponds to "one" degree of freedom. The number of degrees of freedom is found by substituting and solving as

$$DF = (r - 1)(c - 1)$$
$$= (2 - 1)(2 - 1)$$
$$= 1 \times 1 = 1$$

where the number of rows, r, equals 2 and the number of columns, c, equals 2.

Step 4. Establish the Decision Rule and Draw a Conclusion

The decision rule for the test is

$$\text{Accept } H_0 \text{ if } \chi^2 < 3.841$$
$$\text{Reject } H_0 \text{ if } \chi^2 \geq 3.841$$

Since the calculated value $\chi^2 = 3.353$ is less than the critical value $\chi^2_{.05} = 3.841$, we *accept* the null hypothesis and *conclude* there is no association or relationship between the cure and the drug; the two characteristics are independent. The results of the test are summarized in the following diagram:

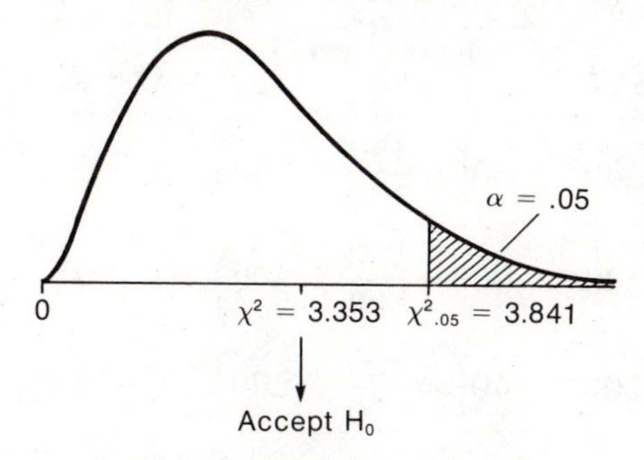

Example 2 Consider the results of a poll of 200 teenagers and 300 adults concerning the beverage that they prefer most.

BEVERAGES PREFERRED BY TEENAGERS AND ADULTS
(NUMBER OF RESPONDENTS)

	Teenagers	*Adults*	*Total*
Coffee	60	90	150
Tea	25	75	100
Soft drinks	85	90	175
Other	30	45	75
Total	200	300	500

The results are in the form of a 4 × 2 table (i.e., 4 rows and 2 columns) and correspond to an overall sample of 500 individuals. The problem is to determine whether a relationship exists between status as an adult or a teenager and perference for beverages. Assume a 1 percent level of significance to perform the test.

In addition to the joint and marginal frequencies provided by the table, we are given the following information:

$$n = 500$$
$$r = 4$$
$$c = 2$$
$$\alpha = 0.01$$

The null and alternate hypotheses are

H_0: There is no relationship between status and beverage preference; they are independent

H_1: A relationship exists between status and preference; preference for beverages is dependent upon status as a teenager or an adult.

Computation of expected frequencies and the value of chi-square appear in the following table.

COMPUTATION OF EXPECTED FREQUENCIES AND CHI-SQUARE IN BEVERAGE PREFERENCE EXAMPLE

O	R	C	$E\left(=\frac{R}{n} \times C\right)$	O − E	$(O-E)^2$	$(O-E)^2/E$
60	150	200	$60\left(=\frac{150}{500} \times 200\right)$	0	0	0.0000
90	150	300	$90\left(=\frac{150}{500} \times 300\right)$	0	0	0.0000
25	100	200	$40\left(=\frac{100}{500} \times 200\right)$	−15	255	5.6250
75	100	300	$60\left(=\frac{100}{500} \times 300\right)$	15	255	3.7500
85	175	200	$70\left(=\frac{175}{500} \times 200\right)$	15	225	3.2143
90	175	300	$105\left(=\frac{175}{500} \times 300\right)$	−15	225	2.1429
30	75	200	$30\left(=\frac{75}{500} \times 200\right)$	0	0	0.0000
45	75	300	$45\left(=\frac{75}{500} \times 300\right)$	0	0	0.0000
500			500			$\chi^2 = 14.7322$

Based upon the above table, the observed value of chi-square is determined as

$$\chi^2 = \sum \frac{(O - E)^2}{E}$$
$$= 14.732 \text{ (rounded)}$$

The number of degrees of freedom for the test is found as

$$\begin{aligned} DF &= (r - 1)(c - 1) \\ &= (4 - 1)(2 - 1) \\ &= 3 \times 1 \\ &= 3 \end{aligned}$$

The critical value of chi-square is found from Table 5 for 3 degrees of freedom and a 1 percent level of significance (i.e., $\alpha = 0.01$) as

$$\begin{aligned} CV &= \chi^2_{\alpha} \\ &= \chi^2_{.01} \\ &= 11.345 \end{aligned}$$

The corresponding decision rule becomes

$$\text{Accept } H_0 \text{ if } \chi^2 < 11.345$$
$$\text{Reject } H_0 \text{ if } \chi^2 \geq 11.345$$

Since the computed value of χ^2 equal to 14.732 is greater than the critical value $\chi^2_{.01}$ equal to 11.345, we *reject* the null hypothesis and *conclude* that there is a relationship between preference for beverages and status as a teenager or an adult.

Example 3 Consider part of the results of a study concerned with people's trust in political leaders. In response to a direct question asking whether respondents trust political leadership, the following results were obtained based upon a breakdown according to geographical location within the country.

TRUST IN POLITICAL LEADERSHIP BY REGION
(NUMBER OF PEOPLE)

	North	*East*	*South*	*West*	*Total*
Trust	105	90	85	70	350
Do not trust	100	115	110	125	450
Do not know	45	45	55	55	200
Total	250	250	250	250	1000

An equal number of people were sampled from each of the four regions, resulting in a total sample size of 1000. The table has three rows and four columns. The problem is to determine whether a relationship exists between location and trust in leadership. We shall use a 5 percent level of significance to perform the test. In addition to the tabled data, we are given the following information:

$$n = 1000$$
$$r = 3$$
$$c = 4$$
$$\alpha = 0.05$$

H_0: Trust in leadership is independent of location
H_1: Trust in leadership is dependent upon, or related to, location

The computation of the expected frequencies and chi-square appear in the following table.

COMPUTATIONS REQUIRED IN TRUST IN LEADERSHIP EXAMPLE

O	$E\left(=\frac{R}{n}\times C\right)$	O − E	$(O-E)^2$	$(O-E)^2/E$
105	$87.5\left(=\frac{350}{1000}\times 250\right)$	17.5	306.25	3.5000
90	87.5	2.5	6.25	.0714
85	87.5	−2.5	6.25	.0714
70	87.5	−17.5	306.25	3.5000
100	$112.5\left(=\frac{450}{1000}\times 250\right)$	−12.5	156.25	1.3889
115	112.5	2.5	6.25	.0556
110	112.5	−2.5	6.25	.0556
125	112.5	12.5	156.25	1.3889
45	$50.0\left(=\frac{200}{1000}\times 250\right)$	−5.0	25.00	.5000
45	50.0	−5.0	25.00	.5000
55	50.0	5.0	25.00	.5000
55	50.0	5.0	25.00	.5000
1000	1000.0			12.0318

The value of chi-square, obtained as the total of the last column is

$$\chi^2 = \sum \frac{(O - E)^2}{E}$$

$$= 12.032 \text{ (rounded)}$$

Recognize, the expected frequencies for all elements in a single row of the original contingency table are identical since the column totals are equal.

The number of degrees of freedom is found as

$$\begin{aligned} DF &= (r - 1)(c - 1) \\ &= (3 - 1)(4 - 1) \\ &= 2 \times 3 \\ &= 6 \end{aligned}$$

The critical value of chi-square found from Table 5 based upon 6 degrees of freedom and a 5 percent level of significance (i.e., $\alpha = 0.05$) is

$$\begin{aligned} CV &= \chi^2_\alpha \\ &= \chi^2_{.05} \\ &= 12.592 \end{aligned}$$

Based upon this value, the decision rule for the test is

$$\text{Accept } H_0 \text{ if } \chi^2 < 12.592$$
$$\text{Reject } H_0 \text{ if } \chi^2 \geq 12.592$$

Since the computed value of χ^2 equal to 12.032 is less than the critical value $\chi^2_{.05}$ equal to 12.592, we *accept* the null hypothesis and *conclude* that attitudes regarding trust in leadership are the same regardless of location, or that they are not related to location within the country.

DISCUSSION

The chi-square test of association applies to contingency tables with two characteristics and any number of categories, or any number of rows and columns. This is true as long as the categories are not constructed on the basis of the actual observations in the sample. Furthermore, the test is independent of the way the marginal totals are generated. The chi-square test is limited to a certain extent by the overall sample size and the size of the marginal frequencies. In this section we shall discuss some of the problems of interpretation of results, limitations, and applications of the chi-square test of association.

The Expected Frequencies. When we perform a test for association we compare expected and observed frequencies using the formula for χ^2. We do this in order to determine whether the discrepancy between the two can be attributed to chance or whether the disparity is large enough to conclude that a real difference exists. The formula used to compute expected frequencies is based upon the assumption that conditional probabilities among all categories should be equal when the characteristics are independent. Hence, the formula for E considers the differences in the size of the marginal totals in order to obtain expected frequencies that correspond to conditional probabilities that are equal for each category.

For example, consider Exhibits 6 and 7 both of which relate to the drug illustration of Example 1. Exhibit 6 presents the observed frequencies based upon the original sample data on the left and the corresponding expected frequencies using the formula for E in the table on the right. Notice the marginal totals are the same in both tables. Exhibit 7 presents the same frequencies expressed as proportions of the column totals. Each of the proportions represents a conditional probability. In the case of the observed values the conditionals vary among the columns, whereas they are the same for the expected frequencies. In other words, the formula for E provides absolute frequencies whose ratio, based upon the different marginal totals, is the same for all columns. Therefore, we would *expect* these frequencies to be produced by chance under the assumption that the drug is not related to or is independent of the cure.

Expected frequencies represent ones that are expected to be produced by chance and are calculated such that their ratio to the marginal totals are the same for all columns.

Interpreting Results of Tests of Association. In the preceding sections we focused on the concept of association and how to perform the appropriate test when the data is in the form of a sample. In one instance, we specified that the column totals were fixed in advance; however, we did not place much emphasis upon the way the marginal totals were generated.

There are three distinct ways in which marginal totals can be generated: (1) all margins are free to vary at random and are not fixed in advance, (2) one set of margins is fixed in advance and the other set is free to vary at random, and (3) all margins, column and row totals, are fixed in advance. The third case where all margins are fixed is somewhat rare in practice and will not be discussed here. In order to understand

EXHIBIT 6
COMPARISON OF OBSERVED AND EXPECTED FREQUENCIES IN DRUG EXAMPLE

	Observed				*Expected*		
	Drug	*No Drug*	*Total*		*Drug*	*No Drug*	*Total*
Cured	65	80	145	Cured	58	87	145
Not Cured	35	70	105	Not Cured	42	63	105
Total	100	150	250	Total	100	150	250

EXHIBIT 7
COMPARISON OF CONDITIONAL PROBABILITIES BASED ON OBSERVED AND EXPECTED FREQUENCIES IN DRUG EXAMPLE

	Observed				*Expected*		
	Drug	*No Drug*	*Total*		*Drug*	*No Drug*	*Total*
Cured	.65	.53	.58	Cured	.58	.58	.58
Not Cured	.35	.47	.42	Not Cured	.42	.42	.42
Total	1.00	1.00	1.00	Total	1.00	1.00	1.00

EXHIBIT 8
ADDITION OF THIRD SAMPLE TO BEVERAGE PREFERENCE EXAMPLE

	Pre-Teens	*Teenagers*	*Adults*	*Total*
Coffee	55	60	90	205
Tea	100	25	75	200
Soft drinks	105	85	90	280
Other	90	30	45	165
Total	350	200	300	850

the other two cases, let us reexamine the beverage preference example. If a sample of individuals is selected and then classified as teenagers and adults according to preference, both the column and row totals are free to vary. A problem of this sort is said to be classified as a *double dichotomy*, and is truly a problem of independence of two characteristics.

The chi-square test of association can be used regardless of the way the marginal totals are generated although the interpretation of the result varies. There are additional theoretical differences among the three cases that are not considered here.

If, however, we select a designated number of teenagers and adults so that one set of marginal totals is fixed before we select the actual sample, we refer to the problem as one of *homogeneity*. In this case, we are comparing two *separate samples* in order to determine whether the corresponding populations of teenagers and adults are similar with respect to their preferences.

When the chi-square test for association is applied to problems of homogeneity, it is nothing more than an *extension* of the two-sample test for the difference between proportions presented in Chapter 13. The advantage of the chi-square test is that it accommodates more than two categories. Furthermore, the chi-square test can be used when more than two samples are compared.

When one set of marginal totals is fixed in advance, the chi-square test for a 2 × 2 table is equivalent to the large sample test for the difference between two proportions and represents an extension of this test when there are more categories or more samples. In the latter case, conclusions resulting from the test are more difficult to make.

In order to elaborate upon the last point, consider the beverage preference example that is modified to include a third sample of pre-teens in addition to teenagers and adults. The results are shown in Exhibit 8. Assuming the column totals are fixed, we now are comparing three samples in order to determine whether a difference exists regarding preference.

Based upon the data presented in Exhibit 8, the value of chi-square is 51.2487 which, at 6 degrees of freedom, is significant at any of the tabulated probability levels. Although we can conclude that there is a difference in preference among the three groups, the single value of χ^2 does not tell us what contributes to the significant result. We can say that a relationship exists but further analysis is necessary in order to locate the source.

Chi-Square and Small Samples. The chi-square test for association is an approximation that, generally speaking, becomes better as the sample increases. The fact that it is an approximation is not a serious limitation since it performs well enough in most practical applications. There are a number of rules that have been developed specifying conditions under which the chi-square test should or should not be applied. The most general of these specifies that the chi-square test for association should not be applied when any of the expected frequencies is less than 5.

The chi-square test is an approximation. A general rule to follow is that it can be used when the expected frequencies are not less than 5.

An exact test exists that can be used as alternative when chi-square does not apply. This is referred to as *Fisher's exact test,* named after its originator, and applies to tables with small frequencies. It is not necessary to consider the details of this test, since we are interested in the overall concept of association here. It is important to recognize that an alternative exists when the chi-square test does not apply.

Other Considerations. Very little has been said about data collection and how it relates to the tests presented in this part of the text. Although there are many considerations about data collection that generally apply to all tests of hypotheses, it is of interest to touch on them now since they can be related easily to crossclassified data.

Example 2 refers to the results of a poll of 500 people regarding preferences for beverages, or specifically, the beverage they prefer most. Four preference categories appear in the table: (1) coffee, (2) tea, (3) soft drinks, and (4) other. Because the data are used to illustrate the procedure for the chi-square test, nothing was said about the way the data was generated. As in the case of all methods of statistical inference, the sample elements must be chosen randomly. If the number of teenagers and adults are specified in advance, then separate random samples from appropriately defined universes of teenagers and adults must be selected.

Although the individuals in the sample may be chosen randomly, responses to the questions asked of them may not be valid since they may depend upon the way the questions are posed. For example, if the four beverage categories actually are listed on a questionnaire for the respondent to see, there may be a tendency to choose the category at the top of the list, or the first one seen by the respondent. This may introduce a bias in the results, whereas a completely voluntary response based upon an open question regarding the most preferred beverage may not. The test for association does not explicitly account for this type of bias, which represents a non-sampling problem requiring specially designed survey techniques.

Another type of problem can be illustrated by reexamining the table used in Example 3. The frequencies in the table correspond to categories related to trust in leadership, broken down by location. One of the categories in the table is given as "Do Not Know." Depending upon the original purpose or goal of the study, a problem could arise if a large number of frequencies correspond to this category.

If it is of specific interest to determine whether people have no attitude or are unaware of their attitude toward trust in leadership, then information can be imparted by this category. However, if the problem is to distinguish between "trust" and "no trust," a large number of responses in the "do not know" category may force us to reassess our conclusion since this category may contribute to a significant result. By performing a chi-square test without the "do not know" category, a significant result is obtained that leads to the conclusion that a relationship exists between trust and location; this was not the case when the "do not know" category was included in the calculations. Hence, the manner in which the original problem is defined is an important consideration when assessing the final results.

A closely related problem arises when information is elicited, say, by mail questionnaire and there is a high degree of non-response. In

EXHIBIT 9
AN EXAMPLE OF A CONTINGENCY TABLE WITH THREE CHARACTERISTICS

	East		*West*		
	Male	*Female*	*Male*	*Female*	*Total*
Trust	175	20	100	55	350
Do not trust	155	60	90	145	450
Do not know	70	20	60	50	200
Total	400	100	250	250	1000

such a case, a test generally is performed with the available data. Implicit in the conclusion that is reached is that non-respondents demonstrate the same pattern as those responding. If this is not so, the conclusion will be biased by some unknown amount. The chi-square test, or any other test, does not account for this type of problem.

The test for association presented in this chapter was illustrated with contingency tables involving two characteristics, regardless of the number of categories for each characteristic. Problems exist where more than two characteristics are under investigation. For example, reconsider the trust in leadership example regarding trust against location within the country; it might have been of interest to consider sex as a third characteristic in addition to geographic location and trust. Assuming two locations instead of four for simplicity, the contingency table might appear like the one in Exhibit 9.

Based upon the exhibit, we can see that two categories for sex correspond to each of the location categories. Hence, what would be a table with 3 rows and 2 columns when sex is not considered, now is a table with 3 rows and 4 columns. We could perform a chi-square test of association on the data assuming 3 rows and 4 columns. If this were done, we would find the result is highly significant. This result alone, however, would *not* disclose which of the characteristics, sex or location, contributes to the rejection of the null hypothesis and the conclusion that a relationship exists. Modifications of the chi-square test appearing in more advanced texts can be used to determine which characteristics are significant.

Another point can be made about the effect of additional characteristics. If additional characteristics are not considered, in some cases relationships among those used in the test may be masked or hidden depending upon the amount of data and the way the data are collected. The choice of variables that is included in an analysis is based upon an investigator's understanding of a particular problem and the way it is defined initially. Moreover, as more characteristics are considered, it is necessary to use a larger overall sample size in order to establish relationships; otherwise, the frequencies in some of the cells may be too small to be meaningful.

No test of a hypothesis accounts for non-sampling errors or problems. More about these problems is considered here because they are easier to understand when related to cross-classified data; however, similar considerations can be applied to other tests presented in other chapters.

SHORTCUTS, CALCULATORS, AND THE COMPUTER

When the number of categories in a crossclassified table is large, computation of the chi-square statistic becomes tedious and time-consuming. Some saving in time can be achieved by using a shortcut formula similar to the one used to compute a variance. Furthermore, computer programs are available to compute chi-square. An added advantage of many of the existing programs is that the raw data can be tallied in crossclassified form before chi-square is computed.

Programmable hand-held calculators are available with a stored program feature that provides routines for calculating χ^2. Generally, these are limited to contingency tables of a certain size. Calculators with continuously stored program libraries have an added feature that makes it possible to determine whether a result is significant by calling a program that computes the "p-value" or probability of observing the calculated value or greater under the null hypothesis.

Shortcut. The shortcut formula considered here avoids computing each deviation between the observed and expected frequencies, which provides a moderate saving in time and avoids some errors in calculation. The shortcut formula for χ^2 is given as

$$\chi^2 = \sum \frac{O^2}{E} - n$$

Shortcut formula for the chi-square test of association.

With this formula, it is necessary to square each observed frequency, divide by the expected frequency and add the results. The only subtraction required is at the end where the sample size is subtracted from the sum of the squared observations divided by the expected frequencies. These must be found with the aid of the formula for E presented earlier.

To illustrate the use of the formula, let us recompute χ^2 for Example 1. The observed and expected frequencies together with the computation of the summation are provided in the following table:

COMPUTATIONS TO BE USED IN SHORTCUT FOR TEST OF ASSOCIATION

O	E	O^2	O^2/E
65	58	4225	72.8448
80	87	6400	73.5632
35	42	1225	29.1667
70	63	4900	77.7778
Total			253.3525

The total of the last column of the table is substituted into the shortcut formula,

$$\chi^2 = \sum \frac{O^2}{E} - n$$

$$= 253.3525 - 250$$

$$= 3.3525$$

where the sample size, n, equals 250. The value of χ^2 equal to 3.3525 is the same as the one found earlier.

EXHIBIT 10
AN EXAMPLE OF COMPUTER OUTPUT OF A CROSSTAB WITH CHI-SQUARE CALCULATION

```
***********************CROSSTAB***********************
                  NUMBER OF RESPONDENTS
*****************************************************

           COUNT     .
           ROW PCT   .
           COL PCT   .    NORTH     EAST      SOUTH     WEST        ROW
           TOT PCT   .                                              TOTAL
                     .        1.        2.        3.        4.
           ..........:........................................
                  1. .    105   .    90   .    85   .    70        350
                     .   30.0   .   25.7  .   24.3  .   20.0       35.0
TRUST                .   42.0   .   36.0  .   34.0  .   28.0
                     .   10.5   .    9.0  .    8.5  .    7.0
                   ..:........................................
                  2. .    100   .   115   .   110   .   125        450
                     .   22.2   .   25.6  .   24.4  .   27.8       45.0
NOT TRUST            .   40.0   .   46.0  .   44.0  .   50.0
                     .   10.0   .   11.5  .   11.0  .   12.5
                   ..:........................................
                  3. .     45   .    45   .    55   .    55        200
                     .   22.5   .   22.5  .   27.5  .   27.5       20.0
DO NOT KNOW          .   18.0   .   18.0  .   22.0  .   22.0
                     .    4.5   .    4.5  .    5.5  .    5.5
                   ...........................................
COLUMN               .    250   .   250   .   250   .   250
TOTAL                    25.0      25.0      25.0      25.0

CHI-SQUARE EQUALS 12.0318 WITH 6 DEGREES OF FREEDOM
```

Output from the Computer. The format of the output from a computer varies depending upon a user's needs and the particular program employed. All, however, provide the same basic information. A typical output appears in Exhibit 10.

The sample output corresponds to Example 3, which dealt with respondents' trust in political leadership by geographic location. The information needed to understand the table is printed along with the numerical results. Frequently, the output contains more information than necessary for a particular user's needs.

In the example presented, column and row headings are printed along with numerical codes. Some programs allow for the codes only. The information in the upper left corner provides the contents of *each* cell of the table. In this case 4 items are given. The first number is the observed joint frequency similar to the values presented in earlier examples. At the option of the user, the remaining 3 quantities may be obtained. The second is the *row percent* which represents the joint frequency as a percent of the row total. The third is the *column percent* that represents the joint frequency as a percent of the column total. The last, or *total percent,* represents the joint frequency as a percent of the overall sample size, which in the case illustrated is 1000. Recognize, all percents correspond to observed relative frequencies, or estimated probability values, of different types depending upon the total used as a base; typically, percents are used instead of decimal values. The numbers under the row and column totals or marginal frequencies represent the percentage of the overall sample size. Printed at the end of the output is the computed value of the test statistic, χ^2, with the corresponding number of degrees of freedom.

Some programs also provide other quantities that are used to measure the degree of association. Although χ^2 is useful to determine significance, it is influenced by the size of the sample in addition to differences in frequencies. Consequently, larger values of chi-square result from larger sample sizes even in cases where the conditional probabilities in the sample remain the same. Measures of association that are independent of the influence of the sample size are useful as indicators of the degree to which characteristics are related and as a basis for making comparisons among different sets of data. Unfortunately, no single commonly accepted measure exists, and special knowledge is required in order to obtain a valid interpretation.

QUESTIONS AND PROBLEMS

DEFINITIONAL QUESTIONS

1. What is a crossclassified or contingency table?

2. What is meant by the term *association*? How is it related to the concept of a relationship as it is presented in this chapter?

3. What is the meaning of the following terms?

   (a) Joint Frequency
   (b) Marginal Frequency
   (c) Joint Probability
   (d) Marginal Probability
   (e) Conditional Probability

4. What is the meaning of the terms *dependent* and *independent*? How are they related to the concept of association and contingency tables?

5. What are the null and alternate hypotheses for tests of association or independence?

6. What is the test statistic that is used to test for association? How does it measure association between two characteristics?

7. What is the meaning of the following terms?

   (a) Chi-Square Statistic
   (b) Critical Value of Chi-Square
   (c) Chi-Square Distribution
   (d) Degrees of Freedom

   Relate each of the above terms to the test for association.

8. Describe the procedure used to test for association among two characteristics.

9. Distinguish between the terms *homogeneity* and *double dichotomy*.

COMPUTATIONAL PROBLEMS

10. Based upon the data in the following table, determine whether or not there is a relationship between the two characteristics.

	Characteristic 1		
Characteristic 2	A	B	*Total*
C	50	50	100
D	40	60	100
Total	90	110	200

11. The table below presents the number of men and women faculty members at a large university who were promoted or not during the previous academic year.

NUMBER OF FACULTY MEMBERS

	Men	*Women*	*Total*
Promoted	302	103	405
Not promoted	98	47	145
Total	400	150	550

Using the chi-square test, determine whether there is a relationship between the university's promotion policy and sex.

12. From a list of registered voters, 400 individuals are randomly selected and asked their opinion about a pending bond issue. The results are tabulated below, classified in terms of property ownership.

NUMBER OF REGISTERED VOTERS

	For	*Against*	*Undecided*	*Total*
Owns property	115	53	42	210
Does not own property	67	77	46	190
Total	182	130	88	400

Determine whether there is a relationship between opinion and property ownership.

13. As part of a study of serious crimes, the number of arrests is recorded by type of crime and four key police districts during a six-month period.

NUMBER OF ARRESTS

Type of Crime	*District 1*	*District 2*	*District 3*	*District 4*	*Total*
Larceny	30	96	104	72	302
Robbery	26	13	29	22	90
Burglary	90	23	5	57	175
Auto theft	34	76	84	51	245
Aggravated assault	9	44	53	7	113
Other	4	35	25	11	75
Total	193	287	300	220	1000

Based upon the data provided in the table, determine whether a relationship exists between police district and type of crime.

14. Use the data presented in the following table to determine whether a relationship exists between number of accidents and age of drivers.

NUMBER OF DRIVERS

Number of Accidents	*Age* Under 21	21–30	31–45	46–60	Over 60	*Total*
0	137	203	179	75	83	677
1	72	119	85	33	32	341
2	53	84	64	20	26	247
More than 2	20	47	35	9	64	175
Total	282	453	363	137	205	1440

CONCEPTUAL QUESTIONS AND PROBLEMS

15. Why is a test for association necessary when comparison of conditional probabilities indicates whether or not characteristics are dependent?

16. What are the advantages and limitations of the chi-square test of association? What is the general rule governing the applicability of the chi-square test?

17. What is the relationship between the chi-square test for association presented in this chapter and the tests in Chapter 13 concerning proportions and percentages? What are the similarities and differences between the procedures presented in the two chapters?

18. Refer to Problem 12. The result of the chi-square test applied to this problem is significant.
    (a) Calculate the conditional probabilities associated with voter opinion conditional upon the property ownership categories.
    (b) Based upon your calculations in part (a) and the fact that chi-

square is significant, what conclusions can be drawn about opinions toward the bond issue as they relate to property ownership?

(c) What percentage of the sample of voters is undecided? What is the effect of the undecideds on your conclusions in part (b)?

19. What are the problems of interpreting the results of a test of hypothesis for association? Discuss.

20. Does the chi-square test for association account for non-sampling errors? Discuss. In your answer define the term *non-sampling error*.

21. Refer to Problem 11. Suppose the data presented in the table are used to determine whether the university discriminates against women faculty members with respect to promotions.

(a) Is the data provided in the problem sufficient to determine whether or not discrimination exists? Discuss.

(b) What other information should be collected in order to determine whether discrimination exists against women regarding promotions? Can the chi-square test be used? Explain.

22. Suppose a claim is made that a particular toothpaste called Brand A has been tested which results in fewer cavities than other leading brands. Assume the claim is made on the basis of the following results:

NUMBER OF PEOPLE

Amount of Cavities	*Brand A*	*Other Brands*	*Total*
None	185	10	195
Few	115	16	131
Many	100	24	124
Total	400	50	450

(a) Use the chi-square test to determine whether a relationship exists. Is the claim made about Brand A valid on the basis of the above evidence? Explain.

Suppose that at the beginning of the study, 1000 people were in the group using brand A and 1000 people were in the group using other brands. In other words, there were 600 dropouts in the first group and 950 in the second. Assume that data is available on the dropouts during the study period, which is presented in the following table:

NUMBER OF DROPOUTS

Amount of Cavities	*Brand A*	*Other Brands*	*Total*
None	100	425	525
Few	130	395	525
Many	140	360	500
Total	370	1180	1550

(b) Explain how it is possible that the sum of the number of dropouts using Brand A and the number using Brand A in the original table does not equal 1000.

(c) Combine the two tables presented in this problem by adding the frequencies in each cell and perform a chi-square test for association. What conclusion about the original claim should be reached based upon this test?

(d) Explain why there is a difference in the conclusions reached in part (a) and part (c). Based upon the additional evidence, can you say that the conclusion reached in part (a) is wrong? Explain.

(e) Is it reasonable to assume that information about dropouts in studies of this kind would be available? Explain.

(f) In order to determine whether the study was conducted properly, what additional information beside the data on dropouts would you need to know? Why?

CHAPTER 16

REGRESSION AND CORRELATION ANALYSIS

You tell me war a man gits his corn pone, en I'll tell you what his pinions is.

Mark Twain

. . . don't be too cock-a-hoop about your .95 correlation with 7 cases.

W. Sealy Gosset

We often turn to accepted relationships in order to reach solutions to problems. Through our experience supplementary information is digested, relationships are formed, conclusions are reached, and, when necessary, these conclusions are applied to similar problems in the form of projections about the future. For example, consider the following situations concerning some very ordinary problems: (1) Will it rain today; is it cloudy? (2) Do I need gasoline; how low is the fuel gauge? (3) How much cash do I need for a trip; how long am I to stay away? The answer to the second part of each set of questions, which is based upon past experience, provides information that helps to answer the first. The relationships we form are not perfect, however; occasionally we run out of gas, need more cash, or get caught in the rain. In other words, factors that are not considered and chance may contribute to the outcome in the form of error.

In the last chapter we worked with relationships in the form of associations among characteristics that were broken down into various categories. There, we dealt with testing for a certain kind of relationship based upon frequencies or counts occurring in the cells of a contingency table. In this chapter we consider relationships among variables that can be measured, with the overall aim of using the relationship to estimate what has occurred in the past or to make predictions about the

EXHIBIT 1
EXAMPLES OF RELATIONSHIPS BASED UPON SAMPLE DATA

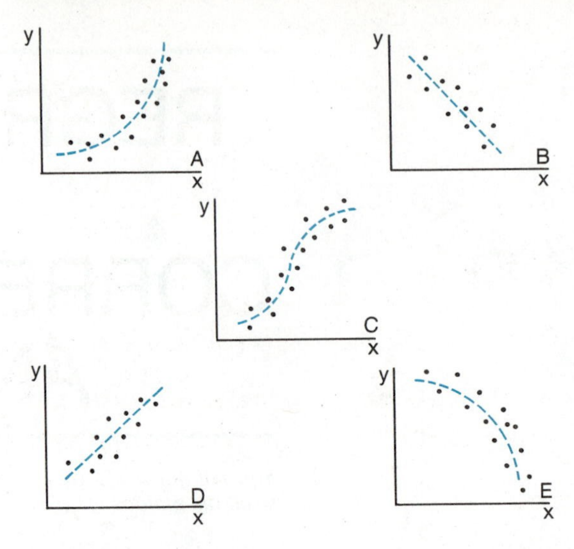

occurrence of future events. Unlike the examples cited at the beginning, we will develop relationships in the form of an *equation* that can be used to estimate the value of one variable based upon the value of another variable. The equation is developed on the basis of observations from a sample. In general, the variable to be estimated is referred to as the **dependent variable.** The variable used to estimate the dependent variable is called the **independent variable.**

Exhibit 1 presents examples of different kinds of relationships based upon samples of data. The variable on the vertical axis indicated by the symbol y is the dependent variable, whereas x on the horizontal axis is the independent variable. Each point in the diagrams corresponds to a pair of observations or values of y and x. A dotted line appears in each diagram to describe the overall relationship between the two variables.

The patterns depicted in B and D of the exhibit represent linear relationships since the overall patterns are described by straight lines. D is referred to as a *direct* or positive relationship since y tends to increase as x increases, whereas B is an *indirect* or negative relationship since y becomes smaller as x increases.

The remaining patterns shown in the exhibit are curvilinear since overall they are described by curves rather than straight lines. The pattern in C illustrates a situation that is not as common as the others since it is "S-shaped." Relationships of this sort often apply to dosage response, where the proportion of a sample of test animals either cured or killed is related to varying dosages of a drug under study. Market share related to advertising outlay is another example.

Situations described by the other four patterns are numerous. A or D, for example, may correspond to flood intensity as a function of the amount of paved surface in a given locality, or to the amount of air pollution based upon increased industrial development. B or E could describe declining gas mileage corresponding to increased car size.

The particular form of a relationship depends upon the nature of an underlying problem and the type of data that is available. Each type of relationship, however, can be described by an equation relating the dependent variable to the independent variable. In this chapter we shall concentrate on linear relationships, or ones that can be described by a straight line. These are the most commonly used in basic applications and provide the simplest way to understand the concepts presented.

The statistical term for the equation used to relate one variable to another variable is a **regression line.** The entire process of developing an estimating equation from sample data and drawing inferences about the results is referred to as **regression analysis.** When the strength of the relationship is considered, we have a problem of **correlation.** The two types of analysis frequently are combined and are called *regression and correlation analysis*. Although related, each has a special meaning.

In order to obtain a better understanding of the material presented in this chapter, consider a specific problem. Suppose it is of interest to establish a relationship or regression equation between the math aptitude test score and grade point average of college students. The overall objective is to use the equation to predict grade point average given a student's math aptitude score before entering college. With respect to this example, math aptitude score is the independent variable and grade point average is the dependent variable. Values of these variables for a sample of 12 students are presented in Exhibit 2.

EXHIBIT 2

EXAMPLE OF DATA USED TO FIND A REGRESSION EQUATION

Student	*Math Aptitude, x*	*Grade Point Average, y*
1	425	1.9
2	700	3.9
3	500	2.9
4	375	1.8
5	450	2.9
6	625	3.2
7	575	2.7
8	350	2.0
9	700	3.0
10	475	2.4
11	750	3.6
12	575	3.4

The first point to note about the exhibit is that there are two observations for each student in the sample. In order for a regression analysis to be valid, values of the two variables must come from the same sample element. This common sample element, which is the student in our example, is called the **unit of association.** The unit of association represents the link between values of the dependent and independent variables. When two observations correspond to a single unit of association, we refer to them as *paired observations*.

If we look at the individual values in the exhibit, we will observe that differences occur in the values of each variable. Although some math aptitude scores are repeated, they correspond to different grade point averages. Conversely, some grade point averages are repeated and correspond to different math aptitudes. In order to use math aptitude scores to predict grade point averages, there must be a pattern in the way the two variables vary together. Such a pattern is detected in the graph presented in Exhibit 3.

EXHIBIT 3
SCATTER DIAGRAM ILLUSTRATING RELATIONSHIP BETWEEN DEPENDENT AND INDEPENDENT VARIABLE

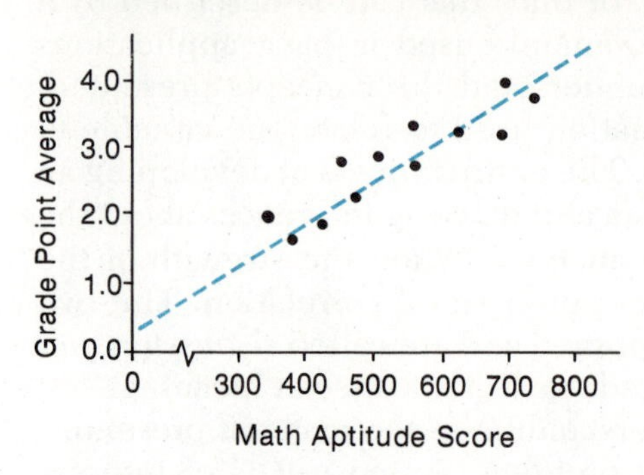

The graph presents the sample grade point averages plotted on the vertical axis against math aptitude scores plotted on the horizontal axis. Each point on the graph represents an individual pair of values corresponding to a single sample item, or student. In statistical terms, a graph of this kind is referred to as a *scatter diagram* due to the visual display of the "scatter" of points. Although the points are dispersed about the graph, there is a tendency for grade point averages to become larger as math aptitudes increase. This tendency is described by the dotted line in the exhibit.

Notice that most of the points do not fall on the line. In other words, there is variability about the line. Some of this variability can be attributed to chance, since no observed relationship is perfect. Other factors beside math aptitude may be related to grade point average, and if these factors are considered, they may reduce the amount of variability. The line, however, provides a description of the overall relationship between the two variables and can be used to obtain a general or average estimate of grade point average, based upon math aptitude. This estimate would be better than one using the average of the grade points alone.

In general, lines similar to the one depicted in Exhibit 3 can be represented by an equation of the form

$$y = a + bx$$

where y and x represent the dependent and independent variables, respectively. This equation is the *regression line* where a and b are referred to as *regression coefficients.* The main goals of this chapter are (1) to show how to establish a regression line in a sample by estimating the regression coefficients, (2) to show how to use the regression line to estimate or predict a dependent variable on the basis of an independent variable, (3) to provide measures of variability of the dependent variable about the regression line, (4) to provide a way of incorporating a measure of variability into the estimate of the dependent variable, (5) to measure the degree of relationship that exists between the dependent and independent variables in a sample, and (6) to provide a test of a hypothesis to determine whether a linear relationship actually exists between the variables.

Since there are many goals in this chapter, the format has been modified and is not the same as most of the chapters in the text. There are three main sections before the discussion. The first deals with finding and using a regression line, the second is concerned with measuring and applying variability, and the third presents methods for measuring and testing the relationship in terms of correlation. The order in which different items are presented within each of these sections is similar to the format of earlier chapters. General considerations associated with regression and correlation analysis are presented in the discussion section. Some attention is given to multiple regression and correlation analysis that is an extension of the material for two variables covered in detail in this chapter.

The symbols used in this chapter are summarized as follows:

y = a value of a dependent variable in the sample
x = a value of an independent variable in the sample
n = the sample size, or the number of pairs of x and y values
y' = an estimated value of the dependent variable for a given value of the independent variable
a = a regression coefficient, representing the estimated value of y when x equals zero
= the y-intercept
b = a regression coefficient, representing the change in the estimated value of y corresponding to a one-unit change in x
= the slope of the line
$\hat{S}^2_{y \cdot x}$ = the estimated variance of the y-values about the regression line
$\hat{S}_{y \cdot x}$ = the estimated standard deviation of the y-values about the regression line
= the estimated standard error of estimate
t = a value of the Student's t-distribution obtained from Table 4 in the Appendix
= the t-statistic (this represents an alternative use of the symbol t)
DF = degrees of freedom
α = the level of significance, or the probability of committing a Type I error when testing a hypothesis
$t_{\alpha/2}$ = a value of the t-distribution corresponding to half the level of significance, α, in *each* tail, which is obtained from Table 4 in the Appendix
CV = the critical value defining the decision rule used to perform a test of a hypothesis
r = the sample coefficient of correlation
r^2 = the sample coefficient of determination
ρ = the universe coefficient of correlation (read as Rho)
R^2 = the sample coefficient of multiple determination
R = the sample coefficient of multiple correlation

FINDING AND USING A REGRESSION LINE

When interested in describing a relationship between two variables on the basis of sample data, it is necessary to decide on the form of the equation to use. Once this is done, a procedure is needed to "fit" the equation to the data due to the variability or scatter of points that is present. In this section we shall introduce one method for fitting a straight line to a set of data and show how the line can be used to make point estimates of the dependent variable.

Finding a Regression Line. In order to determine the regression line relating a dependent variable, y, to an independent variable, x, it is necessary to estimate the coefficients a and b in the equation

$$y' = a + bx$$

where y′ represents an estimated value of the dependent variable based upon x. Estimates of the coefficients are found by substituting into the following formulas:

$$a = \frac{(\Sigma x^2)(\Sigma y) - (\Sigma x)(\Sigma xy)}{n(\Sigma x^2) - (\Sigma x)^2}$$

$$b = \frac{n(\Sigma xy) - (\Sigma x)(\Sigma y)}{n(\Sigma x^2) - (\Sigma x)^2}$$

Formulas for finding regression coefficients.

These formulas are based upon a general procedure for fitting equations to data called the method of *least squares*. More is said about this method in the discussion section. Before substituting into the above formulas, *preliminary calculations* must be made to find the cumulative terms Σx^2, Σx, Σy, and Σxy. Some saving in computational effort can be achieved by recognizing that the denominator in both formulas is the same and can be computed once.

Example 1 Consider the example concerned with relating grade point average to math aptitude test score. Using symbols, we have

y = grade point average
x = math aptitude score
n = sample size
= 12

The original data is reproduced in Exhibit 4 along with the preliminary calculations needed to find the values of the regression coefficients, a and b. The totals from the table are substituted into the formulas

EXHIBIT 4
PRELIMINARY CALCULATIONS FOR TWO-VARIABLE REGRESSION AND CORRELATION ANALYSIS

Grade Point Average Problem

Student	*Math Aptitude, x*	*Grade Point, y*	*xy*	x^2	y^2
1	425	1.9	807.5	180625	3.61
2	700	3.9	2730.0	490000	15.21
3	500	2.9	1450.0	250000	8.41
4	375	1.8	675.0	140625	3.24
5	450	2.9	1305.0	202500	8.41
6	625	3.2	2000.0	390625	10.24
7	575	2.7	1552.5	330625	7.29
8	350	2.0	700.0	122500	4.00
9	700	3.0	2100.0	490000	9.00
10	475	2.4	1140.0	225625	5.76
11	750	3.6	2700.0	562500	12.96
12	575	3.4	1955.0	330625	11.56
Total	6500 (Σx)	33.7 (Σy)	19115.0 (Σxy)	3716250 (Σx^2)	99.69 (Σy^2)

for a and b as follows:

$$a = \frac{(\Sigma x^2)(\Sigma y) - (\Sigma x)(\Sigma xy)}{n(\Sigma x^2) - (\Sigma x)^2}$$

$$= \frac{(3716250)(33.7) - (6500)(19115.0)}{12(3716250) - (6500)^2}$$

$$= \frac{125237625 - 124247500}{44595000 - 42250000}$$

$$= \frac{990125}{2345000}$$

$$= 0.4222$$

$$b = \frac{n(\Sigma xy) - (\Sigma x)(\Sigma y)}{n(\Sigma x^2) - (\Sigma x)^2}$$

$$= \frac{12(19115.0) - (6500)(33.7)}{12(3716250) - (6500)^2}$$

$$= \frac{10330}{2345000}$$

$$= 0.0044$$

Based upon these values, the *regression line* relating grade point average to math aptitude test score in the sample is

$$y' = a + bx$$

$$= 0.4222 + 0.0044x$$

Using the Regression Line. Once the coefficients are estimated, the regression line can be used to make a point estimate of the dependent variable by substituting a value of x into the regression equation and solving for y′.

Assume we are interested in using the equation found above to estimate the grade point average of a student with a math aptitude test score of 550. This is found by substituting 550 for x into the regression equation and solving as follows:

$$\begin{aligned} y' &= 0.4222 + 0.0044(550) \\ &= 0.4222 + 2.42 \\ &= 2.8422 \text{ or } 2.8 \text{ (rounded)} \end{aligned}$$

Since the original grade points are given to one decimal place, the rounded figure of 2.8 is a satisfactory estimate of a student's grade point average whose math aptitude test score is 550.

Extrapolation should be done with caution and is based upon the assumption that the observed relationship applies outside the range of the observed data.

Based upon this example we can see that it is a simple matter to obtain an estimate of the dependent variable using the regression line. There are, however, some problems associated with this approach. One problem is called extrapolation. An **extrapolated value** is an estimate of the dependent variable that corresponds to a value of the independent variable beyond the values observed in the sample. For example, in the grade point problem, math aptitude test scores in the sample range between 350 and 750. If we use the regression line to estimate a student's grade point average outside this range, the estimate is an extrapolated value. Whenever we extrapolate, we do so under the assumption that the same relationship described by the regression equation exists for x-values that are lower *or* higher than those observed in the sample. In general, the further we extrapolate beyond the observed data, the more likely that the projection is incorrect.

An estimate of the dependent variable that is obtained by substituting into the regression equation is a point estimate since the result is in the form of a single value.

The purpose of the regression line is to account for variability or error in the dependent variable so that it can be estimated or predicted better. By substituting a value of x into the regression equation, we obtain a point estimate of y. In other words, the projection is in the form of a single value. Since no line fitted to a set of data fits the points exactly, point projections are subject to error. Furthermore, by substituting into the regression equation to obtain an estimate, it is assumed that once fitted the equation is ready to use. Whenever we fit an equation to data, we should consider how well the line fits the data before applying it. These problems are considered in the remaining sections of the chapter.

The regression coefficient "a" literally represents the estimated value of the dependent variable when the independent variable equals zero. "b" represents the estimated change in the dependent variable for a unit change in the independent variable.

The Regression Coefficients. Particular values of a and b define a specific linear relationship between y and x, based upon the sample observed. The coefficient "a" is termed the *y-intercept* and represents the estimated value of y when x equals zero. "b" is the *slope* of the line, and is a measure of the change in the estimated value of y for a one-unit change in x.

In order to understand these interpretations, consider the result of the numerical example presented earlier. There, we obtained the line

$$y' = 0.4222 + 0.0044x$$

where y′ represents the estimated grade point average and x math aptitude test score. The value of b equal to 0.0044 can be interpreted in the following manner: if the math aptitude test score of two students differs by one point, their estimated grade point average will differ by 0.0044. The coefficient "a" equal to 0.4222 is interpreted literally as the grade point average of a student scoring zero on the math aptitude test.

Clearly, the above interpretation of "a" has no practical meaning since a student would not be admitted to college with a zero math score. Furthermore, it would seem highly unlikely that anyone would obtain a zero score. Since a math score of zero was not included in the original sample and lies below the lowest observed math score, the value of "a" is an extrapolated value. In this example, it merely positions the line so that it may properly describe the relationship between the variables.

Whether or not the regression coefficients have a practical meaning depends upon a particular problem. It is not necessary that they have a practical meaning for the regression line to be useful.

There are cases where the coefficients have practical meaning and are of interest in their own right. As an example, consider a linear cost function

$$C = 250000 + 200\ Q$$

where C represents the total cost of production in dollars and Q corresponds to total output, or the quantity produced, in terms of the number of units. The symbols C and Q are used instead of y and x to make the point more clearly. Assume the coefficients were estimated on the basis of a sample of production runs ranging in size between 500 and 1500 units.

The a-term, or intercept, equal to 250,000 is interpreted literally as the production cost when nothing (i.e., Q = 0) is produced. Unlike the grade point example, this interpretation has a practical meaning since the component of total cost corresponding to zero output is defined as overhead or fixed cost. Since the coefficient was based on production run data between 500 and 1500 units that did not include zero output, estimating overhead as $250,000 should be done with caution; this value should not be used as an estimate of fixed cost *unless* it can be safely assumed that the specific linear cost function applies at this level.

The term "200 Q" of the cost function represents the variable cost component, since total cost increases by $200 for every unit produced. Hence, the intercept or b-term can be viewed as the marginal cost, or variable cost per unit. Finding the regression line, therefore, provides a way of estimating fixed and marginal cost, which could be the goal in performing the analysis rather than predicting total cost.

MEASURING AND APPLYING ERROR

When we use regression analysis to relate variables, one goal is to account for as much variability in the dependent variable as possible. We have seen, however, that observations scatter about a regression line. Since the line does not explain everything about the dependent variable, variability or error exists. This error can result from several sources.

If we knew the effect of all relevant factors, we should be able to account for the variability in the dependent variable and predict it exactly. Since this is not possible in reality, variability in the dependent variable will occur due to the fact that complete knowledge about the dependent variable is not available. Furthermore, virtually all observable phenomena are not completely predictable due to the presence of a purely random element that is associated with any relationship. Generally, these two forms of error representing incomplete knowledge of the behavior of a dependent variable are treated together and are considered as random error. A third source of error about the regression line can result when the algebraic form of the relationship between y and x is not properly specified. For example, if we employ a straight line when a curvilinear relationship exists, another form of error is added. Finally, errors can arise owing to imprecise measuring instruments even when the form of the relationship is specified properly.

The material presented in this chapter is based on the assumption that the relationship between the dependent and independent variable is a straight line. We shall ignore errors of measurement and assume that observations are measured properly. Consequently, we will focus our attention on variability about the regression line that is the result of chance.

When specific assumptions are made about the way the points distribute about the line, we can make inferences like the ones presented in earlier chapters. These inferences are based upon the amount of variation about the regression line. Variability about the regression line is measured in terms of residuals. A **residual** is defined as the difference between an observed value of y and an estimated value, y′, determined by the regression equation. In order to understand the concept of a residual, consider Exhibit 5, which presents the scatter diagram of students' grade point averages (y) and math aptitude test scores (x) together with the regression line found in the previous chapter. A dotted vertical line extends from each of the observed points to the regression line shown in the exhibit. The length of the vertical line in each case represents a residual.

As an example, consider the observed and estimated values of grade point averages corresponding to a math aptitude test score, x, equal to 450. The observed value of y equals 2.9, whereas the estimated value, y′, on the regression line equals 2.2. Recall, the estimated value is obtained by substituting the value of x into the regression equation and solving for y′. The residual is the difference between y and y′, or the observed value, y, and the estimated value, y′, for a given value of x.

$$\text{Residual} = y - y'$$

Residual about the regression line.

In the above example, the residual is found as

$$\begin{aligned} y - y' &= 2.9 - 2.2 \\ &= 0.7 \end{aligned}$$

Hence, the residual corresponding to x equal to 450 is 0.7, which is shown in Exhibit 5.

There is a different residual computed like the one above for every observed pair of x and y values. These values represent the variability in the dependent variable, y, that is not explained by the independent variable, x, in terms of the regression line. Since the sizes of the residuals

EXHIBIT 5
SCATTER DIAGRAM WITH REGRESSION LINE TO ILLUSTRATE CONCEPT OF RESIDUALS

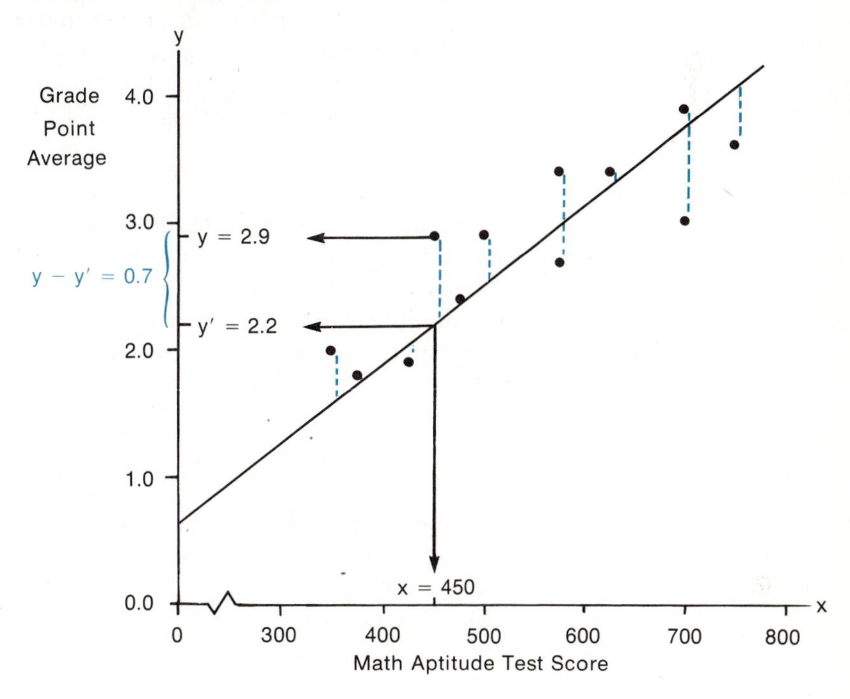

differ, it is useful to employ an overall measure of variability in the dependent variable about the line.

The measures we shall use are in the form of a variance and standard deviation similar to those presented in earlier chapters. They are, however, based upon the residuals for varying values of y′ rather than the same mean value used in earlier formulas. The variance about the regression line can be defined by the formula

$$\hat{S}^2_{y \cdot x} = \frac{\Sigma(y - y')^2}{n - 2}$$

Definition of the estimated variance about the regression line.

where $\hat{S}^2_{y \cdot x}$ is the symbol for the variance and the subscript "y·x" indicates that the variance is of the dependent variable y, given or conditional upon, x; n is the sample size. The denominator, n − 2, is referred to as the number of *degrees of freedom*. We can see from the formula that the variance about the line is defined as the sum of the squared residuals, or squared deviations of the observed y-values about the line, divided by the degrees of freedom. Each value of y′ is determined from the regression equation based upon the values of x in the sample.

The standard deviation is obtained in the usual way by taking the square root of the variance. This measure, which is the standard deviation about the regression line, is given a special name in statistics. It is referred to as the **standard error of estimate.** Both this and the variance measure the variability about the line, just as the variance and standard deviation used in earlier chapters measure variability about the mean.

Note. Since the above formula is based upon the degrees of freedom, n − 2, rather than upon the sample size, n, directly, it actually represents an estimated value similar to the estimated variance introduced in Chapter 11 concerning estimation of means. For large samples, it is not necessary to subtract 2 since the estimated and sample values are almost the same.

Calculating the Estimated Variance and Standard Error. The formula given above can be used directly to find the value of the variance about the regression line. In order to use the formula, it is necessary to calculate values of y′ for all values of x in the sample and compute each squared residual. A simpler formula that provides the same answer is given by the following expression.

$$\hat{S}^2_{y \cdot x} = \frac{\Sigma y^2 - a\Sigma y - b\Sigma xy}{n - 2}$$

Formula for finding the estimated variance about the regression line.

In order to find the variance about the line using this formula, we substitute the values for the cumulative terms, the regression coefficients and n, and solve for $\hat{S}^2_{y \cdot x}$. The standard error of estimate is found by taking the square root of $\hat{S}^2_{y \cdot x}$.

$$\hat{S}_{y \cdot x} = \sqrt{\hat{S}^2_{y \cdot x}}$$

Formula for finding the estimated standard error of estimate.

Example 2 Using the formulas just presented, let us find the variance and standard error of estimate of grade point average conditional upon math aptitude test score. Based upon the preliminary calculations in Exhibit 4 and our earlier work, we have the following information:

$$\begin{aligned} n &= 12 \\ \Sigma y &= 33.7 \\ \Sigma xy &= 19115.0 \\ \Sigma y^2 &= 99.69 \\ a &= 0.4222 \\ b &= 0.0044 \end{aligned}$$

Using this information, the variance of the y-values about the line is found by substituting and solving as follows:

$$\begin{aligned} \hat{S}^2_{y \cdot x} &= \frac{\Sigma y^2 - a\Sigma y - b\Sigma xy}{n - 2} \\ &= \frac{99.69 - (.4222)(33.7) - (.0044)(19115)}{12 - 2} \\ &= \frac{99.69 - 14.23 - 84.11}{10} \\ &= \frac{1.35}{10} \\ &= 0.135 \end{aligned}$$

The standard error of estimate, based upon the variance, is found as

$$\begin{aligned}\hat{S}_{y \cdot x} &= \sqrt{\hat{S}^2_{y \cdot x}} \\ &= \sqrt{.135} \\ &= 0.37 \text{ (rounded)}\end{aligned}$$

Hence, the variance of student grade point average based upon math aptitude test score is 0.135. The standard error of estimate is 0.37.

Meaning of $\hat{S}^2_{y \cdot x}$ and $\hat{S}_{y \cdot x}$. In general, the concept of a variance and standard deviation are difficult to understand, especially when seen for the first time. The variance and standard error of estimate about the regression line are complicated further by the fact that their formulas are more complex than the ones used earlier. Conceptually, however, they have a similar interpretation. With respect to regression analysis a large variance (and standard error of estimate) indicates a large amount of scatter or dispersion of the points about the line. This means that the smaller the variance and the standard error of estimate, the better the regression line fits the data and describes the relationship between y and x.

The smaller the standard error of estimate, the closer the points fall about the regression line and the better the line fits the data. When all the points fall on the line, the standard error of estimate equals zero.

Since one of the goals in regression analysis is to find a variable, x, that accounts for variation of the dependent variable, y, the smaller the variance about the line, the more useful the regression line may be in predicting y. When working with actual problems, there generally is some variability left over about the line. It is important to measure this for a number of reasons: (1) it provides a way of determining the usefulness of the regression line in prediction, (2) it can be used to construct interval estimates of the dependent variable, and (3) inferences can be made about other components of the problem.

Sample vs. Universe Values. Up to this point we have considered data and measures that are based upon a particular sample of paired observations. No mention was made about a universe. In order to make inferences in regression problems, we must consider certain assumptions about the universe from which the sample was drawn and the way the observations are generated. The basic assumptions underlying simple regression problems are listed as follows:

1. A *relationship* between the dependent and independent variables exists in the universe.
2. For every value of the independent variable, there is an arithmetic *mean* of the values of the dependent variable.
3. The arithmetic mean of the dependent variable varies with the independent variable according to a *linear* equation.
4. *Variability* in the dependent variable exists about the line for every value of the independent variable; residual values of the dependent variable exist about the line.

There are certain assumptions that form the basis for inferences that are made in typical regression problems.

5. The residuals about the line are distributed randomly according to a *normal* probability distribution.
6. There is a *constant variance* and standard deviation of the dependent variable about the line that are the same for each value of the independent variable.

In order to understand the meaning of these assumptions, consider Exhibit 6. Symbols that are not used in any calculations are presented in the exhibit in order to convey the ideas more clearly. Capital letters are used to signify universe rather than sample values.

The symbols X and Y appearing in the exhibit represent the independent and dependent variables. Particular values of the independent variable, X, appear on the horizontal axis and are distinguished with individual subscripts. The symbol $\overline{Y}_X$ represents the arithmetic mean of y-values *conditional* upon a particular value of X. The relationship between the conditional means and X represents the true regression line in the universe. This is depicted in terms of the equation $\overline{Y}_X = A + BX$, where A and B are the universe regression coefficients that are estimated by "a" and "b" from a sample. Each value on the line corresponds to a different mean value of Y that is dependent upon X. These means are noted on the Y-axis. Hence, the regression obtained from the sample, $y' = a + bx$, provides estimates of the conditional mean values, $\overline{Y}_X$.

For a given value of X, Y-values are distributed about the line or conditional mean, $\overline{Y}_X$, according to a normal curve. Since the mean, $\overline{Y}_X$, changes with the value of X, there is a different normal distribution of the Y-values for every value of X. Each of these normal distributions has the same shape, which means that each has the same standard deviation. This represents the universe standard error of estimate, which is estimated by the sample value $\hat{S}_{y \cdot x}$. When a relationship exists between X

When a relationship exists, the variability about the regression line is less than the total variation in the dependent variable.

EXHIBIT 6
DESCRIPTION OF ASSUMPTIONS UNDERLYING REGRESSION ANALYSIS

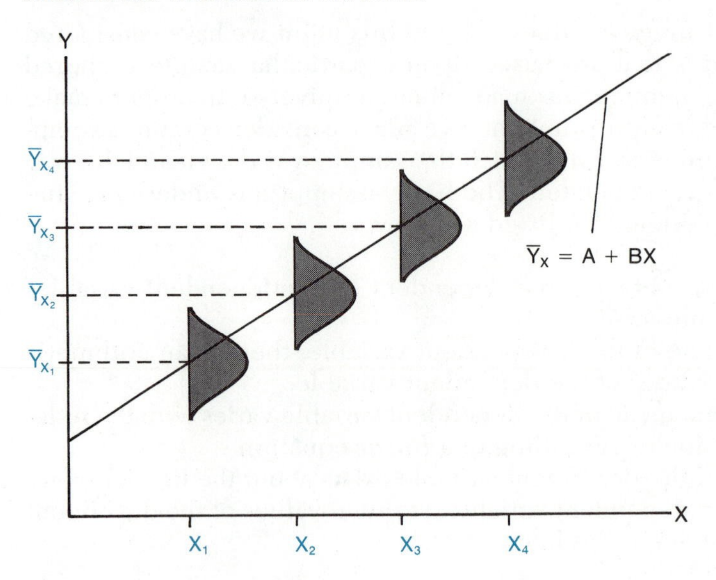

and Y, the standard error of estimate is less than the standard deviation of all the Y-values in the universe computed about their mean without regard to X.

Like all sample characteristics, a, b, and $\hat{S}_{y \cdot x}$ vary from sample to sample. Hence, the estimates obtained by the computations performed earlier depend upon the particular sample selected. Furthermore, different regression lines are obtained from different samples. Based upon the above assumptions and the results of statistical theory, we are able to describe the behavior or sampling properties of the various sample estimates as they vary from sample to sample. These sampling properties are useful in making predictions and drawing inferences about the relationship between X and Y.

Prediction Intervals Using the Standard Error of Estimate. In an earlier part of the chapter we showed how to estimate or predict a value of the dependent variable by substituting a value of x into the regression equation and solving for y′. This prediction can be modified by constructing an interval estimate that accounts for the error about the line. Under the given list of assumptions the distribution of the y-values about the sample regression line is a Student's t-distribution.

Note. The t-distribution that applies to regression problems is the same one that is used for means in Chapters 11 and 12. The distribution is tabulated in the Appendix as **Table 4.** Recall, a particular t-distribution is defined in terms of the number of degrees of freedom. In the case of regression problems involving two variables, the number of degrees of freedom is found as

$$DF = n - 2$$

Degrees of freedom for determining prediction intervals in two-variable regression problems.

The degrees of freedom is found by subtracting 2 from the sample size. For large samples the normal distribution can be used as an approximation, and the degrees of freedom need not be calculated.

An approximate interval estimate of the value of the dependent variable for a given value of the independent variable available, x, is obtained by substituting into the following expression:

$$y' \pm t\hat{S}_{y \cdot x}$$

Approximate interval estimate of the dependent variable.

where y′ is determined by substituting a given value of x into the regression equation; $\hat{S}_{y \cdot x}$ is the estimated standard error of estimate; and t is found from Table 4, based upon a *chosen probability level.* This probability level is similar to a level of confidence described in Chapter 11. The interval estimate found by the above expression is called a **prediction interval.**

Example 3 In order to illustrate the procedure for constructing a prediction interval, consider a different problem in which the cost of production is related to output according to the following regression equation:

$$y' = 109.477 + 217.647x$$

where y' represents estimated production cost in \$1000 units and x is the size of a production run in terms of 100 machine units. The equation is based upon a sample of 9 runs. Assume that we want to construct an interval estimate of the cost of production when 600 machines are produced using a probability level of 0.95. Assume further that the standard error of estimate, $\hat{S}_{y \cdot x}$, already has been found and equals 346.26. Let us break the problem down into 3 parts, first finding y', then t, and finally the prediction interval:

1. Find y' given output equal to 600 machines. Since x is expressed in units of 100, the output figure must be converted, which is accomplished by dividing by 100. Hence,

$$x = \frac{600}{100}$$
$$= 6$$

The value of y' is obtained by substituting this value of x into the regression equation, or

$$\begin{aligned} y' &= 109.477 + 217.647x \\ &= 109.477 + 217.647(6) \\ &= 109.477 + 1305.882 \\ &= 1415.359 \end{aligned}$$

2. Since Table 4 is based upon "two-tailed" probabilities, the value of t obtained from the table corresponds to the given probability level subtracted from one, or 0.05. The degrees of freedom is found as

$$\begin{aligned} DF &= n - 2 \\ &= 9 - 2 \\ &= 7 \end{aligned}$$

Hence, the value of t from Table 4 corresponding to a two-tailed probability of 0.05 and 7 degrees of freedom is

$$t = 2.365$$

3. Using the values obtained, the prediction interval is found

$$y' \pm t\hat{S}_{y \cdot x}$$
$$1415.359 \pm 2.365(346.26)$$
$$\pm 818.905$$
$$596.454 - 2234.264$$

This is an interval estimate in units of $1000; multiplying both limits by 1000 yields an estimate in terms of the original units of cost. Hence, our estimate of the cost of producing 600 machines at the 0.95 probability level lies between $596,454 and $2,234,264.

There is one point that should be made about interval estimates obtained by using the formula given above. The formula accounts for the variability about the line but does not consider the variability of the regression line from sample to sample. Consequently, the interval estimate that is presented is an approximation and is interpreted under the assumption that the regression is estimated without error.* Hence, we can say that a y-value will fall within the prediction limits with a probability equal to the chosen level under the assumption that the regression has been estimated correctly. The approximation gets better as the sample size increases.

Prediction intervals are associated with values of individual elements in a universe, whereas confidence intervals are used to estimate universe parameters. The prediction interval used here is an approximation that gets better for larger sample sizes.

CORRELATION ANALYSIS

There are two problems related to regression analysis that have not been covered. One problem is to determine whether an actual relationship exists between the dependent and independent variables before the regression line is used to make estimates or predictions. The second

*A precise prediction interval can be obtained by using a measure of variability that considers the variability about the line and the variability of the line from sample to sample. The measure of variability, called the standard error of forecast, S_F, is given by the formula

$$S_F = \hat{S}_{y \cdot x}\sqrt{1 + \frac{1}{n} + \frac{(x - \bar{x})^2}{\Sigma(x - \bar{x})^2}}$$

where $\hat{S}_{y \cdot x}$ is the standard error of estimate and n is the sample size. The important term to consider is the one involving the independent variable, x. The standard error of forecast varies depending upon the value of x appearing in the numerator. When the forecast is conditional on the mean of the x-values, $\bar{x}$, the forecast error is smallest since the last term equals zero. Forecasts based upon values of x farther from $\bar{x}$ are subject to greater error since the last term becomes larger. Based upon the standard error of forecast, the exact prediction interval for the dependent variable given x is given by the expression

$$y' \pm tS_F$$

which has the same form as the interval estimate used above except S_F is used in place of $\hat{S}_{y \cdot x}$.

problem involves measuring the degree to which the two variables are related. In some problems, the regression line is not the main interest. Sometimes, the main goal of the analysis is just to establish whether or not variables are related.

In this section we focus on the concept of *correlation*. The term correlation is used in a variety of ways. In the most general sense, it denotes a relationship between variables. We shall focus on a specific kind of correlation, however, that corresponds to linear relationships between variables like the ones considered in the last two chapters. Therefore, the goal of this section is to provide a specific measure of the degree to which variables are *linearly* related and how to use this measure to determine whether a linear relationship actually exists.

The amount of correlation in a sample is measured either by the sample *coefficient of determination*, denoted as r^2, or by its square root, which is referred to as the sample *coefficient of correlation*, r. Later in the discussion we shall see that the coefficient of determination is easier to interpret; however, both measures impart the same basic information. They measure the variability in y that is explained by x. When no correlation exists in the sample, r^2 and r assume a numerical value of *zero*, whereas perfect correlation yields a value of *one*. When a perfect inverse relationship exists, all points fall on a line with a negative slope and r assumes a value of *minus* one. The limits of the two measures can be written in symbol form as

$$-1 \leq r \leq 1$$

$$0 \leq r^2 \leq 1$$

Limits of coefficients of correlation and determination.

Since r^2 and r are sample statistics and subject to sampling variability, values that are not equal to zero could be generated from populations where there is no linear relationship. Consequently, a test of a hypothesis usually is performed to determine whether it is reasonable to conclude that a relationship does exist. To perform this test we must introduce the concept of correlation in the universe. This is reflected in a parameter called the *universe coefficient of correlation*, denoted by the symbol ρ (pronounced Rho). When ρ equals zero there is no correlation in the universe, whereas correlation exists when ρ is not equal to zero.

Formulas for computing r^2 and r are given in the next section. A method for testing whether correlation exists is provided in the section that follows. More about the interpretation of the two measures then is presented.

Calculating the Coefficients of Determination and Correlation. The coefficient of determination, r^2, of a particular sample is found by substituting into the following formula:

$$r^2 = \frac{[n\Sigma xy - \Sigma x\Sigma y]^2}{[n\Sigma x^2 - (\Sigma x)^2][n\Sigma y^2 - (\Sigma y)^2]}$$

Formula for finding the coefficient of determination.

The cumulative terms in the formula are the same as those used to find regression coefficients and the standard error of estimate. If the cumulative terms have not been calculated, preliminary computations are needed to find them before substituting into the formula for r^2.

The coefficient of correlation is obtained as the square root of the coefficient of determination. That is,

$$r = \pm\sqrt{r^2}$$

Formula for finding the coefficient of correlation.

The value of r is either positive or negative, which is indicated by the "plus-minus" sign appearing before the square root. If the above formula for r^2 is used to find r, the sign of r must be determined in terms of the sign of the slope of the regression line, b. If b is positive r is positive, and if b is negative r is negative.

Note. By taking the square root of the formula for r^2 *before* substituting numbers, the resulting formula for r automatically will retain the proper sign. This alternative takes the form

$$r = \frac{n\Sigma xy - \Sigma x\Sigma y}{\sqrt{n\Sigma x^2 - (\Sigma x)^2}\sqrt{n\Sigma y^2 - (\Sigma y)^2}}$$

Formula for r automatically providing proper sign.

If the value of b is not available, this formula for r can be used if the sign of r is important. The coefficient of determination then can be found by squaring the value of r. This formula for r is slightly more difficult to compute since it involves two square roots or the square root of a large number in some cases.

Example 4 Consider the regression analysis of grade point average and math aptitude test score. We already have the following information based upon our earlier work:

$$\begin{aligned} n &= 12 \\ \Sigma x &= 6500 \\ \Sigma x^2 &= 3716250 \\ \Sigma y &= 33.7 \\ \Sigma y^2 &= 99.69 \\ \Sigma xy &= 19115 \end{aligned}$$

Using this information, let us find the coefficient of determination and correlation, r^2 and r, between grade point average and math aptitude test score.

The coefficient of determination is found by by substituting and solving as follows:

$$r^2 = \frac{[n\Sigma xy - \Sigma x\Sigma y]^2}{[n\Sigma x^2 - (\Sigma x)^2][n\Sigma y^2 - (\Sigma y)^2]}$$

$$= \frac{[12(19115) - 6500(33.7)]^2}{[12(3716250) - (6500)^2][12(99.69) - (33.7)^2]}$$

$$= \frac{(10330)^2}{(44595000 - 42250000)(1196.28 - 1135.56)}$$

$$= \frac{(10330)^2}{(2345000)(60.59)}$$

$$= \frac{106708900}{142083550}$$

$$= 0.751 \text{ (rounded)}$$

The coefficient of correlation is found by taking square root of the above result:

$$r = \sqrt{r^2}$$
$$= \sqrt{0.751}$$
$$= 0.867$$

Hence, the coefficient of determination between grade point average and math aptitude test score is 0.751 and the coefficient of correlation equals 0.867. Since the sign of the regression coefficient, b, equal to 0.0044 is positive, r is positive.

Testing for Correlation. In order to determine whether correlation exists, we test the hypothesis

$$H_0: \rho = 0 \text{ (No correlation)}$$

Null hypothesis in test for correlation.

The null hypothesis specifies that the coefficient of correlation in the universe, Rho, equals zero and that there is no correlation. The alternative hypothesis is given as

$$H_1: \rho \neq 0 \text{ (Correlation exists)}$$

Alternate hypothesis in test for correlation.

which specifies that Rho is not equal to zero and that there is correlation between the dependent and independent variables. The test of the hypothesis is performed by conducting the following three steps.

Step 1. Based upon the value of r^2 and r computed according to the above formulas, compute the value of the t-statistic using the formula

$$t = r\sqrt{\frac{n - 2}{1 - r^2}}$$

The test statistic used to test for correlation.

Note. This test statistic is a function of the coefficient of correlation, which equals zero when r equals zero and becomes larger as r increases toward one. Hence, unusually large values of the test statistic lead to the rejection of the null hypothesis.

When the null hypothesis is true or when there is no correlation in the universe, the test statistic is distributed according to Student's t-distribution with degrees of freedom equal to

$$DF = n - 2$$

Formula for degrees of freedom in test for correlation between two variables.

The degrees of freedom that identifies the particular t-distribution that applies in this case equals the sample size minus two. The t-distribution is tabulated in the Appendix as **Table 4.** When the number of degrees of freedom is "large," the normal distribution can be used as a satisfactory approximation to the t-distribution.

Step 2. Based upon a chosen level of significance, α, and the number of degrees of freedom, DF, find the critical values from Table 4. The critical values are given as

$$CV_1 = -t_{\alpha/2}$$
$$CV_2 = t_{\alpha/2}$$

Critical values for test of correlation between two variables.

Note. Two critical values are required since the test is two-tailed. Recall, Table 4 is unlike the tables for other probability distributions since it is given in terms of *two-tailed* probabilities. Consequently, positive values of $t_{\alpha/2}$ are given for the right tail of the distribution, which corresponds to a two-tailed level of significance equal to α. In other words, the upper critical value, CV_2, equal to $t_{\alpha/2}$ is found directly from Table 4 but is found for a probability value in the table equal to α. Because the t-distribution is symmetric, the lower critical value, CV_1, is the negative of the value found in the table, $-t_{\alpha/2}$.

Step 3. Based upon the critical values and the computed t-statistic, apply the following decision-rule:

Accept H_0 when $CV_1 < t < CV_2$
Reject H_0 when $t \leq CV_1$
or
$t \geq CV_2$

Decision-rule for test of correlation between two variables.

The decision rule tells us to accept the null hypothesis of no correlation when the t-statistic falls between the tabulated critical values, and reject the null hypothesis and conclude correlation exists when the t-statistic is either less than or equal to the lower critical value, $-t_{\alpha/2}$, or greater than or equal to the upper critical value, $t_{\alpha/2}$.

Example 5 In order to determine whether the sample results in the grade point problem reflect an actual correlation between grade point average and math aptitude test scores for the entire universe of students, we test the null hypothesis:

$H_0: \rho = 0$ (No correlation exists between grade point average and math aptitude test score)

The alternate hypothesis is

$H_1: \rho \neq 0$ (Correlation exists between grade point average and math aptitude test score)

The test is performed as follows:

Step 1. Compute the t-Statistic and the Number of Degrees of Freedom.

$$\begin{aligned} t &= r\sqrt{\frac{n-2}{1-r^2}} \\ &= .867\sqrt{\frac{12-2}{1-.751}} \\ &= .867\sqrt{\frac{10}{.249}} \\ &= .867\sqrt{40.1606} \\ &= .867(6.3372) \\ &= 5.4944 \end{aligned}$$

where

$$\begin{aligned} DF &= n - 2 \\ &= 12 - 2 = 10 \end{aligned}$$

Step 2. Find the Critical Value of t. Assuming a 5 percent level of significance, $\alpha = 0.05$, the critical value based upon 10 degrees of freedom is obtained from Table 4 as

$$\begin{aligned} CV_1 &= -t_{\alpha/2} \\ &= -t_{.05/2} \\ &= -t_{.025} \\ &= -2.228 \end{aligned}$$

$$\begin{aligned} CV_2 &= t_{\alpha/2} \\ &= t_{.025} \\ &= 2.228 \end{aligned}$$

Remember, the probability level of 0.05 is used in Table 4 to obtain a value of t with half the level of significance in each tail.

Step 3. Construct the Decision Rule and Apply the Test. The decision rule, based upon Step 2 is

$$\text{Accept } H_0 \text{ if } -2.228 < t < 2.228$$
$$\text{Reject } H_0 \text{ if } t \leq -2.228 \quad \textit{or} \quad t \geq 2.228$$

Since t equal to 5.4944 is greater than the critical value equal to 2.228, we *reject* the null hypothesis and *conclude* that correlation exists between grade point average and math aptitude test score. Hence, we may conclude that a linear relationship exists between the two variables.

The Meaning of r^2 and r. The formula given in the procedure section is useful for computing the numerical value of r^2, but is of little help in understanding its meaning. Of the many formulas available, one exists that is easier to understand and can be written as

$$r^2 = \frac{\text{Total variance of y-values} \quad \textit{minus} \quad \text{Variance of y-values from the regression line}}{\text{Total variance of y-values}}$$

This formula is given in words rather than symbols since we are interested in the interpretation rather than the calculation at this point.*

By looking at the above formula we can see that the coefficient of determination is expressed in terms of two variances. The total variance of the y-values measures the variability of the sample values of the dependent variable about the *mean* of the y's. It can be computed by using the formulas for the variance presented in Chapter 3, and represents the maximum amount of variation or error associated with the dependent variable.

The variance of the y-values about the regression line is similar to

---

*The same formula that can be used to calculate r^2 is given in terms of symbols as

$$r^2 = \frac{S_y^2 - S_{y \cdot x}^2}{S_y^2}$$

where each term in the formula is given as

$$S_y^2 = \frac{\Sigma(y - \bar{y})^2}{n} = \frac{\Sigma y^2}{n} - \bar{y}^2$$

$$S_{y \cdot x}^2 = \frac{\Sigma(y - y')^2}{n} = \frac{\Sigma y^2 - a\Sigma y - b\Sigma xy}{n}$$

For each term, two formulas are given: The definition and the shortcut computational form. The first is the variance of the y-values, and is similar to the formula found in Chapter 3 on variability. The second represents the variance of the sample y-values about the regression line. It is similar to the variance presented in the last section except the denominator is the sample size *without* adjustment for degrees of freedom.

the measure presented in the last chapter; the only difference lies in the denominator, which is not adjusted for degrees of freedom. Hence, the variance about the line appearing in the formula for r^2 is a measure of the variability of the y-values about the regression line of the sample. Consequently, it represents the variability in the dependent variable that is unaccounted for, or not explained, by x or the regression line; it corresponds to the variability in y left over after using x to explain the variability in y.

The difference between the total variability and the variability that is unexplained equals the variability that is explained.

The difference between the total variance and the unexplained variance appears in the numerator of r^2, and represents the variability in y that is *explained* by x or the regression line. That is, if we subtract the unexplained variability from the total, the result is the amount of variability that is explained. Hence, we can express r^2 in the following way:

$$r^2 = \frac{\text{Explained variability in y}}{\text{Total variability in y}}$$

The coefficient of determination, r^2, can be interpreted as the proportion of the total variance in the dependent variable that is explained by the independent variable, or the regression line.

In other words, the coefficient of determination is the ratio of the explained variability to the total variability in the dependent variable. Therefore, r^2 can be interpreted as the *proportion of the total variance of the dependent variable, y, that is explained by the independent variable, x.*

When the variability about the regression line is large, little is explained and r^2 is close to zero, whereas a small amount of variability about the line leads to a large amount explained and a value of r^2 that is close to one. The closer r^2 is to one, the stronger the relationship between x and y in the sample.

When all the points fall on the regression line, there is perfect correlation in the sample. An upward sloping line indicates a direct relationship and a downward sloping line indicates an indirect relationship.

Because of the properties of r^2, it is easier to interpret than r. Both measures, however, impart similar information about the strength of the regression in the sample, although numerically they are different. Also, a sign is attached to r, whereas r^2 always is positive. The sign of r is based upon the sign of the regression coefficient, b. When b is positive the sign of r is plus, and when b is negative the sign of r is minus. A positive r indicates a direct linear relationship, and a negative r indicates one that is indirect.

Consider the diagrams presented in Exhibit 7. Each of the diagrams depicts a different degree of correlation and corresponding values of r^2 and r. Perfect correlation is described by the diagrams in the upper left and lower right corners. In both cases, r^2 equals one since all the points fall on the line. The value of r, however, in the case of an inverse relationship is negative. When no correlation exists, the values of r^2 and r equal zero and the regression line is horizontal with a slope equal to zero also. This is shown in the diagram in the upper right portion of the exhibit.

The last diagram to be mentioned appears in the lower left corner of Exhibit 7. It corresponds to the grade point example used throughout this chapter. Although the slope of the regression line is positive, there is scatter or variability about the line. Based upon earlier remarks, we can interpret the value of r^2 equal to 0.751 as follows: 0.751 or 75.1 percent of the variance in grade point average is explained by math

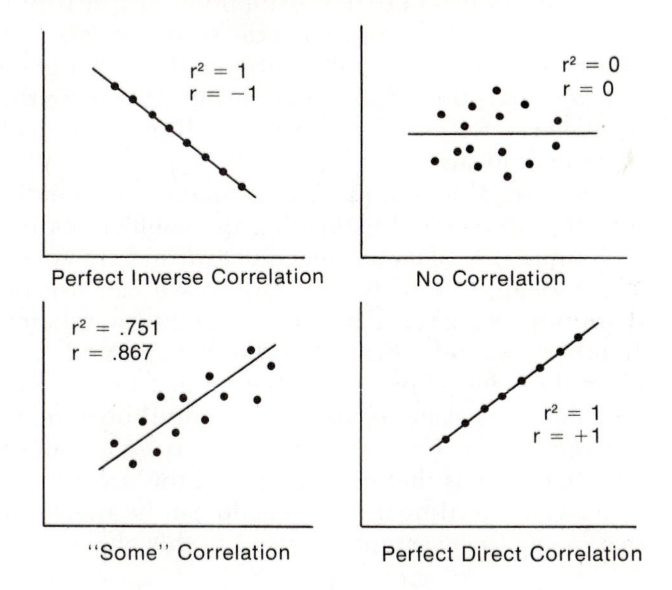

aptitude test score. In other words, the regression line accounts for 75.1 percent of the variance of grade point averages.

We already have performed a statistical test and determined that there is correlation between students' grade point average and math aptitude test scores. Recognize, the hypothesis we tested was that the population correlation is zero against the alternative that it is not zero. By concluding that correlation exists, or that correlation is not zero, we are not able to conclude very much more about the *magnitude* of the correlation in the universe.

The values of r^2 and r represent measures of correlation in the sample. Both measures are a reflection of the correlation in the universe, but each is masked by the fact that they are sample values. If a value of r^2 is "close" to one, we are able to say that correlation is high in the sample. In order to draw the same conclusion about the universe, we must perform a test of a hypothesis that ρ equals some number, other than zero, that we assume or define to be high. A test that ρ equals any number beside zero is available; however, we shall not consider it here.

DISCUSSION

In the preceding portions of this chapter we have covered the basic procedures associated with regression and correlation analysis. Before concluding, however, it is useful to provide some additional insight about these procedures and the way they should be applied.

Equation Form. Whenever an equation is fitted to data, we strive to choose the simplest algebraic formula that adequately describes the relationship between the variables. The straight line considered in the preceding section is one of the simplest forms.

Emphasis in this chapter is placed upon linear relationships. Other types that are curvilinear exist and are used also.

Because of its simplicity, the straight line or linear equation has been used more often than other formulas that are available. In many practical applications the assumption of linearity is justified and leads to a satisfactory description of the overall relationship between the variables. In other cases, different equations that are curvilinear provide a better description of the relationship. It is more difficult to estimate the coefficients for curvilinear relationships since mathematically they are more complicated.

Statistical techniques are available for deciding on the form of the equation to use and estimating the coefficients in the equation. These techniques are of an advanced nature and are not considered in this text. When dealing with two variables, we can use the scatter diagram to determine whether a linear or curvilinear relationship is appropriate. In other words, by inspecting the data plotted on a graph we can determine the general pattern of the points. In some cases the formulas for straight lines can be applied to curvilinear relationships by using a transformation. In other words, by using a mathematical function of a variable such as the logarithm, ln y, the square root, $\sqrt{y}$, or the reciprocal, 1/y, a curvilinear relationship can be transformed to a straight line that meets the assumptions used in this chapter.

Fitting Equations and Least Squares. Most observed data exhibit a scatter of points like the ones described earlier. Consequently, a simple equation with few coefficients rarely passes through the observed points. When fitting an equation to a set of data the goal is to represent the general or "average" relationship between the dependent and independent variables such that the variation of the points about the line is small.

The formulas for estimating the regression coefficients are based upon a procedure called the method of least squares. This is a general procedure for fitting equations that is based upon minimizing the sum of the squared deviations of the values of the dependent variable about the line.

We could, of course, draw a line through the points plotted on a scatter diagram in a "freehand" manner. Such a procedure, however, is not objective since no two individuals would fit the same line to a given set of data. The theory of statistics provides us with objective procedures for fitting regression equations to data. The procedure underlying the formulas used in this chapter is called the *method of least squares*.

Of all possible lines that can be drawn through a set of points, the method of least squares provides a unique line that minimizes the sum of the squares of the deviations, or vertical distances, between the observed y-values and the values of y on the regression line.* It is equivalent to finding a regression that yields the smallest "variance"

---

*Based upon the deviations or residuals about the regression line, the sum of the squared deviations can be computed, i.e.,

$$\Sigma(y - y')^2 \text{ or } \Sigma(y - a - bx)^2$$

The least squares estimates of a and b are the ones that make the above quantity smaller than any other values. The least squares formulas are determined by employing methods from calculus which are used to minimize $\Sigma(y - y')^2$ and result from the simultaneous solution of the two equations

$$\Sigma y = na + b\Sigma x$$
$$\Sigma xy = a\Sigma x + b\Sigma x^2$$

which are referred to as the *normal equations*.

of the y-values about the line. The general formulas for the regression coefficients, a and b, provided earlier are a result of the least squares principle. Consequently, they can be applied to any set of data and will yield estimates that minimize the variability of the points about the resulting regression line.

Correlation and Casuality. Often, when correlation is found, it is assumed that there is a "cause and effect" relationship between the variables. Although there should be some logical basis for relating variables, cause is not demonstrated with a statistical technique.

Statistical methods can be used to find reasons for things that happen and as aids in a decision-making process. Variables are correlated in order to reach a conclusion in response to a specific question or to obtain a tool that is useful in making a more effective decision. When we conclude that two variables are correlated, we can say that they are associated and that more information about one of the variables is available when we have knowledge about the occurrence of the other variable.

Cause is not demonstrated with correlation analysis.

It should be recognized that it is possible to find data providing a "high" correlation for two variables that are not directly connected in any logical way. The classic example of this used in statistics courses is a high correlation coefficient (greater than 0.9) found between the batting average of a famous baseball player and the yearly rainfall in a foreign nation. Another example involves a high correlation observed between the number of births and the size of the stork population in a small Italian town. Results of this type are referred to as nonsense or *spurious* correlation since there is no logical basis for relating the two variables. In cases where a logical basis for relating variables does exist, a relationship can be useful, but we still cannot conclude that one variable causes the other variable to assume certain values.

There are many problems where a dependent and an independent variable are correlated to a *third variable* that can be used as a possible reason for the original two variables to be related. For example, the existence of swamp-like conditions has been related to the prevalence of yellow fever. If swamps are eliminated, it does not automatically follow that the "cause" of the disease has been eliminated. Since the presence of fever-carrying mosquitos is associated with swamp-like conditions, elimination of swamps also removes mosquitos, which is a third factor related to the incidence of the disease.

Additional Considerations. As the sample size increases, the values of r^2 and r become more reliable as measures of the actual correlation, provided the assumptions of randomness, linearity, and constant error are met. The values of r^2 and r are less subject to variability from sample to sample when the sample size is large. In the case of small samples, rejection of the null hypothesis of no correlation does not allow us to conclude very much about the magnitude of the actual correlation.

In many practical applications, "high" values of r^2 are equated with excellent predictability. As indicated, this may not be the case since r^2 is a sample value and is subject to sampling variability. In addition, the value of r^2 may be high, but the assumption of a linear regression may be incorrect. The straight line observed in the sample may represent a portion of a relationship that actually is in the form of a curve. Hence, extrapolated projections may be far from the actual values. Ultimately, if prediction is the goal, the real test of a regression model is whether it works. The value of r^2 does not provide the final answer. It serves as an aid in determining what to do.

Relation Between r^2 and b. We already have seen that a relationship exists between the coefficient of determination, r^2, and the slope of the regression line, b. For fixed units of x and y, b increases as r^2 becomes larger. The magnitude of b can be changed, however, by changing the units of either x or y. The magnitude of r^2 is the same regardless of the units of the two variables.

When the slope of the regression line is zero, there is no correlation in the sample. When the slope is not zero, some correlation in the sample exists.

When there is no correlation in the sample, both r^2 and b are zero. When no relationship between the variables exists, the regression line is horizontal and its slope is zero. Furthermore, lack of correlation in the universe would imply that the slope of the true regression line is zero also. Hence, testing for the significance of an observed regression can be performed in terms of the slope of the line. The test for correlation presented in this chapter is a special case of the test for the significance of the slope. In the case of two variables, the two tests are equivalent under the null hypothesis of no relationship.

Additional independent variables can be included in a regression equation in order to explain more of the variability in the dependent variable. The methods and concepts are similar to the two-variable case; however, the computations are more complex.

Multiple Regression and Correlation *(Optional)*. The concepts of regression and correlation presented in the last three sections can be extended to more than two variables. By including more than one independent variable in the regression equation it is possible to account for more of the variability in the dependent variable. By doing so, we can establish an estimating equation that may predict more reliably. Furthermore, by introducing additional variables it is possible to obtain better explanations about the behavior of the dependent variable.

Methods similar to the ones in this chapter are available that can be used to relate a dependent variable to two or more independent variables. These methods are considered under the general heading of *multiple regression and correlation analysis.* Due to the addition of more variables the calculations become quite complicated. Serious work in multiple regression and correlation is done with the aid of a computer. Consequently, some of the important similarities and differences between the two-variable case and situations involving more than two variables are highlighted in this section in order to provide some awareness of the problem. A sample computer output appears at the end of the chapter and displays the kind of results that are obtained in actual regression and correlation problems.

In order to present the basic ideas of multiple regression and correlation, consider a problem where the amount of family savings is to be related to annual income *and* family size based upon a sample of family data. Data based upon a sample of 9 families appears in Exhibit 8.

Since three variables are involved, three observations for each family representing the unit of association must be obtained. When more than three variables are considered, observations for each addi-

EXHIBIT 8
EXAMPLE OF DATA USED IN MULTIPLE REGRESSION AND CORRELATION ANALYSIS INVOLVING THREE VARIABLES

Family	*Annual Income, x_1 ($1000 Units)*	*Family Size, x_2*	*Annual Savings, y ($100 Units)*
1	15	3	9.0
2	12	6	4.0
3	20	4	9.5
4	10	5	3.5
5	13	3	7.8
6	14	4	8.2
7	19	2	10.5
8	17	3	8.5
9	16	4	9.3

tional variable must be obtained from the same unit of association. In the problem we are considering, savings is the dependent variable and annual income and family size are the independent variables. The regression equation for this problem is written as

$$y' = a + b_1x_1 + b_2x_2$$

Multiple regression equation involving three variables.

where y′ represents estimated savings measured in $100 units, x_1 represents annual income in $1000 units, and x_2 stands for family size in terms of the number of people. Subscripts are attached to the independent variables so that they can be distinguished. The regression coefficient "a" represents the estimated value of the dependent variable, y′, when x_1 *and* x_2 equal zero. The symbol "b" is used again to represent a regression coefficient "attached" to an independent variable; however, each has a subscript that indicates the variable to which it corresponds. b_1 and b_2 are called net regression coefficients and differ slightly in interpretation relative to the two-variable case. If more than three variables are considered, there would be a net regression coefficient for each additional independent variable.

The method of least squares can be used to estimate the coefficients in regression problems involving any number of variables. In general, the results of the least squares method are expressed in terms of a simultaneous system of equations called the "normal equations" since individual equations for each coefficient are too complex. In the case of three variables the normal equations are given as

$$\begin{aligned}\Sigma y &= na + b_1\Sigma x_1 + b_2\Sigma x_2\\ \Sigma x_1y &= a\Sigma x_1 + b_1\Sigma x_1^2 + b_2\Sigma x_1x_2\\ \Sigma x_2y &= a\Sigma x_2 + b_1\Sigma x_1x_2 + b_2\Sigma x_2^2\end{aligned}$$

Normal equations used to find least squares estimates of coefficients in a regression equation involving three variables.

By substituting the cumulative terms based upon preliminary computations, this system of equations can be solved simultaneously for a, b_1, and b_2. The result represents the least squares estimates of the regression coefficients when three variables are involved. For multiple regression problems involving more than three variables, there is a separate equation for each additional variable and an added term in each equation for each additional regression coefficient.

EXHIBIT 9
PRELIMINARY CALCULATIONS NEEDED IN THREE-VARIABLE REGRESSION PROBLEM INVOLVING SAVINGS, INCOME, AND FAMILY SIZE

Family	*Income, x_1 ($1000)*	*Size, x_2*	*Savings, y ($100)*	x_1x_2	x_1y	x_2y	x_1^2	x_2^2	y^2
1	15	3	9.0	45	135.0	27.0	225	9	81.00
2	12	6	4.0	72	48.0	24.0	144	36	16.00
3	20	4	9.5	80	190.0	38.0	400	16	90.25
4	10	5	3.5	50	35.0	17.5	100	25	12.25
5	13	3	7.8	39	101.4	23.4	169	9	60.84
6	14	4	8.2	56	114.8	32.8	196	16	67.24
7	19	2	10.5	38	199.5	21.0	361	4	110.25
8	17	3	8.5	51	144.5	25.5	289	9	72.25
9	16	4	9.3	64	148.8	37.2	256	16	86.49
(n = 9)	136 (Σx_1)	34 (Σx_2)	70.3 (Σy)	495 (Σx_1x_2)	1117 (Σx_1y)	246.4 (Σx_2y)	2140 (Σx_1^2)	140 (Σx_2^2)	596.57 (Σy^2)

Consider the savings example and let us find estimates of the coefficients in the regression equation relating family savings to annual income and family size. Based upon preliminary calculations of the cumulative terms that appear in Exhibit 9, we substitute into the normal equations as follows:

$$\begin{aligned} 70.3 &= 9a + 136b_1 + 34b_2 \\ 1117 &= 136a + 2140b_1 + 495b_2 \\ 246.4 &= 34a + 495b_1 + 140b_2 \end{aligned}$$

By solving this system of equations simultaneously for the unknown values, a, b_1, and b_2, we get

$$\begin{aligned} a &= 4.88709 \textit{ or } 4.89 \text{ (rounded)} \\ b_1 &= 0.43264 \textit{ or } 0.43 \text{ (rounded)} \\ b_2 &= -0.95657 \textit{ or } -0.96 \text{ (rounded)} \end{aligned}$$

Based upon the *rounded* estimates, the regression equation is

$$y' = 4.88 + 0.43x - 0.96x$$

y' = Estimated family savings measured in \$100 units
x_1 = Annual income measured in \$1000 units
x_2 = Family size

The a-term equal to 4.88 is the value of y' when x_1 and x_2 are zero. Literally, this is interpreted to mean that estimated savings equals $488 for a family with no income and no people. Therefore, this represents an extrapolated value that has no practical meaning in this particular problem.

The interpretation of the regression coefficients in a multiple regression equation is different from the two-variable case.

In general, the b-terms are of more interest. Earlier we mentioned that these are referred to as *net* regression coefficients. This means that they represent a change in the dependent variable for a unit change in one of the independent variables net of the effect of the other variable, or under the assumption that the other variable is held constant.

The value of b_1 equal to 0.43 can be interpreted in terms of the original units to mean that $43 more is saved by a family whose income is $1000 more than the income of another family that is of the same size. The value of b_2 equal to -0.96 can be interpreted to mean that $96 less is saved by a family with one more member than another family with the same income. In general, the results indicate that families with greater income save more but will save less when the size of the family is larger. This conclusion is based upon the signs of the coefficients. Sometimes, there can be a problem associated with attaching importance to estimates of net regression coefficients that does not occur when there is one independent variable.

Measurement of the amount of error in the dependent variable not explained by the independent variables of a multiple regression equation is similar to the two-variable case. The measures used are the variance and standard error of estimate of the dependent variable about the regression equation. The estimated conditional variance is defined by the formula

$$\hat{S}^2_{y\cdot 12} = \frac{\Sigma(y - y')^2}{n - 3}$$

Estimated variance about regression equation involving three variables.

The subscript used here indicates that the variance is of the dependent variable y conditional upon the variables having the subscripts 1 and 2, or x_1 and x_2. The numerator represents the sum of the squared deviations of the observed y-values from the estimated value, y′, obtained from the regression equation for given values of x_1 and x_2. The denominator of the formula represents the degrees of freedom equal to the sample size minus 3, where the 3 corresponds to the number of coefficients in the regression equation that are estimated from the sample. In general, the estimated variance of the dependent variable about the regression equals the sum of the squared deviations of the sample y-values from the estimated values divided by the sample size *minus* the number of coefficients in the regression equation. The standard error of estimate is obtained as the square root of the variance similar to the two-variable case. For the savings problem, the variance and standard error of estimate about the regression are calculated as

$$\hat{S}^2_{y\cdot 12} = 0.90792 \text{ or } 0.91 \text{ (rounded)}$$

$$\hat{S}_{y\cdot 12} = 0.95285 \text{ or } 0.95 \text{ (rounded)}$$

Hence, the standard error of estimate of savings in original units conditional upon income and family size approximately equals $95. This value can be used to construct an interval estimate similar to the two-variable case using the t-distribution. The number of degrees of freedom, however, equals the denominator of the estimated variance.

For problems of multiple regression and correlation, correlation is measured in terms of the *coefficient of multiple determination*, R^2, and the *coefficient of multiple correlation*, R. Capital letters are used to distinguish between two-variable and multiple regression problems.

In general, these are defined similarly to the two-variable case. The coefficient of the multiple determination is defined by the expression*

$$R^2 = \frac{\text{Total variance of y} \;\; \textit{minus} \;\; \text{Variance of y conditional on the independent variables}}{\text{Total variance of y}}$$

$$= \frac{\text{Variance in y explained by the independent variables}}{\text{Total variance of y}}$$

Definition of the coefficient of multiple determination.

The coefficient of multiple determination represents the proportion of the total variance in the dependent variable that is explained by the independent variables simultaneously. The coefficient of multiple correlation is the positive square root of the coefficient of multiple determination; its sign always is plus.

The coefficient of multiple determination, R^2, can be interpreted as the proportion of the total variance in the dependent variable that is explained by all of the independent variables appearing in the regression equation.

The value of R^2 in the savings problem equals 0.88519 or 0.89 when rounded. Hence, we can say that 89 percent of the variance in family savings of the sample is explained by annual income *and* family size. If family income alone were used, 74 percent of the variability in savings is explained, and 67 percent is explained if family size alone were used.

It is possible to perform a test of a hypothesis to determine whether correlation exists between the dependent variable and any number of independent variables. A general test is available that is based upon the F-distribution. This is the distribution that is used in Chapter 14 to perform tests for the equality of means using the analysis of variance. The test for correlation is based upon the following statistic:

$$F = \frac{\Sigma(y' - \bar{y})^2/(k - 1)}{\Sigma(y - y')^2/(n - k)}$$

F-statistic used to test for correlation.

The expression $\Sigma(y' - \bar{y})^2$ appearing in the numerator of F represents the sum of the squared deviations of the estimated values, y′, from the mean of the y-values, $\bar{y}$. This sum of squared deviations, or "explained sum of squares," is divided by the "number of degrees of freedom," $k - 1$, to obtain the variance in the dependent variable *explained* by the regression equation. k equals the number of coefficients in the regression equation.

The expression $\Sigma(y - y')^2$ appearing in the denominator of F represents the sum of the squared deviations of the observed y-values from the estimated values based upon the regression equation, y′. This sum of squared deviations is referred to as the residual or "error sum of squares." When the error sum of squares is divided by the "denominator degrees of freedom," $n - k$, the result is the variance in the dependent

*In this form, the coefficient of determination can be expressed in terms of a formula as

$$R^2 = \frac{s_y^2 - s_{y\cdot12}^2}{s_y^2}$$

where s_y^2 is the variance of the y-values about the mean of the y's in the sample and $s_{y\cdot12}^2$ is the variance of the y's about the regression, where neither measure is adjusted for degrees of freedom.

variable that is *unexplained* by the regression equation. The term in the denominator of F is the general expression for the estimated variance about the regression, which we denote as $\hat{S}^2_{y \cdot 12}$ in the case of three variables.

A general test is available for testing whether overall correlation exists in terms of an F-statistic. This is the ratio of the explained variability to the unexplained variability.

The F-statistic, therefore, represents the ratio of the explained variance to the unexplained variance. The unexplained or residual variance is a measure of chance variability. If the explained variability is "close" to this measure of chance variation, it can be concluded that no correlation exists. If, however, the F-statistic is "unusually" large, or much greater than one, it can be concluded that there is correlation between the dependent and the independent variables.

The F-statistic given above can be written in another form in terms of the coefficient of multiple determination, R^2:

$$F = \frac{(n - k)R^2}{(k - 1)(1 - R^2)}$$

F-statistic used to test for correlation expressed in terms of the coefficient of multiple determination.

This also is a general expression that applies to any regression problem. It can be shown that it equals the square of the value of "t" presented earlier in order to test for correlation between two variables, or when k equals 2 and R^2 is replaced with r^2.

When performing the test with the F-statistic, Table 6 in the Appendix is used. If the computed value of the F-statistic exceeds the value in the table for a given level of significance, one may conclude that correlation exists.

When it is concluded that correlation exists based upon the test outlined, it is not possible to determine which of the variables in the multiple regression equation is correlated with the dependent variable. The conclusion is that *all* of the independent variables combined are correlated with the dependent variable. It is possible that only one of the independent variables is correlated, and it is this variable that creates a significant result.

Instead of testing for overall correlation using the F-statistic given above, it is possible to use a test to determine whether each independent variable has a significant effect upon the dependent variable. A common practice is to test the b-values individually to determine whether they are significantly different from zero. When a b-value is significantly different from zero, it is concluded that the corresponding variable has a significant effect upon the dependent variable. Tests that are applied individually to the regression coefficients are based upon the t-distribution and are similar to the ones for means. The difference lies in the way the standard errors of the sampling distributions of the b-values are computed.

Tests associated with each of the net regression coefficients can be performed in order to determine whether each of the independent variables has a significant effect on the dependent variable.

Whenever there is more than one independent variable in a regression equation, there is the possibility that correlation exists among the independent variables. If the goal is to use the regression equation to predict the dependent variable, there is no real problem since R^2 is not affected. There is a problem, however, if the goal is to measure the separate effects of the independent variables upon the dependent variable directly in terms of the regression coefficients. When the independent variables are correlated, or when *multicolinearity* exists, the regression coefficients are affected. More specifically, the estimates of the

Relationships can exist among the independent variables. When this occurs, interpretation of the net regression coefficients becomes a problem.

regression coefficients become increasingly unreliable as the correlation among the independent variables increases. Consequently, more variability in the values of b_1 and b_2 exists from sample to sample. This means that an estimate of b_1 or b_2 for a specific sample cannot be trusted to reflect the true values when multicolinearity is present.

CALCULATORS AND COMPUTERS

By examining the formulas in this and the last three chapters, it is easy to see that computations become increasingly complex when more variables are introduced into the problem. Although the amount of computation in a three-variable regression and correlation problem is manageable, there is a lot of work involved.

An increasing number of hand-held and desk-top programmable calculators with a stored program feature can be used to solve regression problems. Many calculators are preprogrammed to calculate the regression coefficients and r^2 for two-variable problems. Optional program libraries are available for some of these calculators that can be used to solve three-variable problems.

Although specific programs for calculators are not available for problems involving more than three variables, some programmable calculators include programs that can be used to solve simultaneous equation systems. These can be adapted in order to solve regression problems that include up to eight or nine variables at once. Large scale problems, however, are solved with a computer.

There are many computer programs available that can be used to solve regression problems. These programs are general and can be used on a variety of computer systems to solve regression problems involving two or many variables. When there are a large number of variables, the computer is the only reasonable way to solve a regression problem. A sample of the output from a computer is presented in Exhibit 10.

The format of the output from a computer varies depending upon the program used. The output illustrated in the exhibit gives an idea of the kind of information provided in a computer output. Some programs provide options so that all the information is not printed and only what is necessary can be chosen by the individual using the program.

The output in the exhibit is associated with the savings example presented in this chapter. The variables in the problem have been renumbered in terms of the above output. Savings is denoted by 1., whereas income is represented by 2. and family size by 3. Some programs provide the option of using actual alphabetic labels instead of numbers to identify the variables. Some discrepancies exist in the results presented in the exhibit due to rounding since the results are printed to 5 decimal places. The computer carries the actual calculations to many more places.

Most of the output in the exhibit is self-explanatory since each set of results is labeled. There is, however, information provided in the sample output that has not been discussed or has been treated lightly in this chapter. The regression coefficient that we designated as "a" is given alongside the savings variable. The standard error column (i.e.,

EXHIBIT 10
AN EXAMPLE OF THE OUTPUT OF A COMPUTER FOR PROBLEMS IN REGRESSION AND CORRELATION

```
                    MULTIPLE LINEAR REGRESSION ANALYSIS

NO OF VARIABLES  3
DEPENDENT VARIABLE  1
NUMBER OF OBSERVATIONS  9

REGRESSION COEFFICIENTS

        VARIABLE       COEFFICIENT       STD ERROR         BETA
           1.            4.88709
           2.            0.43264          0.12917        0.57868
           3.           -0.95657          0.35009       -0.47206

COEFFICIENT OF DETERMINATION        0.88528
COEFFICIENT OF CORRELATION          0.94089
STANDARD ERROR OF ESTIMATE          0.95247

ANALYSIS OF VARIANCE TABLE

        SOURCE             DF      SUM SQUARES      MEAN SQUARES          F
          REGRESSION        2       42.00568         21.00284         23.15123
               ERROR        6        5.44321          0.90720

CORRELATION MATRIX

                1.            2.            3.
        1.    1.00000
        2.    0.86171       1.00000
        3.   -0.81901      -0.59955       1.00000

OBSERVATION   NO.        ACTUAL        PREDICTED        RESIDUAL
              1.          9.0           8.50704          0.49296
              2.          4.0           4.33940         -0.33940
              3.          9.5           9.71369         -0.21369
              4.          3.5           4.43068         -0.93068
              5.          7.8           7.64175          0.15825
              6.          8.2           7.11782          1.08218
              7.         10.5          11.19418         -0.69418
              8.          8.5           9.37233         -0.87233
              9.          9.3           7.98311          1.31689
```

STD ERROR) under the regression coefficients section presents the standard deviations of the b-values mentioned earlier. These can be used to test the separate effects of the independent variables. Generally, the standard error of the a-term is not always printed since it is not used frequently.

A new concept is introduced in the same section. The last column is entitled BETA. The values in this column represent values of the b-values that are converted into standard deviation units and are called *Beta coefficients*. By converting to standard deviation units the relative effects of each of the variables can be compared since the independent variables often are expressed in different units. For example, the value of 0.57868 means that savings changes by 0.57868 standard deviations when variable 2. or income changes by one standard deviation. A similar interpretation is attached to −0.47206. These coefficients should not be compared when the independent variables are correlated.

The results appearing in the section entitled ANALYSIS OF VARIANCE TABLE correspond to the earlier discussion regarding the F-test for overall correlation. The "regression" row corresponds to the numerator of the F-statistic and the "error" row corresponds to the denominator. Hence, in the degrees of freedom column (i.e., DF), the value of 2 corresponds to $k - 1$ and 6 corresponds to $n - k$. The regression sum of squares equal to 42.00568 represents the quantity $\Sigma(y' - \bar{y})^2$ and the error sum of squares equal to 5.44321 represents the value of $\Sigma(y - y')^2$. The mean squares are obtained by dividing the sum of squares by the degrees of freedom.

The value of F in the last column is the value of the F-statistic or the ratio of the mean squares. This result can be used to determine whether there is overall correlation between the dependent variable and the independent variables. Table 6 is used to perform the test.

A common practice is to report the correlation coefficients or "r-values" corresponding to *all pairs* of variables. This is given in terms of the CORRELATION MATRIX appearing in the exhibit. Each value in the table represents the coefficient of correlation between the variables corresponding to the row and column numbers associated with each value. For example, the value of 0.86171 is the coefficient of correlation, $r_{y \cdot 2}$, between variable 1., or savings, and variable 2., or income. Notice the correlation coefficient between income and family size, r_{23}, equals −0.59955, which indicates that there is some correlation that is negative between the independent variables. The values of 1.00000 merely indicate that there is perfect correlation for a variable that is correlated with itself.

Printed at the end of Exhibit 10 are the original y-values in the sample that appear in the column entitled ACTUAL. Corresponding to each of these values are the values of y′ based upon the regression line for the sample values of the independent variables. These appear in the PREDICTED column. The last column entitled RESIDUAL provides the difference between the actual and predicted values, $y - y'$.

The information presented in this part of the printout can be used to compare the observed values of the dependent variable with the values obtained from the regression equation. Furthermore, other methods of analysis not discussed here may be applied to the residuals in order to determine how closely underlying assumptions are met and to determine whether a better explanation of the dependent variable can be achieved using additional techniques.

The computer output shown in the above exhibit is an example of a multiple regression problem where the independent variables are chosen in advance and the analysis considers these variables all at one time. Frequently, many independent variables may be considered as candidates to be included in the regression equation when the overall goal is to choose the combination that will provide the best explanation of the dependent variable. A key advantage of the computer is that it can be used as an aid in the selection of variables using other statistical methods that are too difficult or virtually impossible to accomplish with other means of computation.

QUESTIONS AND PROBLEMS

DEFINITIONAL QUESTIONS

1. What is the meaning of the following terms?

   (a) Relationship between variables
   (b) Unit of association
   (c) Scatter diagram
   (d) Regression line

2. Distinguish between a linear and a curvilinear regression relationship. What kind of relationship is emphasized in this chapter?

3. What is meant by the term *fitting a regression line* to a set of data? What is the name of the method used in this chapter to fit a regression line?

4. What is the purpose of fitting a regression line to a set of data?

5. What are regression coefficients? In general, how are they interpreted?

6. What is a point estimate? How is a regression line used to find a point estimate?

7. What is meant by the term *extrapolation?* What is the basic assumption that must be made in order to use a regression line to extrapolate?

8. What is the meaning of the following terms?

   (a) Variability in the dependent variable
   (b) Variability about the regression line
   (c) Variance about the regression line
   (d) Standard error of estimate

9. What are residuals? Why are they important in regression problems?

10. List the possible sources of error in a regression problem. Of these, which is considered in this chapter?

11. What measures are used to measure variability about a regression line?

12. What are the definitional and computational forms of the estimated variance about the regression line?

13. List the assumptions used when making inferences in regression problems?

14. What is the difference between the universe regression line and the sample regression equation? What is meant by the term *conditional mean of the dependent variable?*

15. Describe the procedure for establishing an interval estimate for the dependent variable. What is meant by the term *degrees of freedom*, and how is it related to this procedure?

16. In general, what is meant by the term *correlation?* How is the term used in this chapter?

17. What is the purpose of correlation analysis?

18. What is the meaning of the following terms?

    (a) No correlation
    (b) Some correlation
    (c) Perfect correlation

19. In words, how is the coefficient of determination, r^2, interpreted?

20. What are the numerical limits of r^2 and r? What does it mean when r equals one? Zero? Minus one?

21. How can the algebraic sign of r be determined? What does it mean when r is positive? Negative?

22. What is the relationship between the coefficient of correlation, r, and the regression coefficient representing the slope of the regression line, b?

23. What is the difference between ρ and r?

24. Specify the null and alternate hypotheses for a test of correlation.

25. What is the t-statistic that is used in a test for correlation? What is meant by the number of degrees of freedom in a test for correlation and how is it used?

26. Describe the procedure for determining whether or not correlation between two variables exists.

27. What is the relationship between the coefficient of determination and the coefficient of correlation?

COMPUTATIONAL PROBLEMS

28. Given the following set of observations where y represents the dependent variable and x the independent variable:

x	y
30	47
30	50
40	52
50	52
100	59
20	47
90	59
30	52
70	53
60	59
60	56
40	50

(a) Find the least squares estimates of the coefficients in the equation $y = a + bx$ relating y to x.
(b) Use the results in part (a) to estimate y when x equals 30.

29. For the regression equation $y = a + bx$, estimate the coefficients based upon the following information:

$$n = 10$$
$$\Sigma x = 50$$
$$\Sigma x^2 = 284$$
$$\Sigma y = 110$$
$$\Sigma xy = 618$$

Use the results to estimate y when x equals 4.

30. A firm administers a test to sales trainees before they go into the field. Management of the firm is interested in determining the relationship between test scores and sales made by the trainees at the end of one year in the field. The following data were collected for 10 salesmen who have been in the field one year:

Salesman	*First Year Sales ($1000 Units)*	*Test Score*
1	112	2.50
2	104	2.37
3	125	2.64
4	136	3.05
5	129	2.85
6	158	4.12
7	140	3.38
8	177	4.91
9	169	4.46
10	147	3.73

(a) Find the least squares regression line relating first year sales to trainee test scores.

(b) Based upon the estimated relationship, predict the first year sales of a trainee who receives a test score of 4.75.

31. A regional unemployment office is interested in determining the relationship between the number of weeks benefits are collected as a function of the amount of unemployment compensation received. A sample of records of 9 claimants who have stopped receiving benefits and returned to work is selected. The following data is recorded.

Claimant	*Weekly Benefit (Dollars)*	*Number of Weeks*
A	93	17
B	80	10
C	85	13
D	65	8
E	96	14
F	75	11
G	107	15
H	72	6
I	84	18

(a) Use the data in the table to determine estimates of the coefficients in a linear regression equation relating the number of weeks benefits are received to the amount of the benefit collected.
(b) Based upon the results in part (a), estimate the number of weeks compensation is received for a claimant who receives $87.

32. A production manager of a firm manufacturing a piece of heavy equipment is interested in establishing an equation relating the cost of production to the size of a production run in terms of the number of units produced. The following information is available:

y = production cost per run ($1000 units)
x = size of production run (number of units)
x is observed between 142 and 189 units for 6 production runs (i.e., $n = 6$)

$$\Sigma x = 965 \qquad \Sigma y = 1489$$
$$\Sigma x^2 = 156609 \qquad \Sigma xy = 241293$$

(a) Based upon the information given, determine the regression line relating cost of production to output.
(b) Use the line found in part (a) to estimate the cost associated with producing 175 units.

33. Based upon a sample of 9 paired observations of the two variables x and y, the following information is available:

$$y' = 9.9 - 0.11x$$
$$\Sigma y = 41$$
$$\Sigma y^2 = 212$$
$$\Sigma xy = 1833$$

Use the information to calculate the values of $\hat{S}^2_{y \cdot x}$ and $\hat{S}_{y \cdot x}$.

34. Refer to Problem 33. Assuming $\hat{S}_{y \cdot x}$ equals 1.05, construct a 95 percent interval estimate of the dependent variable when x equals 47.

35. The following information was collected about two variables, x and y, based upon a sample of 11 items.

x	y
33	534
24	272
7	127
13	213
37	581
21	365
46	683
13	174
28	482
42	614
26	416

(a) Find the coefficients in the regression equation $y' = a + bx$.
(b) Find the estimated variance and standard error of estimate associated with the regression line found in part (a).

36. Trainees learning to keypunch are given a standard test each week of a 10 week training program in order to determine their progress. In order to establish a relationship between keypunching speed as a function of the week of training, one trainee is selected at random within each week of the training period. The following results are available for the sample:

x = Number of weeks in the training program
y = Number of minutes to keypunch standard assignment

$n = 10$ $\Sigma y = 119$
$\Sigma x = 55$ $\Sigma y^2 = 1799$
$\Sigma x^2 = 385$ $\Sigma xy = 486$

$$y' = 23.13 - 2.04x$$

Using the information given, calculate the estimated variance and standard error of estimate associated with the time to keypunch the standard assignment based upon the week in the training period.

37. Refer to Problem 36. Assume the standard error of estimate equals 2 minutes. Find a 98 percent interval estimate of the time

it takes a trainee who is in the fifth week of training to keypunch the standard assignment.

38. Given below are the monthly food expenditures and annual incomes for a random sample of 10 families.

Family	*Annual Income ($100 Units)*	*Food Expenditure (Dollars)*
1	95	135
2	75	140
3	110	167
4	122	246
5	140	207
6	170	325
7	165	268
8	105	118
9	135	186
10	183	250

(a) Assuming the relationship is linear, find the coefficients in the regression equation relating monthly food expenditure to income.
(b) Find the estimated variance and standard error of estimate of monthly food expenditure based upon annual family income.

39. Given two variables x and y. Determine the values of r^2 and r based upon the following information:

$$n = 50 \qquad \Sigma y = 1540$$
$$\Sigma x = 167 \qquad \Sigma y^2 = 57460$$
$$\Sigma x^2 = 693.1 \qquad \Sigma xy = 6286$$

The regression line relating the variables is $y' = 2.6 + 8.4x$.

40. The following information is associated with two variables x and y:

$$n = 20 \qquad \Sigma y = 133$$
$$\Sigma x = 83 \qquad \Sigma y^2 = 960$$
$$\Sigma x^2 = 463 \qquad \Sigma xy = 503$$

(a) Find the coefficient of determination and the coefficient of correlation.
(b) Use the results found in part (a) to determine if correlation exists in the universe.

41. Given the following information based upon a sample of 9 paired observations associated with the variables x and y:

x	y
2.8	41.6
2.3	48.2
1.3	17.4
4.6	68.3
2.1	36.5
3.7	58.1
1.3	21.3
0.7	12.7
2.4	27.2

Use the information given to determine whether or not correlation exists.

42. Part of a crime study report contains information regarding the percentage change in the amount of crime between 1975 and 1977 for a sample of 12 townships adjacent to a major city. The average population for each of the townships during the three year period also is contained in the report. The following information is available:

y = Percentage change in crime between 1975 and 1977
x = Average population size during 1975 and 1977
x is in units of 1000 people

$n = 12$ $\Sigma y = 547$
$\Sigma x = 470$ $\Sigma y^2 = 41485$
$\Sigma x^2 = 22294$ $\Sigma xy = 16639$

Use the information given to calculate the coefficient of determination associated with the percentage change in crime and township population size.

43. As part of a study of the operations of a medium sized bank, management is interested in determining whether or not an effective relationship exists between the amount of money loaned to business organizations and the firms' current operating ratio, or the ratio of current assets to current liabilities. A random sample of 6 loan accounts is selected from the bank's current records that contains the following information:

Loan Amount ($1000 Units)	*Current Ratio*
1150	5.25
800	3.25
1200	5.75
100	1.50
1100	4.50
300	2.00

(a) Find the coefficient of determination and the coefficient of correlation for the sample data.
(b) Using the results of part (a), determine whether actual

correlation exists between the current operating ratio and the amount of loans given by the bank.

44. A firm manufacturing a variety of over-the-counter products is interested in determining whether or not a relationship exists between its market share for a certain mouthwash and the amount of advertising in print media. Estimates of last year's market share, or the percentage of industry sales attributed to the firm's mouthwash, are available for 27 marketing areas throughout the country. The amount of advertising for each of these areas is available in terms of the dollar amount spent on print media per thousand population. The following results are available:

x = dollar amount of advertising per thousand population
y = market share

$n = 27$	$\Sigma y = 348$	$a = 5.85$
$\Sigma x = 44$	$\Sigma y^2 = 4626$	$b = 4.32$
$\Sigma x^2 = 77$	$\Sigma xy = 590$	

(a) Find the coefficient of determination and the coefficient of correlation in the sample.
(b) Use the results of part (a) to determine whether there is a relationship between print media advertising expenditure and market share.

CONCEPTUAL QUESTIONS AND PROBLEMS

45. Assuming a relationship exists between two variables, why is it necessary to "fit" an equation to data associated with these variables? What essential property of the data makes this necessary?

46. In order to fit a regression line to a set of data, why is it necessary to observe a pair of observations from the same unit of association?

47. Given that there is variability or scatter among the observations, how does a regression line describe a relationship between two variables?

48. Refer to Problem 30. The least squares regression line fitted to the data in that problem is $y = 49.6 + 26.5x$, where y represents first year sales and x represents a trainee's test score.

(a) In terms of the problem, interpret the coefficients in the regression equation.
(b) Does the interpretation of the "a-term" have any practical meaning in this problem? Explain. Is this necessary in order to use the equation to predict a trainee's first year sales based upon his test score?
(c) Based upon the regression line given above, estimate first

year sales for a trainee who scores 3.5 on the test. Is it likely that the actual sales for this trainee will equal this estimate exactly? Why or why not?

49. Refer to Problem 32. The regression line relating cost of production to output is given as $y = 40.7 + 1.29x$.

(a) In terms of the problem, interpret the coefficients in the regression equation.
(b) Does the interpretation of the "a-term" have any practical meaning in the problem? Explain. Under what circumstances is this value useful?
(c) Use the regression line given above to predict the cost to produce 150 units. Is it likely that the actual cost will equal this value exactly? Explain.

50. Given below are the total number of accidents attributed to trucks in various classes of size in terms of average weight.

Average Weight (Pounds)	*Number of Accidents*
53400	3319
27200	2370
12700	720
21300	1290
58100	3650
68300	4572
36500	2140
17400	1340
48200	2764
61400	4200
41600	2623
75000	4760

(a) Calculate the least squares regression coefficients corresponding to a linear equation fitted to the data. *Note.* The amount of computations can be reduced if the units of the weights are changed to 100 pounds.
(b) Interpret the coefficients found in part (a) in terms of the problem. Does the "a-term" have a practical meaning in this case? Explain. Is this necessary for the resulting equation to be useful?
(c) Plot the original data on a graph. Also, trace the regression line based upon the values in part (a) through the points. Does a straight line describe the relationship between truck weight and accidents satisfactorily? Explain.

51. Incomes of residents of a certain state are taxed at a standard rate of 8 percent. Since this percentage is applied to net income after deductions are taken, the percentage of gross income varies below the stipulated rate. In order to determine the relationship between high bracket incomes and the actual percent of gross income paid in taxes, a sample of 90 taxpayer records are examined, resulting in the following information:

x = gross income ($1000 units)
y = percent of gross income paid in taxes
x is observed between $30,000 and $74,000

$$\Sigma x = 4500 \qquad \Sigma y = 410$$
$$\Sigma x^2 = 245160 \qquad \Sigma xy = 18328$$

(a) Assuming the relationship is linear, determine the regression line relating the actual tax rate as a function of gross income.
(b) In terms of the problem, interpret the "b-term" or the slope found in part (a). Does the result seem reasonable? Explain.
(c) Use the regression line found in part (a) to estimate the percent of gross income paid in taxes by a taxpayer earning $35,000 and one who earns $65,000. Are the results consistent with the estimate of the slope of the regression line? Explain.
(d) Would you be willing to use the regression line to estimate the actual tax rate for a taxpayer who earns $80,000? Discuss what assumption is necessary in order to make a valid estimate?
(e) Use the regression line found in part (a) to estimate the tax rate paid by a taxpayer earning $100,000. Is the result a valid estimate? Explain. Does the result mean that the regression line is useless? Would a curvilinear relationship fitted to the data provide a more reasonable description of the relationship? Explain.

52. What information is provided by the variance computed about the regression line? By the standard error of estimate?

53. What is meant when it is said that the smaller the standard error of estimate, the better the regression line "fits" the data? How is the standard error of estimate related to the possible usefulness of a regression line?

54. What is the relationship between a residual and the standard error of estimate? If all residuals about a regression line are zero, what is the value of the standard error of estimate? What would the scatter diagram look like in such a case?

55. Suppose the ratio of profit to assets, or the rate of return on assets, for a certain type of small manufacturer is linearly related to assets according to the relationship

$$y' = 1 + .00002x$$

where y represents the rate of return on assets and x represents dollar assets. The standard error of estimate corresponding to the rate of return is 4.8. Construct a 95 percent interval estimate of the rate of return for a firm with 1.5 million dollars in assets. Interpret the result.

56. Given below are the total sales of 10 firms for one year along with the amount spent on advertising:

Firm	*Advertising ($100,000 Units)*	*Sales ($1,000,000 Units)*
A	2.6	5.1
B	1.3	2.1
C	1.7	3.5
D	3.1	6.6
E	1.9	4.0
F	2.8	5.8
G	3.7	7.1
H	1.0	1.4
I	2.3	4.6
J	1.6	2.7

(a) Calculate the least squares estimates of the coefficients in a linear regression equation relating sales to advertising.
(b) Based upon the values found in part (a), estimate the sales of a firm that spends $200,000 for advertising.
(c) Compute the standard error of estimate of annual sales based upon advertising expenditure.
(e) Using a 98 percent probability level, establish an interval estimate of the sales of a firm that spends $200,000 for advertising. How is this estimate interpreted?

57. Listed below are the midterm and final exam grades for a class of 15 students:

Midterm Grade	*Final Exam Grade*
96	97
61	54
80	80
73	70
78	80
67	70
65	67
88	93
76	77
74	73
91	87
50	62
72	77
85	82
83	86

(a) Assuming the relationship is linear, determine the regression equation relating final exam grades to midterm grades.
(b) Interpret the coefficients found in part (a) in terms of the problem.
(c) Assume the relationship found in part (a) applies to another class taking the same course. If 60 is a passing grade, can

a student who failed the midterm with a grade of 50 "expect" to pass the final exam?

(d) Compute the standard error of estimate and use it to construct a 95 percent interval estimate of a student's final exam grade who fails the midterm with a grade of 50. Does this estimate provide any additional information about the student's final exam grade than the estimate found in part (c)? Explain.

(e) Would the interval estimate found in part (d) be more informative if the standard error of estimate were smaller? Explain.

58. Based upon a random sample of 102 sales representatives of a large firm, a regression analysis was conducted relating annual commission to a representative's age. The results of the analysis are given below:

x = age in years

y = annual commission in \$1000 units
x is observed between 30 and 50 years

$y' = -10 + x$

$\hat{S}_{y \cdot x} = 3$

Based upon these results, should a sales representative who is 40 years old and receives \$17,000 in annual commissions be considered a "poor performer"? Justify your conclusion statistically and explain why you reached that conclusion.

59. What information is provided by the coefficient of correlation of a sample? Why is it necessary to perform a test of a hypothesis for correlation?

60. When the result of a test for correlation is significant, what conclusion is drawn if r is positive? If r is negative?

61. Briefly, but clearly, explain how the coefficient of correlation and the standard error of estimate are related.

62. Criticize or explain the following statement. A relationship exists between the coefficient of correlation and the slope of the regression line; however, the magnitude of the slope does not indicate anything about the strength of the relationship between the variables unless the slope is zero.

63. What is meant when it is said that a "high" value of r^2 indicates a "good fit"?

64. How is the value of r^2 related to the use of a regression line to make estimates or predictions?

65. Criticize or explain the following statement. When the value of r^2 is high and the regression line fits the data well, extrapolation is not a problem.

66. Refer to Problem 42. The coefficient of determination between y and x is 0.356. In terms of the problem, interpret this value.

67. Refer to Problem 44. The coefficient of determination between y and x is 0.703. Interpret this value in terms of the problem.

68. The school board of a moderately large city undertakes a study of student failures throughout the school system during the previous year. Part of the study involves an examination of the relationship between failure to be promoted and the grade level of students. The following information is available:

y = percentage of students not promoted
x = grade level
x ranges between 1 and 12

$n = 12 \qquad \Sigma y = 117$
$\Sigma x = 76 \qquad \Sigma y^2 = 2118$
$\Sigma x^2 = 646 \qquad \Sigma xy = 1040$

(a) Find the coefficient of determination between the failure rate and grade level.
(b) Interpret the value found in part (a) in terms of the problem.
(c) Perform a test to determine whether correlation exists between failure rate and grade level. What is your conclusion? Why did you reach this conclusion?

69. The number of units of two computer programming manuals, MM and VV, that were sold for a five year period are given below:

NUMBER OF MANUALS SOLD (HUNDREDS)

Year	*MM*	*VV*
1972	24.0	22.2
1973	26.0	24.4
1974	39.1	21.2
1975	46.0	17.3
1976	60.7	16.4
1977	59.1	14.6

(a) Determine the least squares regression line relating VV sales to those of MM.
(b) Interpret the slope, b, found in part (a) in terms of the problem.
(c) Find the coefficient of determination, r^2, for the sales of the two manuals.

(d) In terms of the problem, interpret the value found in part (c).
(e) Using a 5 percent level of significance, determine whether a relationship exists between the sales of the two manuals.

70. Based upon a random sample of 8 families, the following information is available regarding annual income and family insurance expenditures.

y = Family insurance expenditure (dollar units)
x = Family income ($1000 units)
x varies between $12,000 and $33,000

$n = 8 \qquad \Sigma y = 9519$
$\Sigma x = 160 \qquad \Sigma y^2 = 13205311$
$\Sigma x^2 = 3562 \qquad \Sigma xy = 213617$

(a) Find the least squares regression line relating family insurance expenditure to annual income.
(b) Using the information provided, find the coefficient of determination and interpret the result in terms of the problem.
(c) Test a hypothesis in order to determine whether a relationship exists between income and insurance payments.
(d) Use the line found in part (a) to estimate the amount spent on insurance for a family whose annual income is $16,000.
(e) Calculate the standard error of estimate. Establish a 95 percent interval estimate of the amount of money spent on insurance by a family with a $16,000 income.
(f) Would you consider $2000 spent on insurance to be an unusually high expenditure for a family with an annual income of $16,000? Explain.

71. A publishing house undertakes a study regarding the market characteristics of a particular line of textbooks. One of these is sold in both a clothbound and a paperback version. The number of units of each version sold during the period 1972 to 1977 is given in the following table:

NUMBER OF UNITS SOLD (THOUSANDS)

Year	*Clothbound*	*Paperback*
1972	5.8	82.8
1973	4.3	61.3
1974	4.2	55.5
1975	4.3	49.1
1976	4.4	40.6
1977	6.0	38.7

(a) Based upon this data, determine whether a relationship exists between sales of the two versions of the textbook.
(b) Plot the data presented on a scatter diagram. Does the in-

formation in the graph support your conclusion in part (a)? Explain.

(c) On another graph, plot the sales of each version against time, which is scaled on the horizontal axis. Does this graph provide any information about the conclusions made above? Explain.

(d) On the basis of your conclusion in part (a), can you automatically conclude that the two versions of the textbook are sold in different markets since their sales are "not related"? Explain.

72. What is multiple regression and correlation analysis?

73. Define the following terms:

(a) Multiple regression equation
(b) Net regression coefficient
(c) Standard error of estimate
(d) Coefficient of multiple determination
(e) Coefficient of multiple correlation

74. What are "normal equations" and how are they used in multiple regression analysis?

75. What is the meaning of the term *multicolinearity?*

76. A sample of 20 college students is selected at the end of their freshman year in order to determine the relationship between freshman grade point average and the students' high school grade point average and the aptitude test score on the college entrance examination. The following information is available:

x_1 = Aptitude test score
x_2 = High school grade point average
y = Freshman year grade point average

x_1 observed between 465 and 688
x_2 observed between 1.6 and 3.9

The regression equation that applies to this problem is $y' = -0.9 + 0.002x_1 + 0.7x_2$ where $R^2 = 0.5$. In terms of the problem, interpret the coefficients of the independent variables, or "b-values," and the value of R^2.

77. What goals are satisfied by performing a multiple regression and correlation analysis? What kinds of problems can be "solved" using multiple regression and correlation?

78. What are the similarities and differences between two-variable and multiple regression and correlation analysis?

79. What is the main goal associated with including additional

independent variables in a regression equation? How is this related to problems of prediction of the dependent variable?

80. (a) What is accomplished by performing the F-test outlined in the last section of this chapter? What is the main disadvantage of the test?
(b) Why may it be of interest to perform tests of hypotheses for the net regression coefficients, or "b-terms," in multiple regression problems?

81. The following table presents the weekly wage of a sample of 10 workers of a particular firm. Also given are the number of years of service with the firm and a rating indicating the level of skill required on the part of the workers when performing their jobs.

Skill Rating, x_1	*Number of Years of Service*, x_2	*Weekly Wage*, y
2	7	$215
3	9	$210
2	10	$193
3	11	$242
4	12	$321
4	14	$261
3	14	$282
5	17	$343
6	17	$400
5	18	$325

Cumulative terms associated with these data are given as follows:

$n = 10$ $\quad\Sigma y = 2792$

$\Sigma x_1 = 37$ $\quad\Sigma y^2 = 820098$

$\Sigma x_1^2 = 153$ $\quad\Sigma x_1x_2 = 517$

$\Sigma x_2 = 129$ $\quad\Sigma x_1y = 11086$

$\Sigma x_2^2 = 1789$ $\quad\Sigma x_2y = 37922$

(a) Determine the regression equation relating weekly wage rate to skill and years of service with the firm. Interpret the regression coefficients in terms of the problem.
(b) Estimate the weekly wage for a worker with a skill rating of 3 and 13 years of service.

82. Refer to the computer output associated with the savings example presented at the end of this chapter. Use the results presented in the analysis of variance table to determine whether there is correlation between savings and income and family size.

83. Under the null hypothesis that the true net regression coefficients are zero, the following test statistic

$$t = \frac{b}{s_b}$$

is distributed according to Student's t-distribution with $n - k$ degrees of freedom, where k equals the number of coefficients in the regression equation and s_b represents the estimated standard error of the regression coefficient. Use this fact to test whether the coefficients are zero in the savings example based upon the results presented in the computer output at the end of the chapter. Perform the tests at a 5 percent level of significance. What conclusions can be reached on the basis of these tests about the original problem?

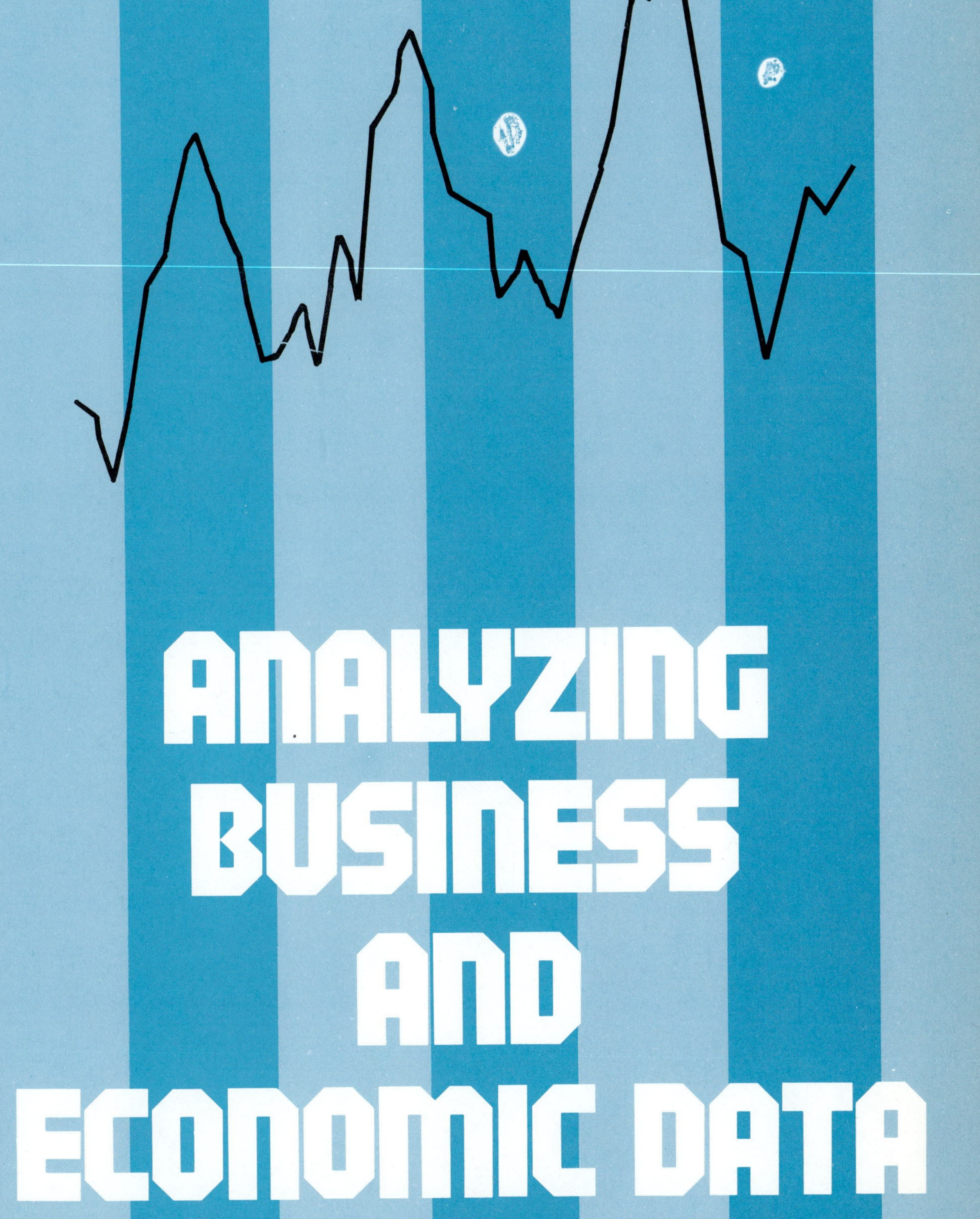
ANALYZING
BUSINESS
AND
ECONOMIC DATA

CHAPTER 17

TIME SERIES ANALYSIS AND THE SECULAR TREND

You can never plan the future by the past.
Edmund Burke

I have but one lamp by which my feet are guided, and that is the lamp of experience. I know of no way of judging the future but by the past.
Patrick Henry

Most data are observed over a period of time, yet time has not been considered as a relevant variable in our work in earlier chapters. There are many problems where time plays an important role and should be considered since it helps to understand the past and to predict events relating to decisions made about the future. This is especially true in business and economics where planning and decision-making for a partially uncertain future are a necessity. For example, a manufacturer may be interested in its long-term profit picture based upon potential sales of its product. This may be necessary to determine whether to expand production facilities in order to maintain its share of the market. Of related interest is the overall state of the economy. If financing is to be sought, the rate of interest prevailing at crucial decision points in the future must be considered.

In the short run, cash balances must be anticipated in order to meet payrolls, acquire raw materials, pay taxes, control inventories, and maintain facilities for production and the administration of the firm's activities. In an economy in which the activities of all institutions, public and private, are interdependent, there is a constant need to understand what has happened in the past and to anticipate or forecast the future.

EXHIBIT 1

EXAMPLES OF DATA RECORDED OVER TIME

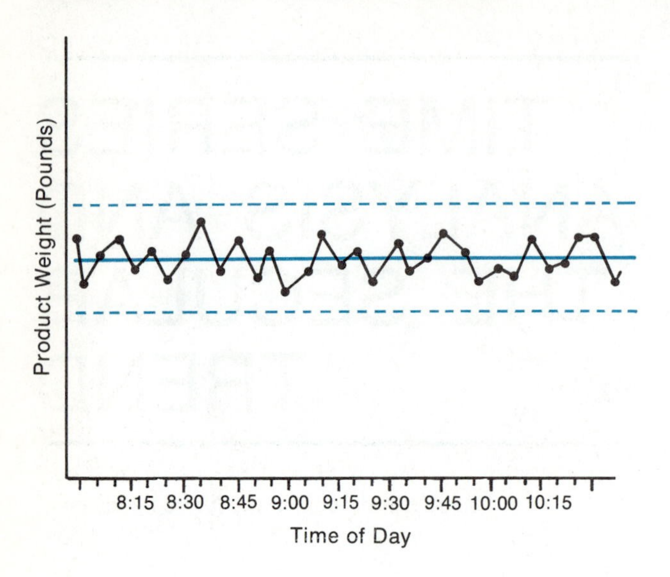

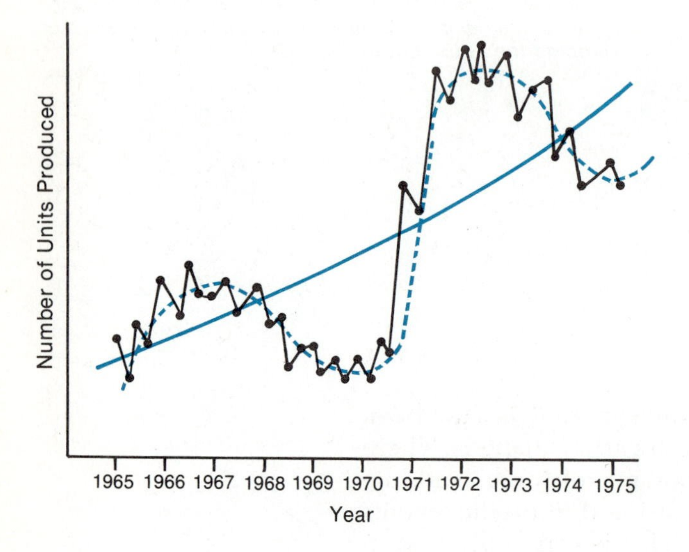

Although data used to illustrate earlier material may have been collected over time, the passage of time did not play an important role in the problems considered. However, as we just indicated there are situations where time does play an important part. In order to emphasize this point, consider the graphs presented in Exhibit 1. The graph at the top presents the weight of a product produced by a continuous manufacturing process plotted against the time that an item is produced. An item is produced every five minutes. The graph at the bottom of the exhibit presents the total number of units of the same product produced annually. Individual points or observations on each graph are connected by straight lines to provide a better picture of the possible patterns in the data.

In both cases, time appears on the horizontal axis. Recognize, however, the graphs appear very different. The one at the top displays the weights of individual items that vary at random about the arithmetic mean weight of the process. This is described by the horizontal line. The pattern of variability about the mean appears stable and is confined within a relatively fixed band. Although the weights were measured at different points in time, the only variability present can be attributed to chance. As long as the production process is in a state of control, the same basic pattern of variability will be observed. The pattern of variability can be described satisfactorily with a frequency distribution like the one presented in Chapter 1, which is the basis of the material presented in the chapters to this point in the text. Time, therefore, is of little importance in describing the output of the process as long as control is maintained. Changes in the process can be detected with methods similar to the ones already presented.

The situation is different with respect to the pattern of variation in total annual output. One difference is that the length of time over which the observations were made is much longer; that is, ten years compared with two and a half hours. Another difference is that there are patterns in the data that cannot be ascribed to chance alone. First, we can see that there is a general pattern of overall increase in the series during the entire ten-year period that is described by the smooth colored curve. Second, we can see that there is a "wavelike" movement about the smooth curve that is described by the dotted line. Lastly, there are fluctuations about the wavelike pattern. Since patterns besides chance are present in the second series, a frequency distribution will not provide a satisfactory description and will mask the effect of the different components of the series.

We can explain the differences between the two series very simply at this point. The weights associated with individual items of output are subject to greater control and are dependent exclusively upon the characteristics of the manufacturing process. Total output, however, is based upon the demand for the firm's product, which is influenced by factors relating to the economy and mostly beyond the direct control of the firm. Hence, time plays a role in describing total output since these economic factors change and exhibit certain distinct patterns of variation over time.

Data that are recorded or generated at successive points in time are defined as a **time series.** Analysis of the patterns in data recorded over time is referred to as *time series analysis* and is the topic covered in this part of the text. Although the material can be applied to other fields, emphasis is placed upon business and economic data. This results from the fact that we encounter such data more routinely than other types. Furthermore, business and economic data typically exhibit distinct patterns that require special attention.

There are three basic reasons for analyzing a time series: (1) to describe the past, (2) to understand the past, and (3) to forecast the future. Sometimes a time series is analyzed in order to determine what has occurred in the past without regard to the future. Typically, however, the goal of time series analysis is to forecast, and the more we understand about the way in which things have occurred in the past the better we are able to forecast the future. Many methods of forecasting exist, but the success with which these methods forecast the future varies.

A method that is successful in one problem may not be successful in another. Actually, the area of forecasting can become very complex and difficult to understand. The approach taken in this text is quite simple. Emphasis is placed upon describing a time series in order for us to become aware of the components that are considered in more complicated methods of analysis. Approaches to forecasting are discussed but are not treated in detail.

A simple description of a time series rests on the assumption that individual components can be isolated. These components are listed as (1) trend, (2) cycle, (3) seasonal, and (4) irregular and random. The *trend* is defined as a smooth, irreversible movement in the series. It can be increasing or decreasing; the smooth curve representing the overall increase in annual production shown in Exhibit 1 is an example of a trend. Typically, in the case of an economic time series, the trend is defined as a long-term movement that occurs over a large number of years. It is difficult to specify an exact number of years, since different series exhibit trend-like patterns for varying numbers of years. Moreover, the faster changes occur in different segments of the underlying economy, the quicker a trend in a series may change.

Both the cycle and the seasonal components are wavelike movements that reoccur periodically. The wavelike pattern depicted in the second diagram in Exhibit 1 is an example of a *cycle*. Cycles associated with economic time series have a shorter duration than trends but last for a period of more than one year. *Seasonal* variations represent reoccurrences at given time points within a year. Typically, they are associated with events that continue to occur year after year. The *random* or irregular component of a time series corresponds to movements in a series that are completely unpredictable.

The remainder of this chapter is devoted to the trend component. The procedure section deals with "fitting" a trend equation to a set of data. The procedure is similar to fitting a regression equation except time is used in place of the independent variable. Only straight line or linear trend equations are considered in detail. More about the use of trend equations in general is presented in the discussion section. Seasonal variation is considered in the next chapter. The cyclical and random components are discussed in Chapter 19 along with an overview of the area of forecasting.

The following symbols are used in this chapter:

- N = the number of observations, or the number of time points for which data are observed
- t = a point in time, or the time at which an observation is generated (time can be measured in terms of weeks, months, quarters, years, etc.)
- Y = a value of a variable observed at a particular point in time
- A = a coefficient in a linear trend equation representing the value of the trend when t = 0, or at the base period
- B = a coefficient in a linear trend equation representing the slope
- Y_T = a value of the trend corresponding to the variable Y

PROCEDURE FOR FITTING A LINEAR TREND EQUATION

A method for estimating the coefficients in a linear trend equation is presented in this section. A linear trend equation takes the form

$$Y_T = A + Bt$$

where Y is a variable of interest, t represents time, and A and B are coefficients to be estimated based upon N observed values of Y. In order to fit the equation to data, we must estimate the coefficients A and B, which is accomplished by performing the following steps:

Step 1. In order to simplify the amount of calculation, *code* the time variable in terms of a successive set of integer values (e.g., $t = 0, 1, 2, \ldots, N$)

Step 2. Substitute the coded values of time and the values of Y into the following formulas* and solve for A and B:

$$A = \frac{(\Sigma t^2)(\Sigma Y) - (\Sigma t)(\Sigma tY)}{N(\Sigma t^2) - (\Sigma t)^2}$$

$$B = \frac{N(\Sigma tY) - (\Sigma t)(\Sigma Y)}{N(\Sigma t^2) - (\Sigma t)^2}$$

Formulas for estimating coefficients in linear trend equation.

Step 3. Accompany the resulting trend line with information explaining the time (month, quarter, year, etc.) corresponding to the base period or the value of t equal to zero and the units of time associated with values of t.

Note. Preliminary computations should be performed to find the cumulative terms that must be substituted into the above equations. These computations are similar to the ones performed in regression and correlation analysis.

NUMERICAL EXAMPLES

Example 1 Appearing in the table following on page 390 are the annual gross sales figures of a firm for ten years beginning with 1969. Sales are given in units of one million dollars.

*These formulas are based upon the method of *least squares* and are similar to the ones used for estimating the coefficients of a two-variable linear regression equation. Since one variable is time, which is measured successively, it can be coded to simplify the calculations. One form of coding is used in this section; a simpler version is given at the end of the chapter.

Year	Y Sales (Million dollars)
1969	4
1970	9
1971	8
1972	7
1973	14
1974	17
1975	16
1976	13
1977	20
1978	19

Step 1. Code the years by using the successive values of 0 through 9 for the years 1969 through 1978. Based upon the coded years and the original data, the preliminary computations appear in the following table:

PRELIMINARY COMPUTATIONS FOR LINEAR TREND EXAMPLE

Year	t (Coded years)	Y	t^2	tY
1969	0	4	0	0
1970	1	9	1	9
1971	2	8	4	16
1972	3	7	9	21
1973	4	14	16	56
1974	5	17	25	85
1975	6	16	36	96
1976	7	13	49	91
1977	8	20	64	160
1978	9	19	81	171
Total	45 (Σt)	127 (ΣY)	285 (Σt^2)	705 (ΣtY)

Step 2. Estimate the coefficients by substituting into the formulas for A and B and solve as follows:

$$A = \frac{(\Sigma t^2)(\Sigma Y) - (\Sigma t)(\Sigma tY)}{N(\Sigma t^2) - (\Sigma t)^2}$$

$$= \frac{285(127) - (45)(705)}{10(285) - (45)^2}$$

$$= \frac{36195 - 31725}{2850 - 2025}$$

$$= \frac{4470}{825}$$

$$= 5.4$$

$$B = \frac{N(\Sigma tY) - (\Sigma t)(\Sigma Y)}{N(\Sigma t^2) - (\Sigma t)^2}$$

$$= \frac{10(705) - 45(127)}{10(285) - (45)^2}$$

$$= \frac{7050 - 5715}{2850 - 2025}$$

$$= \frac{1335}{825}$$

$$= 1.6$$

Step 3. Based upon the estimated coefficients, the trend equation along with the base period and time units is

$$Y_T = 5.4 + 1.6t$$

$t = 0$ in 1969

t is in one-year intervals

Example 2 Consider the production data presented in the following table:

Year	*Output (1000 units)*
1960	110
1965	185
1970	430
1975	750

The data represent a time series corresponding to a 15-year period, where output is recorded every 5 years. If we were to plot the output figures on a graph against time, we would find the points follow a curve rather than a straight line. It happens, however, that the logarithms of the output values plot very closely to a straight line. This becomes evident by examining the following graphs, where output is labeled with the symbol Y.

GRAPHIC COMPARISON OF OUTPUT DATA AND LOGARITHMS OF OUTPUTS PLOTTED AS A TIME SERIES

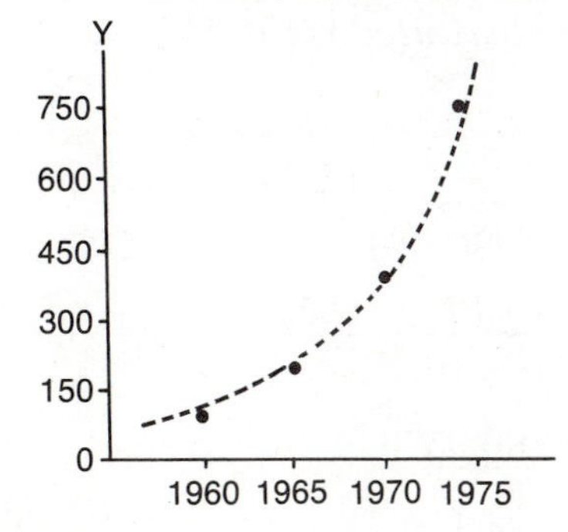

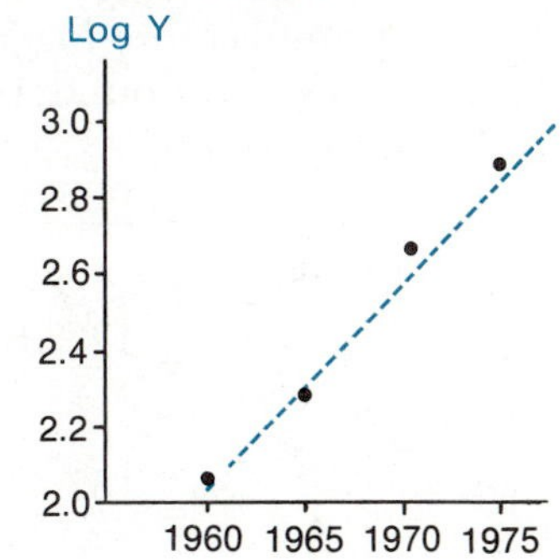

The logarithms used in the graph on the right are to the base 10; the same pattern would be observed if natural logarithms were used, but the numerical scale would differ. Notice how closely the logarithms scatter about a straight line. Since this is the case for this set of data, it is possible to use the formulas provided in the procedure section. Instead of applying them directly to the output figures, we must first take the logarithm of each and then apply the formulas with the logarithm of output substituted for the individual values. This is an example of the use of a *transformation* that was mentioned in previous chapters.

Appearing in the table below are the original data, the coded time values, the logarithms of the output values, and the preliminary computations needed to obtain the cumulative terms; the logarithms are to the base 10 and are obtained from Table 11.

PRELIMINARY INFORMATION NEEDED TO FIT A LINEAR TREND TO THE LOGARITHMS OF OUTPUT

Year	t	Y	logY	t^2	tlogY
1960	0	110	2.0414	0	0.0000
1965	1	185	2.2672	1	2.2672
1970	2	430	2.6335	4	5.2670
1975	3	750	2.8751	9	8.6253
Total	6 (Σt)		9.8172 $(\Sigma logY)$	14 (Σt^2)	16.1595 $(\Sigma t\,logY)$

The trend equation is of the same general form as the one introduced earlier, *except* that it is in terms of the logarithm of Y. The equation for trend, therefore, takes the form

$$\log Y_T = A + Bt$$

Estimates of A and B are found using the same formulas, but log Y is *substituted* for Y. The estimates are found as

$$A = \frac{(\Sigma t^2)(\Sigma \log Y) - (\Sigma t)(\Sigma t \log Y)}{N(\Sigma t^2) - (\Sigma t)^2}$$

$$= \frac{14(9.8172) - 6(16.1595)}{4(14) - (6)^2}$$

$$= \frac{137.4408 - 96.9570}{56 - 36}$$

$$= \frac{40.4838}{20}$$

$$= 2.0242$$

$$B = \frac{N(\Sigma t \log Y) - (\Sigma t)(\Sigma \log Y)}{N(\Sigma t^2) - (\Sigma t)^2}$$

$$= \frac{4(16.1595) - 6(9.8172)}{4(14) - (6)^2}$$

$$= \frac{64.6380 - 58.9032}{56 - 36}$$

$$= \frac{5.7348}{20}$$

$$= 0.2867$$

Therefore, the trend equation in terms of logarithms becomes

$$\log Y_T = 2.0242 + 0.2867t$$
$t = 0$ in 1960
t is in 5-year intervals

Note. The original output values are given every 5 years beginning with 1960. Consequently, whenever a value of t changes by one unit it corresponds to a change of 5 years in time due to the way time is recorded in this example. This is reflected in the information accompanying the above trend equation. Also, note that the estimates of the coefficients in the equation are carried to 4 decimal places. This is not related to the decimal place accuracy of the original output figures, but to the fact that the equation is in terms of logarithms. In order to obtain enough accuracy with logarithms, a large number of decimal places should be used. Since logarithms are given to 4 decimal places in Table 11, this is the number of places used in the results. Greater accuracy can be obtained with a scientific calculator.

DISCUSSION

This chapter introduces the concept of variation over time in terms of the components of a time series. Emphasis was placed on the trend or overall long-term component. The procedure for estimating the coefficients in a linear trend equation is illustrated in the last section. There, the approach is applied to trends fitted to actual observations and to the logarithms of the actual observations. The two applications are the simplest trend models to fit and interpret. Here, we shall examine each in detail and compare the situations where they can be applied.

Based upon the sales data given in Example 1, we obtained a linear trend equation

$$Y_T = 5.4 + 1.6t$$

where

Y = sales in million dollar units

t = 0 in 1969

t is in one-year intervals

Trend estimates for a given year using a linear trend equation are obtained by substituting the corresponding coded value of t into the trend equation and solving for Y_T.

In order to estimate the trend component for any year during the period observed, we must substitute the coded value of t into the trend equation and solve for Y_T. For example, the trend estimate of sales for 1976 is obtained by substituting the coded value of t equal to 7 into the equation and solving for Y_T as follows:

$$\begin{aligned} Y_T &= 5.4 + 1.6(7) \\ &= 5.4 + 11.2 \\ &= 16.6 \end{aligned}$$

The trend estimate of sales in 1976, therefore, is 16.6 million dollars, or $16,600,000.

A trend forecast beyond the range of the observed data is performed in the same way, except a value of t for the year forecasted must be determined. In general, the value of t for *any* period can be found by the formula

$$t = \frac{\text{Given year } \textit{minus} \text{ Base year}}{\text{Size of interval}}$$

Formula for determining the value of t for any given year.

The given year represents the year to be forecasted, the base year corresponds to the point where t equals zero, and the size of the interval represents the number of units associated with t. For example, suppose we want a trend forecast for 1985. The value of t for 1985 is found as

$$\begin{aligned} t &= \frac{1985 - 1969}{1} \\ &= 16 \end{aligned}$$

which leads to a trend forecast of 31 by substituting into the trend equation and solving for Y_T as follows:

$$\begin{aligned} Y_T &= 5.4 + 1.6(16) \\ &= 5.4 + 25.6 \\ &= 31 \end{aligned}$$

Therefore, the trend forecast of sales for 1985 is 31 million dollars, or $31,000,000.

Now, let us use the logarithmic trend equation to make a trend forecast. The trend equation found in Example 2 is

$$\log Y_T = 2.0242 + 0.2867t$$

where

Y = Output in 1000 units
t = 0 in 1960
t is in 5-year intervals

Suppose we want to obtain the trend estimate for 1975. The value of t corresponding to 1975 is 3. This can be obtained by using the formula for t as

$$t = \frac{1975 - 1960}{5} = \frac{15}{5} = 3$$

By substituting this value into the trend equation, we obtain

$$\log Y_T = 2.0242 + 0.2867(3) = 2.0242 + 0.8601 = 2.8843$$

This result is the logarithm of the trend value for 1975 and must be converted by taking the antilog using Table 11. The antilog provides the estimate of trend in 1975, which is given as

Trend estimates using a log-linear trend equation are obtained by substituting a coded value of t for a given year into the trend equation, solving for log Y_T, and taking the antilog of the result to obtain Y_T.

$$Y_T = \text{antilog } 2.8843 = 766$$

The trend estimate of production in 1975 is 766,000 units. Trend estimates and forecasts for any year using the log-linear trend equation are found in the same way, based upon the appropriate value of t.

Now that we have seen how to obtain a trend estimate, we shall discuss situations described by the two equations. Consider the trend equation, $Y_T = A + Bt$. The A-term represents the trend estimate when t equals zero. Consequently, the numerical value of A depends upon the coding scheme and the year designated with a zero. Of the two coefficients, B is of greater interest. It is the slope of the line and represents the change in the trend for a one unit change in t; the number of years represented by this change depends upon the time units used. Reconsider the trend line obtained in Example 1, where A equals 5.4 and B equals 1.6. Hence, the trend estimate in the base year, 1969, is 5.4, or $5,400,000. The estimate of trend changes by 1.6, or $1,600,000, for every change of one year in time.

In general, the values of A and B depend upon the coding scheme used. A represents the estimated value of the trend in the base period, or when t equals zero. B represents the change in the trend estimate for a change in time corresponding to one unit of t.

The interpretations of A and B for the equation $\log Y_T = A + Bt$ are the same in terms of the *logarithm* of the trend estimate. The critical

For log-linear trend equations, the interpretations of A and B are expressed in terms of logarithms. The antilog of B represents the multiple that a particular trend value is of a previous period. The antilog of A represents the trend estimate in the base period in terms of the original units of Y.

difference lies in the interpretation of the antilog of B, which is in terms of the units of the *original* observations. Recall from basic algebra that constant amounts of change in the logarithms of a series of values represent constant *rates* of change in the original values. Hence, the value of B equal to 0.2867 represents the change in the logarithm of the trend estimate of output for every 5-year period. The antilog of the B-value is 1.94. This means that for each 5-year change in time, output is 1.94 times the previous trend value, or that the trend increases by 94 percent every 5 years. The antilog of A corresponds to the trend estimate in the base period expressed in terms of the original units of Y. Hence, the antilog of A, equal to 2.0242, equals 106 and corresponds to a trend estimate of 106,000 units of output in 1960.

Both the linear trend and log-linear trend models correspond to situations where there are constant changes over time. The linear trend fitted to the actual observations represents constant *amounts* of change, whereas the log-linear trend is used to describe constant *rates* or constant percentage changes from period to period.*

Other types of equations besides the ones considered in this chapter can be used to describe trends depending upon the nature of the data described.

The trend lines described in this chapter are only two of many possible equation forms available. Different series of data may exhibit patterns that can be represented by curvilinear expressions similar to the ones described in Chapter 16 concerning regression analysis. A scatter diagram can be used to determine the general pattern of the trend in a time series. Procedures for determining which equation describes the trend best are complicated by the presence of the other time series components and are not as clearly defined as in regression analysis.

When there is a small or relatively stable cyclical pattern and little random variation, it is safer to assume that the trend has been estimated satisfactorily. There is no guarantee, however, that the same trend pattern will continue in the future.

In general, if the cyclical effect is small or if its behavior is somewhat regular and the random effect is small, we can feel more confident that a trend has been satisfactorily described historically. Whether the same pattern will continue into the future is a problem. For example, consider the graph presented in Exhibit 2. The wavy line appearing in the exhibit represents an actual series observed during a particular time period. The solid line (color) corresponds to a trend line fitted to the observed points. We can see that there is a small amount of variation of the actual series about the fitted line and that the line appears to fit the series well. The dotted curve also appears to describe the series satisfactorily.

For the particular time series depicted in the exhibit, both the straight line and the curve appear to describe the observed data from the past satisfactorily. A real difference is evident when we project the two into the future. This can be seen by looking at the dotted extensions of the two equations starting at the end of the observed series. Since these represent projections into the future, it is impossible to know which is correct. Consequently, we can see that an adequate description of the past trend does not necessarily guarantee a correct projection of the future.

Any projection of a trend into the future is based upon the *assumption* that the same pattern described in the past will continue into the

---

*Series of data that change by constant amounts are referred to as *arithmetic,* whereas series that change by a constant rate are referred to as *geometric.*

EXHIBIT 2
PROJECTING A TREND

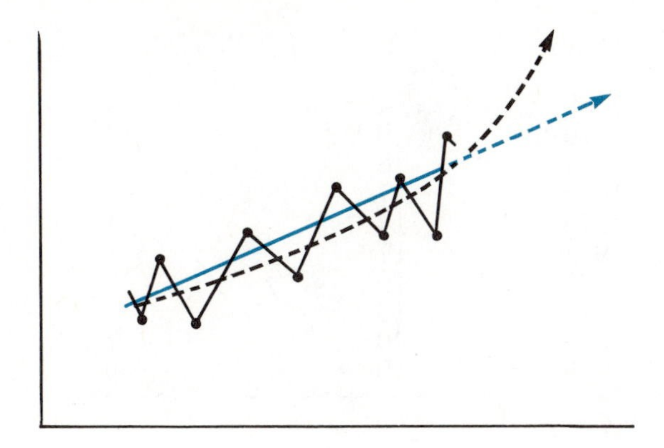

future. In general, the larger the number of years described in the past, the safer one feels that the pattern will continue. If a trend is fitted to a series that is too short, it is possible that part of a cycle is described instead of a trend. Furthermore, projections or forecasts for short periods of time generally will be more accurate than ones for long periods. It is less reasonable to assume that a similar trend pattern will continue for periods further into the future.

Projections of trends into the future are based on the assumption that the same pattern described in the past will continue into the future. It is safer to make this assumption when past trends are based upon a lot of data and projections are not made very far into the future.

In this chapter we have seen how a trend equation can be used to estimate the trend component in the past and forecast it into the future. Since there are other components in a time series, an accurate forecast of an actual observation generally is not possible unless these components also are considered. Furthermore, we have considered yearly data only. Consequently, seasonal effects could not influence our estimates of the trend. If time intervals within a year are used, the seasonal effect must be considered also. Seasonal variation is considered in the next chapter. Ways in which all time series components can be handled together are presented at the end of Chapter 19.

SIMPLIFYING CALCULATIONS FOR TREND

The procedure for finding a trend line presented earlier is fairly simple based upon the coding approach used. This coding procedure was illustrated since it enables us to use the same approach that was used for two-variable linear regression and is easier to understand. Furthermore, programmable calculators with either stored programs or continuous memory generally use a similar coding scheme. This coding scheme also is used in computer programs that are available to perform an analysis of a time series.

A simpler coding procedure is available that saves some time and effort when performing a trend analysis by hand computation. In order to illustrate the procedure, consider the data in Example 1. These appear in the following table along with preliminary calculations used in the formulas that are modified to account for the change in coding scheme.

ORIGINAL DATA AND PRELIMINARY CALCULATIONS FOR SHORTCUT CODING METHOD FOR TREND

Year	*Y*	*Re-coded Years, t*	*tY*	t^2
1969	4	−9	−36	81
1970	9	−7	−63	49
1971	8	−5	−40	25
1972	7	−3	−21	9
1973	14	−1	−14	1
1974	17	1	17	1
1975	16	3	48	9
1976	13	5	65	25
1977	20	7	140	49
1978	19	9	171	81
Total	127	0	267	330

The first two columns in the table contain the original data. The coded values for each year appear in the third column. These codes are assigned in such a way that their sum equals *zero*. This is accomplished by assigning a value of zero to the middle time period in the series. Since the series in this example contains an *even* number of years, the zero does not appear in the list of coded values and corresponds to the time period midway between 1973 and 1974. The remaining codes are assigned in such a way that there is a constant difference between successive values. This is done to reflect the constant difference in time between successive years.

Minus-one and plus-one are assigned to the middle two years of 1973 and 1974. The absolute difference between these codes is 2. Hence, every difference in time of one year must be represented by a difference of 2 in codes corresponding to successive years. Plus and minus signs are used to indicate whether a year is after or before the base period, which corresponds to a value of t equal to zero.

Because the sum of the coded time values equals zero, the least squares formulas for the coefficients A and B are simplified since all terms containing Σt vanish. The estimated coefficients, therefore, are found by the formulas

$$A = \frac{\Sigma Y}{N}$$

$$B = \frac{\Sigma tY}{\Sigma t^2}$$

Simplified formulas for estimates of trend coefficients.

Using these formulas, the trend estimates in the example are found as

$$A = \frac{127}{10} = 12.7$$

$$B = \frac{267}{330} = 0.8$$

The resulting trend equation, therefore, is

$$Y_T = 12.7 + 0.8t$$
$$t = 0 \text{ in } 1973\tfrac{1}{2}$$
$$t \text{ is in } \tfrac{1}{2}\text{-year intervals}$$

As in the earlier case, the trend equation must be accompanied by information specifying the base period and the meaning of the coded intervals. In this example, the base period is midway between 1973 and 1974 and is designated as "1973½." Since a two-unit change in t represents a one-year change in time, a one-unit change in t corresponds to a one-half-year change in time.

The values of A and B in the equation are different than the ones found earlier using the other coding scheme. This is due to the difference in the coding scheme used which results in a different base period and different units of t. Since the coded time unit in this case equals half the earlier value, the value of B of 0.8 is half the earlier value of 1.6. Trend estimates for a given period, however, are the same using either equation except for rounding discrepancies.

> *Note.* In the case of an *odd* number of years, the shortcut procedure basically is the same. The formulas are identical; however, the coded value for the middle year is *zero* and successive values of t change by *one* unit instead of two. In both cases, even and odd years, trend estimates and forecasts are based upon the formula for t given in the discussion section. The base period used in the formula always corresponds to the value of t equal to zero. For an odd number of years, the shortcut coding scheme yields the same value of B as the procedure presented earlier.

QUESTIONS AND PROBLEMS

DEFINITIONAL QUESTIONS

1. What is a time series? Why is it important to consider the analysis of time series as a special topic?

2. List the main components of a time series.

3. What is the meaning of the following terms?

   (a) Trend
   (b) Trend equation
   (c) Linear trend
   (d) Log-linear trend
   (e) Curvilinear trend

4. How are the coefficients A and B in a linear trend equation interpreted?

5. What are the uses of a trend equation in general?

COMPUTATIONAL PROBLEMS

6. Find a linear trend equation fitted to the time series given in the following table:

Year	*Y*
1967	52.1
1968	50.4
1969	60.0
1970	65.5
1971	67.7
1972	70.8
1973	76.0
1974	80.1
1975	83.0
1976	85.6
1977	89.7
1978	97.0

7. Given below are the annual number of residences in a suburban community that are delinquent regarding payment of the annual trash collection levy between the years 1968 and 1977:

Year	*Number of Residences*
1968	6415
1969	6309
1970	6776
1971	7039
1972	7966
1973	8048
1974	8367
1975	8342
1976	8669
1977	9003

(a) Find the linear trend equation for the number of delinquent residences.
(b) Use the equation found in part (a) to obtain a trend estimate for 1975.
(c) Assuming the same pattern of delinquency continues in the future, forecast the trend value of the number of defaulting residences in 1984.

8. The annual cash reserves of a moderate sized loan company for a 16-year period are given in the following table:

Year	*Cash Reserve ($1000 Units)*
1963	376
1964	464
1965	388
1966	440
1967	393
1968	371
1969	390
1970	376
1971	327
1972	385
1973	432
1974	378
1975	331
1976	330
1977	322
1978	336

(a) Fit a linear trend equation to the data in the table.
(b) Use the equation obtained in part (a) to forecast the trend component of the company's cash reserves for 1980.

9. The dollar amount of a company's inventory of small motors is registered every 5 years for a 30-year period beginning in 1945.

Year	*Inventory ($1000 Units)*
1945	239
1950	269
1955	262
1960	295
1965	297
1970	325
1975	306

(a) Plot the data as a scatter diagram, where time appears on the horizontal axis.
(b) Based upon your work in part (a), does a straight line describe the trend in the data? Explain.
(c) Fit a trend line to the data presented in the table. Plot the result on the scatter diagram developed in part (a).
(d) Project the trend line on the scatter diagram to 1980 in order to obtain a trend forecast for that year. Verify the result by using the actual trend equation found in part (c).

10. Fit a linear trend equation to the logarithms of the Y-values in the following time series:

Year	*Y*
1970	108
1971	203
1972	397
1973	770
1974	1485
1975	2871
1976	5556

11. The following table presents the population of the United States every decade between 1880 and 1970:

Year	*Population (Millions)*
1880	50
1890	63
1900	76
1910	92
1920	106
1930	123
1940	132
1950	151
1960	179
1970	203

(a) Fit a log-linear trend equation to the population figures provided in the table.
(b) Use the equation found in part (a) to forecast the U.S. population in 1965.

CONCEPTUAL QUESTIONS AND PROBLEMS

12. What are the main reasons for analyzing a time series? Why shouldn't a frequency distribution be used?

13. What is the relationship between a time series and forecasting? How is the concept of a trend equation related?

14. Under what circumstances can a trend equation be used to forecast a value in a series in the future? Explain.

15. Criticize or explain the following statement. Although a log-linear trend equation is a straight line, it is used to describe a particular type of trend which is curvilinear.

16. What is the basic assumption that is made when a trend is extrapolated in order to make a trend forecast?

17. An administrator of a state's employment office is interested in the pattern of the number of unemployment benefit claims made in the state's largest metropolitan area. The following information is available:

Year	*Number of Claims (Thousands)*
1973	763
1974	842
1975	1493
1976	1234
1977	1882

(a) Fit a linear trend equation to the number of unemployment compensation claims.
(b) What information is provided by the slope of the line found in part (a)?
(c) Based upon the information provided in the problem, is it reasonable to assume that the result in part (a) is a trend line? Is it possible that the result is part of a cycle? Explain.

18. The following table presents the total number of army military personnel on active duty between 1940 and 1975:

Year	*Military Personnel (Thousands)*
1940	270
1945	8250
1950	590
1955	1100
1960	870
1965	965
1970	1320
1975	780

(a) By examining the data given, can you account for the pattern of variability in the series? Explain.
(b) Is it possible to describe the overall long-term pattern in the series with a trend equation? What kind of trend equation would you use? Explain.

19. The table below presents the average number of men and the average number of women per year that were employed in a large metropolitan area:

Year	**Number Employed**	
	Men	*Women*
1971	1,336,400	806,300
1972	1,253,000	780,700
1973	1,294,200	808,900
1974	1,247,300	779,500
1975	1,277,100	798,200

(a) Is there any evidence of a trend in either series? Explain.
(b) Is there evidence of a cycle in either series? Explain.
(c) In order to answer the questions in parts (a) and (b), would it be helpful to have data for more years? Explain.
(d) What would you expect to happen to the trends in the two series for the years following the ones reported in the table? Explain.

20. The table following on page 404 presents the total U.S. energy consumption between 1945 and 1970:

Year	*Energy Consumption (Quadrillion BTU)*
1945	33.0
1950	36.5
1955	41.7
1960	47.4
1965	58.3
1970	72.1

(a) Which would better describe the trend in the series, a linear or log-linear equation? Explain.
(b) Fit a linear trend equation to the logarithms of the energy consumption series.
(c) Interpret the slope of the line found in part (b).
(d) Find the antilog of the slope found in (b). What information is provided by this number?
(e) Use the equation found in part (b) to obtain a forecast of the trend in energy consumption in 1985. Should the actual energy consumption in 1985 equal this value? Explain.

21. The trend in the size of the population of a small town is described by the following equation:

$$Y_T = 40 + t + 0.1t^2$$

Y = Population in units of 1000 people
t = 0 in 1970
t is in one-year intervals

(a) Is the trend equation linear? Explain.
(b) In words explain the way the trend in population is changing.
(c) Use the trend equation to forecast the town's population in 1983.

22. Criticize or explain the following comments. A trend equation is like a regression equation since they both describe general patterns in a set of data. Forecasts can be made with both types of equations. A forecast can be made using a regression equation without extrapolating, whereas extrapolation must be used in order to forecast a trend.

23. The coefficient of determination, r^2, can be used to determine how well a regression line fits a set of data. Why shouldn't it be used in general to determine how well a trend line fits a time series?

24. The compound interest formula used in basic finance is given by the formula

$$A = P(1 + i)^N$$

where

A = the accumulated amount
P = the principal
i = the rate of interest
N = the number of compounding periods

Using basic algebra, express the logarithm of A in terms of P, i, and N. What is the relationship between the result and a linear trend equation fitted to the logarithms of a set of data?

25. Criticize or explain the following comment. Since a linear trend and a log-linear trend are the only two equations considered in this chapter, these are the only types of equations that can be used to describe a trend in a time series.

26. Criticize or explain the following comment. Since codes are used instead of the actual years, the equations for estimating the coefficients in a trend equation cannot be applied to the actual years.

27. Refer to Problem 11. In that problem you are asked to fit a log-linear trend equation to the series. The result is

$$\log Y_T = 1.7406 + 0.0647t$$

$t = 0$ in 1880

t is in 10-year intervals

The linear trend equation fitted to the actual observations is

$$Y_T = 43.8 + 16.4t$$

$t = 0$ in 1880

t is in 10-year intervals

(a) Estimate the trend value of the population in 1950 using the log-linear equation. Do the same using the linear trend equation.
(b) Forecast the trend value of the population in the year 2000 using the log-linear equation. Do the same using the linear trend equation.
(c) How can you account for the differences in the two sets of results in parts (a) and (b)? What general conclusions can be made about the trend in U.S. population on the basis of your results?

28. The concept of a trend is treated in this chapter in terms of problems in business and economics and represents a component of a time series that is measurable. Other types of trends exist in our society that represent prevailing tendencies or preferences.

(a) Is there a relationship between the trend in a time series and social or cultural tendencies and preferences? Explain.

(b) Is it possible that changes in social or cultural tendencies can have an impact on a trend in an economic time series? Explain. Prepare an original example in order to illustrate the point.

CHAPTER 18

SEASONAL VARIABILITY

Thus with the year
Seasons return
John Milton

Seasonal variability has been defined as a movement in a time series during a particular time of year that reoccurs year after year. Consequently, it is not possible to observe seasonal variation in an economic time series expressed in terms of annual data of the kind considered in the previous chapter. In order to observe or measure the seasonal component, time must be measured in terms of units such as days, weeks, months, or quarters. Typical units for business and economic problems are months or quarters. Seasonal patterns have been observed in many series to occur with great regularity so that they can be measured and predicted satisfactorily.

Seasonal patterns develop as the result of natural changes in the seasons throughout the year or they may result from habits, customs, or events that reoccur at the same time every year. For example, toy sales in December are greater than in other months owing to the presence of holidays in that month. Although toys are sold throughout the year, year after year there is a marked increase in sales during the month of December.

The seasonal rise in toy sales is based upon the occurrence of certain events in December. Examples of product sales that are tied to the natural seasons of the year are air conditioner and beach wear sales. These are associated with warmer months in climates experiencing weather changes. With respect to beach apparel, a second season exists, since a cruise-wear market has been developed in colder months by travelers to warmer climates.

A seasonal pattern that has undergone change in recent years is associated with ice cream sales. Traditionally, ice cream sales soared during warmer periods. With the emergence of ice cream parlor chains, ice cream sales have increased in other parts of the year although more still is sold in warmer months.

There are three reasons for studying seasonal variability. First, *description* of the seasonal effect provides a better understanding of the impact of this component upon a particular series. Second, it is possible to eliminate the seasonal effect from a time series in order to observe the effect of the other components. Eliminating the seasonal effect from a series is referred to as *deseasonalizing* or *seasonally adjusting* the data. Third, knowledge of the seasonal effect can be used to include this component into a *forecast.*

Seasonal variability is measured with a *seasonal index.* A seasonal index is an average that indicates the percentage of an actual observation relative to what it would be if no seasonal variation in a particular period is present. For example, if the seasonal index for the number of people unemployed in the second quarter is 110, it means that unemployment is 10 percent higher during the months of April, May, and June than it would be if there were no seasonal effect during that period. More is said about the interpretation of a seasonal index later.

Seasonal variation is measured in terms of an index attached to each period of a time series within a year. Hence, if monthly data are considered, there are twelve separate seasonal indexes, one for each month. For quarterly data there are four. The most common procedure for determining a seasonal index is based upon a special kind of average called a *moving average,* which is used to eliminate the seasonal effect from a time series. The ratio of the original observations to the corresponding moving average provides a rough measure of seasonal variability for a given period. Refined values of these ratios represent the seasonal index.

The procedure briefly outlined above is referred to as the *ratio to moving average method.* The procedure is not very difficult but involves a great deal of computation, especially when there are a large number of years, and time units of a month or less are used. The next section describes the procedure for determining seasonal indexes when data are in the form of quarters or months. The procedure is illustrated for quarterly data in the section that follows. The discussion section provides the rationale for each of the steps in the procedure and ways that seasonal indexes are used.

The symbols used in this chapter are summarized as follows:

N = the number of observations or the number of time periods considered
Y = a value of a variable observed at a particular period of time
O = an *alternative* symbol for Y typically used in connection with seasonal indexes, which stands for an *original* observation
MA = a moving average
SI = a seasonal index

THE RATIO TO MOVING AVERAGE METHOD

This section presents a general procedure for determining a seasonal index for quarterly or monthly data. In order to apply this procedure, quarterly or monthly data must be available for at least two years, but preferably for more than two. Seasonal indexes are obtained by performing the following steps:

Step 1. Eliminate seasonal variation from the original series by calculating a moving average, MA. A moving average is an arithmetic mean that is computed by successively dropping the first value in a series and including the next. The moving average for determining a seasonal index is performed in two stages.

(a) If *quarterly* data are used, the first value of a 4-quarter moving average is calculated by adding the first four observations and dividing by 4. Remaining values of the 4-quarter moving average are obtained by successively dropping the first value in the previously computed average and including the next value in the series. The process is continued until four observations are not available to form the average. The first value computed is placed midway between the second and third quarter observations of the first year.

If *monthly* data are used, the same process is followed except 12 observations are averaged at a time. The process is continued until 12 observations are not available to form the average. The first moving average is placed midway between the sixth and seventh month observations of the first year.

(b) The moving averages computed in the first stage do not correspond to the quarters or months in the series. A second moving average is necessary in order to center the averages on the quarters or months in the series. This is accomplished by successively computing the arithmetic mean of the original moving averages, *two* at a time. In the case of *quarterly* data, the first of these moving averages is placed alongside the third quarter of the first year in the series. For *monthly* data, the first of these moving averages is placed beside the seventh month of the first year in the series. The results at this stage are the MA values.

Step 2. For all periods, quarters or months, with a corresponding moving average, calculate the ratio of the original observation to the moving average, or O/MA values.

Step 3. For each quarter or each month, isolate the O/MA values corresponding to all years in the series.

Step 4. Based upon the O/MA values for each quarter or each month, calculate the *positional mean* and convert to percents by multiplying by 100. The positional mean is obtained by ordering the O/MA values, eliminating an equal number of high and low extremes, and computing the arithmetic mean of the remaining values. The number of extremes eliminated depends upon the of years of data observed in the original series. If the original series contains 4, 5, or 6 years of data, one high and one low extreme should be eliminated. It is satisfactory to eliminate two high and two low extremes when more than 6 years of data are used.

Step 5. If *quarterly* data are used, adjust the positional means for all quarters so their sum equals 400. This is accomplished by summing the positional means and multiplying each positional mean by 400 divided by this sum.

If *monthly* data are used, adjust the positional means for all months so their sum equals 1200. This is done by summing the positional means and multiplying each by 1200 divided by this sum.

The results of this step multiplied by 100 provide the seasonal indexes. If quarterly data are used, there are 4 indexes, one for each quarter. If monthly data are used, there are 12 indexes, one for each month.

NUMERICAL EXAMPLE

Quarterly department store sales for a particular standard metropolitan statistical area are presented in the following table for a four-year period:

DEPARTMENT STORE SALES (MILLION DOLLAR UNITS)

Year	*Quarter*	*Sales*	*Year*	*Quarter*	*Sales*
1975	Jan–Mar	204	1977	Jan–Mar	242
	Apr–Jun	254		Apr–Jun	327
	Jul–Sep	248		Jul–Sep	305
	Oct–Dec	385		Oct–Dec	465
1976	Jan–Mar	227	1978	Jan–Mar	263
	Apr–Jun	288		Apr–Jun	344
	Jul–Sep	266		Jul–Sep	330
	Oct–Dec	432		Oct–Dec	507

Our goal is to find a seasonal index for each of the four quarters based upon the data given in the table. Quarters are used here so that the example remains as simple as possible. Furthermore, only four years of data are presented which limits the averaging procedure at the end.

Step 1. Eliminate seasonal variability from the original sales series by calculating a moving average MA.

(a) The 4-quarter moving averages are presented in the fourth column of Exhibit 1. In order to understand how these are obtained, consider the first two values. The first value, equal to 273, is found by taking the arithmetic mean of the first four sales values as

$$\frac{204 + 254 + 248 + 385}{4} = \frac{1091}{4}$$

$$= 273$$

EXHIBIT 1
RESULTS OF MOVING AVERAGE AND O/MA CALCULATIONS TO FIND SEASONAL INDEXES

Year	Quarter	*Y* Sales	*4-Quarter Moving Average*	*MA* *2 of 4-Quarter Centered Moving Average*	*O/MA* *Ratio of Original Sales to Moving Average*
1975	Jan–Mar	204			
	Apr–Jun	254			
			273		
	Jul–Sep	248		276	.899
			279		
	Oct–Dec	385		283	1.360
			287		
1976	Jan–Mar	227		290	.783
			292		
	Apr–Jun	288		298	.996
			303		
	Jul–Sep	266		305	.872
			307		
	Oct–Dec	432		312	1.385
			317		
1977	Jan–Mar	242		322	.752
			327		
	Apr–Jun	327		331	.988
			335		
	Jul–Sep	305		338	.902
			340		
	Oct–Dec	465		342	1.360
			344		
1978	Jan–Mar	263		348	.756
			351		
	Apr–Jun	344		356	.966
			361		
	Jul–Sep	330			
	Oct–Dec	507			

This value is placed midway between the four values in the average.

The second moving average, equal to 279, is obtained by dropping the first sales value, adding the fifth, and taking the arithmetic mean:

$$\frac{254 + 248 + 385 + 227}{4} = \frac{1114}{4}$$
$$= 279$$

This value follows immediately after the average calculated above and is midway between the values comprising the second average.

The remaining values in the 4-quarter moving average column are calculated in the same way by successively dropping the first value in the previous average and including the next value in the series. When actually performing these calculations, some find it convenient to include a column with 4-quarter *moving totals* and then to divide these totals by 4.

(b) In order to obtain a moving average that corresponds to the actual quarterly observations in the series, a centered moving average of two of the 4-quarter averages is calculated. These appear in the fifth column of Exhibit 1 and represent the moving average, MA, used to find the seasonal indexes. For example, the first of these centered moving averages, equal to 276, is found as

$$\frac{273 + 279}{2} = \frac{552}{2}$$
$$= 276$$

The remaining values in the same column are obtained by successively dropping the first 4-quarter average and including the next in the series of 4-quarter averages. The first centered moving average is placed between the two values comprising the average. Consequently, it corresponds to the third quarter (Jul–Sep) in the actual series. The remaining MA values follow successively and correspond to an actual quarter in the original series. Due to the averaging procedure, there are no MA values for the first two and last two quarters in the original series.

Step 2. The ratio of the original sales figures given in the third column are divided by the moving average or MA values to obtain the O/MA values in the last column in the exhibit. For example, the first value in the O/MA column equal to 0.899 is found as

$$\frac{248}{276} = .899$$

The remaining O/MA values are found by dividing each value in the original series by the corresponding MA value for each quarter where an MA value is available.

Step 3. Isolate the O/MA values for each quarter. This is done in the following table:

O/MA Values Isolated By Quarter

	1975	*1976*	*1977*	*1978*
Jan–Mar		.783	.752	.756
Apr–Jun		.966	.988	.966
Jul–Sep	.899	.872	.902	
Oct–Dec	1.360	1.385	1.360	

Step 4. Based upon Step 3, calculate the positional mean of the O/MA values for each quarter. This is done in the table below:

Calculation of Positional Means

	Ordered O/MA Values			*Positional Mean*	*Converted to Percents*
Jan–Mar	.752	.756	.783	.756	75.6
Apr–Jun	.966	.966	.988	.966	96.6
Jul–Sep	.872	.899	.902	.899	89.9
Oct–Dec	1.360	1.360	1.385	1.360	136.0
					398.1

The ordered values in the above table represent the O/MA values for each quarter in Step 3 ordered by increasing magnitude. For each quarter, the highest and lowest O/MA values are eliminated. In this case the positional mean is the middle value since there is only one value remaining after eliminating the extremes. If more than one middle value remains, the arithmetic mean of these values is the positional mean. The last column of the table provides the positional means multiplied by 100.

Step 5. The positional means expressed as percents must be adjusted so their sum equals 400. Since the sum of the positional means equals 398.1, each must be adjusted so their sum equals 400. This is done by multiplying each positional mean by 400 divided by 398.1, or 1.005. The results appear in the following table:

ADJUSTING POSITIONAL MEANS

	Positional Mean	*SI* *Multiply By 1.005*
Jan–Mar	75.6	76.0
Apr–Jun	96.6	97.1
Jul –Sep	89.9	90.3
Oct–Dec	136.0	136.7
	398.1	400.1

The positional means multiplied by 1.005 represent the final seasonal indexes, which appear in the last column of the table. The sum of the seasonal indexes equals 400.1 instead of 400 due to rounding error.

DISCUSSION

The procedure for developing a seasonal index is rather long and involved and is more so when more than four years of monthly data are used. The procedure is included in this chapter because it provides a stronger basis for understanding the meaning of a seasonal index and the analysis of time series. We can do this by examining each component of the procedure.

Like any other average, a moving average is a single value that represents a set of observations that exhibit variability. Previously, however, we were not concerned with using an average to smooth a particular type of variability in cases where different sources of variability exist. In general, a moving average is used to smooth a time series in order to isolate one or more components in the series.

When developing a seasonal index, a moving average is used primarily to eliminate the seasonal variation from a time series. Since seasonal variability is defined as variation within a year, an average over an entire year period removes the seasonal component. This is the reason for using a 4-quarter or a 12-month moving average. Depending on the type of data (i.e., quarterly or monthly), they represent averages over the course of an entire year that remove seasonal variation from a series. If weekly data are used to obtain a weekly seasonal index, a 52-week moving average would be used. The series of moving averages

A moving average is used here to eliminate seasonal variability from the series. The resulting moving averages over a year period mainly represent the trend and cyclical components when they are present.

EXHIBIT 2
MOVING AVERAGE USED TO SMOOTH DEPARTMENT STORE SALES

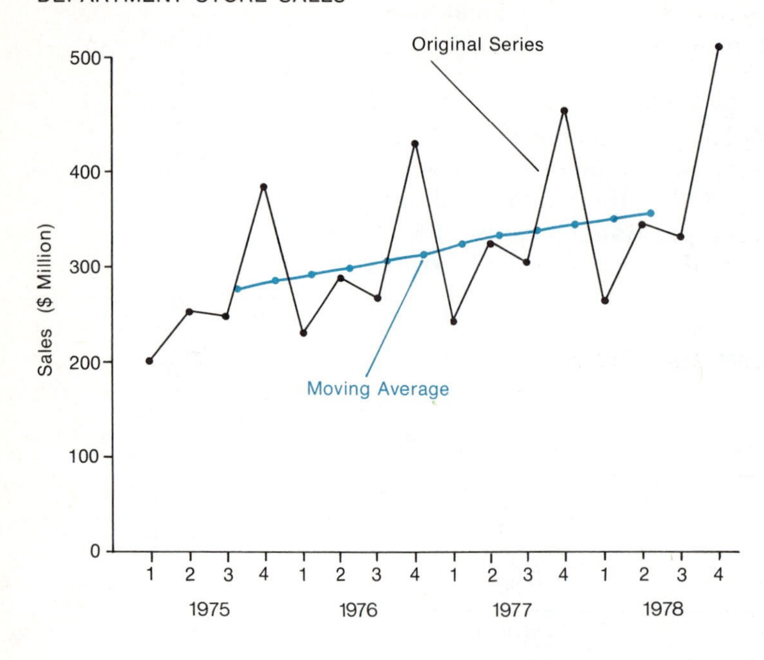

over a year period mainly represents the *trend and cyclical components* of the original series.

The department store sales data used in the numerical example and the centered moving average are plotted in Exhibit 2. The jagged high and low points in the original series correspond mostly to seasonal and some random variability. The moving average is much smoother and eliminates the high and low points. The pronounced peaks occur in the fourth quarters and correspond to a seasonal index of 136.7, which is somewhat higher than the other indexes. Since there does not appear to be any wavelike oscillation in the moving average, it mostly represents a trend in this particular case.

The O/MA values are crude measures of seasonal variability. Variability among the O/MA values for the same period primarily is due to random effects that are eliminated by the positional mean.

It was mentioned earlier that a seasonal index is a ratio of what is actually observed to the value that would be obtained if there were no seasonal variability. Hence, the O/MA values represent crude measures of seasonal variability since the original observation, O, contains the seasonal component and the moving average, MA, does not.

The O/MA values exhibit variability for the same quarter (or month) that primarily is due to the effect of random variation. By eliminating extremes and calculating the positional mean, the effects of the random element on the index are eliminated.

No seasonal variation in a particular period corresponds to a seasonal index of 100. Consequently, the average of the seasonal indexes for all periods within a year should equal 100 since an average over an entire year removes the seasonal effect.

Since a seasonal index is expressed as a percentage, *no* seasonal variation is indicated by an index equal to 100, or one-hundred percent. Consequently, the arithmetic mean of the seasonal indexes for all quarters or all months of the year should equal 100. In other words, seasonal variation should "average out" over the course of a year. This forms the basis for the final adjustment of the positional means.

If quarterly data are used, the sum of the four seasonal indexes should equal 400 and if monthly data are used, the sum of the twelve seasonal indexes should equal 1200. Otherwise, the average of the indexes would not equal 100. Therefore, the final adjustment increases each index proportionately if the sum of the positional means is less than 400 or 1200, and proportionately decreases them if the sum is greater.

Meaning of a Seasonal Index. Although the meaning of a seasonal index has been indicated in earlier remarks, let us summarize our earlier comments and then discuss this in more detail. A seasonal index corresponding to a particular period can be defined in the following way:

$$\text{Seasonal index, SI} = \text{Aver}\left[\frac{\text{Actual value}}{\text{Deseasonalized value}}\right] \times 100$$

Hence, a seasonal index is an average measure of the ratio of the actual observation to the deseasonalized value, or value containing no seasonal. A seasonal index, therefore, is a descriptive measure that indicates the percent of a series that is attributable to seasonal variability.

A seasonal index is a descriptive measure that indicates the percent of a series that is attributable to seasonal effects.

Consider the seasonal indexes in the following table:

SEASONAL INDEX—BEACHWEAR SALES

Quarter	*SI* *Seasonal Index*
Jan – Mar	105
Apr – Jun	95
Jul – Sep	125
Oct – Dec	75

The table provides quarterly indexes for department store beachwear sales. Consider the seasonal index for the third quarter, July–September. The value of 125 is interpreted to mean that beachwear sales are 125 percent of the sales that would be realized if there were no seasonal effect in this quarter. Another way to interpret this value is that beachwear sales in the third quarter are 25 percent higher than average yearly sales resulting from a seasonal effect in that quarter. A seasonal increase in sales is described by the third quarter index.

Notice the index for the fourth quarter is 75, which is less than 100. This means that a seasonal drop in sales occurs in this quarter resulting in sales that are 75 percent of the yearly average. Except for the specific numbers, similar interpretations can be given to the remaining values in the above table.

A seasonal index greater than 100 indicates that there is an increase in the series due to seasonal factors, and a seasonal index less than 100 indicates that there is a drop in the series due to seasonal effects.

By using months instead of quarters in our example, it would be possible to obtain more detail regarding the time of year when seasonal increases and decreases occur. Regardless of the time units that are used, the basic interpretation of a seasonal index is the same.

Deseasonalizing Time Series Data. In addition to describing the seasonal effect in the past, a seasonal index can be used to adjust or remove the seasonal effect from a time series. The process of removing the seasonal effect is referred to as deseasonalizing. By deseasonalizing a set of data, we are able to observe the series as it would appear if no seasonal variation were present.

Deseasonalizing an observation or a series means that it is seasonally adjusted such that the seasonal effect is removed.

An observation is deseasonalized by using the following formula:

$$\text{Deseasonalized value} = \frac{\text{Actual value}}{\text{SI}} \times 100$$

Seasonal effects are removed by dividing an observation for a given period by the seasonal index and multiplying by 100.

The seasonal effect is removed from an observation by *dividing* the actual value by the seasonal index and multiplying by 100. The same thing can be accomplished by moving the decimal place of an index two places to the left and dividing this result into the actual observation.

In order to illustrate the process, consider quarterly beachwear sales for 1978 that appear in the following table:

Quarter	*Sales*
1	$206,000
2	$192,000
3	$262,000
4	$156,000

First quarter sales are deseasonalized based upon the index for that quarter, 105, as follows:

$$\begin{aligned}\text{Deseasonalized first quarter sales} &= \frac{\text{Actual sales}}{\text{SI}} \times 100 \\ &= \frac{206{,}000}{105} \times 100 \\ &= 1961.9 \times 100 \\ &= 196{,}190\end{aligned}$$

A deseasonalized value can be interpreted as the value that would be obtained if no seasonal effect were present during the corresponding period.

We can interpret this value to mean that beachwear sales in the first quarter would be $196,190 if there were no seasonal effect present in that quarter. Each of the remaining sales figures can be deseasonalized similarly based upon the appropriate seasonal index. A summary of the results together with the sales figures and seasonal indexes appears in Exhibit 3. The seasonally adjusted or deseasonalized sales figures appear in the last column of the table. Notice, actual sales fluctuate over the four quarters with an especially large jump in the third quarter, whereas the seasonally adjusted values tend to increase steadily.

By deseasonalizing a set of data, we are able to better observe the other components. In the example just introduced, sales for 1978 is part of an upward movement of the series. Whether this movement is

EXHIBIT 3
DESEASONALIZING DATA WITH A SEASONAL INDEX

Quarter	*Beachwear Sales*	*SI Seasonal Index*	*Deseasonalized Sales = (Sales/SI) × 100*
1	$206,000	105	$196,190
2	$192,000	95	$202,105
3	$262,000	125	$209,600
4	$156,000	75	$208,000

due to a trend or a cycle cannot be determined without having data for additional years.

Including Seasonal Effects in a Forecast. A seasonal effect can be incorporated into a forecast by reversing the process of deseasonalizing. This is accomplished by *multiplying* an estimate of a future value of a time series by the appropriate seasonal index and dividing by 100. For example, assume we estimate the trend *and* cyclical component for beachwear sales in the third quarter of 1980 to be $350,000; the seasonal index for that quarter is 125. Then,

$$\text{Third quarter forecast, 1980} = \frac{(\text{Estimate of trend and cycle}) \times \text{SI}}{100}$$

$$= \frac{350{,}000 \times 125}{100}$$

$$= \frac{43{,}750{,}000}{100}$$

$$= 437{,}500$$

The sales forecast for the third quarter of 1980 containing trend, cycle, *and* seasonal effects is $437,500.

A seasonal effect can be included in a forecast by multiplying estimates of the other components by the seasonal index and dividing by 100.

The procedure just described applies to time series forecasts regardless of the way the forecasts of the other components are obtained. We shall see in the next chapter that forecasting the cycle is not an easy matter. The cyclical component is considered in the above example to illustrate the way a seasonal index can be used to make a forecast. If no cycle is present, the seasonal component can be included in a forecast simply by multiplying the trend forecast by the seasonal index for the period forecasted.

Changing Seasonal Patterns. The procedure for developing a seasonal index presented in this chapter applies in cases where the same seasonal pattern occurs year after year. This is referred to as a *stable seasonal* pattern. Other cases exist where seasonal increases and decreases occur within the same period of the year, year after year; however, the level of increase or decrease varies. For example, if toy sales not only increase in December but the relative volume of toy sales becomes larger and larger every year, the seasonal pattern is not the same. Hence, this kind of movement is referred to as a *changing seasonal* pattern.

Changing seasonal patterns can be detected by plotting the O/MA values for a particular period against the years corresponding to these values. If the O/MA values fluctuate about a horizontal line, no changing seasonal pattern is present. If, however, a trend is apparent, there is a changing seasonal pattern. One way to measure a changing seasonal pattern is to fit a trend line to the O/MA value for each period within a year similar to the way a trend is fitted to any time series. A value of a seasonal index is obtained by substituting the code for a quarter or month of a particular year into the trend equation. Regardless of the method of developing a seasonal index, the way an index is used and interpreted is the same.

The seasonal index based upon the procedure in this chapter applies to stable seasonal patterns, which are ones that recur at the same levels year after year. Some series exhibit changing seasonal patterns that must be measured with the aid of other methods. Regardless of the method that is used to obtain a seasonal index, the interpretation is the same.

QUESTIONS AND PROBLEMS

DEFINITIONAL QUESTIONS

1. What is seasonal variability? How is it measured?

2. What kind of time units are necessary in order to measure seasonal variability?

3. How is a seasonal index interpreted? What is the meaning of a seasonal index equal to 100?

4. What are the three uses of a seasonal index?

5. What is meant by the following terms?

   (a) Moving average
   (b) Centered moving average
   (c) 4-Quarter moving average
   (d) 12-Month moving average

6. How is a moving average used to develop a seasonal index?

7. What is an O/MA value? A positional mean? How are they used to find a seasonal index?

8. Distinguish between a stable and a changing seasonal pattern. Which type of seasonal pattern can be measured by the procedure presented in this chapter?

COMPUTATIONAL PROBLEMS

9. The number of departing passengers for the first six years of a specially built international airport are presented in the following table:

Year	*Quarter*	*Number of Passengers*	*Year*	*Quarter*	*Number of Passengers*
1970	1	372	1973	1	519
	2	413		2	630
	3	480		3	749
	4	375		4	559
1971	1	454	1974	1	645
	2	479		2	750
	3	583		3	921
	4	460		4	692
1972	1	530	1975	1	780
	2	602		2	903
	3	669		3	1070
	4	495		4	778

Determine the seasonal indexes for each quarter for the passenger data given in the table.

10. Regional department store sales on a monthly basis for a 4-year period are presented in the following table:

REGIONAL DEPARTMENT STORE SALES (MILLION DOLLAR UNITS)

Month	*1975*	*1976*	*1977*	*1978*
Jan	63	70	80	84
Feb	59	68	67	77
Mar	82	89	95	102
Apr	79	96	109	107
May	90	98	113	128
Jun	86	94	105	110
Jul	77	80	94	100
Aug	78	87	101	112
Sep	93	98	110	117
Oct	94	103	117	127
Nov	114	137	142	159
Dec	176	192	206	221

Determine the seasonal index for each month for the department store sales data given in the table.

11. The number of building permits issued by a suburban township for each quarter of 1978 and corresponding seasonal indexes are given in the following table:

Quarter	*Number of Permits*	*Seasonal Index*
Jan – Mar	317	77.6
Apr – Jun	498	116.7
Jul – Sep	443	109.3
Oct – Dec	396	96.4

Find the number of permits that would have been issued in each quarter if no seasonal variation were present in the data.

12. The total number of hospital admissions in a tri-county area during October of 1978 equals 2750 patients. The seasonal index corresponding to the number of patients for the month of October is 109. How many patients would be admitted in October 1978 if there were no seasonal effect in that month?

13. A city's budget director needs to estimate the city's cash requirements for the month of June 1979. A trend estimate of cash needs for the month is available that equals $500 million. Seasonal influences on cash requirements are summarized by an index number that equals 115 for the month of June. Assuming no

cyclical variability in cash needs exist, forecast the cash requirement for the month of June 1979.

CONCEPTUAL QUESTIONS AND PROBLEMS

14. What effect does seasonal variability have on a time series? What is the basis for this variability for an economic time series?

15. What is measured by a moving average? Why are 4-quarter and 12-month moving averages used to develop a seasonal index? Under what circumstances would a moving average based upon a different number of periods be used?

16. Criticize or explain the following statement. An O/MA value is a crude measure of seasonal variability but should be modified to account for other effects in order to provide a better description of seasonal variation.

17. Suppose you were to record the monthly population of permanent residents of a particular geographic location. In general, would you expect to find any seasonal variability present in the data? Why or why not?

18. Refer to Problem 9. The seasonal indexes for the four quarters are given as follows:

Quarter	*Seasonal Index*
1	93.6
2	103.3
3	117.2
4	85.9

(a) Is there any seasonal variability in the number of departing passengers? Explain.
(b) Interpret the seasonal index for the third quarter.
(c) Interpret the seasonal index for the first quarter.
(d) Which quarters exhibit seasonal highs and which exhibit seasonal lows? Why?

19. Refer to Problem 9.

(a) Plot the original quarterly data on a graph as a time series. How does the variability appearing in the graph relate to the seasonal indexes given in Problem 18?
(b) Use the seasonal indexes in Problem 18 to deseasonalize the original airline passenger series. Plot the deseasonalized values on a graph and compare the result with the graph in part (a).

20. The total number of persons unemployed on a quarterly basis for an eight-county labor market area in northeastern U.S. is

presented for 3 years in the following table:

NUMBER OF PERSONS UNEMPLOYED (THOUSANDS)

	1975	*1976*	*1977*
Jan – Mar	376	324	380
Apr – Jun	422	371	430
Jul – Sep	392	372	367
Oct – Dec	329	326	372

Is there evidence of seasonal variation in the unemployment data presented? Explain the reason for reaching your conclusion.

21. Refer to Problem 20. The quarterly seasonal indexes for unemployment are as follows:

	Seasonal Index
Jan – Mar	96
Apr – Jun	109
Jul – Sep	105
Oct – Dec	90

(a) What information is provided by each index in the table?
(b) Suppose you computed similar seasonal indexes for construction workers only. Would you expect to find similar values for these seasonal indexes? Give reasons for possible differences in the two sets of values.

22. A playground administrator has developed a trend equation for the number of children attending special programs. The equation is given as

$$Y_T = 100 + 2t$$

t = 0 in January 1970
t is in one-month intervals

The seasonal index for attendance at special programs during May is 110. Assuming no cyclical variation exists, estimate the attendance at the playground's special programs during May 1979.

23. For any time series with time units less than a year, how does one know if there is no seasonal variability present in the data? Explain.

24. The total number of gardening and plant books sold monthly by a paperback book store is given in the following table:

	Jan	Feb	Mar	Apr	May	Jun	Jul	Aug	Sep	Oct	Nov	Dec
1974	346	328	285	348	416	438	545	520	457	427	362	358
1975	405	355	451	372	525	549	635	516	484	463	454	349
1976	320	380	315	320	409	532	513	544	429	369	365	300
1977	266	327	389	382	413	382	435	396	366			

(a) Plot the data on a graph as a time series. By examining the graph, is it possible to determine whether there is a trend, cyclical, or seasonal variability in the data? Explain.

(b) Monthly seasonal indexes for the above data are given as follows:

Month	Jan	Feb	Mar	Apr	May	Jun	Jul	Aug	Sep	Oct	Nov	Dec
Seasonal Index	81	85	90	82	110	116	131	122	109	100	94	80

Based upon these indexes, what conclusions can be reached about the seasonal sales of the store's gardening and plant books?

(c) Use the seasonal indexes in part (b) to deseasonalize the original data. How are the values interpreted?

(d) Plot the deseasonalized values found in part (c) on a graph as a time series. Is there evidence of a trend or cycle in the deseasonalized values. Explain. What overall conclusion can you make based upon this exercise about the nature of time series data and the use of a seasonal index?

25. Is it reasonable to use one year of monthly or quarterly data to develop seasonal indexes? Why or why not? What is the advantage of using a number of years to calculate seasonal indexes?

26. Why shouldn't the complete procedure presented in this chapter be used to develop an index when a changing seasonal pattern exists?

27. The O/MA values corresponding to June sales of jogging paraphernalia for a sporting goods store are given for a 10-year period in the following table:

Year	1969	1970	1971	1972	1973	1974	1975	1976	1977	1978
O/MA	.89	1.03	.96	1.05	1.17	1.08	1.24	1.19	1.29	1.25

(a) What information is provided by the O/MA values in the table? Is there evidence of a changing seasonal pattern? Explain.

(b) Using the methods of Chapter 29, fit a trend line to the O/MA values in the table.

(c) Use the trend line found in part (b) to estimate the seasonal index for June sales in 1978. Interpret this value.

(d) Assume actual sales of jogging paraphernalia in June of 1978 equals $20,000. What would the sales have been if there were no seasonal variation present in this month?

(e) Why should the approach in this problem be used instead of the one presented in the chapter?

CHAPTER 19

CYCLICAL AND RANDOM VARIABILITY

Forecasting Methods

Civilization exists by geological consent, subject to change without notice.

Will Durant

Man must go up or down, and the odds seem to be all in favor of his going down and out.

H. G. Wells

We have seen how to determine the effect of the trend and seasonal components and how to make projections of these components into the future. Of the two, seasonal influences are easier to understand and to measure since they are external to the underlying economic system that generates a time series. The trend is more difficult to define since it depends upon the nature of each individual situation. The real problem, however, is associated with the cyclical and random components.

Cyclical variability has been defined as a periodic or wavelike movement about the trend. For economic time series, a cycle occurs within a varying number of years, but for a period that is shorter than the trend. When a cycle appears in a series, the series is observed to rise and fall over and over as time passes. It is quite common to hear news reports concerning the current status of the "business cycle" and what is expected to happen to it in the future. Productivity may be "up,"

EXHIBIT 1
CYCLICAL VARIABILITY

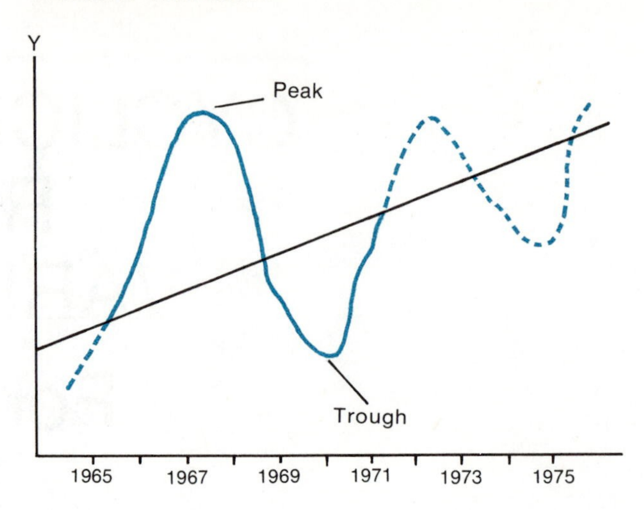

employment "down," interest rates "high," or housing starts expected to be "low" in the near future. Actually, each of these items represent separate parts of overall economic activity that rise and fall in level over time.

Appearing in Exhibit 1 is an example of a typical cycle. The thin black line in the exhibit represents the trend. The actual series fluctuates about the trend in a wavelike manner. Consider the solid portion of the series (color) corresponding to the years 1965 to 1971. This represents one complete cycle. In other words, a complete cycle begins at the trend, rises to a high point or *peak*, falls to a low point or *trough*, and rises again reaching the trend where the cycle terminates. In this particular case, it takes six years for the cycle to execute completely. This period, or the time it takes to return to the trend, is called the *duration* of the cycle. The distance between the highest and lowest points, or the distance between the peak and the trough, is referred to as the *amplitude*.

A continuation of the cyclical pattern appears in the exhibit as a dotted line. This indicates that the cycle is a recurring phenomenon. Notice, the second time the series completes a cycle, the duration and amplitude are different. There are series experienced in the physical sciences where the duration and amplitude occur with near-perfect regularity. When this occurs, cycles can be described and forecasted using mathematical equations with no difficulty.

Series in economics and business do not exhibit the same kind of regularity. In other words, the duration and amplitude generally vary each time a series undergoes a new cycle. Because of this, it is difficult to forecast the behavior of the cycle for an economic time series. The main problem is to determine the turning points when the cycle changes direction.

Actual time series exhibit fluctuations about the cycle. These fluctuations frequently are attributed to irregular movements in the series. Simpler approaches to the analysis of time series generally do not distinguish between an irregular component and a random component. These approaches view irregular movements as disconnected episodes that occur infrequently and with no regularity. Such reasons as strikes, wars, floods, famines, and now oil embargoes are offered to explain irregular changes in a time series. The effect of each of these can be identified after they occur but whether they will occur is not known in advance. Although such occurrences can be viewed as the result of random forces, they are not treated as random in the simpler methods in terms of probability as we know of it from our earlier work.

Disconnected episodes actually represent one kind of random phenomenon present in a time series. In general, we can define a random component of a time series as that part of the series that cannot be explained and is distributed according to some probability distribution. In general, a random component is one that does not occur with any regularity and cannot be explained or predicted. In many cases it is difficult to separate random and cyclical movement in a time series.

Although there is general agreement regarding the components of a time series, there is no single method of isolating all the components or of forecasting a series that is agreed upon by all. This is due to the difficulties mainly associated with the cyclical and random components. Consequently, the primary goal of this chapter is to provide some awareness of the problems associated with analyzing time series and of the existence of techniques that are available. A simple way of describing a cycle is presented in the next section and then is discussed. The last section brings together the ideas presented up to that point and provides a brief survey of the area of forecasting.

DESCRIBING CYCLICAL VARIABILITY

The simplest way to describe cyclical variation is in terms of a residual about the trend. This can be done in absolute terms or as a relative, or percent. In absolute terms, we simply determine the portion of the series left over after eliminating the trend. This is given by the following expression:

$$\text{Residual} = Y - Y_T$$

Absolute cyclical residual.

The absolute cyclical residual is the difference between the observed value of a series, Y, and the corresponding trend value, Y_T.

Another way to eliminate the trend effect is to express the actual value as a percent of the trend:

$$\text{Percent of trend} = \frac{Y}{Y_T} \times 100$$

Cycle described as a percent.

Here we calculate the ratio of the actual observation, Y, to the trend value, Y_T, and multiply by 100. Differences between the result and 100 represent the residual effect expressed as a percentage.

In order to illustrate the meaning of the two formulas, consider data representing plant and equipment expenditures presented in the following table:

Year	*Plant & Equipment Expenditures ($ Billion)*
1965	16.9
1966	16.0
1967	16.4
1968	19.0
1969	21.0
1970	16.4
1971	17.1
1972	19.5
1973	18.7
1974	19.7

The trend line fitted to the above series is

$$Y_T = 16.7 + 0.3t$$
$$t = 0 \text{ in } 1965$$
$$t \text{ is in 1-year intervals}$$

If, for example, we are to find the absolute cyclical residual for 1968, we first find the corresponding trend estimate as

$$\begin{aligned} Y_T &= 16.7 + 0.3(3) \\ &= 16.7 + 0.9 \\ &= 17.6 \end{aligned}$$

where the value of t equal to 3 substituted above corresponds to 1968. The absolute cyclical residual for this year is obtained as

$$\begin{aligned} \text{Residual} &= Y - Y_T \\ &= 19.0 - 17.6 \\ &= 1.4 \end{aligned}$$

where 19.0 represents the actual plant and equipment expenditure for 1968. The value of 1.4, or $1.4 billion, represents the amount left over in the series after eliminating trend for 1968.

Using the same values, the percent of trend is

$$\begin{aligned} \text{Percent of trend} &= \frac{Y}{Y_T} \times 100 \\ &= \frac{19.0}{17.6} \times 100 \\ &= 1.08 \times 100 \\ &= 108 \end{aligned}$$

In 1968, the actual value in the series is 108 percent of the trend or 8 percent higher.

The same procedure can be used to find the absolute cyclical residuals and the percent of trend for the remaining values of plant and equipment expenditures. A summary of the results appears in the following table:

Year	Plant & Equipment Expenditure (Y)	Trend (Y_T)	Absolute Cyclical Residual ($Y - Y_T$)	Percent of Trend $\left(\frac{Y}{Y_T} \times 100\right)$
1965	16.9	16.7	.2	101
1966	16.0	17.0	−1.0	94
1967	16.4	17.3	−.9	95
1968	19.0	17.6	1.4	108
1969	21.0	17.9	3.1	117
1970	16.4	18.2	−1.8	90
1971	17.1	18.5	−1.4	92
1972	19.5	18.8	.7	104
1973	18.7	19.1	−.4	98
1974	19.7	19.4	.3	102

The absolute cyclical residuals for the original series are presented in the next to the last column of the table. A minus sign indicates that an observed value is below the trend whereas positive values indicate that it lies above the trend. With respect to the percent of trend figures in the last column, values above 100 percent indicate that an observation is above the trend and ones below 100 percent correspond to observations below the trend.

Meaning of Residual Measures. In order to understand the meaning of the residual measures illustrated above, it is instructive to plot the results for the entire series on a graph. This is done in Exhibit 2. Three graphs are presented in the exhibit. The first corresponds to the original time series of plant and equipment expenditures; the dotted line represents the trend. The second graph represents the absolute cyclical residuals, and the third corresponds to the percent of trend.

The absolute cyclical residuals are plotted about a horizontal line at zero, since an observed value equal to the trend yields zero or no residual. A wavelike or cyclical pattern is evident in the graph. Roughly, two completed cycles are apparent, one ending around 1969 and the other about 1973.

The same basic pattern is evident by examining the third graph, although the pattern is more pronounced. By using relative rather than absolute values, the importance of the cyclical movement relative to the trend is described more effectively. In general, if the trend has a steep slope and the swings in the cycle are small, the percents of trend would be less pronounced than in a situation where there are large swings about a gradually rising trend.

Notice that the cyclical patterns in the graphs do not appear very smooth, especially in 1969 where the peak is much more pronounced than other high and low points. The residual measures used here also contain random and irregular movements. Consequently, these measures represent the part of the time series attributed to cyclical and random factors after the effect of trend is removed.

It is possible to smooth the residual measures with a moving average in order to remove some of the random component. For annual data, however, the choice of the number of years to use in the average is arbitrary. In other words, there is no overall guide to determine the number of years to be used in the averaging procedure; the result of the averaging procedure will vary depending upon the number of years

EXHIBIT 2
GRAPHS OF ACTUAL DATA, ABSOLUTE CYCLICAL RESIDUALS, AND PERCENT OF TREND

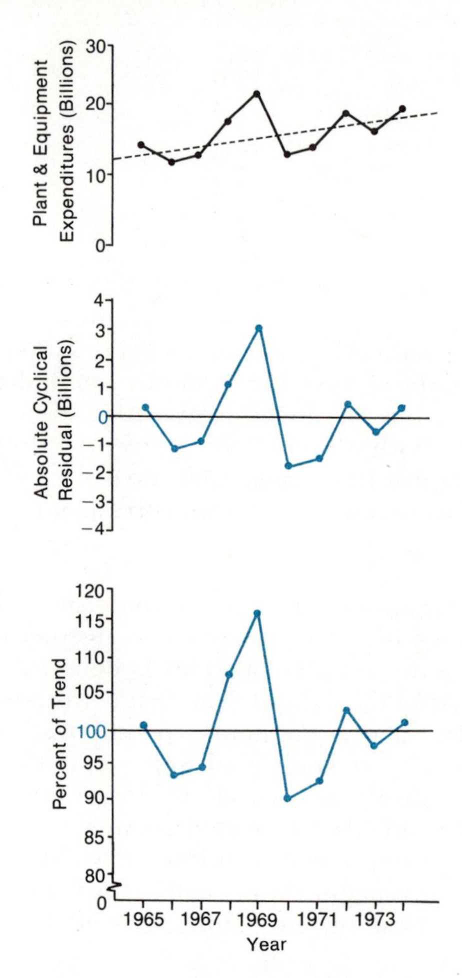

used and the nature of the cyclical and random effects in a particular time series.

We have just seen the most basic way to describe the cycle. Although a cycle can be observed directly from the original time series when the random effect is not too great, the use of residual measures eliminates the trend component which can obscure the cyclical pattern to some extent. By using residual measures it is possible to obtain an idea of the behavior of the cycle in the past with respect to duration and amplitude. Because of the way these change each time a new cycle occurs, residual measures are not useful to forecast the future behavior of the cycle, primarily the turning points. More sophisticated techniques are necessary for this purpose. Simple extensions similar to trend projections of upswings and downswings, however, do a satisfactory job between the peaks and troughs in some cases.

Monthly and Quarterly Data. In our discussion of cyclical variability to this point, it has been assumed that the data are in terms of years. When monthly or quarterly data are used, the data must be *de-*

seasonalized before any analysis of the cyclical effect is undertaken. Otherwise, the effects of the cyclical and seasonal components could not properly be separated.

For example, suppose department store sales for a particular region equal $190 million in December 1978. The seasonal index for the month of December is 125 and the trend in department store sales is given by the expression

$$Y_T = 62 + 2t$$

t = 0 in January 1975

t is in 1-month intervals

Y is in million dollar units

In order to determine either the absolute cyclical residual or the percent of trend due to cycle, we first must deseasonalize the sales for December. This is accomplished by dividing the actual sales by the seasonal index and multiplying by 100:

$$\begin{aligned}\text{Deseasonalized December sales} &= \frac{190}{125} \times 100 \\ &= 152\end{aligned}$$

The trend estimate of sales is found by substituting into the trend equation

$$\begin{aligned}Y_T &= 62 + 2(35) \\ &= 132\end{aligned}$$

where t equal to 35 corresponds to December 1978, given that the base period is January 1975.

On the basis of the above two results, the absolute cyclical residual is found as

$$\begin{aligned}\text{Absolute cyclical residual} &= \\ \text{Deseasonalized December sales } &\textit{minus} \text{ Trend} \\ &= 152 - 132 \\ &= 20\end{aligned}$$

In terms of the original units, $20 million is attributed to cyclical and random factors. The deseasonalized sales also can be expressed as a percent of trend as

$$\begin{aligned}\text{Percent of trend} &= \frac{\text{Deseasonalized December sales}}{\text{Trend}} \times 100 \\ &= \frac{152}{132} \times 100 \\ &= 115\end{aligned}$$

Cyclical and random factors are 115 percent of the trend or 15 percent higher.

Based upon our work in this example, we are able to establish the amount of department store sales in December 1978 attributable to each of the components. We are provided with the following information:

$$
\begin{aligned}
\text{Actual sales} &= \$190 \text{ Million} \\
\text{Deseasonalized sales} &= \$152 \text{ Million} \\
\text{Trend in sales} &= \$132 \text{ Million} \\
\text{Residual sales} &= \$20 \text{ Million}
\end{aligned}
$$

Sales attributable to seasonal effects for the month are obtained by subtracting the deseasonalized sales from the actual sales.

$$
\begin{aligned}
\text{Seasonal sales} &= 190 - 152 \\
&= \$38 \text{ Million}
\end{aligned}
$$

Hence, an increase of $38 million is attributed to seasonal effects, $132 million to the trend, and $20 million is attributed to cyclical and random factors. Recognize, these three figures should sum to total sales. We have

$$
\begin{aligned}
\text{Actual sales} &= \text{Trend} + \text{Seasonal} + \text{Residual} \\
&= 132 + 38 + 20 \\
&= \$190 \text{ Million}
\end{aligned}
$$

The main point made in this section is that data in units of less than a year should be adjusted for seasonal effects before measuring the effects of the other components in a time series. Otherwise, it is not possible to determine which component contributes to the result. By illustrating this point we also can see that the procedures we have used allow us to *decompose* a time series into amounts attributable to each of the components.

OVERVIEW OF FORECASTING

At the beginning of Chapter 17 it was mentioned that the analysis of time series data is related to the problem of forecasting. The procedures that are considered were included primarily to understand the nature of time series data and some of the problems associated with it. These methods of analysis centered on isolating the individual components of a time series: trend, seasonal, and cycle plus random. This process is referred to as **classical decomposition,** or simply as decomposition.

Suppose we are interested in using classical decomposition for the purpose of forecasting in the department store example. Assume it is February 1979 and we want to forecast sales in million dollar units for December of the same year. Based upon the trend equation from past work, $Y_T = 62 + 2t$ with base period equal to January of 1975, we forecast the trend component as $156 million dollars. Further, assume that the residual effect due to cyclical factors will be 115 percent of the trend. The seasonal index for December is 125.

We can represent the problem in the form of a "model," or an equation, describing the way the forecast is made on the basis of the information given. This can be written as

$$Y = T \times C \times S$$

where Y represents an actual value, T is the trend effect, C stands for cycle expressed as a percent of trend divided by 100, and S represents the seasonal effect in terms of a seasonal index divided by 100. Using the information given and substituting into the equation we get a forecast of Y as

$$\begin{aligned} Y &= T \times C \times S \\ &= 156 \times \left(\frac{115}{100}\right) \times \left(\frac{125}{100}\right) \\ &= 224.25 \end{aligned}$$

Hence, our forecast of department store sales for December 1979 is $224.25 million or $224,250,000.

A forecast of this kind is based on the assumption that each of the effects has been measured correctly in the past and will continue in the same way in the future. It already has been indicated that the assumption will hold for seasonal effects and generally it will hold for the trend. The real problem is that the cyclical component does not recur in a similar manner in the future. Furthermore, the model used above does not consider the random element.

When a time series is considered in terms of the classical decomposition method it appears as if each of the components act separately and can be identified distinctly. By viewing a time series in this way it is possible to understand it better and in some instances the distinction is useful when forecasting. However, in general the distinction among the components may not be clear.

For example, a set of data may exhibit a smooth overall movement for a seven-year period that is described as a trend. It is possible, however, that this represents one portion of a cycle that executes completely over a much longer period of time. Whether the movement described is defined as a trend or part of a cycle depends upon the amount of data available and the purpose of analyzing the data. If forecasting is the goal, projection as a trend may be satisfactory for a short period into the future. If the goal is to understand what actually happens to a particular series of data, treating it as a trend may not be sufficient.

In other instances, it is possible that a trend is present, but it is obscured because of large and erratic swings in the cycle. Further, a large and changing random element can obscure the other components of a series. Overall, we can say that in some cases it is useful to identify the four time series components. In other cases, this is not sufficient and other methods of analysis are required in order to explain the past and predict the future.

Other Methods. Numerous modifications of the decomposition method have been developed in order to overcome some of its weaknesses. One of the most popular of these versions is called the **Census II** method which was developed at the U.S. Census Bureau. This is a computerized system that provides trading-day adjustments to monthly data to account for the fact that all months do not have the same number of days. In addition, the method adjusts the data for outliers, which are extreme observations that do not conform to any recognizable patterns and affect the measurement of the other components. Some clues relating to turning points in the cycle are provided also.

Procedures such as the basic decomposition method and Census II do not deal with the "reasons" why a particular time series behaves the way it does. Such methods are referred to as **naive** since they do not anticipate changes in the underlying economic system when making a forecast. Other methods, which we can call **regression methods,** frequently are used that do consider the reasons for time series variation by explicitly incorporating other variables beside time into the model.

We already have been exposed to regression analysis in an earlier chapter. However, we did not deal explicitly with problems where time is the unit of association linking the variables. When variables are related over time, forming relationships can help to explain the cyclical component and some of the remaining residual that is not truly random. These relationships often are useful for forecasting purposes. Furthermore, multiple regression equations can be modified to include variables that describe individual time series components. For example, consider the following equation:

$$Y_t = 750 - 0.2x_t + 0.1t$$

where

$$\begin{aligned} Y_t &= \text{company sales (million dollar units)} \\ x_t &= \text{average price (dollar units)} \\ t &= \text{time measured in quarters} \\ &= 0 \text{ in Jan–Mar 1976} \end{aligned}$$

The equation represents quarterly company sales as a function of price, x, and a trend variable, t, where the trend variable assumes coded values similar to trend equations considered earlier.

Suppose we want to forecast sales in the third quarter of 1980 assuming the average price for that quarter will be \$5. We obtain a forecast by substituting into the above equation as follows:

$$Y_t = 750 - 0.2(5) + 0.1(18)$$

The value of t equal to 18 corresponds to the trend for the third quarter of 1980. If we solve for y in the above expression we get

$$\begin{aligned} y &= 750 - 1.0 + 1.8 \\ &= 750.8 \end{aligned}$$

Hence, the third quarter forecast of sales in 1980 equals \$750.8 million. Notice, the trend contributes \$1.8 million to forecasted sales.

It is possible to include variables describing seasonal effects into regression equations in addition to the trend. For example, we can define a seasonal variable, Q_1, corresponding to the first quarter, which is defined in the following way:

$$Q_1 = 1 \text{ in the first quarter}$$
$$= 0 \text{ in other quarters}$$

A 1 indicates the presence of the first quarter and a 0 indicates that the first quarter is not present. Variables corresponding to the other quarters can be defined in the same way. By including these variables in a regression equation it is possible to obtain coefficients corresponding to the seasonal effects.

In general, a variable that indicates the presence or absence of a particular condition in terms of 0's and 1's is called a *dummy variable.* Dummy variables can be used to include the effect of any condition that assumes one of two possibilities. For example, the presence of consumers in a particular age group can be designated by a 1 and by a 0 if the group is not present. An irregular effect such as wartime versus peacetime conditions also could be included in terms of a dummy variable. Anticipated increases in promotional efforts too can be considered with dummy variables. One must be careful about the number of dummy variables that are used since breakdowns of the least squares method can result if too many are considered in particular problems.

We have just seen a simple example where the concepts of time series and regression are combined into a single equation. In the example it is assumed that price is an independent variable and is subject to the control of the firm. Instead, suppose that price is dependent upon a number of variables in the economy so that price is not subject to complete control by the firm. In this case both price and sales are dependent variables such that price must be predicted with another equation before the sales equation can be used to forecast sales. Situations of this kind where two or more regression equations are needed generally are considered in the field of econometrics.

Econometrics is a field of study in which statistical methods, mathematics, and economic theory are used to solve economic problems. Although the field is quite diverse and involves many types of activities, the primary output is in the form of econometric models. **Econometric models** are systems of equations of interrelated variables. Each equation describes a different aspect of an economic problem and can be used to obtain forecasts of different variables that are required. Such models are useful for making policy and planning decisions since the impact of changes of one or more variables on the others in the system can be evaluated.

Cyclical Indicators. The National Bureau of Economic Research and the Bureau of Economic Analysis of the Department of Commerce have studied a large number of time series, resulting in a list of series that are useful as indicators of change in the cyclical movements in the economy. The approach is based on the fact that some time series exhibit changes in direction before turning points occur in overall economic activity. Series of this kind are referred to as *leading series.* For example, changes in the money supply, changes in the price of selected stocks, and changes in the layoff rate in manufacturing tend to occur before changes appear in the overall economy. Consequently, time series of this kind can be used to obtain indications about cyclical turns in other series. Moreover, they can be incorporated into regression type models discussed above to provide more predictability.

Forecasting Criteria. Actual forecasting done in practice is based upon considerations taken from both the naive and the regression approaches. Combinations and extensions of these methods are based upon a number of factors associated with a particular decision-making problem and the many individual techniques available.

The factors that relate to the choice of a method of forecasting are listed as follows: (1) the time horizon, (2) the frequency of forecasting, (3) the number of forecasts, (4) the level of detail, (5) past data patterns, (6) availability of data, (7) cost, (8) accuracy, (9) ease of understanding and application, and (10) the experience of the forecaster. The time horizon represents the length of time into the future that a forecast is to be made. The broadest way to classify the time horizon is in terms of a *short term* and a *long term.* Exact boundaries for the two do not exist and vary with a particular problem. The forecast period can range between as short a time as one month and as long as five or ten years or more. In general, the shorter the time horizon of a forecast, the easier it is to forecast "correctly." Long-term forecasts would involve the trend component whereas short-term forecasts would be concerned with seasonal and cyclical movements also.

Forecasts made far into the future are for planning purposes and are not made frequently. At the other extreme, there are situations when a large number of forecasts are necessary on a month-to-month basis for the purpose of operational control. For example, an operation that requires an inventory of a large number of individual parts may require forecasts of the monthly demand for each part. In such a case, a method that is inexpensive, quick, and easy to implement is necessary rather than a sophisticated forecasting procedure.

Within an individual firm or organization, forecasts may be needed at different levels of management with varying amounts of accuracy and detail. The vice president of sales at the home office may be interested in overall, or aggregate, company sales whereas the director of a regional sales office may be interested only in sales in a particular region. If, for example, the firm sells a very large and expensive piece of equipment, the only forecasting requirement may be at the aggregate level since too few are sold at each of the regional offices. Here, efforts would be concentrated on sales and service rather than planning. If, on the other hand, a large number of smaller units are sold, the individual sales manager may be required to submit a forecast of sales anticipated in a coming period. Furthermore, the needs at the home office may be greater since they relate not only to current sales but to the long-term growth of the firm. Since productivity and many related factors depend upon current sales, the shorter term forecast might have to be better than the long-term projection. In other words, the requirements for accuracy may vary depending upon the type of decision to be made. In the short term the requirement might be that the forecast falls within 3 percent of actual sales whereas 10 or 20 percent may be satisfactory for a longer term projection.

In addition to considerations associated with the type of decision, there are others. The pattern of past data is important to consider when building a forecasting technique. For example, if a time series is observed to be stable over a long period of time with a marked trend with little variability, a simple trend projection may do a satisfactory job. If, however, large cyclical swings of changing duration and amplitude occur about the trend, a more sophisticated approach may have to be used. Further, the approach that is used is dependent upon the kind of data that are available. If a regression method is employed, data regarding other variables beside the one forecasted must be available. To a great extent, the final forecasting approach that is chosen depends upon one's experience regarding the success of various methods as they relate to different types of data and decision-making goals.

QUESTIONS AND PROBLEMS

DEFINITIONAL QUESTIONS

1. What is cyclical variability? What is a cycle?

2. Define the following terms relating to the concept of a cycle:

   (a) Peak
   (b) Trough
   (c) Duration
   (d) Amplitude

3. What is meant by random variation as it relates to a time series?

4. What measures of cyclical variability are used in this chapter? Do they measure anything else beside cyclical variability? Explain.

5. Distinguish between an absolute cyclical residual and a percent of trend.

6. If cyclical residual measures are calculated on the basis of monthly or quarterly data, what adjustment to the data must be made? Why?

7. What is meant by the term *classical decomposition* of a time series?

8. Distinguish between naive methods and regression methods of forecasting.

9. List some alternative approaches to forecasting.

10. List the criteria that relate to the choice of a forecasting method.

CONCEPTUAL QUESTIONS AND PROBLEMS

11. What information is provided by an absolute cyclical residual and the percent of trend?

12. Why is it necessary to examine residual measures for an entire time series in order to determine whether cyclical variability is present? Under what circumstances is information about cyclical variability unclear when using residual measures?

13. Why is it difficult to forecast the cyclical component of a time series?

14. The trend in cash reserves for a certain company for 10 years between 1967 and 1976 is given by the equation

$$Y_T = 51 + 4t$$

t = 0 in 1967
t is in 1-year intervals
(Y is in \$10,000 units)

Actual cash reserves for 1974 equal \$930,000. Find the absolute cyclical residual and the percent of trend for 1974. What information is provided by these results?

15. Company profits for a fifteen-year period are given in the following table:

Year	*Company Profits (\$1000 Units)*
1963	475
1964	550
1965	650
1966	775
1967	650
1968	600
1969	575
1970	550
1971	663
1972	752
1973	800
1974	750
1975	700
1976	685
1977	750

The series of profits is described by a trend equation

$$Y_T = 568.492 + 13.312t$$

t = 0 in 1963
t is in 1-year intervals

Based upon the information given, determine whether there is evidence of a cycle in company profits.

16. Refer to the department store sales problem given in the numerical example of Chapter 18. The trend equation that applies to the data is given as

$$Y_T = 262.9 + 7.3t$$

t = 0 in January 1975
t is in one-quarter-year intervals

(a) Is there evidence of a cycle in the department store sales data? Explain.
(b) Would it be helpful to have data for more years in order to answer part (a)? Explain.

17. Refer to the airline passenger data in Problem 9 of Chapter 18. The quarterly seasonal indexes corresponding to the data are given as follows:

Quarter	*Seasonal Index*
1	93.6
2	103.3
3	117.2
4	85.9

(a) Fit a trend line to the deseasonalized data in Problem 9.
(b) Is there evidence of cyclical variability in the passenger data? Explain.
(c) Based upon your work and the information given, calculate the number of passengers corresponding to each of the time series components for the second quarter of 1973.
(d) On the basis of the information available, is it possible to make a meaningful forecast of the number of passengers for the second quarter of 1977? What additional information or assumptions are necessary to make the forecast? Explain.

18. Under what circumstances is it reasonable to use a description of the cycle in the past as a basis for forecasting it in the future?

19. Refer to Exhibit 2 in Chapter 18. By examining the moving average of quarterly department store sales shown in the exhibit, can you draw any conclusions about the presence of a cycle? Explain. In general, how can a moving average be used as an alternative to describe cycles in time series data? Relate your answer to annual data, and monthly and quarterly data.

20. Criticize or explain the following statements. Basically, there are two goals of time series analysis: (1) to decompose a series into its various components, and (2) to forecast. By decomposing a series it is possible to understand what has occurred in the past, which is useful in reconstructing a series so that it can be predicted in the future.

CHAPTER 20

INDEX NUMBERS

You have only . . . to multiply and divide at discretion; and you can pay the National Debt in half an hour. Calculation is nothing but cookery.

Lord Brougham

And in such indexes . . . , there is seen the baby figure of the giant mass of things to come.

William Shakespeare

In this chapter we devote our attention to a particular type of variability over time. Here, we are interested in changes in the values of a *group* of items between two time periods. Since more than one item is to be considered, it is necessary to use an average measure of change. Such a measure is referred to as an *index number*.

In order to understand the problem, consider the question of whether a consumer must pay more today to live than five years ago. If we are interested in the change in price of a single item consumed, say air conditioners, we could simply compare the prices of a similar brand between the two time periods. If, however, we address ourselves to changes in price for many different items essential to a consumer, the problem is not as simple. We hear about this problem repeatedly in terms of changes in the "cost of living."

On the one hand, we are interested in price changes faced by a "typical" consumer, since different people do not purchase exactly the same group of items, although some similarities occur. On the other hand, a particular individual will not purchase an identical group of items consistently over a five-year period, or the quantity purchased may vary. Furthermore, some items will increase in price, while others will decrease over the given time period. Sometimes, an item purchased in one year may not be available in a later period. Consequently, an average measure is needed to reflect the various changes in the prices of a variety of products purchased by a large number of consumers.

The problem outlined above is not restricted to the cost of living. Manufacturers or construction firms, for example, may be interested in determining how raw material costs change over time. If they purchase a large number of different raw materials, an overall measure may be more useful than examining the cost of each individual item. Actually, overall price changes are not the only type of changes of interest. Changes in quantities used or consumed are of interest also.

Whenever changes in a group of items are measured, whether price or quantity, an average measure or index is used. Since price changes are considered more frequently, this chapter focuses on price indexes rather than on quantity indexes. The underlying principles are the same in both cases, except for simple modifications in the formula used to calculate each type of index. Consequently, the procedure and example sections deal exclusively with price indexes. A short section at the end of the discussion is devoted to quantity changes.

A **price index** can be defined as a relative measure describing the average change in the prices of a group of items between two time periods. Generally, the time period used as a basis for constructing a price index is a year. An index is constructed as a *ratio* of prices prevailing in one year to those representing an earlier period. Prices associated with the earlier period always appear in the denominator of the ratio. The period or year corresponding to the denominator of the index is called the **base period.**

A price index is expressed as a *percentage*. The value of the index in the base period, therefore, always is defined equal to 100. Since an index is an average measure, both the numerator and denominator of the index are *sums* associated with the items considered. Because changes in price are being measured, prices of the same item for the two periods are allowed to vary.

Each of the prices is *weighted* to reflect *typical* patterns of consumption or usage of the items included in the index. The weights used must be held *constant* between the periods compared so that changes observed can be attributed to changes in price. Weights used are of two kinds, quantity weights and value weights.

There are two price indexes that commonly are in use. These are the *weighted relative of aggregates* and the *weighted average of price relatives*. The procedures for computing both are presented and illustrated in the next two sections.

The symbols used in this chapter are summarized as follows:

P_B = the price of an item in the base period
P_G = the price of an item in a given year other than the base period
P_T = a typical price of an item
Q_B = the quantity of an item used or consumed in the base period
Q_G = the quantity of an item used or consumed in a given year other than the base period
Q_T = a typical quantity of an item used or consumed

PROCEDURES FOR CONSTRUCTING A PRICE INDEX

Weighted Relative of Aggregates Index. Based upon prices of a group of items corresponding to a base period and a given year, the weighted relative of aggregates index is found by substituting into the formula

$$\text{Weighted relative of aggregates} = \frac{\Sigma Q_T P_G}{\Sigma Q_T P_B} \times 100$$

In words, this index is found by multiplying each price in the given year, P_G, by the corresponding typical quantity, Q_T, and adding the results; this is divided by the sum of the base period prices, P_B, multiplied by the *same* typical quantities. This answer is multiplied by 100 to express the index as a percentage.

Weighted Average of Price Relatives Index. The weighted average of price relatives index is found by averaging the ratios of prices in the given year to those in the base period according to the formula

$$\text{Weighted average of price relatives} = \frac{\Sigma (Q_T P_T)\,\dfrac{P_G}{P_B}}{\Sigma Q_T P_T} \times 100$$

The weights used in this index are *value* weights found by multiplying a typical price, P_T, by a typical quantity, Q_T. The ratio or relative of the price in the given year to the price in the base period, P_G/P_B, is found and multiplied by the value weight, $Q_T P_T$. These results are summed to obtain the numerator of the index that is then divided by the sum of the value weights. Multiplying by 100 converts the index to a percentage.

NUMERICAL EXAMPLE

Appearing in the table below are the prices and typical quantities of standard stationery supplies used by one department of a large company for two years, 1970 and 1978.

Stationery Item	Typical Quantity Used Q_T	Unit Price 1970 P_B	Unit Price 1978 P_G
Ball Pens	20 dozen	$11.35	$10.65
Pencils	325 dozen	1.35	2.75
Transparent Tape	400 rolls	1.40	1.75
Tablets	225 dozen	5.65	7.15
Staples	125 boxes	1.75	2.50
Typing Ribbon	75 boxes	3.00	4.70
Typing Paper	120 boxes	3.30	5.00

Our problem is to construct a price index reflecting the overall change in price of the stationery items between the two years.

Weighted Relative of Aggregates. Before using the formula, it is convenient to perform a preliminary set of calculations. This is done in the following table:

PRELIMINARY CALCULATIONS FOR WEIGHTED AGGREGATES INDEX

Q_T	P_B	P_G	Q_TP_B	Q_TP_G
20	11.35	10.65	227.00	213.00
325	1.35	2.75	438.75	893.75
400	1.40	1.75	560.00	700.00
225	5.65	7.15	1271.25	1608.75
125	1.75	2.50	218.75	312.50
75	3.00	4.70	225.00	352.50
120	3.30	5.00	396.00	600.00
			3336.75	4680.50

The last two columns of the table present the prices in the two periods multiplied by the same typical quantities. Based upon the totals of these two columns, the aggregates index is found as

$$\text{Weighted relative of aggregates} = \frac{\Sigma Q_TP_G}{\Sigma Q_TP_B} \times 100$$

$$= \frac{4680.50}{3336.75} \times 100$$

$$= 1.40 \times 100$$

$$= 140$$

Based upon the weighted relative of aggregates index equal to 140, we can say that prices of the standard stationery items used in 1978 were 140 percent of those in the base period of 1970. Another interpretation is that, overall, prices increased on the average by 40 percent between the two periods.

Weighted Average of Price Relatives. Using the same data presented at the beginning of this section, the preliminary computations for the weighted average of relatives index are presented in the following table:

PRELIMINARY CALCULATIONS FOR WEIGHTED AVERAGE OF PRICE RELATIVES

Q_T	P_T	P_B	P_G	Q_TP_T	P_G/P_B	$(Q_TP_T)(P_G/P_B)$
20	11.00	11.35	10.65	220.00	.94	206.80
325	2.05	1.35	2.75	666.25	2.04	1359.15
400	1.58	1.40	1.75	632.00	1.25	790.00
225	6.40	5.65	7.15	1440.00	1.27	1828.80
125	2.13	1.75	2.50	266.25	1.43	380.74
75	3.85	3.00	4.70	288.75	1.57	453.34
120	4.15	3.30	5.00	498.00	1.52	756.96
				4011.25		5775.79

Recall, the typical quantities, Q_T, were included in the original information provided at the beginning of this section; however, typical prices,

P_T, were not. For purposes of this example, assume the typical prices are given; more is said about this and the weighting problem in the discussion section. The value weights, Q_TP_T, are obtained by multiplying the typical quantities by the typical prices. These value weights are multiplied by the ratio of the prices between the given and base year, which results in the figures in the last column. Based upon the totals obtained in the table, we compute

$$\text{Weighted average of price relatives} = \frac{\Sigma(Q_TP_T)\dfrac{P_G}{P_B}}{\Sigma Q_TP_T} \times 100$$

$$= \frac{5775.79}{4011.25} \times 100$$

$$= 1.44 \times 100$$

$$= 144$$

Based upon the weighted average of relatives equal to 144 we can say that prices of the standard stationery items used in 1978 were 144 percent of those in the base period of 1970. Or, we can say that prices increased, on the average, by 44 percent between the two periods.

DISCUSSION

We stated that a price index describes the way prices of a group of items change over a period of time. A number of different methods for constructing index numbers exist. The most commonly used methods, however, are the weighted relative of aggregates and the weighted average of price relatives. Regardless of the method of construction, the interpretation of any price index is the same. A value of 100 means there is no overall change in prices; a value exceeding 100 means there is an overall increase; and a value less than 100 means there is an overall decrease. Since an index is an average measure, some prices in the group may increase while others decrease.

A price index equal to 100 means that there is no overall change in price. A value greater than 100 means there is an overall increase, and a value less than 100 means there is an overall decrease.

Recognize that a price index is a comparatively rough measure of change and is subject to many of the limitations associated with an average as a representative measure. In part, this is evident from the numerical example presented earlier where the values of the two indexes are different, even though they are based upon the same data. In general, the weighted relative of aggregates and the weighted average of price relatives will not assume the same value, but the values generally will be close for the same set of data. The difference between the two depends upon the choice of a typical price used in the value weight of the weighted average of price relatives index.

When base period prices are used in the value weight of the average of relatives index, the method of relatives yields the same numerical answer as the method of aggregates. This easily is seen by examining the formula for the relatives index with base period prices in the weights:

$$\text{Weighted average of price relatives with base period prices in weights} = \frac{\Sigma(Q_T P_B)\frac{P_G}{P_B}}{\Sigma Q_T P_B} \times 100$$

The base period prices divide out in the numerator leaving us with the result

In general, some difference exists in the values of the weighted aggregates index and the average of relatives index. When base period prices are used in the weights of the average of relatives, the values of the two indexes are identical.

$$\text{Weighted average of price relatives with base period prices in weights} = \frac{\Sigma Q_T P_G}{\Sigma Q_T P_B} \times 100 = \text{Weight relative of aggregates}$$

Hence, the two indexes are identical when base period prices are used in the value weight of the relatives index.

A considerable amount of effort has been applied toward refining index numbers, especially by government agencies, so that index numbers describe changes in price as accurately as possible. Moreover, the two indexes presented have been studied from time to time in order to determine which is better under varying circumstances. From a practical viewpoint, it does not really matter. The one advantage, however, of the aggregates index is that it is easier to compute than the relatives index.

Using a Price Index. We have noted that a price index can be used to *describe* the overall change in prices of a group of items. As a descriptive measure, price indexes are useful to private companies as well as to government agencies. By examining the movement of price indexes of different segments of a firm's operations or of the economy, administrators can assess the impact of price changes. Since many aspects of economic activity are related to price movements, price indexes can be used as *indicators* of change in various segments of the economy. The usefulness of indicators in forecasting was cited in the last chapter. The kinds of index numbers available as indicators can be found in various publications produced by the government.

A price index can be used to describe the overall change in prices of a group of items. Price indexes also may be used as indicators of change in various segments of the economy.

Index numbers also can be used to make *comparisons*. For example, consider the numerical illustration in the last section where we found the value of an index number for standard stationery items used by a particular department within a firm. Suppose there are a large number of departments in the firm, each of which does its own purchasing of stationery supplies. Top management may be interested in cutting costs in a number of areas including supplies for stationery. An item by item comparison of prices paid by each department is complicated and does not provide an overall picture of prices paid by the various departments, all of which may not purchase exactly the same items. By comparing price indexes among departments using the same base period, it is possible to determine whether disparities in prices paid, say for stationery, exist. If this is the case, a decision may be made to centralize purchasing of supplies so that lower prices are paid.

Price indexes are useful for making comparisons of prices paid by different units purchasing the same basic group of items. They also are useful for estimating current or replacement costs of a group of commodities.

Price indexes are useful in *estimating* current costs of projects or of replacement costs, say for insurance purposes. The cost of a group of items obtained from company records can be adjusted to current dollar values by simply *multiplying* by an appropriate price index and dividing by 100. The appropriate index is one that is representative of the items in question and is determined for a base period corresponding to the purchase data appearing in the company records.

Current vs. Real Values. Frequently, we hear that the dollar is not worth as much as it was in the past. Whether or not this is true depends upon the relative quantities of various items that can be purchased over time by an individual for a given amount of money. An important use of a price index is to adjust *current* dollar values into *real* values in order to eliminate the effect of price changes upon purchasing power.

In order to understand the problem, suppose a worker earns $4.20 an hour in 1969. If an item cost $1.00 in the same year, the worker could purchase 4.2 units of the item with one hour's wages. That is,

$$\begin{aligned}\text{Number of units purchased with an hour's wage} &= \frac{\text{Hourly wage}}{\text{Price of item}} \\ &= \frac{\$4.20}{\$1.00} \\ &= 4.2 \text{ Units}\end{aligned}$$

If, however, in 1977 the same worker earns $5.60 an hour but the item costs $1.23, an hour's wage buys 4.6 units of the same item (i.e., $5.60/ $1.23 = 4.6). Hence, the real benefit of an hour's wage relative to the particular item is greater in 1977 than it was in 1969 since more units can be purchased with current earnings. Despite the fact that the price of the item increased, more of the item can be purchased with current earnings. By dividing by the item price in both cases, we eliminated the effect of price and determined the real purchasing power for that item.

In general, when questions of purchasing power arise, they are not associated with a particular individual but with an entire class or group. Furthermore, it is not meaningful to consider a single item since individuals purchase many different items in order to live. Consequently, earnings of a group of people or a class of workers must be adjusted with a price index that provides an overall picture of the purchasing power for the group. The process of converting current earnings to real terms basically is the same as illustrated above except that a price index is used instead of the price of an individual item; an adjustment also is necessary since a price index is expressed as a percentage.

Typically, when real values are computed, the base period is earlier than are the given years for which these values are determined. In such a case, the adjustment of current values to real terms is referred to as *deflating a value series* since prices typically increase over time. In order to adjust current or money values into real terms, each value in a series is *divided* by the corresponding price index and the result multiplied by 100. This is equivalent to moving the decimal point in the price index two places to the left and dividing into the value in the series. The process of adjusting current earnings into real terms is illustrated in Exhibit 1.

Current or money values are converted into real terms by dividing by a price index and multiplying by 100.

EXHIBIT 1
ADJUSTMENT OF CURRENT EARNINGS TO REAL VALUES

Year	(1) Average Hourly Earnings Current Values	(2) Price Index (1969 = 100)	$(3) = \frac{(1)}{(2)} \times 100$ Real Earnings (1969 Prices)
1969	$3.21	100.0	$3.21
1970	$3.25	101.3	$3.21
1971	$3.43	104.5	$3.28
1972	$3.50	107.2	$3.26
1973	$3.66	110.6	$3.31
1974	$3.83	113.4	$3.38
1975	$3.91	116.3	$3.36
1976	$4.06	119.7	$3.39
1977	$4.22	122.4	$3.45
1978	$4.31	125.2	$3.44

Column (1) in the exhibit presents the average hourly earnings of a specific group of workers. These earnings represent the actual wage paid in the given years. The price index that applies to this group is presented in Column (2). In order to adjust the actual hourly earnings figures for the effect of price, the current values in column (1) are divided by the corresponding price index and the result is multiplied by 100. The adjusted earnings in real terms expressed in terms of 1969 prices appears in column (3) of the exhibit.

By examining column (1) we can see that earnings in money terms steadily rose from year to year. Overall, this is true of the real values also. In other words, there is an increase in purchasing power between 1969 and 1978 since the adjusted hourly wage in 1978 is greater than the value in 1969. Within the nine-year period, however, the pattern varies. For example, a slight drop in purchasing power occurs in 1972 since the real value of $3.26 is less than the value of $3.28 in 1971.

Relatively, real values have the same relationship to one another regardless of the base year used.

The real values appearing in column (3) of Exhibit 1 are based upon price indexes computed on a 1969 base year. Consequently, the real earnings figures for intervening years can be compared directly to determine the effect of price upon purchasing power for the specific group of workers considered. As long as the price indexes are computed on the same base, regardless of the year, similar comparisons can be made. If a different base period were used, the numerical values in the real earnings column would be different; however, the *relative* differences among the real values will be the same no matter what base period is used.

The adjustment of a value series in this section is illustrated in terms of hourly earnings. The concept, however, can be applied to other series of data. For example, the gross national product (GNP) of the economy represents the total amount of goods and services produced for a given year. When expressed in the actual dollar amounts of the goods and services for the year, it is not possible to determine whether an increase is due to changes in price or to changes in the actual volume of output in the economy. By adjusting GNP with a price index that converts it into real terms, it is possible to measure changes in the volume of real output.

Weights Used in a Price Index. When computing a price index, prices are weighted to reflect the relative importance of the individual items. If weights are not used, a price index would be a haphazard measure of change based upon the units of the items included in the index. No completely satisfactory system of weighting exists, however.

Quantity weights should reflect the pattern of consumption or utilization of a group of items in both periods compared. The quantities used, however, must be held constant in order to measure changes in price. The major problem with any set of weights lies in the fact that patterns of consumption or utilization change over time. In some cases, products used in one period are not available in another. In other cases, the composition of a product or item is different so that there is a problem whether changes are measured for the same item.

Weights must be held constant in order to measure variations in price and should reflect typical patterns of consumption in the periods compared. A problem exists, however, since patterns of consumption change over time.

When establishing quantity weights, or typical quantities, there are three basic choices. Quantities reflecting use in the base period or the current period can be used, or an average of quantities in some period of time ranging between the two periods can be used. Regardless of the choice made, the weights must be periodically revised in order to reflect more current behavior.

When value weights are used as in the case of the weighted average of price relatives, the choice of a typical price to include is subject to the same basic considerations as in the case of quantity weights. Typical prices should reflect prices paid in a "typical" period. For the numerical example presented earlier, the arithmetic mean of base period and current year prices was used as the typical price. The resulting index was calculated to be 144. If base year prices were used, the result would be 141, and if current year prices were used in the weights the index would equal 146. Actually, there is not a big difference among the three results. In general, it does not matter very much which weights are chosen as long as they are used consistently.

The Base Period. The base period associated with an index number is a period of time that is used as a basis for comparing changes observed. In the numerical examples, we used a period of one year as a base period. However, it can be any period of time. No matter what period of time is used, the value of the index number for the base period is 100. The period chosen should be one that corresponds to a time from which relative changes are to be measured.

Since patterns of utilization and production do not remain constant over time, the period chosen as a base should *not* be too far in the past. Consequently, published indexes such as those developed by government agencies are updated periodically. Generally, this is done when the weighting scheme or the items included in an index are changed.

A base period should reflect a period of relative stability. A base period can be a year or a period that equals more than one year.

The base period of an index should represent a period of relative economic *stability*. In other words, it should not reflect a period in which unusual fluctuations in prices occur. For this reason, some indexes are based upon a number of years, usually three. The average price over the three-year period then is used as a basis for comparison.

In order to make proper comparisons between two index numbers or two series of indexes, the index numbers must be related to the same base period. In some cases it is necessary to change the base period of an index. This process is referred to as *shifting* the base. Without actually recomputing the entire index, shifting can be accomplished by adjusting a series of indexes based upon the fact that the base period value must equal 100. Hence the base of a series of index numbers is shifted by dividing each index in the series by the value of the index corresponding to the newly desired base period.

Items Included in an Index. When an index is computed, it is based upon a group of items associated with a particular problem or the reason for developing the index. For indexes used internally within private organizations, complete records of item quantities and prices generally are available. Consequently, all items related to a particular problem such as in the stationery example may be included when computing an index.

When many items are included in an index or all necessary data are not available, a sample of items is used. Many considerations are required in order to choose a representative sample of the items to be included.

In other cases, especially indexes developed and published by governmental bodies, either too many items must be considered or the necessary data are not available. When this occurs, a *sample* of items is used to compile an index. The sample must be representative of the group described in terms of the actual items included and the weights used. For example, the Bureau of Labor Statistics compiles a consumer price index for goods and services purchased by urban clerical workers and wage earners. On the one hand, it would be virtually impossible to include every item and service purchased by all such workers. On the other hand, to determine the pattern of consumption for all such wage earners would be a difficult and time consuming, if not an impossible, task. Consequently, a representative sample of items is used.

The Consumer Price Index. Of the many index numbers currently in use, one that is cited by the media and is heard of more frequently than the others is the Consumer Price Index, or CPI. This index, which is related to economic activity, is strongly tied to labor negotiations involving workers' bargaining for pay increases that will maintain their real purchasing ability. The main uses of the CPI are to formulate economic policy, escalate income payments, and measure real earnings and purchasing power of the consumer dollar. Because of its importance, some of the main features of this index are provided here. The material also illustrates the type of considerations involved in assembling an index number.

The Consumer Price Index is published monthly by the Bureau of Labor Statistics of the U.S. Department of Labor. This index measures changes in prices of goods and services in major expenditure groups such as housing, food, clothing, recreation, health, and transportation for all urban consumers and urban wage earners and clerical workers. It is based upon a modification of the weighted average of price relatives with base period weights, where the base period is 1967.

Previously the Consumer Price Index applied to urban wage earners and clerical workers, but now as a result of a current revision there are two indexes, the Consumer Price Index for All Urban Consumers, CPI-U, and the Consumer Price Index for Urban Wage Earners and Clerical Workers, CPI-W. The CPI-W covers the same population as the old index, which includes urban wage and clerical worker families and single individuals living alone where at least one family member must be employed for at least 37 weeks during the survey year in a wage or clerical occupation such that half the total family income is derived from this occupation. The CPI-U considers all urban residents including salaried workers, self-employed workers, retirees, unemployed individuals, and wage earners and clerical workers; there are no restrictions on earnings and family income.

Weights used in the indexes are based upon average annual expenditures established from the 1972–73 Consumer Expenditure Survey conducted in 216 areas throughout the country. Items considered in the indexes include necessities and luxuries purchased for consumption and real estate, sales, and excise taxes, where life insurance payments and income and personal property taxes are excluded. About 400 items are included in the indexes and are selected on the basis of importance in family spending for the individuals covered. The items are subdivided into various levels of expenditure classes, which are reported separately along with the overall indexes.

Prices for the revised indexes are based upon a full probability sample of retail outlets based upon the results of a point-of-purchase survey. More than 20,000 families across the country provided information regarding retail outlets in which they did their shopping. Sampling was done in 85 areas, which included cities, suburbs, and urban regions within a fixed limit of selected counties. Price reports were elicited mainly by combinations of personal visits and telephone interviews, except for some items that were based upon mailed reports and secondary sources. Reports were obtained from about 60,000 units that include food store outlets, rental units, and housing units. Over one million price quotations were obtained.

Quantity Indexes. A quantity index measures the relative changes in the volume of a group of items used or produced between two time periods. Two examples are the Index of Department Store Sales and the Index of Industrial Production. The considerations associated with constructing quantity indexes are basically the same as those related to price indexes. Methods of construction are the same except quantities are varied from period to period. The two most common quantity indexes are the weighted relative of aggregates and the weighted average of quantity relatives index. The formulas are the same as those for the corresponding price indexes except that prices are used as weights and are held constant in the aggregates index.

QUESTIONS AND PROBLEMS

DEFINITIONAL QUESTIONS

1. What is an index number? A price index?

2. Why are index numbers important?

3. Why is a "weight" used when computing an index number?

4. What is a weighted aggregates index? A weighted average of price relatives?

5. What is the meaning of the following terms?

(a) Base year	(e) Base year quantity
(b) Given year	(f) Given year quantity
(c) Base year price	(g) Typical quantity
(d) Given year price	(h) Typical price

6. What are the main uses of a price index?

7. What is meant by the term *deflating a value series?*

8. What is meant when it is said that output or consumption is expressed in "real" terms rather than in "money" terms?

9. What does it mean when a price index is equal to 100? Greater than 100? Less than 100?

10. What is a quantity index?

COMPUTATIONAL PROBLEMS

11. Presented in the following table are the prices of four items for 1970 and 1978 with typical quantities of utilization:

Item	Price *1970*	*1978*	Q_T
A	$10.70	$14.50	10
B	8.30	6.95	14
C	15.47	16.30	6
D	2.24	3.90	12

Compute the weighted aggregates price index for the group of items using 1970 as a base year.

12. Refer to Problem 11. Based upon the data given calculate a weighted average of price relatives. Use current year prices in the value weights.

13. Given below are the prices of breakfast foods for 1972 and 1978 along with quantities that reflect the relative consumption of each item:

Item	Price *1972*	*1978*	Quantity
Bacon (lb)	$1.29	$1.72	$1\frac{1}{2}$
Butter (lb)	1.12	1.50	1
Eggs (doz)	.70	.92	2
Cereal (box)	.60	.89	2
Bread (loaf)	.37	.56	$2\frac{1}{2}$
Coffee (lb)	1.20	2.50	$\frac{1}{2}$
Milk (qt)	.40	.55	4
Sugar (lb)	.29	.48	$\frac{1}{2}$

Measure the overall change in the price of a breakfast between 1972 and 1978 using a weighted average of price relatives index. Use 1978 prices in the weights.

14. Repeat Problem 13 using a weighted aggregates index.

15. Given below are the prices of four types of pocket calculators in 1973 and 1978. The typical quantities given in the table represent the average number of each type sold.

Type	Price 1973	Price 1978	Typical Quantities
Four function	$50	$11	21
Single memory	65	15	14
Scientific	140	29	11
Programmable	250	65	6

Measure the change in the price paid for pocket calculators between the two periods.

CONCEPTUAL QUESTIONS AND PROBLEMS

16. How does an index number measure changes in price?

17. What are the similarities and differences between the weighted aggregates index and the weighted average of price relatives index?

18. Refer to Problem 11. The price index using a 1970 base for the group of items equals 113. Interpret this result.

19. Refer to Problem 13. The price index for that problem equals 146. Interpret this result in terms of the problem.

20. Refer to Problem 15. Discuss the problems associated with the measurement of changes in prices paid by consumers over time for pocket calculators.

21. The following table presents the dollar amount of assets in the form of heavy machinery purchased by a company for each of three years; price indexes for the three years are given also.

Year	Asset Purchases (*$1000 Units*)	Price Index (*1970 = 100*)
1976	503	172.2
1977	517	187.1
1978	523	211.8

(a) Express the assets purchased for each year in terms of 1970 prices.

(b) Has there been an increase or a decrease in the amount of real assets in terms of heavy machinery purchased each year between 1976 and 1978? Explain.

22. Criticize or explain the following statement. Since wages of a certain group of workers increased 40 percent between two periods and prices increased 60 percent for the same period of time, the workers received 20 percent less in terms of real wages.

23. Discuss the problems associated with the construction of an index number in terms of weights used, the base period, and the items included in the index.

Bibliography

General Appreciation

Campbell, S. K., *Flaws and Fallacies in Statistical Thinking.* Englewood Cliffs, N.J.: Prentice-Hall, Inc., 1974.
Epstein, R. A., *The Theory of Gambling and Statistical Logic.* New York: Academic Press, 1967.
Huff, D., *How to Lie with Statistics.* New York: W. W. Norton, 1954.
Larsen, R. J., and D. F. Stroup, *Statistics in the Real World.* New York: Macmillan Publishing Co., Inc., 1976.
Levinson, N. C., *Chance, Luck, and Statistics.* New York: Dover Publications, Inc., 1963.
Moroney, M. J., *Facts from Figures.* London: Pelican Books, 1956.
Mosteller, F., R. Rourke, and G. Thomas, Jr., *Probability and Statistics.* Reading, Mass.: Addison-Wesley Publishing Co., Inc., 1961.
Sellers, G. R., *Elementary Statistics.* Philadelphia: W. B. Saunders Company, 1977.
Tanur, J. M., et al., *Statistics: A Guide to the Unknown.* San Francisco: Holden-Day, Inc., 1962.
Weaver, W., *Lady Luck.* New York: Doubleday and Company, Inc., 1963.
Zeisel, H., *Say It With Figures.* New York: Harper and Row, 1957.

Probability

Feller, W., *An Introduction to Probability Theory and Its Applications.* Vol. 1, 2nd ed. New York: John Wiley & Sons, Inc., 1971.
Goldberg, S., *Probability: An Introduction.* Englewood Cliffs, N.J.: Prentice-Hall, Inc., 1960.
Parzen, E., *Modern Probability Theory and Its Applications.* New York: John Wiley & Sons, Inc., 1960.
Thorp, E. O., *Elementary Probability.* New York: John Wiley & Sons, Inc., 1966.

General Statistics

Anderson, T. W., and S. L. Sclove, *An Introduction to the Statistical Analysis of Data.* Boston: Houghton Mifflin Company, 1978.
Dixon, W., and F. Massey, Jr., *Introduction to Statistical Analysis,* 3rd ed. New York: McGraw-Hill Book Company, 1969.
Gulezian, R. C., *Statistics for Decision Making.* Philadelphia: W. B. Saunders Company, 1979.
Hamburg, M., *Statistical Analysis for Decision Making.* New York: Harcourt Brace Jovanovich, Inc., 1977.
Hoel, P., *Introduction to Mathematical Statistics,* 4th ed. New York: John Wiley & Sons, Inc., 1971.
Mason, R. D., *Statistical Techniques in Business and Economics.* Richard D. Irwin, Inc.: Homewood, Illinois, 1974.
Neter, J., W. Wasserman, and G. A. Whitmore, *Applied Statistics.* Boston: Allyn and Bacon, Inc., 1978.

Special Topics

Butler, W. F., and R. A. Kavesh, *How Business Economists Forecast.* Englewood Cliffs, N. J.: Prentice-Hall, Inc., 1966.
Chisholm, R. K., and G. R. Whitaker, *Forecasting Methods.* Homewood, Ill.: Richard D. Irwin, Inc., 1971.
Cochran, W. G., *Sampling Techniques,* 2nd ed. New York: John Wiley & Sons, Inc., 1963.
Cochran, W. G., and G. M. Cox, *Experimental Designs.* New York: John Wiley & Sons, Inc., 1957.
Conover, W. J., *Practical Non-Parametric Statistics.* New York: John Wiley & Sons, Inc., 1971.
Draper, N. R., and H. Smith, *Applied Regression Analysis.* New York: John Wiley & Sons, Inc., 1966.
Fleiss, J. L., *Statistical Methods for Rates and Proportions.* New York: John Wiley & Sons, Inc., 1973.
Gilchrist, W., *Statistical Forecasting.* New York: John Wiley & Sons, Inc., 1976.
Hammersley, J. M., and D. C. Handscomb, *Monte Carlo Methods.* London: Methuen and Company, Ltd., 1964.
Kish, L., *Survey Sampling.* New York: John Wiley & Sons, Inc., 1965.
Klein, L. R., *An Introduction to Econometrics.* Englewood Cliffs, N. J.: Prentice-Hall, Inc., 1962.
Lewis, J. P., and R. C. Turner, *Business Conditions Analysis.* New York: McGraw-Hill Book Company, 1967.
McNeil, D. R., *Interactive Data Analysis.* New York: John Wiley & Sons, Inc., 1977.
Morgan, B. W., *An Introduction to Bayesian Statistical Decision Processes.* Englewood Cliffs, N. J.: Prentice-Hall, Inc., 1968.
Nelson, C. R., *Applied Time Series Analysis.* San Francisco: Holden Day, Inc., 1973.
Quenouille, M. H., *Rapid Statistical Calculations.* New York: Hafner Publishing Company, 1972.
Raiffa, H., *Decision Analysis, Introductory Lectures on Choices Under Uncertainty.* Reading, Mass.: Addison-Wesley Publishing Co., Inc., 1968.
Schlaifer, R., *Probability and Statistics for Business Decisions.* New York: McGraw-Hill Book Company, 1959.
Siegel, S., *Nonparametric Statistics for the Behavioral Sciences.* New York: McGraw-Hill Book Company, 1956.
Winkler, R. L., *Introduction to Bayesian Inference and Decision.* New York: Holt, Rinehart and Winston, Inc., 1972.
Wonnacott, R. J., and T. H. Wonnacott, *Econometrics.* New York: John Wiley & Sons, Inc., 1970.

Data Sources

Coman, E. T., Jr., *Sources of Business Information,* 2nd ed. Berkeley: University of California Press, 1964.
Directory of International Statistics. New York: United Nations Publications.
National Referral Center for Science and Technology, *A Directory of Information Resources in the United States, Social Sciences.* Washington, D.C.: U. S. Government Printing Office, 1967.
Silk, L. S., and M. L. Curley, *A Primer on Business Forecasting with a Guide to Sources of Business Data.* New York: Random House, Inc., 1970.
Wasserman, P., E. Allen, A. Kruzas, and C. Georgi, *Statistics Sources,* 4th ed. Detroit: Gale Research Company, 1971.

Tables

Beyer, W. H., *Handbook of Tables for Probability and Statistics,* 2nd ed. Cleveland: Chemical Rubber Company, 1968.
Burington, R. S., and D. C. May, *Handbook of Probability and Statistics with Tables,* 2nd ed. New York: McGraw-Hill Book Company, 1970.
Hald, A., *Statistical Tables and Formulas.* New York: John Wiley & Sons, Inc., 1952.
Owen, D., *Handbook of Statistical Tables.* Reading, Mass.: Addison-Wesley Publishing Company, Inc., 1962.

Dictionaries

Freund, J., and F. Williams, *Dictionary/Outline of Basic Statistics*. New York: McGraw-Hill Book Company, 1966.

Kendall, Maurice, and W. R. Buckland, *Dictionary of Statistical Terms*, 3rd ed. New York: Hafner Press, 1971.

History

David, F. N., *Games, Gods, and Gambling*. London: Charles Griffin and Company, Ltd., 1962.

Pearson, E. S., and M. G. Kendall, *Studies in the History of Statistics and Probability*. Darien, Conn.: Hafner Publishing Company, 1970.

APPENDIX

Table 1
Binomial Probabilities

This table provides probabilities of f(x) for values of x based upon selected values of n and P for the binomial distribution

$$f(x) = \binom{n}{x} P^x (1 - P)^{n-x};\ x = 0, 1, \ldots, n$$

The table considers values of n between 1 and 20 and values of P between 0.05 and 0.50 in intervals of 0.05. When P exceeds 0.5, probabilities for values of x can be found by using n − x in place of x and 1 − P in place of P.

Example In order to find the probability that x equals 5 given n equals 8 and P equals 0.35, locate n = 8 in the left margin and P = .35 at the top. The probability is found below P across from x = 5 associated with n.

$$P(x = 5 \mid n = 8, P = .35) = .0808$$

Example Suppose n equals 10 and P equals 0.8 and we want to find the probability that x equals 7. Since the table does not consider P greater than 0.5 directly, we must locate n − x = 10 − 7 = 3 and 1 − P = 1 − .8 = .2 corresponding to n = 8. That is

$$P(x = 7 \mid n = 10, P = .8) = P(n - x = 3 \mid n = 10, 1 - P = .2)$$

$$= .2013$$

Note. Some probabilities are recorded as 0.0000. This means that the actual probability is very small and results in zero when rounded to four decimal places.

Table 1. Binomial Probabilities

n	x	P .05	.10	.15	.20	.25	.30	.35	.40	.45	.50
1	0	.9500	.9000	.8500	.8000	.7500	.7000	.6500	.6000	.5500	.5000
	1	.0500	.1000	.1500	.2000	.2500	.3000	.3500	.4000	.4500	.5000
2	0	.9025	.8100	.7225	.6400	.5625	.4900	.4225	.3600	.3025	.2500
	1	.0950	.1800	.2550	.3200	.3750	.4200	.4550	.4800	.4950	.5000
	2	.0025	.0100	.0225	.0400	.0625	.0900	.1225	.1600	.2025	.2500
3	0	.8574	.7290	.6141	.5120	.4219	.3430	.2746	.2160	.1664	.1250
	1	.1354	.2430	.3251	.3840	.4219	.4410	.4436	.4320	.4084	.3750
	2	.0071	.0270	.0574	.0960	.1406	.1890	.2389	.2880	.3341	.3750
	3	.0001	.0010	.0034	.0080	.0156	.0270	.0429	.0640	.0911	.1250
4	0	.8145	.0561	.5220	.4096	.3164	.2401	.1785	.1296	.0915	.0625
	1	.1715	.2916	.3685	.4096	.4219	.4116	.3845	.3456	.2995	.2500
	2	.0135	.0486	.0975	.1536	.2109	.2646	.3105	.3456	.3675	.3750
	3	.0005	.0036	.0115	.0256	.0469	.0756	.1115	.1536	.2005	.2500
	4	.0000	.0001	.0005	.0016	.0039	.0081	.0150	.0256	.0410	.0625
5	0	.7738	.5905	.4437	.3277	.2373	.1681	.1160	.0778	.0503	.0312
	1	.2036	.3280	.3915	.4096	.3955	.3602	.3124	.2592	.2059	.1562
	2	.0214	.0729	.1382	.2048	.2637	.3087	.3364	.3456	.3369	.3125
	3	.0011	.0081	.0244	.0512	.0879	.1323	.1811	.2304	.2757	.3125
	4	.0000	.0004	.0022	.0064	.0146	.0284	.0488	.0768	.1128	.1562
	5	.0000	.0000	.0001	.0003	.0010	.0024	.0053	.0102	.0185	.0312
6	0	.7351	.5314	.3771	.2621	.1780	.1176	.0754	.0467	.0277	.0156
	1	.2321	.3543	.3993	.3932	.3560	.3025	.2437	.1866	.1359	.0938
	2	.0305	.0984	.1762	.2458	.2966	.3241	.3280	.3110	.2780	.2344
	3	.0021	.0146	.0415	.0819	.1318	.1852	.2355	.2765	.3032	.3125
	4	.0001	.0012	.0055	.0154	.0330	.0595	.0951	.1382	.1861	.2344
	5	.0000	.0001	.0004	.0015	.0044	.0102	.0205	.0369	.0609	.0938
	6	.0000	.0000	.0000	.0001	.0002	.0007	.0018	.0041	.0083	.0156
7	0	.6983	.4783	.3206	.2097	.1335	.0824	.0490	.0280	.0152	.0078
	1	.2573	.3720	.2960	.3670	.3115	.2471	.1848	.1306	.0872	.0547
	2	.0406	.1240	.2097	.2753	.3115	.3177	.2985	.2613	.2140	.1641
	3	.0036	.0230	.0617	.1147	.1730	.2269	.2679	.2903	.2918	.2734
	4	.0002	.0026	.0109	.0287	.0577	.0972	.1442	.1935	.2388	.2734
	5	.0000	.0002	.0012	.0043	.0115	.0250	.0466	.0774	.1172	.1641
	6	.0000	.0000	.0001	.0004	.0013	.0036	.0084	.0172	.0320	.0547
	7	.0000	.0000	.0000	.0000	.0001	.0002	.0006	.0016	.0037	.0078
8	0	.6634	.4305	.2725	.1678	.1001	.0576	.0319	.0168	.0084	.0039
	1	.2793	.3826	.3847	.3355	.2670	.1977	.1373	.0896	.0548	.0312
	2	.0515	.1488	.2376	.2936	.3115	.2965	.2587	.2090	.1569	.1094
	3	.0054	.0331	.0839	.1468	.2076	.2541	.2786	.2787	.2568	.2188
	4	.0004	.0046	.0185	.0459	.0865	.1361	.1875	.2322	.2627	.2734
	5	.0000	.0004	.0026	.0092	.0231	.0467	.0808	.1239	.1719	.2188
	6	.0000	.0000	.0002	.0011	.0038	.0100	.0217	.0413	.0703	.1094
	7	.0000	.0000	.0000	.0001	.0004	.0012	.0033	.0079	.0164	.0312
	8	.0000	.0000	.0000	.0000	.0000	.0001	.0002	.0007	.0017	.0039

Table 1. Binomial Probabilities

n	x	P .05	.10	.15	.20	.25	.30	.35	.40	.45	.50
9	0	.6302	.3874	.2316	.1342	.0751	.0404	.0207	.0101	.0046	.0020
	1	.2985	.3874	.3679	.3020	.2253	.1556	.1004	.0605	.0339	.0176
	2	.0629	.1722	.2597	.3020	.3003	.2668	.2162	.1612	.1110	.0703
	3	.0077	.0446	.1069	.1762	.2336	.2668	.2716	.2508	.2119	.1641
	4	.0006	.0074	.0283	.0661	.1168	.1715	.2194	.2508	.2600	.2461
	5	.0000	.0008	.0050	.0165	.0389	.0735	.1181	.1672	.2128	.2461
	6	.0000	.0001	.0006	.0028	.0087	.0210	.0424	.0743	.1160	.1641
	7	.0000	.0000	.0000	.0003	.0012	.0039	.0098	.0212	.0407	.0703
	8	.0000	.0000	.0000	.0000	.0001	.0004	.0013	.0035	.0083	.0176
	9	.0000	.0000	.0000	.0000	.0000	.0000	.0001	.0003	.0008	.0020
10	0	.5987	.3487	.1969	.1074	.0563	.0282	.0135	.0060	.0025	.0010
	1	.3151	.3874	.3474	.2684	.1877	.1211	.0725	.0403	.0207	.0098
	2	.0746	.1937	.2759	.3020	.2816	.2335	.1757	.1209	.0763	.0439
	3	.0105	.0574	.1298	.2013	.2503	.2668	.2522	.2150	.1665	.1172
	4	.0010	.0112	.0401	.0881	.1460	.2001	.2377	.2508	.2384	.2051
	5	.0001	.0015	.0085	.0264	.0584	.1029	.1536	.2007	.2340	.2461
	6	.0000	.0001	.0012	.0055	.0162	.0368	.0689	.1115	.1596	.2051
	7	.0000	.0000	.0001	.0008	.0031	.0090	.0212	.0425	.0746	.1172
	8	.0000	.0000	.0000	.0001	.0004	.0014	.0043	.0106	.0229	.0439
	9	.0000	.0000	.0000	.0000	.0000	.0001	.0005	.0016	.0042	.0098
	10	.0000	.0000	.0000	.0000	.0000	.0000	.0000	.0001	.0003	.0010
11	0	.5688	.3138	.1673	.0859	.0422	.0198	.0088	.0036	.0014	.0005
	1	.3293	.3835	.3248	.2362	.1549	.0932	.0518	.0266	.0125	.0054
	2	.0867	.2131	.2866	.2953	.2581	.1998	.1395	.0887	.0513	.0269
	3	.0137	.0710	.1517	.2215	.2581	.2568	.2254	.1774	.1259	.0806
	4	.0014	.0158	.0536	.1107	.1721	.2201	.2428	.2365	.2060	.1611
	5	.0001	.0025	.0132	.0388	.0803	.1321	.1830	.2207	.2360	.2256
	6	.0000	.0003	.0023	.0097	.0268	.0566	.0985	.1471	.1931	.2256
	7	.0000	.0000	.0003	.0017	.0064	.0173	.0379	.0701	.1128	.1611
	8	.0000	.0000	.0000	.0002	.0011	.0037	.0102	.0234	.0462	.0806
	9	.0000	.0000	.0000	.0000	.0001	.0005	.0018	.0052	.0126	.0269
	10	.0000	.0000	.0000	.0000	.0000	.0000	.0002	.0007	.0021	.0054
	11	.0000	.0000	.0000	.0000	.0000	.0000	.0000	.0000	.0002	.0005
12	0	.5404	.2824	.1422	.0687	.0317	.0138	.0057	.0022	.0008	.0002
	1	.3413	.3766	.3012	.2062	.1267	.0712	.0368	.0174	.0075	.0029
	2	.0988	.2301	.2924	.2835	.2323	.1678	.1088	.0639	.0339	.0161
	3	.0173	.0852	.1720	.2362	.2581	.2397	.1954	.1419	.0923	.0537
	4	.0021	.0213	.0683	.1329	.1936	.2311	.2367	.2128	.1700	.1208
	5	.0002	.0038	.0193	.0532	.1032	.1585	.2039	.2270	.2225	.1934
	6	.0000	.0005	.0040	.0155	.0401	.0792	.1281	.1766	.2124	.2256
	7	.0000	.0000	.0006	.0033	.0115	.0291	.0591	.1009	.1489	.1934
	8	.0000	.0000	.0001	.0005	.0024	.0078	.0199	.0420	.0762	.1208
	9	.0000	.0000	.0000	.0001	.0004	.0015	.0048	.0125	.0277	.0537
	10	.0000	.0000	.0000	.0000	.0000	.0002	.0008	.0025	.0068	.0161
	11	.0000	.0000	.0000	.0000	.0000	.0000	.0001	.0003	.0010	.0029
	12	.0000	.0000	.0000	.0000	.0000	.0000	.0000	.0000	.0001	.0002

Table 1. Binomial Probabilities

n	x	.05	.10	.15	.20	.25	.30	.35	.40	.45	.50
13	0	.5133	.2542	.1209	.0550	.0238	.0097	.0037	.0013	.0004	.0001
	1	.3512	.3672	.2774	.1787	.1029	.0540	.0259	.0113	.0045	.0016
	2	.1109	.2448	.2937	.2680	.2059	.1388	.0836	.0453	.0220	.0095
	3	.0214	.0997	.1900	.2457	.2517	.2181	.1651	.1107	.0660	.0349
	4	.0028	.0277	.0838	.1535	.2097	.2337	.2222	.1845	.1350	.0873
	5	.0003	.0055	.0266	.0691	.1258	.1803	.2154	.2214	.1989	.1571
	6	.0000	.0008	.0063	.0230	.0559	.1030	.1546	.1968	.2169	.2095
	7	.0000	.0001	.0011	.0058	.0186	.0442	.0833	.1312	.1775	.2095
	8	.0000	.0000	.0001	.0011	.0047	.0142	.0336	.0656	.1089	.1571
	9	.0000	.0000	.0000	.0001	.0009	.0034	.0101	.0243	.0495	.0873
	10	.0000	.0000	.0000	.0000	.0001	.0006	.0022	.0065	.0162	.0349
	11	.0000	.0000	.0000	.0000	.0000	.0001	.0003	.0012	.0036	.0095
	12	.0000	.0000	.0000	.0000	.0000	.0000	.0000	.0001	.0005	.0016
	13	.0000	.0000	.0000	.0000	.0000	.0000	.0000	.0000	.0000	.0001
14	0	.4877	.2288	.1028	.0440	.0178	.0068	.0024	.0008	.0002	.0001
	1	.3593	.3559	.2539	.1539	.0832	.0407	.0181	.0073	.0027	.0009
	2	.1229	.2570	.2912	.2501	.1802	.1134	.0634	.0317	.0141	.0056
	3	.0259	.1142	.2056	.2501	.2402	.1943	.1366	.0845	.0462	.0222
	4	.0037	.0349	.0998	.1720	.2202	.2290	.2022	.1549	.1040	.0661
	5	.0004	.0078	.0352	.0860	.1468	.1963	.2178	.2066	.1701	.1222
	6	.0000	.0013	.0093	.0322	.0734	.1262	.1759	.2066	.2088	.1833
	7	.0000	.0002	.0019	.0092	.0280	.0618	.1082	.1574	.1952	.2095
	8	.0000	.0000	.0003	.0020	.0082	.0232	.0510	.0918	.1398	.1833
	9	.0000	.0000	.0000	.0003	.0018	.0066	.0183	.0408	.0762	.1222
	10	.0000	.0000	.0000	.0000	.0003	.0014	.0049	.0136	.0312	.0611
	11	.0000	.0000	.0000	.0000	.0000	.0002	.0010	.0033	.0093	.0222
	12	.0000	.0000	.0000	.0000	.0000	.0000	.0001	.0005	.0019	.0056
	13	.0000	.0000	.0000	.0000	.0000	.0000	.0000	.0001	.0002	.0009
	14	.0000	.0000	.0000	.0000	.0000	.0000	.0000	.0000	.0000	.0001
15	0	.4633	.2059	.0874	.0352	.0134	.0047	.0016	.0005	.0001	.0000
	1	.3658	.3432	.2312	.1319	.0668	.0305	.0126	.0047	.0016	.0005
	2	.1348	.2669	.2856	.2309	.1559	.0916	.0476	.0219	.0090	.0032
	3	.0307	.1285	.2184	.2501	.2252	.1700	.1110	.0634	.0318	.0139
	4	.0049	.0428	.1156	.1876	.2252	.2186	.1792	.1268	.0780	.0417
	5	.0006	.0105	.0449	.1032	.1651	.2061	.2123	.1859	.1404	.0916
	6	.0000	.0019	.0132	.0430	.0917	.1472	.1906	.2066	.1914	.1527
	7	.0000	.0003	.0030	.0138	.0393	.0811	.1319	.1771	.2013	.1964
	8	.0000	.0000	.0005	.0035	.0131	.0348	.0710	.1181	.1647	.1964
	9	.0000	.0000	.0001	.0007	.0034	.0116	.0298	.0612	.1048	.1527
	10	.0000	.0000	.0000	.0001	.0007	.0030	.0096	.0245	.0515	.0916
	11	.0000	.0000	.0000	.0000	.0001	.0006	.0024	.0074	.0191	.0417
	12	.0000	.0000	.0000	.0000	.0000	.0001	.0004	.0016	.0052	.0139
	13	.0000	.0000	.0000	.0000	.0000	.0000	.0001	.0003	.0010	.0032
	14	.0000	.0000	.0000	.0000	.0000	.0000	.0000	.0000	.0001	.0005
	15	.0000	.0000	.0000	.0000	.0000	.0000	.0000	.0000	.0000	.0000

Table 1. Binomial Probabilities

n	x	.05	.10	.15	.20	.25	P .30	.35	.40	.45	.50
16	0	.4401	.1853	.0743	.0281	.0100	.0033	.0010	.0003	.0001	.0000
	1	.3706	.3294	.2097	.1126	.0535	.0228	.0087	.0030	.0009	.0002
	2	.1463	.2745	.2775	.2111	.1336	.0732	.0353	.0150	.0056	.0018
	3	.0359	.1423	.2285	.2463	.2079	.1465	.0888	.0468	.0215	.0085
	4	.0061	.0514	.1311	.2001	.2252	.2040	.1553	.1014	.0572	.0278
	5	.0008	.0137	.0555	.1201	.1802	.2099	.2008	.1623	.1123	.0667
	6	.0001	.0028	.0180	.0550	.1101	.1649	.1982	.1983	.1684	.1222
	7	.0000	.0004	.0045	.0197	.0524	.1010	.1524	.1889	.1969	.1746
	8	.0000	.0001	.0009	.0055	.0197	.0487	.0923	.1417	.1812	.1964
	9	.0000	.0000	.0001	.0012	.0058	.0185	.0442	.0840	.1318	.1746
	10	.0000	.0000	.0000	.0002	.0014	.0056	.0167	.0392	.0755	.1222
	11	.0000	.0000	.0000	.0000	.0002	.0013	.0049	.0142	.0337	.0667
	12	.0000	.0000	.0000	.0000	.0000	.0002	.0011	.0040	.0115	.0278
	13	.0000	.0000	.0000	.0000	.0000	.0000	.0002	.0008	.0029	.0085
	14	.0000	.0000	.0000	.0000	.0000	.0000	.0000	.0001	.0005	.0018
	15	.0000	.0000	.0000	.0000	.0000	.0000	.0000	.0000	.0001	.0002
	16	.0000	.0000	.0000	.0000	.0000	.0000	.0000	.0000	.0000	.0000
17	0	.4181	.1668	.0631	.0225	.0075	.0023	.0007	.0002	.0000	.0000
	1	.3741	.3150	.1893	.0957	.0426	.0169	.0060	.0019	.0005	.0001
	2	.1575	.2800	.2673	.1914	.1136	.0581	.0260	.0102	.0035	.0010
	3	.0415	.1556	.2359	.2393	.1893	.1245	.0701	.0341	.0144	.0052
	4	.0076	.0605	.1457	.2093	.2209	.1868	.1320	.0796	.0411	.0182
	5	.0010	.0175	.0668	.1361	.1914	.2081	.1849	.1379	.0875	.0472
	6	.0001	.0039	.0236	.0680	.1276	.1784	.1991	.1839	.1432	.0944
	7	.0000	.0007	.0065	.0267	.0668	.1201	.1685	.1927	.1841	.1484
	8	.0000	.0001	.0014	.0084	.0279	.0644	.1134	.1606	.1883	.1855
	9	.0000	.0000	.0003	.0021	.0093	.0276	.0611	.1070	.1540	.1855
	10	.0000	.0000	.0000	.0004	.0025	.0095	.0263	.0571	.1008	.1484
	11	.0000	.0000	.0000	.0001	.0005	.0026	.0090	.0242	.0525	.0944
	12	.0000	.0000	.0000	.0000	.0001	.0006	.0024	.0081	.0215	.0472
	13	.0000	.0000	.0000	.0000	.0000	.0001	.0005	.0021	.0068	.0182
	14	.0000	.0000	.0000	.0000	.0000	.0000	.0001	.0004	.0016	.0052
	15	.0000	.0000	.0000	.0000	.0000	.0000	.0000	.0001	.0003	.0010
	16	.0000	.0000	.0000	.0000	.0000	.0000	.0000	.0000	.0000	.0001
	17	.0000	.0000	.0000	.0000	.0000	.0000	.0000	.0000	.0000	.0000
18	0	.3972	.1501	.0536	.0180	.0056	.0016	.0004	.0001	.0000	.0000
	1	.3763	.3002	.1704	.0811	.0338	.0126	.0042	.0012	.0003	.0001
	2	.1683	.2835	.2556	.1723	.0958	.0458	.0190	.0069	.0022	.0006
	3	.0473	.1680	.2406	.2297	.1704	.1046	.0547	.0246	.0095	.0031
	4	.0093	.0700	.1592	.2153	.2130	.1681	.1104	.0614	.0291	.0117
	5	.0014	.0218	.0787	.1507	.1988	.2017	.1664	.1146	.0666	.0327
	6	.0002	.0052	.0301	.0816	.1436	.1873	.1941	.1655	.1181	.0708
	7	.0000	.0010	.0091	.0350	.0820	.1376	.1792	.1892	.1657	.1214
	8	.0000	.0002	.0022	.0120	.0376	.0811	.1327	.1734	.1864	.1669
	9	.0000	.0000	.0004	.0033	.0139	.0386	.0794	.1284	.1694	.1855

Table 1. Binomial Probabilities

n	x	.05	.10	.15	.20	.25	.30	.35	.40	.45	.50
						P					
18	10	.0000	.0000	.0001	.0008	.0042	.0149	.0385	.0771	.1248	.1669
	11	.0000	.0000	.0000	.0001	.0010	.0046	.0151	.0374	.0742	.1214
	12	.0000	.0000	.0000	.0000	.0002	.0012	.0047	.0145	.0354	.0708
	13	.0000	.0000	.0000	.0000	.0000	.0002	.0012	.0045	.0134	.0327
	14	.0000	.0000	.0000	.0000	.0000	.0000	.0002	.0011	.0039	.0117
	15	.0000	.0000	.0000	.0000	.0000	.0000	.0000	.0002	.0009	.0031
	16	.0000	.0000	.0000	.0000	.0000	.0000	.0000	.0000	.0001	.0006
	17	.0000	.0000	.0000	.0000	.0000	.0000	.0000	.0000	.0000	.0001
	18	.0000	.0000	.0000	.0000	.0000	.0000	.0000	.0000	.0000	.0000
19	0	.3774	.1351	.0456	.0144	.0042	.0011	.0003	.0001	.0000	.0000
	1	.3774	.2852	.1529	.0685	.0268	.0093	.0029	.0008	.0002	.0000
	2	.1787	.2852	.2428	.1540	.0803	.0358	.0138	.0046	.0013	.0003
	3	.0533	.1796	.2428	.2182	.1517	.0869	.0422	.0175	.0062	.0018
	4	.0112	.0798	.1714	.2182	.2023	.1491	.0909	.0467	.0203	.0074
	5	.0018	.0266	.0907	.1636	.2023	.1916	.1468	.0933	.0497	.0222
	6	.0002	.0069	.0374	.0955	.1574	.1916	.1844	.1451	.0949	.0518
	7	.0000	.0014	.0122	.0443	.0974	.1525	.1844	.1797	.1443	.0961
	8	.0000	.0002	.0032	.0166	.0487	.0981	.1489	.1797	.1771	.1442
	9	.0000	.0000	.0007	.0051	.0198	.0514	.0980	.1464	.1771	.1762
	10	.0000	.0000	.0001	.0013	.0066	.0220	.0528	.0976	.1449	.1762
	11	.0000	.0000	.0000	.0003	.0018	.0077	.0233	.0532	.0970	.1442
	12	.0000	.0000	.0000	.0000	.0004	.0022	.0083	.0237	.0529	.0961
	13	.0000	.0000	.0000	.0000	.0001	.0005	.0024	.0085	.0233	.0518
	14	.0000	.0000	.0000	.0000	.0000	.0001	.0006	.0024	.0082	.0222
	15	.0000	.0000	.0000	.0000	.0000	.0000	.0001	.0005	.0022	.0074
	16	.0000	.0000	.0000	.0000	.0000	.0000	.0000	.0001	.0005	.0018
	17	.0000	.0000	.0000	.0000	.0000	.0000	.0000	.0000	.0001	.0003
	18	.0000	.0000	.0000	.0000	.0000	.0000	.0000	.0000	.0000	.0000
	19	.0000	.0000	.0000	.0000	.0000	.0000	.0000	.0000	.0000	.0000
20	0	.3585	.1216	.0388	.0115	.0032	.0008	.0002	.0000	.0000	.0000
	1	.3774	.2702	.1368	.0576	.0211	.0068	.0020	.0005	.0001	.0000
	2	.1887	.2852	.2293	.1369	.0669	.0278	.0100	.0031	.0008	.0002
	3	.0596	.1901	.2428	.2054	.1339	.0716	.0323	.0123	.0040	.0011
	4	.0133	.0898	.1821	.2182	.1897	.1304	.0738	.0350	.0139	.0046
	5	.0022	.0319	.1028	.1746	.2023	.1789	.1272	.0746	.0365	.0148
	6	.0003	.0089	.0454	.1091	.1686	.1916	.1712	.1244	.0746	.0370
	7	.0000	.0020	.0160	.0545	.1124	.1643	.1844	.1659	.1221	.0739
	8	.0000	.0004	.0046	.0222	.0609	.1144	.1614	.1797	.1623	.1201
	9	.0000	.0001	.0011	.0074	.0271	.0654	.1158	.1597	.1771	.1602
	10	.0000	.0000	.0002	.0020	.0099	.0308	.0686	.1171	.1593	.1762
	11	.0000	.0000	.0000	.0005	.0030	.0120	.0336	.0710	.1185	.1602
	12	.0000	.0000	.0000	.0001	.0008	.0039	.0136	.0355	.0727	.1201
	13	.0000	.0000	.0000	.0000	.0002	.0010	.0045	.0146	.0366	.0739
	14	.0000	.0000	.0000	.0000	.0000	.0002	.0012	.0049	.0150	.0370
	15	.0000	.0000	.0000	.0000	.0000	.0000	.0003	.0013	.0049	.0148
	16	.0000	.0000	.0000	.0000	.0000	.0000	.0000	.0003	.0013	.0046
	17	.0000	.0000	.0000	.0000	.0000	.0000	.0000	.0000	.0002	.0011
	18	.0000	.0000	.0000	.0000	.0000	.0000	.0000	.0000	.0000	.0002
	19	.0000	.0000	.0000	.0000	.0000	.0000	.0000	.0000	.0000	.0000
	20	.0000	.0000	.0000	.0000	.0000	.0000	.0000	.0000	.0000	.0000

Table 2
Areas of the Normal Curve

This table is used to find probabilities associated with normal variables in terms of areas under the standardized normal distribution:

$$f(z) = \frac{1}{\sqrt{2\pi}} e^{-\frac{1}{2}\left(\frac{x-\mu}{\sigma}\right)^2}; \quad -\infty < x < \infty$$

which is a normal curve with a mean of zero and variance of one. The area tabulated is shown in the following diagram.

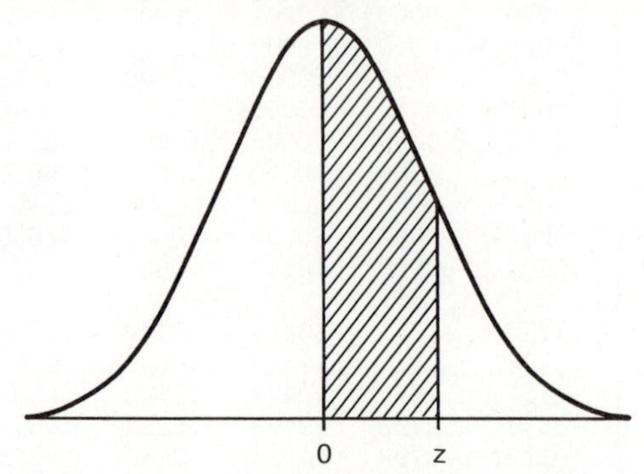

The area tabulated lies between the mean equal to zero and a value z of the standardized normal variate. Probabilities for any normal variate with mean μ and standard deviation σ are found in terms of z and the area provided in the table. The value of z is obtained with the formula

$$z = \frac{x - \mu}{\sigma}$$

The table is given in terms of positive values of z since the distribution is symmetric; however, probabilities in other regions of the curve also can be obtained.

Example To find the probability that x falls between the mean and 16.4 of a normal distribution with a mean of 10 and standard deviation of 5, we first compute z as

$$z = \frac{x - \mu}{\sigma} = \frac{16.4 - 10}{5} = \frac{6.4}{5} = 1.28$$

Values of z are given in the extreme left column of the table where the second decimal place is provided at the top row. Hence, we locate 1.2 in the left column and the second decimal place 0.08 at the top. The intersection of these values provides the required probability

$$P(10 \leq x \leq 16.4) = .3997$$

All other probabilities are found in terms of positive values of z corresponding to areas between the mean and a given number.

Example The table can be used in reverse to find values of x when a probability is given. Using the same mean and standard deviation in the above example, let us find the value of x such that 30 percent of the area falls between x and the mean. This is found by substituting into the formula for z and solving for x:

$$z = \frac{x - \mu}{\sigma}$$

$$0.84 = \frac{x - 10}{5} \rightarrow x = 10 + (.84)(5)$$

$$= 14.2$$

The value of z is determined by finding the given probability in the body of the table and adding the corresponding values in the left column and top row. In this case 30 percent or 0.3000 is not included so the closest value of .2995 is used. It is not necessary to interpolate. When x is to the left of the mean, a minus sign must be attached to z.

Table 2. Areas of the Normal Curve

Z	.00	.01	.02	.03	.04	.05	.06	.07	.08	.09
0.0	.0000	.0040	.0080	.0120	.0160	.0199	.0239	.0279	.0319	.0359
0.1	.0398	.0438	.0478	.0517	.0557	.0596	.0636	.0675	.0714	.0753
0.2	.0793	.0832	.0871	.0910	.0948	.0987	.1026	.1064	.1103	.1141
0.3	.1179	.1217	.1255	.1293	.1331	.1368	.1406	.1443	.1480	.1517
0.4	.1554	.1591	.1628	.1664	.1700	.1736	.1772	.1808	.1844	.1879
0.5	.1915	.1950	.1985	.2019	.2054	.2088	.2123	.2157	.2190	.2224
0.6	.2257	.2291	.2324	.2357	.2389	.2422	.2454	.2486	.2517	.2549
0.7	.2580	.2611	.2642	.2673	.2704	.2734	.2764	.2794	.2823	.2852
0.8	.2881	.2910	.2939	.2967	.2995	.3023	.3051	.3078	.3106	.3133
0.9	.3159	.3186	.3212	.3238	.3264	.3289	.3315	.3340	.3365	.3389
1.0	.3413	.3438	.3461	.3485	.3508	.3531	.3554	.3577	.3599	.3621
1.1	.3643	.3665	.3686	.3708	.3729	.3749	.3770	.3790	.3810	.3830
1.2	.3849	.3869	.3888	.3907	.3925	.3944	.3962	.3980	.3997	.4015
1.3	.4032	.4049	.4066	.4082	.4099	.4115	.4131	.4147	.4162	.4177
1.4	.4192	.4207	.4222	.4236	.4251	.4265	.4279	.4292	.4306	.4319
1.5	.4332	.4345	.4357	.4370	.4382	.4394	.4406	.4418	.4429	.4441
1.6	.4452	.4463	.4474	.4484	.4495	.4505	.4515	.4525	.4535	.4545
1.7	.4554	.4564	.4573	.4582	.4591	.4599	.4608	.4616	.4625	.4633
1.8	.4641	.4649	.4656	.4664	.4671	.4678	.4686	.4693	.4699	.4706
1.9	.4713	.4719	.4726	.4732	.4738	.4744	.4750	.4756	.4761	.4767
2.0	.4772	.4778	.4783	.4788	.4793	.4798	.4803	.4808	.4812	.4817
2.1	.4821	.4826	.4830	.4834	.4838	.4842	.4846	.4850	.4854	.4857
2.2	.4861	.4864	.4868	.4871	.4875	.4878	.4881	.4884	.4887	.4890
2.3	.4893	.4896	.4898	.4901	.4904	.4906	.4909	.4911	.4913	.4916
2.4	.4918	.4920	.4922	.4925	.4927	.4929	.4931	.4932	.4934	.4936
2.5	.4938	.4940	.4941	.4943	.4945	.4946	.4948	.4949	.4951	.4952
2.6	.4953	.4955	.4956	.4957	.4959	.4960	.4961	.4962	.4963	.4964
2.7	.4965	.4966	.4967	.4968	.4969	.4970	.4971	.4972	.4973	.4974
2.8	.4974	.4975	.4976	.4977	.4977	.4978	.4979	.4979	.4980	.4981
2.9	.4981	.4982	.4982	.4983	.4984	.4984	.4985	.4985	.4986	.4986
3.0	.4987	.4987	.4987	.4988	.4988	.4989	.4989	.4989	.4990	.4990
3.1	.4990	.4991	.4991	.4991	.4992	.4992	.4992	.4992	.4993	.4993
3.2	.4993	.4993	.4994	.4994	.4994	.4994	.4994	.4995	.4995	.4995
3.3	.4995	.4995	.4995	.4996	.4996	.4996	.4996	.4996	.4996	.4997
3.4	.4997	.4997	.4997	.4997	.4997	.4997	.4997	.4997	.4997	.4998

Table 3
Poisson Probabilities

This table provides probabilities or f(x) for values of x based upon selected values of λ for the Poisson distribution:

$$f(x) = \frac{\lambda^x e^{-\lambda}}{x!};\ x = 0, 1, 2, \ldots.$$

The table includes values of λ between 0.1 and 20 in units of 0.1.

Example In order to find the probability that x equals 4 when λ equals 3.5, locate 3.5 in the center of the table and look down the column until you reach the row corresponding to x on the left equal to 4. The probability is found as

$$P(x = 4 \mid \lambda = 3.5) = .1888$$

Note. Some probabilities are recorded as 0.0000. This means that the actual probability is very small and results in zero when rounded to four decimal places.

Table 3. Poisson Probabilities

x	λ 0.1	0.2	0.3	0.4	0.5	0.6	0.7	0.8	0.9	1.0
0	.9048	.8187	.7408	.6703	.6065	.5488	.4966	.4493	.4066	.3679
1	.0905	.1637	.2222	.2681	.3033	.3293	.3476	.3595	.3659	.3679
2	.0045	.0164	.0333	.0536	.0758	.0988	.1217	.1438	.1647	.1839
3	.0002	.0011	.0033	.0072	.0126	.0198	.0284	.0383	.0494	.0613
4	.0000	.0001	.0002	.0007	.0016	.0030	.0050	.0077	.0111	.0153
5	.0000	.0000	.0000	.0001	.0002	.0004	.0007	.0012	.0020	.0031
6	.0000	.0000	.0000	.0000	.0000	.0000	.0001	.0002	.0003	.0005
7	.0000	.0000	.0000	.0000	.0000	.0000	.0000	.0000	.0000	.0001

x	λ 1.1	1.2	1.3	1.4	1.5	1.6	1.7	1.8	1.9	2.0
0	.3329	.3012	.2725	.2466	.2231	.2019	.1827	.1653	.1496	.1353
1	.3662	.3614	.3543	.3452	.3347	.3230	.3106	.2975	.2842	.2707
2	.2014	.2169	.2303	.2417	.2510	.2584	.2640	.2678	.2700	.2707
3	.0738	.0867	.0998	.1128	.1255	.1378	.1496	.1607	.1710	.1804
4	.0203	.0260	.0324	.0395	.0471	.0551	.0636	.0723	.0812	.0902
5	.0045	.0062	.0084	.0111	.0141	.0176	.0216	.0260	.0309	.0361
6	.0008	.0012	.0018	.0026	.0035	.0047	.0061	.0078	.0098	.0120
7	.0001	.0002	.0003	.0005	.0008	.0011	.0015	.0020	.0027	.0034
8	.0000	.0000	.0001	.0001	.0001	.0002	.0003	.0005	.0006	.0009
9	.0000	.0000	.0000	.0000	.0000	.0000	.0001	.0001	.0001	.0002

x	λ 2.1	2.2	2.3	2.4	2.5	2.6	2.7	2.8	2.9	3.0
0	.1225	.1108	.1003	.0907	.0821	.0743	.0672	.0608	.0550	.0498
1	.2572	.2438	.2306	.2177	.2052	.1931	.1815	.1703	.1596	.1494
2	.2700	.2681	.2652	.2613	.2565	.2510	.2450	.2384	.2314	.2240
3	.1890	.1966	.2033	.2090	.2138	.2176	.2205	.2225	.2237	.2240
4	.0992	.1082	.1169	.1254	.1336	.1414	.1488	.1557	.1622	.1680
5	.0417	.0476	.0538	.0602	.0668	.0735	.0804	.0872	.0940	.1008
6	.2146	.0174	.0206	.0241	.0278	.0319	.0362	.0407	.0455	.0504
7	.0044	.0055	.0068	.0083	.0099	.0118	.0139	.0163	.0188	.0216
8	.0011	.0015	.0019	.0025	.0031	.0038	.0047	.0057	.0068	.0081
9	.0003	.0004	.0005	.0007	.0009	.0011	.0014	.0018	.0022	.0027
10	.0001	.0001	.0001	.0002	.0002	.0003	.0004	.0005	.0006	.0008
11	.0000	.0000	.0000	.0000	.0000	.0001	.0001	.0001	.0002	.0002
12	.0000	.0000	.0000	.0000	.0000	.0000	.0000	.0000	.0000	.0001

x	λ 3.1	3.2	3.3	3.4	3.5	3.6	3.7	3.8	3.9	4.0
0	.0450	.0408	.0369	.0334	.0302	.0273	.0247	.0224	.0202	.0183
1	.1397	.1304	.1217	.1135	.1057	.0984	.0915	.0850	.0789	.0733
2	.2165	.2087	.2008	.1929	.1850	.1771	.1692	.1615	.1539	.1465
3	.2237	.2226	.2209	.2186	.2158	.2125	.2087	.2046	.2001	.1954
4	.1734	.1781	.1823	.1858	.1888	.1912	.1931	.1944	.1951	.1954
5	.1075	.1140	.1203	.1264	.1322	.1377	.1429	.1477	.1522	.1563
6	.0555	.0608	.0662	.0716	.0771	.0826	.0881	.0936	.0989	.1042
7	.0246	.0278	.0312	.0348	.0385	.0425	.0466	.0508	.0551	.0595
8	.0095	.0111	.0129	.0148	.0169	.0191	.0215	.0241	.0269	.0298
9	.0033	.0040	.0047	.0056	.0066	.0076	.0089	.0102	.0116	.0132

Table 3. Poisson Probabilities

X	λ 3.1	3.2	3.3	3.4	3.5	3.6	3.7	3.8	3.9	4.0
10	.0010	.0013	.0016	.0019	.0023	.0028	.0033	.0039	.0045	.0053
11	.0003	.0004	.0005	.0006	.0007	.0009	.0011	.0013	.0016	.0019
12	.0001	.0001	.0001	.0002	.0002	.0003	.0003	.0004	.0005	.0006
13	.0000	.0000	.0000	.0000	.0001	.0001	.0001	.0001	.0002	.0002
14	.0000	.0000	.0000	.0000	.0000	.0000	.0000	.0000	.0000	.0001

X	λ 4.1	4.2	4.3	4.4	4.5	4.6	4.7	4.8	4.9	5.0
0	.0166	.0150	.0136	.0123	.0111	.0101	.0091	.0082	.0074	.0067
1	.0679	.0630	.0583	.0540	.0500	.0462	.0427	.0395	.0365	.0337
2	.1393	.1323	.1254	.1188	.1125	.1063	.1005	.0948	.0894	.0842
3	.1904	.1852	.1798	.1743	.1687	.1631	.1574	.1517	.1460	.1404
4	.1951	.1944	.1933	.1917	.1898	.1875	.1849	.1820	.1789	.1755
5	.1600	.1633	.1662	.1687	.1708	.1725	.1738	.1747	.1753	.1755
6	.1093	.1143	.1191	.1237	.1281	.1323	.1362	.1398	.1432	.1462
7	.0640	.0686	.0732	.0778	.0824	.0869	.0914	.0959	.1002	.1044
8	.0328	.0360	.0393	.0428	.0463	.0500	.0537	.0575	.0614	.0653
9	.0150	.0168	.0188	.0209	.0232	.0255	.0280	.0307	.0334	.0363
10	.0061	.0071	.0081	.0092	.0104	.0118	.0132	.0147	.0164	.0181
11	.0023	.0027	.0032	.0037	.0043	.0049	.0056	.0064	.0073	.0082
12	.0008	.0009	.0011	.0014	.0016	.0019	.0022	.0026	.0030	.0034
13	.0002	.0003	.0004	.0005	.0006	.0007	.0008	.0009	.0011	.0013
14	.0001	.0001	.0001	.0001	.0002	.0002	.0003	.0003	.0004	.0005
15	.0000	.0000	.0000	.0000	.0001	.0001	.0001	.0001	.0001	.0002

X	λ 5.1	5.2	5.3	5.4	5.5	5.6	5.7	5.8	5.9	6.0
0	.0061	.0055	.0050	.0045	.0041	.0037	.0033	.0030	.0027	.0025
1	.0311	.0287	.0265	.0244	.0225	.0207	.0191	.0176	.0162	.0149
2	.0793	.0746	.0701	.0659	.0618	.0580	.0544	.0509	.0477	.0446
3	.1348	.1293	.1239	.1185	.1133	.1082	.1033	.0985	.0938	.0892
4	.1719	.1681	.1641	.1600	.1558	.1515	.1472	.1428	.1383	.1339
5	.1753	.1748	.1740	.1728	.1714	.1697	.1678	.1656	.1632	.1606
6	.1490	.1515	.1537	.1555	.1571	.1584	.1594	.1601	.1605	.1606
7	.1086	.1125	.1163	.1200	.1234	.1267	.1298	.1326	.1353	.1377
8	.0692	.0731	.0771	.0810	.0849	.0887	.0925	.0962	.0998	.1033
9	.0392	.0423	.0454	.0486	.0519	.0552	.0586	.0620	.0654	.0688
10	.0200	.0220	.0241	.0262	.0285	.0309	.0334	.0359	.0386	.0413
11	.0093	.0104	.0116	.0129	.0143	.0157	.0173	.0190	.0207	.0225
12	.0039	.0045	.0051	.0058	.0065	.0073	.0082	.0092	.0102	.0113
13	.0015	.0018	.0021	.0024	.0028	.0032	.0036	.0041	.0046	.0052
14	.0006	.0007	.0008	.0009	.0011	.0013	.0015	.0017	.0019	.0022
15	.0002	.0002	.0003	.0003	.0004	.0005	.0006	.0007	.0008	.0009
16	.0001	.0001	.0001	.0001	.0001	.0002	.0002	.0002	.0003	.0003
17	.0000	.0000	.0000	.0000	.0000	.0001	.0001	.0001	.0001	.0001

Table 3. Poisson Probabilities

x	λ 6.1	6.2	6.3	6.4	6.5	6.6	6.7	6.8	6.9	7.0
0	.0022	.0020	.0018	.0017	.0015	.0014	.0012	.0011	.0010	.0009
1	.0137	.0126	.0116	.0106	.0098	.0090	.0082	.0076	.0070	.0064
2	.0417	.0390	.0364	.0340	.0318	.0296	.0276	.0258	.0240	.0223
3	.0848	.0806	.0765	.0726	.0688	.0652	.0617	.0584	.0552	.0521
4	.1294	.1249	.1205	.1162	.1118	.1076	.1034	.0992	.0952	.0912
5	.1579	.1549	.1519	.1487	.1454	.1420	.1385	.1349	.1314	.1277
6	.1605	.1601	.1595	.1586	.1575	.1562	.1546	.1529	.1511	.1490
7	.1399	.1418	.1435	.1450	.1462	.1472	.1480	.1486	.1489	.1490
8	.1066	.1099	.1130	.1160	.1188	.1215	.1240	.1263	.1284	.1304
9	.0723	.0757	.0791	.0825	.0858	.0891	.0923	.0954	.0985	.1014
10	.0441	.0469	.0498	.0528	.0558	.0588	.0618	.0649	.0679	.0710
11	.0245	.0265	.0285	.0307	.0330	.0353	.0377	.0401	.0426	.0452
12	.0124	.0137	.0150	.0164	.0179	.0194	.0210	.0227	.0245	.0264
13	.0058	.0065	.0073	.0081	.0089	.0098	.0108	.0119	.0130	.0142
14	.0025	.0029	.0033	.0037	.0041	.0046	.0052	.0058	.0064	.0071
15	.0010	.0012	.0014	.0016	.0018	.0020	.0023	.0026	.0029	.0033
16	.0004	.0005	.0005	.0006	.0007	.0008	.0010	.0011	.0013	.0014
17	.0001	.0002	.0002	.0002	.0003	.0003	.0004	.0004	.0005	.0006
18	.0000	.0001	.0001	.0001	.0001	.0001	.0001	.0002	.0002	.0002
19	.0000	.0000	.0000	.0000	.0000	.0000	.0000	.0001	.0001	.0001

x	λ 7.1	7.2	7.3	7.4	7.5	7.6	7.7	7.8	7.9	8.0
0	.0008	.0007	.0007	.0006	.0006	.0005	.0005	.0004	.0004	.0003
1	.0059	.0054	.0049	.0045	.0041	.0038	.0035	.0032	.0029	.0027
2	.0208	.0194	.0180	.0167	.0156	.0145	.0134	.0125	.0116	.0107
3	.0492	.0464	.0438	.0413	.0389	.0366	.0345	.0324	.0305	.0286
4	.0874	.0836	.0799	.0764	.0729	.0696	.0663	.0632	.0602	.0573
5	.1241	.1204	.1167	.1130	.1094	.1057	.1021	.0986	.0951	.0916
6	.1468	.1445	.1420	.1394	.1367	.1339	.1311	.1282	.1252	.1221
7	.1489	.1486	.1481	.1474	.1465	.1454	.1442	.1428	.1413	.1396
8	.1321	.1337	.1351	.1363	.1373	.1382	.1388	.1392	.1395	.1396
9	.1042	.1070	.1096	.1121	.1144	.1167	.1187	.1207	.1224	.1241
10	.0740	.0770	.0800	.0829	.0858	.0887	.0914	.0941	.0967	.0993
11	.0478	.0504	.0531	.0558	.0585	.0613	.0640	.0667	.0695	.0722
12	.0283	.0303	.0323	.0344	.0366	.0388	.0411	.0434	.0457	.0481
13	.0154	.0168	.0181	.0196	.0211	.0227	.0243	.0260	.0278	.0296
14	.0078	.0086	.0095	.0104	.0113	.0123	.0134	.0145	.0157	.0169
15	.0037	.0041	.0046	.0051	.0057	.0062	.0069	.0075	.0083	.0090
16	.0016	.0019	.0021	.0024	.0026	.0030	.0033	.0037	.0041	.0045
17	.0007	.0008	.0009	.0010	.0012	.0013	.0015	.0017	.0019	.0021
18	.0003	.0003	.0004	.0004	.0005	.0006	.0006	.0007	.0008	.0009
19	.0001	.0001	.0001	.0002	.0002	.0002	.0003	.0003	.0003	.0004
20	.0000	.0000	.0001	.0001	.0001	.0001	.0001	.0001	.0001	.0002
21	.0000	.0000	.0000	.0000	.0000	.0000	.0000	.0000	.0001	.0001

Table 3. Poisson Probabilities

X	λ 8.1	8.2	8.3	8.4	8.5	8.6	8.7	8.8	8.9	9.0
0	.0003	.0003	.0002	.0002	.0002	.0002	.0002	.0002	.0001	.0001
1	.0025	.0023	.0021	.0019	.0017	.0016	.0014	.0013	.0012	.0011
2	.0100	.0092	.0086	.0079	.0074	.0068	.0063	.0058	.0054	.0050
3	.0269	.0252	.0237	.0222	.0208	.0195	.0183	.0171	.0160	.0150
4	.0544	.0517	.0491	.0466	.0443	.0420	.0398	.0377	.0357	.0337
5	.0882	.0849	.0816	.0784	.0752	.0722	.0692	.0663	.0635	.0607
6	.1191	.1160	.1128	.1097	.1066	.1034	.1003	.0972	.0941	.0911
7	.1378	.1358	.1338	.1317	.1294	.1271	.1247	.1222	.1197	.1171
8	.1395	.1392	.1388	.1382	.1375	.1366	.1356	.1344	.1332	.1318
9	.1256	.1269	.1280	.1290	.1299	.1306	.1311	.1315	.1317	.1318
10	.1017	.1040	.1063	.1084	.1104	.1123	.1140	.1157	.1172	.1186
11	.0749	.0776	.0802	.0828	.0853	.0878	.0902	.0925	.0948	.0970
12	.0505	.0530	.0555	.0579	.0604	.0629	.0654	.0679	.0703	.0728
13	.0315	.0334	.0354	.0374	.0395	.0416	.0438	.0459	.0481	.0504
14	.0182	.0196	.0210	.0225	.0240	.0256	.0272	.0289	.0306	.0324
15	.0098	.0107	.0116	.0126	.0136	.0147	.0158	.0169	.0182	.0194
16	.0050	.0055	.0060	.0066	.0072	.0079	.0086	.0093	.0101	.0109
17	.0024	.0026	.0029	.0033	.0036	.0040	.0044	.0048	.0053	.0058
18	.0011	.0012	.0014	.0015	.0017	.0019	.0021	.0024	.0026	.0029
19	.0005	.0005	.0006	.0007	.0008	.0009	.0010	.0011	.0012	.0014
20	.0002	.0002	.0002	.0003	.0003	.0004	.0004	.0005	.0005	.0006
21	.0001	.0001	.0001	.0001	.0001	.0002	.0002	.0002	.0002	.0003
22	.0000	.0000	.0000	.0000	.0001	.0001	.0001	.0001	.0001	.0001

X	λ 9.1	9.2	9.3	9.4	9.5	9.6	9.7	9.8	9.9	10
0	.0001	.0001	.0001	.0001	.0001	.0001	.0001	.0001	.0001	.0000
1	.0010	.0009	.0009	.0008	.0007	.0007	.0006	.0005	.0005	.0005
2	.0046	.0043	.0040	.0037	.0034	.0031	.0029	.0027	.0025	.0023
3	.0140	.0131	.0123	.0115	.0107	.0100	.0093	.0087	.0081	.0076
4	.0319	.0302	.0285	.0269	.0254	.0240	.0226	.0213	.0201	.0189
5	.0581	.0555	.0530	.0506	.0483	.0460	.0439	.0418	.0398	.0378
6	.0881	.0851	.0822	.0793	.0764	.0736	.0709	.0682	.0656	.0631
7	.1145	.1118	.1091	.1064	.1037	.1010	.0982	.0955	.0928	.0901
8	.1302	.1286	.1269	.1251	.1232	.1212	.1191	.1170	.1148	.1126
9	.1317	.1315	.1311	.1306	.1300	.1293	.1284	.1274	.1263	.1251
10	.1198	.1210	.1219	.1228	.1235	.1241	.1245	.1249	.1250	.1251
11	.0991	.1012	.1031	.1049	.1067	.1083	.1098	.1112	.1125	.1137
12	.0752	.0776	.0799	.0822	.0844	.0866	.0888	.0908	.0928	.0948
13	.0526	.0549	.0572	.0594	.0617	.0640	.0662	.0685	.0707	.0729
14	.0342	.0361	.0380	.0399	.0419	.0439	.0459	.0479	.0500	.0521
15	.0208	.0221	.0235	.0250	.0265	.0281	.0297	.0313	.0330	.0347
16	.0118	.0127	.0137	.0147	.0157	.0168	.0180	.0192	.0204	.0217
17	.0063	.0069	.0075	.0081	.0088	.0095	.0103	.0111	.0119	.0128
18	.0032	.0035	.0039	.0042	.0046	.0051	.0055	.0060	.0065	.0071
19	.0015	.0017	.0019	.0021	.0023	.0026	.0028	.0031	.0034	.0037
20	.0007	.0008	.0009	.0010	.0011	.0012	.0014	.0015	.0017	.0019
21	.0003	.0003	.0004	.0004	.0005	.0006	.0006	.0007	.0008	.0009
22	.0001	.0001	.0002	.0002	.0002	.0002	.0003	.0003	.0004	.0004
23	.0000	.0001	.0001	.0001	.0001	.0001	.0001	.0001	.0002	.0002
24	.0000	.0000	.0000	.0000	.0000	.0000	.0000	.0001	.0001	.0001

Table 3. Poisson Probabilities

x	λ 11	12	13	14	15	16	17	18	19	20
0	.0000	.0000	.0000	.0000	.0000	.0000	.0000	.0000	.0000	.0000
1	.0002	.0001	.0000	.0000	.0000	.0000	.0000	.0000	.0000	.0000
2	.0010	.0004	.0002	.0001	.0000	.0000	.0000	.0000	.0000	.0000
3	.0037	.0018	.0008	.0004	.0002	.0001	.0000	.0000	.0000	.0000
4	.0102	.0053	.0027	.0013	.0006	.0003	.0001	.0001	.0000	.0000
5	.0224	.0127	.0070	.0037	.0019	.0010	.0005	.0002	.0001	.0001
6	.0411	.0255	.0152	.0087	.0048	.0026	.0014	.0007	.0004	.0002
7	.0646	.0437	.0281	.0174	.0104	.0060	.0034	.0018	.0010	.0005
8	.0888	.0655	.0457	.0304	.0194	.0120	.0072	.0042	.0024	.0013
9	.1085	.0874	.0661	.0473	.0324	.0213	.0135	.0083	.0050	.0029
10	.1194	.1048	.0859	.0663	.0486	.0341	.0230	.0150	.0095	.0058
11	.1194	.1144	.1015	.0844	.0663	.0496	.0355	.0245	.0164	.0106
12	.1094	.1144	.1099	.0984	.0829	.0661	.0504	.0368	.0259	.0176
13	.0926	.1056	.1099	.1060	.0956	.0814	.0658	.0509	.0378	.0271
14	.0728	.0905	.1021	.1060	.1024	.0930	.0800	.0655	.0514	.0387
15	.0534	.0724	.0885	.0989	.1024	.0992	.0906	.0786	.0650	.0516
16	.0367	.0543	.0719	.0866	.0960	.0992	.0963	.0884	.0772	.0646
17	.0237	.0383	.0550	.0713	.0847	.0934	.0963	.0936	.0863	.0760
18	.0145	.0256	.0397	.0554	.0706	.0830	.0909	.0936	.0911	.0844
19	.0084	.0161	.0272	.0409	.0557	.0699	.0814	.0887	.0911	.0888
20	.0046	.0097	.0177	.0286	.0418	.0559	.0692	.0798	.0866	.0888
21	.0024	.0055	.0109	.0191	.0299	.0426	.0560	.0684	.0783	.0846
22	.0012	.0030	.0065	.0121	.0204	.0310	.0433	.0560	.0676	.0769
23	.0006	.0016	.0037	.0074	.0133	.0216	.0320	.0438	.0559	.0669
24	.0003	.0008	.0020	.0043	.0083	.0144	.0226	.0328	.0442	.0557
25	.0001	.0004	.0010	.0024	.0050	.0092	.0154	.0237	.0336	.0446
26	.0000	.0002	.0005	.0013	.0029	.0057	.0101	.0164	.0246	.0343
27	.0000	.0001	.0002	.0007	.0016	.0034	.0063	.0109	.0173	.0254
28	.0000	.0000	.0001	.0003	.0009	.0019	.0038	.0070	.0117	.0181
29	.0000	.0000	.0001	.0002	.0004	.0011	.0023	.0044	.0077	.0125
30	.0000	.0000	.0000	.0001	.0002	.0006	.0013	.0026	.0049	.0083
31	.0000	.0000	.0000	.0000	.0001	.0003	.0007	.0015	.0030	.0054
32	.0000	.0000	.0000	.0000	.0001	.0001	.0004	.0009	.0018	.0034
33	.0000	.0000	.0000	.0000	.0000	.0001	.0002	.0005	.0010	.0020
34	.0000	.0000	.0000	.0000	.0000	.0000	.0001	.0002	.0006	.0012
35	.0000	.0000	.0000	.0000	.0000	.0000	.0000	.0001	.0003	.0007
36	.0000	.0000	.0000	.0000	.0000	.0000	.0000	.0001	.0002	.0004
37	.0000	.0000	.0000	.0000	.0000	.0000	.0000	.0000	.0001	.0002
38	.0000	.0000	.0000	.0000	.0000	.0000	.0000	.0000	.0000	.0001
39	.0000	.0000	.0000	.0000	.0000	.0000	.0000	.0000	.0000	.0001

Table 4
Values of Student's t-Distribution

This table is used to find values of Student's t-statistic corresponding to a limited selection of probability values needed in problems in inference. Positive values of t are given that correspond to probabilities comprising *both* tails of the distribution, which are shown in the following diagram:

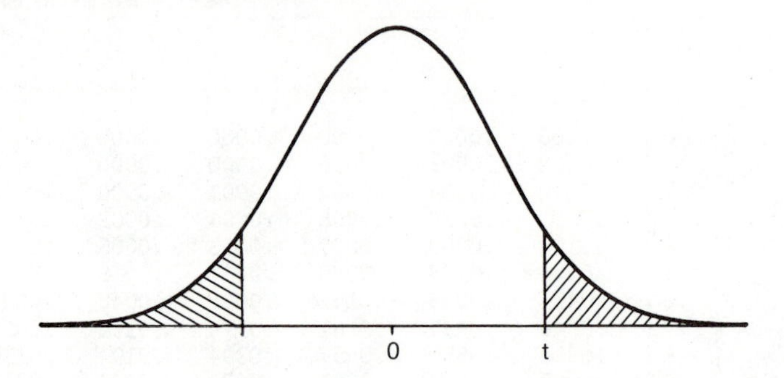

The probabilities in the table represent the sum of the shaded areas and are divided equally between the tails. Since the distribution is symmetric, only positive values of t are given for the right half of the curve.

The t-distribution changes with the number of degrees of freedom, DF, which is determined on the basis of a particular problem. In order to determine t, it is necessary to know the degrees of freedom, the probability level, and whether this is one-tailed or two-tailed. Values of t corresponding to one tail are found by locating the two-tailed probability equal to *twice* the required one-tail probability.

Example Suppose it is of interest to find the values of t such that 2.5 percent of the area lies in *each* tail (or corresponding to 95 percent in the center of the distribution) where the number of degrees of freedom equals 16. The degrees of freedom are given in the leftmost column. Consequently, we locate the two-tailed probability of 0.05 in the top column and find t equal to 2.120 across from 16 degrees of freedom. This is the value of t in the upper tail. The value in the lower tail is the same except a minus sign is attached: −2.120.

Example If we are interested in finding the value of t corresponding to 1 percent or 0.01 in the upper tail for 12 degrees of freedom, we locate a probability of 0.02 at the top. The value of t is in the corresponding column across from 12 degrees of freedom and equals 2.681. If t is required for the lower tail with an area of 0.01 a minus sign is attached to t: −2.681.

Note. For degrees of freedom not considered in the table the standardized normal variate, z, from Table 2 can be used to approximate t. The values on the bottom row ($DF = \infty$) in Table 4 are the same as values of z; however, they correspond to a limited number of probability levels.

Table 4. Values of Student's t-Distribution

DF	Two-Tailed Probabilities 0.2	0.1	0.05	0.02	0.01	0.001
1	3.078	6.314	12.706	31.821	63.657	636.619
2	1.886	2.920	4.303	6.965	9.925	31.598
3	1.638	2.353	3.182	4.541	5.841	12.924
4	1.533	2.132	2.776	3.747	4.604	8.610
5	1.476	2.015	2.571	3.365	4.032	6.869
6	1.440	1.943	2.447	3.143	3.707	5.959
7	1.415	1.895	2.365	2.998	3.499	5.408
8	1.397	1.860	2.306	2.896	3.355	5.041
9	1.383	1.833	2.262	2.821	3.250	4.781
10	1.372	1.812	2.228	2.764	3.169	4.587
11	1.363	1.796	2.201	2.718	3.106	4.437
12	1.356	1.782	2.179	2.681	3.055	4.318
13	1.350	1.771	2.160	2.650	3.012	4.221
14	1.345	1.761	2.145	2.624	2.977	4.140
15	1.341	1.753	2.131	2.602	2.947	4.073
16	1.337	1.746	2.120	2.583	2.921	4.015
17	1.333	1.740	2.110	2.567	2.898	3.965
18	1.330	1.734	2.101	2.552	2.878	3.922
19	1.328	1.729	2.093	2.539	2.861	3.883
20	1.325	1.725	2.086	2.528	2.845	3.850
21	1.323	1.721	2.080	2.518	2.831	3.819
22	1.321	1.717	2.074	2.508	2.819	3.792
23	1.319	1.714	2.069	2.500	2.807	3.767
24	1.318	1.711	2.064	2.492	2.797	3.745
25	1.316	1.708	2.060	2.485	2.787	3.725
26	1.315	1.706	2.056	2.479	2.779	3.707
27	1.314	1.703	2.052	2.473	2.771	3.690
28	1.313	1.701	2.048	2.467	2.763	3.674
29	1.311	1.699	2.045	2.462	2.756	3.659
30	1.310	1.697	2.042	2.457	3.750	3.646
40	1.303	1,684	2.021	2.423	2.704	3.551
60	1.296	1.671	2.000	2.390	2.660	3.460
120	1.289	1.658	1.980	2.358	2.617	3.373
∞	1.282	1.645	1.960	2.326	2.576	3.291

Table 5
Values of the Chi-Square Distribution

This table is used to find values of the chi-square statistic, χ^2, corresponding to selected probabilities required in problems in inference. The values of chi-square correspond to these probabilities which are associated with the right tail of the distribution as shown in the following diagram:

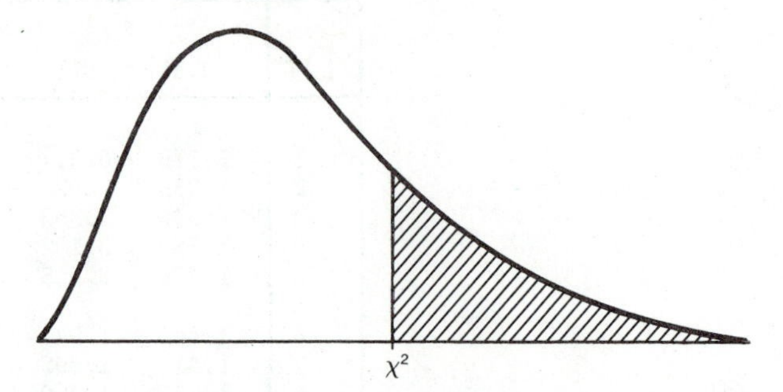

Although probabilities in the table are associated with the right tail, values of χ^2 corresponding to probabilities in the left tail can be obtained by subtracting the tail area from one and locating this probability at the top row on the left half of the table.

The chi-square distribution changes with the number of degrees of freedom, DF, which is determined on the basis of a particular problem. In order to determine χ^2, it is necessary to know the degrees of freedom, the probability level, and to which tail or tails the probability applies.

Example Consider finding the value of χ^2 corresponding to a probability of 0.05 or 5 percent in the upper or right tail when the degrees of freedom equals 14. This is obtained by locating 0.05 on the top row and 14 in the left-most column labeled DF. The corresponding value of chi-square is

$$\chi^2 = \chi^2_{.05}$$

$$= 23.685$$

The subscript is used to indicate the area above the value of χ^2.

Example Suppose we want to find the values of χ^2 corresponding to 0.05 or 5 percent of the area split equally in both tails when the degrees of freedom equals 14. Here we are to find two values of χ^2 corresponding to 0.025 in each tail. The value in the left tail is obtained by locating 0.025 at the top of the table directly and is

$$\chi^2 = \chi^2_{.025}$$

$$= 26.119$$

In order to find the value in the lower or left tail we subtract 0.025 from 1, or .975, and locate the result at the top of the table:

$$\chi^2 = \chi^2_{1-.025}$$

$$= \chi^2_{.975} = 5.629$$

The subscript indicates the area to the right or above the value of χ^2, which is equivalent to this value subtracted from 1 in the lower or left tail.

Table 5. Values of the Chi-Square Distribution

DF	Right-Tail Probability							
	.995	.99	.975	.95	.05	.025	.01	.005
1	.00004	.0002	.001	.004	3.841	5.024	6.635	7.879
2	.010	.020	.051	.103	5.991	7.378	9.210	10.597
3	.072	.115	.216	.352	7.815	9.348	11.345	12.838
4	.207	.297	.484	.711	9.488	11.143	13.277	14.860
5	.412	.554	.831	1.145	11.070	12.832	15.086	16.750
6	.676	.872	1.237	1.635	12.592	14.449	16.812	18.548
7	.989	1.239	1.690	2.167	14.067	16.013	18.475	20.278
8	1.344	1.646	2.180	2.733	15.507	17.535	20.090	21.955
9	1.735	2.088	2.700	3.325	16.919	19.023	21.666	23.589
10	2.156	2.558	3.247	3.940	18.307	20.483	23.209	25.188
11	2.603	3.053	3.816	4.575	19.675	21.920	24.725	26.757
12	3.074	3.571	4.404	5.226	21.026	23.337	26.217	28.300
13	3.565	4.107	5.009	5.892	22.362	24.736	27.688	29.819
14	4.075	4.660	5.629	6.571	23.685	26.119	29.141	31.319
15	4.601	5.229	6.262	7.261	24.996	27.488	30.578	32.801
16	5.142	5.812	6.908	7.962	26.296	28.845	32.000	34.267
17	5.697	6.408	7.564	8.672	27.587	30.191	33.409	35.718
18	6.265	7.015	8.231	9.390	28.869	31.526	34.805	37.156
19	6.884	7.633	8.907	10.117	30.144	32.852	36.191	38.582
20	7.434	8.260	9.591	10.851	31.410	34.170	37.566	39.997
21	8.034	8.897	10.283	11.591	32.671	35.479	38.932	41.401
22	8.643	9.542	10.982	12.338	33.924	36.781	40.289	42.796
23	9.260	10.196	11.689	13.091	35.172	38.076	41.638	44.181
24	9.886	10.856	12.401	13.848	36.415	39.364	42.980	45.558
25	10.520	11.524	13.120	14.611	37.652	40.646	44.314	46.928
26	11.160	12.198	13.844	15.379	38.885	41.923	45.642	48.290
27	11.808	12.879	14.573	16.151	40.113	43.194	46.963	49.645
28	12.461	13.565	15.308	16.928	41.337	44.461	48.278	50.993
29	13.121	14.256	16.047	17.708	42.557	45.722	49.588	52.336
30	13.787	14.953	16.791	18.493	43.773	46.979	50.892	53.672

Table 6(a)
Values of the F-Distribution
5 Percent Level

This table is used to find values of the F-statistic corresponding only to one probability level equal to 0.05 in the upper or right tail of the F-distribution. Hence, values of F in Table 6(a) correspond to the area shown in the following diagram:

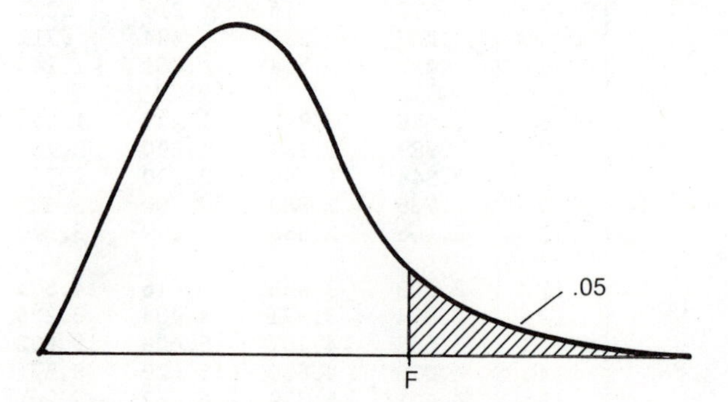

The F-distribution changes with two quantities or parameters, the numerator degrees of freedom, DF_1, and the denominator degrees of freedom, DF_2.

Example Suppose we are to find the value of F corresponding to 0.05 in the upper or right tail where the numerator degrees of freedom equals 9 and the denominator degrees of freedom equals 20. This is found by locating $DF_1 = 9$ at the top row and $DF_2 = 20$ at the left-most column. The corresponding value of F equals 2.39.

Note. Values of F for a probability of 0.01 in the upper tail appear in Table 6(b).

Table 6(a). Values of the F-Distribution—5 Percent Level

Denominator Degrees of Freedom (DF_2)	Numerator Degrees of Freedom (DF_1) 1	2	3	4	5	6	7	8	9	10	12	15	20	24	30	40	60	120	∞
1	161	200	216	225	230	234	237	239	241	242	244	246	248	249	250	251	252	253	254
2	18.5	19.0	19.2	19.2	19.3	19.3	19.4	19.4	19.4	19.4	19.4	19.4	19.4	19.5	19.5	19.5	19.5	19.5	19.5
3	10.1	9.55	9.28	9.12	9.01	8.94	8.89	8.85	8.81	8.79	8.74	8.70	8.66	8.64	8.62	8.59	8.57	8.55	8.53
4	7.71	6.94	6.59	6.39	6.26	6.16	6.09	6.04	6.00	5.96	5.91	5.86	5.80	5.77	5.75	5.72	5.69	5.66	5.63
5	6.61	5.79	5.41	5.19	5.05	4.95	4.88	4.82	4.77	4.74	4.68	4.62	4.56	4.53	4.50	4.46	4.43	4.40	4.37
6	5.99	5.14	4.76	4.53	4.39	4.28	4.21	4.15	4.10	4.06	4.00	3.94	3.87	3.84	3.81	3.77	3.74	3.70	3.67
7	5.59	4.74	4.35	4.12	3.97	3.87	3.79	3.73	3.68	3.64	3.57	3.51	3.44	3.41	3.38	3.34	3.30	3.27	3.23
8	5.32	4.46	4.07	3.84	3.69	3.58	3.50	3.44	3.39	3.35	3.28	3.22	3.15	3.12	3.08	3.04	3.01	2.97	2.93
9	5.12	4.26	3.86	3.63	3.48	3.37	3.29	3.23	3.18	3.14	3.07	3.01	2.94	2.90	2.86	2.83	2.79	2.75	2.71
10	4.96	4.10	3.71	3.48	3.33	3.22	3.14	3.07	3.02	2.98	2.91	2.85	2.77	2.74	2.70	2.66	2.62	2.58	2.54
11	4.84	3.98	3.59	3.36	3.20	3.09	3.01	2.95	2.90	2.85	2.79	2.72	2.65	2.61	2.57	2.53	2.49	2.45	2.40
12	4.75	3.89	3.49	3.26	3.11	3.00	2.91	2.85	2.80	2.75	2.69	2.62	2.54	2.51	2.47	2.43	2.38	2.34	2.30
13	4.67	3.81	3.41	3.18	3.03	2.92	2.83	2.77	2.71	2.67	2.60	2.53	2.46	2.42	2.38	2.34	2.30	2.25	2.21
14	4.60	3.74	3.34	3.11	2.96	2.85	2.76	2.70	2.65	2.60	2.53	2.46	2.39	2.35	2.31	2.27	2.22	2.18	2.13
15	4.54	3.68	3.29	3.06	2.90	2.79	2.71	2.64	2.59	2.54	2.48	2.40	2.33	2.29	2.25	2.20	2.16	2.11	2.07
16	4.49	3.63	3.24	3.01	2.85	2.74	2.66	2.59	2.54	2.49	2.42	2.35	2.28	2.24	2.19	2.15	2.11	2.06	2.01
17	4.45	3.59	3.20	2.96	2.81	2.70	2.61	2.55	2.49	2.45	2.38	2.31	2.23	2.19	2.15	2.10	2.06	2.01	1.96
18	4.41	3.55	3.16	2.93	2.77	2.66	2.58	2.51	2.46	2.41	2.34	2.27	2.19	2.15	2.11	2.06	2.02	1.97	1.92
19	4.38	3.52	3.13	2.90	2.74	2.63	2.54	2.48	2.42	2.38	2.31	2.23	2.16	2.11	2.07	2.03	1.98	1.93	1.88
20	4.35	3.49	3.10	2.87	2.71	2.60	2.51	2.45	2.39	2.35	2.28	2.20	2.12	2.08	2.04	1.99	1.95	1.90	1.84
21	4.32	3.47	3.07	2.84	2.68	2.57	2.49	2.42	2.37	2.32	2.25	2.18	2.10	2.05	2.01	1.96	1.92	1.87	1.81
22	4.30	3.44	3.05	2.82	2.66	2.55	2.46	2.40	2.34	2.30	2.23	2.15	2.07	2.03	1.98	1.94	1.89	1.84	1.78
23	4.28	3.42	3.03	2.80	2.64	2.53	2.44	2.37	2.32	2.27	2.20	2.13	2.05	2.01	1.96	1.91	1.86	1.81	1.76
24	4.26	3.40	3.01	2.78	2.62	2.51	2.42	2.36	2.30	2.25	2.18	2.11	2.03	1.98	1.94	1.89	1.84	1.79	1.73
25	4.24	3.39	2.99	2.76	2.60	2.49	2.40	2.34	2.28	2.24	2.16	2.09	2.01	1.96	1.92	1.87	1.82	1.77	1.71
30	4.17	3.32	2.92	2.69	2.53	2.42	2.33	2.27	2.21	2.16	2.09	2.01	1.93	1.89	1.84	1.79	1.74	1.68	1.62
40	4.08	3.23	2.84	2.61	2.45	2.34	2.25	2.18	2.12	2.08	2.00	1.92	1.84	1.79	1.74	1.69	1.64	1.58	1.51
60	4.00	3.15	2.76	2.53	2.37	2.25	2.17	2.10	2.04	1.99	1.92	1.84	1.75	1.70	1.65	1.59	1.53	1.47	1.39
120	3.92	3.07	2.68	2.45	2.29	2.18	2.09	2.02	1.96	1.91	1.83	1.75	1.66	1.61	1.55	1.50	1.43	1.35	1.25
∞	3.84	3.00	2.60	2.37	2.21	2.10	2.01	1.94	1.88	1.83	1.75	1.67	1.57	1.52	1.46	1.39	1.32	1.22	1.00

Table 6(b)
Values of the F-Distribution
1 Percent Level

This table is used to find values of the F-statistic corresponding only to one probability level equal to 0.01 in the upper or right tail of the F-distribution. Hence, values of F in Table 6(b) correspond to the area shown in the following diagram:

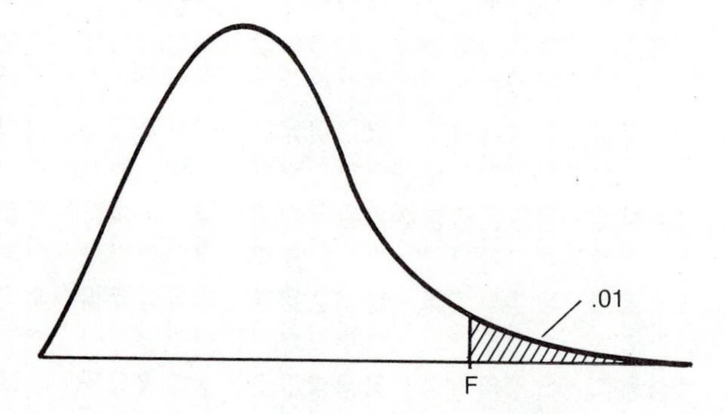

The F-distribution changes with two quantities or parameters, the numerator degrees of freedom, DF_1, and the denominator degrees of freedom, DF_2.

Example Suppose we are to find the value of F corresponding to 0.01 in the upper or right tail where the numerator degrees of freedom equals 4 and the denominator degrees of freedom equals 13. This is found by locating $DF_1 = 4$ at the top row and $DF_2 = 13$ at the left-most column. The corresponding value of F equals 5.21.

Note. Values of F for a probability of 0.05 in the upper tail appear in Table 6(a).

Table 6(b). Values of the F-Distribution—1 Percent level

Denominator Degrees of Freedom (DF_2)	Numerator Degrees of Freedom (DF_1)																		
	1	2	3	4	5	6	7	8	9	10	12	15	20	24	30	40	60	120	∞
1	4052	5000	5403	5625	5764	5859	5928	5982	6023	6056	6106	6157	6209	6235	6261	6287	6313	6339	6366
2	98.5	99.0	99.2	99.2	99.3	99.3	99.4	99.4	99.4	99.4	99.4	99.4	99.4	99.5	99.5	99.5	99.5	99.5	99.5
3	34.1	30.8	29.5	28.7	28.2	27.9	27.7	27.5	27.3	27.2	27.1	26.9	26.7	26.6	26.5	26.4	26.3	26.2	26.1
4	21.2	18.0	16.7	16.0	15.5	15.2	15.0	14.8	14.7	14.5	14.4	14.2	14.0	13.9	13.8	13.7	13.7	13.6	13.5
5	16.3	13.3	12.1	11.4	11.0	10.7	10.5	10.3	10.2	10.1	9.89	9.72	9.55	9.47	9.38	9.29	9.20	9.11	9.02
6	13.7	10.9	9.78	9.15	8.75	8.47	8.26	8.10	7.98	7.87	7.72	7.56	7.40	7.31	7.23	7.14	7.06	6.97	6.88
7	12.2	9.55	8.45	7.85	7.46	7.19	6.99	6.84	6.72	6.62	6.47	6.31	6.16	6.07	5.99	5.91	5.82	5.74	5.65
8	11.3	8.65	7.59	7.01	6.63	6.37	6.18	6.03	5.91	5.81	5.67	5.52	5.36	5.28	5.20	5.12	5.03	4.95	4.86
9	10.6	8.02	6.99	6.42	6.06	5.80	5.61	5.47	5.35	5.26	5.11	4.96	4.81	4.73	4.65	4.57	4.48	4.40	4.31
10	10.0	7.56	6.55	5.99	5.64	5.39	5.20	5.06	4.94	4.85	4.71	4.56	4.41	4.33	4.25	4.17	4.08	4.00	3.91
11	9.65	7.21	6.22	5.67	5.32	5.07	4.89	4.74	4.63	4.54	4.40	4.25	4.10	4.02	3.94	3.86	3.78	3.69	3.60
12	9.33	6.93	5.95	5.41	5.06	4.82	4.64	4.50	4.39	4.30	4.16	4.01	3.86	3.78	3.70	3.62	3.54	3.45	3.36
13	9.07	6.70	5.74	5.21	4.86	4.62	4.44	4.30	4.19	4.10	3.96	3.82	3.66	3.59	3.51	3.43	3.34	3.25	3.17
14	8.86	6.51	5.56	5.04	4.70	4.46	4.28	4.14	4.03	3.94	3.80	3.66	3.51	3.43	3.35	3.27	3.18	3.09	3.00
15	8.68	6.36	5.42	4.89	4.56	4.32	4.14	4.00	3.89	3.80	3.67	3.52	3.37	3.29	3.21	3.13	3.05	2.96	2.87
16	8.53	6.23	5.29	4.77	4.44	4.20	4.03	3.89	3.78	3.69	3.55	3.41	3.26	3.18	3.10	3.02	2.93	2.84	2.75
17	8.40	6.11	5.19	4.67	4.34	4.10	3.93	3.79	3.68	3.59	3.46	3.31	3.16	3.08	3.00	2.92	2.83	2.75	2.65
18	8.29	6.01	5.09	4.58	4.25	4.01	3.84	3.71	3.60	3.51	3.37	3.23	3.08	3.00	2.92	2.84	2.75	2.66	2.57
19	8.19	5.93	5.01	4.50	4.17	3.94	3.77	3.63	3.52	3.43	3.30	3.15	3.00	2.92	2.84	2.76	2.67	2.58	2.49
20	8.10	5.85	4.94	4.43	4.10	3.87	3.70	3.56	3.46	3.37	3.23	3.09	2.94	2.86	2.78	2.69	2.61	2.52	2.42
21	8.02	5.78	4.87	4.37	4.04	3.81	3.64	3.51	3.40	3.31	3.17	3.03	2.88	2.80	2.72	2.64	2.55	2.46	2.36
22	7.95	5.72	4.82	4.31	3.99	3.76	3.59	3.45	3.35	3.26	3.12	2.98	2.83	2.75	2.67	2.58	2.50	2.40	2.31
23	7.88	5.66	4.76	4.26	3.94	3.71	3.54	3.41	3.30	3.21	3.07	2.93	2.78	2.70	2.62	2.54	2.45	2.35	2.26
24	7.82	5.61	4.72	4.22	3.90	3.67	3.50	3.36	3.26	3.17	3.03	2.89	2.74	2.66	2.58	2.49	2.40	2.31	2.21
25	7.77	5.57	4.68	4.18	3.86	3.63	3.46	3.32	3.22	3.13	2.99	2.85	2.70	2.62	2.53	2.45	2.36	2.27	2.17
30	7.56	5.39	4.51	4.02	3.70	3.47	3.30	3.17	3.07	2.98	2.84	2.70	2.55	2.47	2.39	2.30	2.21	2.11	2.01
40	7.31	5.18	4.31	3.83	3.51	3.29	3.12	2.99	2.89	2.80	2.66	2.52	2.37	2.29	2.20	2.11	2.02	1.92	1.80
60	7.08	4.98	4.13	3.65	3.34	3.12	2.95	2.82	2.72	2.63	2.50	2.35	2.20	2.12	2.03	1.94	1.84	1.73	1.60
120	6.85	4.79	3.95	3.48	3.17	2.96	2.79	2.66	2.56	2.47	2.34	2.19	2.03	1.95	1.86	1.76	1.66	1.53	1.38
∞	6.63	4.61	3.78	3.32	3.02	2.80	2.64	2.51	2.41	2.32	2.18	2.04	1.88	1.79	1.70	1.59	1.47	1.32	1.00

Table 7
Critical Values of the Total Number of Runs

This table provides critical values of the total number of runs used to test the hypothesis that a sequence of two kinds of symbols is random against the alternative that the sequence is not random. The symbols, n_1 and n_2, correspond to the number of each kind of symbol, where n_1 represents the smaller of the two values. The probabilities appearing at the top of the table identify critical values corresponding to upper and lower critical values; the right half corresponds to upper critical values and the left half identifies lower critical values. Each probability represents half the level of significance, α. Blank spaces in the table indicate cases in which critical values are impossible.

Example Suppose a sequence of 12 elements has 5 of one kind and 7 of another. The critical values for the test at the 5 percent level of significance are 3 and 11. These are found by letting $n_1 = 5$ and $n_2 = 7$ and locating half the level of significance on the right side of the table equal to 0.025 and one minus half the level of significance equal to 0.975 on the left side of the table. The hypothesis of randomness is accepted if the observed number of runs falls within the critical values in the table.

Table 7. Critical Values of the Total Number of Runs

		Probability							
n_1	n_2	.995	.99	.975	.95	.05	.025	.01	.005
2	2								
	3								
	4								
	5								
	6								
	7								
	8				2				
	9				2				
	10				2				
	11				2				
	12			2	2				
	13			2	2				
	14			2	2				
	15			2	2				
	16			2	2				
	17			2	2				
	18			2	2				
	19		2	2	2				
	20		2	2	2				
3	3								
	4					7			
	5				2				
	6			2	2				
	7			2	2				
	8			2	2				
	9		2	2	2				
	10		2	2	3				
	11		2	2	3				
	12	2	2	2	3				
	13	2	2	2	3				
	14	2	2	2	3				
	15	2	2	3	3				
	16	2	2	3	3				
	17	2	2	3	3				
	18	2	2	3	3				
	19	2	2	3	3				
	20	2	2	3	3				
4	4				2	8			
	5			2	2	9	9	9	

Table 7. Critical Values of the Total Number of Runs

n_1	n_2	Probability .995	.99	.975	.95	.05	.025	.01	.005
4	6		2	2	3	9	9		
	7		2	2	3	9			
	8	2	2	3	3				
	9	2	2	3	3				
	10	2	2	3	3				
	11	2	2	3	3				
	12	2	3	3	4				
	13	2	3	3	4				
	14	2	3	3	4				
	15	3	3	3	4				
	16	3	3	4	4				
	17	3	3	4	4				
	18	3	3	4	4				
	19	3	3	4	4				
	20	3	3	4	4				
5	5		2	2	3	9	10	10	
	6	2	2	3	3	10	10	11	11
	7	2	2	3	3	10	11	11	
	8	2	2	3	3	11	11		
	9	2	3	3	4	11			
	10	3	3	3	4	11			
	11	3	3	4	4				
	12	3	3	4	4				
	13	3	3	4	4				
	14	3	3	4	5				
	15	3	4	4	5				
	16	3	4	4	5				
	17	3	4	4	5				
	18	4	4	5	5				
	19	4	4	5	5				
	20	4	4	5	5				
6	6	2	2	3	3	11	11	12	12
	7	2	3	3	4	11	12	12	13
	8	3	3	3	4	12	12	13	13
	9	3	3	4	4	12	13	13	
	10	3	3	4	5	12	13		
	11	3	4	4	5	13	13		
	12	3	4	4	5	13	13		
	13	3	4	5	5	13			

Table 7. Critical Values of the Total Number of Runs

n_1	n_2	Probability .995	.99	.975	.95	.05	.025	.01	.005
6	14	4	4	5	5	13			
	15	4	4	5	6				
	16	4	4	5	6				
	17	4	5	5	6				
	18	4	5	5	6				
	19	4	5	6	6				
	20	4	5	6	6				
7	7	3	3	3	4	12	13	13	13
	8	3	3	4	4	13	13	14	14
	9	3	4	4	5	13	14	14	15
	10	3	4	5	5	13	14	15	15
	11	4	4	5	5	14	14	15	15
	12	4	4	5	6	14	14	15	
	13	4	5	5	6	14	15		
	14	4	5	5	6	14	15		
	15	4	5	6	6	15	15		
	16	5	5	6	6	15			
	17	5	5	6	7	15			
	18	5	5	6	7	15			
	19	5	6	6	7	15			
	20	5	6	6	7	15			
8	8	3	4	4	5	13	14	14	15
	9	3	4	5	5	14	14	15	15
	10	4	4	5	6	14	15	15	16
	11	4	5	5	6	15	15	16	16
	12	4	5	6	6	15	16	16	17
	13	5	5	6	6	15	16	17	17
	14	5	5	6	7	16	16	17	17
	15	5	5	6	7	16	16	17	
	16	5	6	6	7	16	17	17	
	17	5	6	7	7	16	17		
	18	6	6	7	8	16	17		
	19	6	6	7	8	16	17		
	20	6	6	7	8	17	17		
9	9	4	4	5	6	14	15	16	16
	10	4	5	5	6	15	16	16	17
	11	5	5	6	6	15	16	17	17
	12	5	5	6	7	16	16	17	18

Table 7. Critical Values of the Total Number of Runs

n_1	n_2	Probability .995	.99	.975	.95	.05	.025	.01	.005
9	13	5	6	6	7	16	17	18	18
	14	5	6	7	7	17	17	18	18
	15	6	6	7	8	17	18	18	19
	16	6	6	7	8	17	18	18	19
	17	6	7	7	8	17	18	19	19
	18	6	7	8	8	18	18	19	
	19	6	7	8	8	18	18	19	
	20	7	7	8	9	18	18	19	
10	10	5	5	6	6	16	16	17	17
	11	5	5	6	7	16	17	18	18
	12	5	6	7	7	17	17	18	19
	13	5	6	7	8	17	18	19	19
	14	6	6	7	8	17	18	19	19
	15	6	7	7	8	18	18	19	20
	16	6	7	8	8	18	19	20	20
	17	7	7	8	9	18	19	20	20
	18	7	7	8	9	19	19	20	21
	19	7	8	8	9	19	20	20	21
	20	7	8	9	9	19	20	20	21
11	11	5	6	7	7	17	17	18	19
	12	6	6	7	8	17	18	19	19
	13	6	6	7	8	18	19	19	20
	14	6	7	8	8	18	19	20	20
	15	7	7	8	9	19	19	20	21
	16	7	7	8	9	19	20	21	21
	17	7	8	9	9	19	20	21	22
	18	7	8	9	10	20	20	21	22
	19	8	8	9	10	20	21	22	22
	20	8	8	9	10	20	21	22	22
12	12	6	7	7	8	18	19	19	20
	13	6	7	8	9	18	19	20	21
	14	7	7	8	9	19	20	21	21
	15	7	8	8	9	19	20	21	22
	16	7	8	9	10	20	21	22	22
	17	8	8	9	10	20	21	22	22
	18	8	8	9	10	21	21	22	23
	19	8	9	10	10	21	22	23	23
	20	8	9	10	11	21	22	23	23

Table 7. Critical Values of the Total Number of Runs

n_1	n_2	.995	.99	.975	.95	.05	.025	.01	.005
		Probability							
13	13	7	7	8	9	19	20	21	21
	14	7	8	9	9	20	20	21	22
	15	7	8	9	10	20	21	22	22
	16	8	8	9	10	21	21	22	23
	17	8	9	10	10	21	22	23	23
	18	8	9	10	11	21	22	23	24
	19	9	9	10	11	22	23	24	24
	20	9	10	10	11	22	23	24	24
14	14	7	8	9	10	20	21	22	23
	15	8	8	9	10	21	22	23	23
	16	8	9	10	11	21	22	23	24
	17	8	9	10	11	22	23	24	24
	18	9	9	10	11	22	23	24	25
	19	9	10	11	12	23	23	24	25
	20	9	10	11	12	23	24	25	25
15	15	8	9	10	11	21	22	23	24
	16	9	9	10	11	22	23	24	24
	17	9	10	11	11	22	23	24	25
	18	9	10	11	12	23	24	25	25
	19	10	10	11	12	23	24	25	26
	20	10	11	12	12	24	25	26	26
16	16	9	10	11	11	23	23	24	25
	17	9	10	11	12	23	24	25	26
	18	10	10	11	12	24	25	26	26
	19	10	11	12	13	24	25	26	27
	20	10	11	12	13	25	25	26	27
17	17	10	10	11	12	24	25	26	26
	18	10	11	12	13	24	25	26	27
	19	10	11	12	13	25	26	27	27
	20	11	11	13	13	25	26	27	28
18	18	11	11	12	13	25	26	27	27
	19	11	12	13	14	25	26	27	28
	20	11	12	13	14	26	27	28	29
19	19	11	12	13	14	26	27	28	29
	20	12	12	13	14	27	27	29	29
20	20	12	13	14	15	27	28	29	30

Table 8
Spearman Rank Correlation Coefficient

This table can be used to perform a one-tailed test of the hypothesis that there is no correlation between two sets of rankings using the test statistic

$$r_s = 1 - \frac{6\Sigma d^2}{n^3 - n}$$

where d equals the difference between independent ranks of a similar object and n is the sample size or number of objects. Values in the table represent critical values of r_s for selected sample sizes, n, between 4 and 30.

Example Based upon a sample of 18 objects, the critical value of r_s at the 5 percent level of significance equals .399. No correlation is rejected if r_s is greater than or equal to this value if r_s is positive, or if r_s is less than or equal to it if r_s is negative. The direction of the rejection region is based on the nature of the alternate hypothesis.

Table 8. Spearman Rank Correlation Coefficient

n	One-Tailed Probability .05	.01
4	1.000	
5	.900	1.000
6	.829	.943
7	.714	.893
8	.643	.833
9	.600	.783
10	.564	.746
12	.506	.712
14	.456	.645
16	.425	.601
18	.399	.564
20	.377	.534
22	.359	.508
24	.343	.485
26	.329	.465
28	.317	.448
30	.306	.432

Table 9
Random Digits

This table provides a sequence of integers between 0 and 9 that are generated at random with equal probabilities. Consequently, each of the digits between 0 and 9 has the same chance of occupying each position in the table. The order in which the digits appear in the table is important. The table can be used to generate sequences of values at random and to select random samples. It is possible to begin generating a sequence at any point in the table. Once a starting place is chosen, however, values must be selected consecutively. It does not matter whether the table is used vertically or horizontally.

Example Suppose we are interested in generating 8 values representing the outcomes of 8 rolls of a balanced die. There are 6 possible outcomes, 1 through 6, each with an equal probability of occurrence. Starting at the beginning of the table working horizontally across the page, the sequence is

3 2 3 3 3 1 5 6

This is obtained by recording values between 1 and 6 when they appear in sequence and ignoring values not falling within the prescribed range.

Example Suppose we are to select a random sample of 5 families from a population of 10,000. Each family is assigned a number between 1 and 10,000. Starting at the beginning of the table working vertically, the families in the sample are the ones with the following numbers:

2610 7633 1384 3290 4905

In order to obtain this sample, columns of 5-digit numbers must be considered since the maximum number of families equal to 10,000 contains 5 digits. By examining each number in the first column of five digits, we see that 02610 is the first that falls within the range 1–10000. The zero on the left does not contribute to the value. Hence, the first family in the sample is the one numbered 2610. Remaining numbers are obtained by working successively down the table ignoring those not falling within the desired range. If the first column is exhausted before obtaining the full sample, the next adjacent column of 5 digits is examined, and so on, until the required number in the sample is achieved.

Note. When a number that already has been chosen appears again, it should be ignored if selection or sampling is done without replacement.

Table 9. Random Digits

38233	37175	96866	39089	94736	61380	90458	05453	95172	93666	16125	67117
43130	40343	75278	89548	44005	58983	39772	16209	79469	06890	74283	27617
84903	43656	46791	28316	55508	09819	79956	97381	73127	72270	65239	84032
44696	43241	46124	94993	97925	03283	80883	02641	43499	60871	50818	84164
56914	28546	02030	17028	97370	83488	36378	75833	05283	78546	10998	91909
30074	14617	18690	48142	92617	04706	12193	04771	84140	48038	26780	85503
20186	41895	73296	19029	00017	17641	81484	64558	83465	67338	46930	46501
81313	63535	95939	18038	73134	74555	54081	98885	51761	42759	04733	33750
51911	73119	08690	19557	86258	71465	32769	63188	61768	34779	33360	88759
35822	38244	43411	03660	49072	29825	06263	23218	26840	04273	81301	88710
21200	07495	16847	20245	32743	74882	68751	31859	33004	29646	19058	62772
17417	12765	58163	20019	30924	47313	25911	37882	00192	96388	05688	71174
02610	46858	91521	49759	53071	88082	41613	23758	59294	71824	92870	54956
76214	63470	74371	86432	02076	35952	57370	92730	67174	35663	44502	19696
95452	79732	96859	95538	68381	73400	47510	58040	96700	97139	84996	12714
07633	63263	24574	26378	63566	83696	11401	70684	91342	84727	10146	23469
72042	00641	81750	78377	33868	28119	01150	17941	04988	11946	48036	97893
55017	29963	95676	10470	69049	35296	54500	17695	31023	54065	07728	27436
79373	74634	74692	04302	07451	02186	45172	83942	16003	08011	47475	78049
51264	44022	55247	95868	59779	68605	87917	19981	60743	22596	55917	89792
27403	76388	26201	99006	55602	18836	32165	55814	14742	87484	09982	85667
60098	12629	01946	85056	89691	28961	47774	90510	03251	35537	03149	02248
92874	29696	32607	41121	86929	17624	49885	06170	22467	89743	38396	86827
01384	00359	13967	74694	96676	29890	63198	21550	10503	53028	30119	45589
65038	15306	14625	97516	28773	92337	55423	23585	71623	98086	40452	28854
34178	00711	64194	28906	82587	68647	46919	06417	64283	40403	72294	39812
43857	60911	27444	39676	86000	73531	27995	70736	01360	07415	93227	17228
22682	72727	84258	05945	38458	08768	25120	76183	40021	10244	43034	66001
16118	67929	90658	20236	13861	97215	79892	75693	68930	98897	53307	33544
45477	71051	85585	61148	52906	91412	79272	32953	67456	02582	53693	87317
62687	93308	41231	21079	52014	85361	17064	99581	23395	11456	43531	85369
95291	57926	31321	89373	20798	67556	95510	19295	76399	25342	34494	84500
92790	81492	62506	74903	08086	99645	46788	72726	90034	41450	53635	22932
36075	73713	84367	10676	80894	88053	12374	74655	65485	81645	86568	23826
98531	46896	53996	81931	31164	46261	63045	86942	41462	96916	85889	40630
03290	68482	67799	53067	63728	59944	30196	32109	96288	65025	04731	53148
78103	04530	16513	99222	43653	37343	22291	47259	34643	38773	82291	07756
66372	30789	05371	49879	92604	59276	73829	14774	39016	01309	26079	16144
27992	03779	43576	97076	27486	89095	29984	93535	58244	16233	04470	96479
04905	59642	30039	20427	86340	20760	73340	78022	54618	99610	61963	88360
45256	83348	29872	43056	25182	08510	96386	93812	22892	25673	07773	69867
35618	29606	95609	48843	25573	46922	03378	91542	22102	91005	06780	84021
03465	74932	99918	57257	10335	74548	11548	74584	59835	27434	85148	35165
85648	19233	32351	35552	76958	19304	60048	02535	74722	95768	42093	50214
55909	81536	28816	26003	39209	87926	96746	89508	03941	08420	51650	88066
39831	63766	71470	34381	87796	45784	50744	32883	93233	20975	88307	33284
45153	96739	66624	25320	13057	51228	91404	88157	25574	33774	78890	16494
57331	98914	46826	88573	81418	88299	83879	06773	92077	26997	81559	77863
11014	16511	31929	45930	30669	35548	60438	62509	67195	88450	74572	18858
55067	94103	12842	57212	00579	42647	01978	95511	02975	10540	72882	46012

Table 9. Random Digits

80228	94650	98827	06716	87059	73187	65518	87934	08550	59656	83690	38975
11628	38997	97586	17935	97437	54708	13404	11193	84853	77658	58313	55704
43420	51029	99009	83003	47328	99892	18167	81569	83295	31972	02170	39666
76561	37244	12465	17759	85036	45863	14640	79790	89190	20710	51329	56776
74957	08291	47782	13230	34106	21974	77973	81218	18775	08881	05113	82676
17857	53018	97004	76189	38481	33208	81353	31089	21716	40228	12298	44686
74209	72305	84513	89882	71566	30761	42122	91567	55794	26241	84936	36160
06176	24713	97697	39673	59473	91209	66041	01213	48396	36481	97767	95281
82957	72585	35475	41563	45263	14207	74938	75034	41973	43327	32136	68343
28228	66034	12211	11844	01217	34302	48485	07762	59084	33426	54929	28011
75919	33487	11989	60556	60113	22576	93990	19218	05551	22933	83481	95931
20842	60990	31592	50027	64811	34888	77983	42946	39834	11439	43635	15042
72056	49717	06979	69247	74524	12800	45286	43799	42544	39434	88354	07696
83900	61343	87267	52372	37856	75464	63123	04188	95258	26525	67631	69272
41926	00734	02848	41214	69593	77667	37071	62873	28348	34055	61767	29547
97901	38306	46821	84480	81685	23016	82173	16436	66247	40469	49584	57456
90322	43071	35187	00860	76653	57091	95909	50875	21026	56531	44297	33361
44128	53542	68835	15958	76449	77345	01697	57285	68631	76385	01655	28788
81720	36056	41871	91299	79629	80832	10982	48432	77144	31235	54595	78635
62256	54387	64865	40942	99468	47925	90785	83794	61495	83373	68138	64344
09037	63822	57470	15697	17281	35058	35878	93637	46550	44815	76902	05397
02973	05968	03677	76158	81972	21463	82953	26812	18909	03341	34154	99904
11475	26812	56902	74434	33760	48934	82864	90109	27106	25841	34312	33383
04506	14722	85609	76330	75980	77913	34274	37106	72182	90220	43418	11832
48027	51541	76252	86682	24490	33832	64295	61186	04560	09276	88982	27419
78120	74800	16316	53910	22373	60069	29176	18712	92241	64558	74629	98190
07704	00578	47435	12270	09210	13455	27487	58044	74194	22157	63786	14118
70490	04712	14274	35419	59534	68114	36753	00056	57203	32586	44801	49757
13256	37320	48708	81848	12740	20222	59529	87733	97509	13314	59894	69263
10357	10777	03801	23868	03059	18363	41766	24135	80203	81121	79301	81680
92558	26337	75235	14392	18976	18443	71538	18253	69296	92136	58600	82617
73835	50639	70723	66490	30539	46388	19351	32391	72274	12025	86124	77430
94221	89718	07488	70786	91259	73006	89240	32329	16459	60913	62818	08058
01635	60086	61853	72973	78432	51237	72938	97573	57152	73065	62558	18645
38476	85892	69540	18513	65122	41428	64928	70054	31090	53622	79981	83737
63561	04605	36213	06086	35035	82470	42488	20855	17410	41883	05123	35179
74798	15466	08644	84800	02514	77952	16826	65657	08263	62803	99395	79936
98232	41347	80149	25949	34773	74012	08134	15469	37418	22953	42447	90327
73801	44228	35735	99481	46230	53734	65386	56253	62061	54645	75542	47052
88685	30803	50191	35266	81908	54935	81996	00244	83251	78318	44869	12451
50918	39339	04620	45761	98565	99832	65390	90342	52824	11916	74527	38278
76972	63515	27344	79145	70830	75215	56903	39838	02384	36844	41336	25291
44798	33099	77580	90821	85738	71605	45953	68589	24634	75604	18616	25007
43090	53160	68401	76922	37911	58696	73453	68064	00312	36254	47019	81014
13184	69776	35497	31620	25270	62956	10519	61186	54264	84923	60330	22623
53643	99337	83120	34317	14515	51128	43477	58882	04263	07096	69400	87031
90719	95160	87829	61810	49374	56180	52461	50929	64841	75886	04665	80557
46470	82358	44671	73821	09842	32049	01303	75933	21142	35346	50014	71633
65219	53520	39176	69765	56080	90261	08420	40769	19363	62689	35658	81457
73177	77210	64760	93880	58426	42365	53378	66232	46636	72268	08995	42642

Table 9. Random Digits

33255	77597	82636	01540	41252	23256	57104	88153	73494	92847	06685	43760
52110	36679	33180	57528	88574	21752	64514	36838	29592	32613	43824	44821
49724	02434	79764	42128	01297	46861	75309	50919	84548	12363	51724	55064
63218	28093	06185	94336	42090	92688	20351	87406	08412	92490	04955	42539
20820	81591	10937	10811	22028	45262	15007	78704	81433	45672	11319	91354
38154	69894	76979	74498	26156	98314	63302	64068	28585	36104	54369	21737
04433	14304	52877	96866	41630	13202	15856	99431	34958	17891	47984	86766
73661	21572	30879	55989	93324	71338	58043	08107	42947	92527	37639	60202
18078	00594	27325	94334	90476	68364	12247	77629	31686	58280	67311	41421
73464	68195	53634	06160	44590	91060	06245	28800	90885	75519	26913	63938
78663	54499	58401	23231	51736	91994	09902	64231	18132	54202	37154	49943
64222	96579	76956	97402	11638	25796	52965	10354	29830	90838	93788	84782
79326	34374	78108	64987	43089	88539	76989	10736	59897	64524	84026	16148
85463	05580	69330	02129	82967	05645	05836	67999	45477	94024	98069	26415
03023	96022	62898	60543	37511	18443	92349	94910	97598	83368	00259	42382
21027	99681	35894	61017	34863	64252	17228	72684	98362	67117	52234	06031
11955	75760	34998	17682	35782	46381	08534	38225	82042	89214	97958	51242
56909	36853	16058	85018	56805	00943	95238	58988	38904	32031	26230	92545
29334	33862	00108	23064	44791	25110	58014	16598	83511	33264	40138	54568
10765	00871	07801	60791	78961	58905	32766	13376	33672	08941	03639	93893
24081	69803	82886	91452	20943	73239	70125	42409	12304	76701	18121	10054
04661	47837	08117	73998	12579	82671	65503	80482	89384	81226	71224	25205
13499	42688	42336	60110	21880	87308	44181	90390	42075	82315	67003	22505
62292	42075	53448	22982	51526	17664	24960	17915	57696	17864	37774	64038
91584	59594	36326	61174	55769	64704	32465	42273	40119	64547	06342	12300
56194	24129	96414	81266	30130	84285	00510	88250	34893	71038	63743	49160
70743	21519	09591	13219	71653	69303	23400	57601	23804	14610	53925	87459
57004	25184	35576	75356	14207	53792	41855	99772	63044	56179	37786	25528
30272	19702	52128	99543	94812	77339	25477	82053	09093	48374	93788	47385
64508	52255	64062	40984	66757	40352	21913	56532	04744	24825	02216	21643
03741	26783	98474	27392	27728	32479	55520	16504	31591	84996	44965	19757
64419	01135	66313	58441	99798	88595	60532	62720	58679	40748	04217	15748
26355	48041	80211	28669	90666	62506	46309	47703	30024	49206	44138	93655
76886	30111	93914	82466	78801	17595	26728	55906	19092	93410	47246	80913
69096	47867	69082	01328	01931	34262	49708	88103	56873	45574	19293	85930
13994	55387	12445	23890	46451	02407	99000	83494	42941	61500	29393	81854
50921	86611	61002	31440	50808	61446	97848	04977	58034	59372	20380	85719
30520	71749	86337	03639	48603	00872	42854	30109	13380	95533	64354	69056
94577	99083	44517	72310	81184	54709	70004	85790	53722	66252	55694	27951
00553	67018	93443	52360	06795	59784	52416	43796	25508	78186	86017	51589
83953	87879	22522	03411	38367	96669	13698	36016	75546	60892	74119	94520
14956	72005	15476	23294	38440	48492	23094	16872	54839	82670	27900	49870
03385	75508	71168	77587	71621	72714	59627	00661	40315	91222	59099	56509
06587	21658	50227	48412	31230	95724	08929	90411	05927	60230	14430	39148
07980	32848	64258	46088	51701	52489	62201	33796	21999	90674	27873	60081
04543	65204	48934	32547	78900	78021	02897	87963	32579	19179	28079	81952
30886	06211	94863	87216	11682	06179	82187	35036	05644	61011	19105	99404
25140	88666	13234	35261	83960	20072	60178	63242	90632	46467	16281	51051
37588	13740	53497	36387	52724	94148	86686	74844	82596	88876	24891	84966
55943	62642	48795	16085	64427	33439	58415	59700	36889	20459	96007	53980

Table 10
Values of e^{-x}

This table provides values of the natural constant e = 2.7182818 . . . raised to negative powers, $-x$, where x assumes selected values between 0.05 and 10.00 in increments of 0.05 units.

Example In order to find e raised to the -1.75 power look up 1.75 in the x-column and read the answer in the e^{-x} column. The answer is $e^{-1.75} = 0.17377$.

Table 10. Values of e^{-x}

x	e^{-x}	x	e^{-x}	x	e^{-x}	x	e^{-x}
0.05	0.95123	2.55	0.07808	5.05	0.00641	7.55	0.00053
0.10	0.90484	2.60	0.07427	5.10	0.00610	7.60	0.00050
0.15	0.86071	2.65	0.07065	5.15	0.00580	7.65	0.00048
0.20	0.81873	2.70	0.06721	5.20	0.00552	7.70	0.00045
0.25	0.77880	2.75	0.06393	5.25	0.00525	7.75	0.00043
0.30	0.74082	2.80	0.06081	5.30	0.00499	7.80	0.00041
0.35	0.70469	2.85	0.05784	5.35	0.00475	7.85	0.00039
0.40	0.67032	2.90	0.05502	5.40	0.00452	7.90	0.00037
0.45	0.63763	2.95	0.05234	5.45	0.00430	7.95	0.00035
0.50	0.60653	3.00	0.04979	5.50	0.00409	8.00	0.00034
0.55	0.57695	3.05	0.04736	5.55	0.00389	8.05	0.00032
0.60	0.54881	3.10	0.04505	5.60	0.00370	8.10	0.00030
0.65	0.52205	3.15	0.04285	5.65	0.00352	8.15	0.00029
0.70	0.49659	3.20	0.04076	5.70	0.00335	8.20	0.00027
0.75	0.47237	3.25	0.03877	5.75	0.00318	8.25	0.00026
0.80	0.44933	3.30	0.03688	5.80	0.00303	8.30	0.00025
0.85	0.42741	3.35	0.03508	5.85	0.00288	8.35	0.00024
0.90	0.40657	3.40	0.03337	5.90	0.00274	8.40	0.00022
0.95	0.38674	3.45	0.03175	5.95	0.00261	8.45	0.00021
1.00	0.36788	3.50	0.03020	6.00	0.00248	8.50	0.00020
1.05	0.34994	3.55	0.02872	6.05	0.00236	8.55	0.00019
1.10	0.33287	3.60	0.02732	6.10	0.00224	8.60	0.00018
1.15	0.31664	3.65	0.02599	6.15	0.00213	8.65	0.00018
1.20	0.30119	3.70	0.02472	6.20	0.00203	8.70	0.00017
1.25	0.28650	3.75	0.02352	6.25	0.00193	8.75	0.00016
1.30	0.27253	3.80	0.02237	6.30	0.00184	8.80	0.00015
1.35	0.25924	3.85	0.02128	6.35	0.00175	8.85	0.00014
1.40	0.24660	3.90	0.02024	6.40	0.00166	8.90	0.00014
1.45	0.23457	3.95	0.01925	6.45	0.00158	8.95	0.00013
1.50	0.22313	4.00	0.01832	6.50	0.00150	9.00	0.00012
1.55	0.21225	4.05	0.01742	6.55	0.00143	9.05	0.00012
1.60	0.20190	4.10	0.01657	6.60	0.00136	9.10	0.00011
1.65	0.19205	4.15	0.01576	6.65	0.00129	9.15	0.00011
1.70	0.18268	4.20	0.01500	6.70	0.00123	9.20	0.00010
1.75	0.17377	4.25	0.01426	6.75	0.00117	9.25	0.00010
1.80	0.16530	4.30	0.01357	6.80	0.00111	9.30	0.00009
1.85	0.15724	4.35	0.01291	6.85	0.00106	9.35	0.00009
1.90	0.14957	4.40	0.01228	6.90	0.00101	9.40	0.00008
1.95	0.14227	4.45	0.01168	6.95	0.00096	9.45	0.00008
2.00	0.13534	4.50	0.01111	7.00	0.00091	9.50	0.00007
2.05	0.12873	4.55	0.01057	7.05	0.00087	9.55	0.00007
2.10	0.12246	4.60	0.01005	7.10	0.00083	9.60	0.00007
2.15	0.11648	4.65	0.00956	7.15	0.00078	9.65	0.00006
2.20	0.11080	4.70	0.00910	7.20	0.00075	9.70	0.00006
2.25	0.10540	4.75	0.00865	7.25	0.00071	9.75	0.00006
2.30	0.10026	4.80	0.00823	7.30	0.00068	9.80	0.00006
2.35	0.09537	4.85	0.00783	7.35	0.00064	9.85	0.00005
2.40	0.09072	4.90	0.00745	7.40	0.00061	9.90	0.00005
2.45	0.08629	4.95	0.00708	7.45	0.00058	9.95	0.00005
2.50	0.08208	5.00	0.00674	7.50	0.00055	10.00	0.00005

Table 11
Logarithms of Numbers

A logarithm of a number is the power or exponent of a base value such that the base raised to that power equals the number. This table allows us to find the logarithm of a number to the base 10, or $\log_{10}$. A logarithm is made up of two components, a characteristic and a mantissa. The algebraic sum of the characteristic and the mantissa equals the logarithm. The mantissa is a decimal part and is obtained from Table 11. The characteristic is a whole number. If the original number is at least one, the characteristic equals one less than the number of digits to the left of the decimal. If the number is less than one the characteristic equals the negative of one plus the number of zeros between the decimal point and the first non-zero value. Logarithms of numbers less than one can be expressed in two ways, which is illustrated in the next to the last example. Table 11 provides mantissas for numbers with three non-zero digits.

Example In order to find the logarithm of 255 we first determine the characteristic. Since the number is an integer with three digits, the characteristic is one less, or equal to 2. The mantissa is found from the table by locating the first two digits, or 25, in the left column and the third digit, or 5, at the top row. The corresponding value, or the mantissa, is 4065. The logarithm is written as the sum of 2 and .4065

$$\begin{aligned}\text{logarithm of } 255 &= \log 255\\ &= 2.4065\end{aligned}$$

Example Suppose the number is 3617.4 and we want its logarithm. Using Table 11 we must round the number to three non-zero digits as 3620. The logarithm of this number is 3.5587. The mantissa equal to .5587 is found from the table by looking up 362 and the characteristic is 3 since there are four digits to the right of the decimal in the original number.

Example Let us find the logarithm of a decimal equal to 0.00054. There are three zeros between the decimal point and the first non-zero value. Hence the characteristic is the negative of one plus three and equals -4. The mantissa is found from the table by locating 54 in the left column and 0 at the top. The result is .7324. The logarithm of .00054 is the algebraic sum of the characteristic and mantissa.

$$\begin{aligned}\log .00054 &= -4 + .7324\\ &= -3.2676\end{aligned}$$

The logarithm of a decimal can be expressed in another way by expressing the characteristic in two parts. For example, we can write -4 as $6.0000 - 10$ and insert the mantissa from the table in place of the zeros. Hence,

$$\log .00054 = 6.7324 - 10$$

When the two parts are added algebraically, the result is the same as the one above.

Example Consider a problem where the logarithm is given as 3.4082 and we want to find the original number. We first locate the decimal part in the body of the table and determine the digits of the number from the left column and top row which is 256. Since the characteristic equals three there are four decimal places. The number is 2560.

When the exact value of the mantissa does not appear in the body of the table, using the closest value is satisfactory. Interpolation of the table is possible, however. When a negative logarithm is given, the mantissa used in the table is the decimal part of the result when the logarithm is subtracted from 10.

Table 11. Logarithms of Numbers

n	0	1	2	3	4	5	6	7	8	9
10	0000	0043	0086	0128	0170	0212	0253	0294	0334	0374
11	0414	0453	0492	0531	0569	0607	0645	0682	0719	0755
12	0792	0828	0864	0899	0934	0969	1004	1038	1072	1106
13	1139	1173	1206	1239	1271	1303	1335	1367	1399	1430
14	1461	1492	1523	1553	1584	1614	1644	1673	1703	1732
15	1761	1790	1818	1847	1875	1903	1931	1959	1987	2014
16	2041	2068	2095	2122	2148	2175	2201	2227	2253	2279
17	2304	2330	2355	2380	2405	2430	2455	2480	2504	2529
18	2553	2577	2601	2625	2648	2672	2695	2718	2742	2765
19	2788	2810	2833	2856	2878	2900	2923	2945	2967	2989
20	3010	3032	3045	3075	3096	3118	3139	3160	3181	3201
21	3222	3243	3263	3284	3304	3324	3345	3365	3385	3404
22	3424	3444	3464	3483	3502	3522	3541	3560	3579	3598
23	3617	3636	3655	3674	3692	3711	3729	3747	3766	3784
24	3802	3820	3838	3856	3874	3892	3909	3927	3945	3962
25	3979	3997	4014	4031	4048	4065	4082	4099	4116	4133
26	4150	4166	4183	4200	4216	4232	4249	4265	4281	4298
27	4314	4330	4346	4362	4378	4393	4409	4425	4440	4456
28	4472	4487	4502	4518	4533	4548	4564	4579	4594	4609
29	4624	4639	4654	4669	4683	4698	4713	4728	4742	4757
30	4771	4786	4800	4814	4829	4843	4857	4871	4886	4900
31	4914	4928	4942	4955	4969	4983	4997	5011	5024	5038
32	5051	5065	5079	5092	5105	5119	5132	5145	5159	5172
33	5185	5198	5211	5224	5237	5250	5263	5276	5289	5302
34	5315	5328	5340	5353	5366	5378	5391	5403	5416	5428
35	5441	5453	5465	5478	5490	5502	5514	5527	5539	5551
36	5563	5575	5587	5599	5611	5623	5635	5647	5658	5670
37	5682	5694	5705	5717	5729	5740	5752	5763	5775	5786
38	5798	5809	5821	5832	5843	5855	5866	5877	5888	5899
39	5911	5922	5933	5944	5955	5966	5977	5988	5999	6010
40	6021	6031	6042	6053	6064	6075	6085	6096	6107	6117
41	6128	6138	6149	6160	6170	6180	6191	6201	6212	6222
42	6232	6243	6253	6263	6274	6284	6294	6304	6314	6325
43	6335	6345	6355	6365	6375	6385	6395	6405	6415	6425
44	6435	6444	6454	6464	6474	6484	6493	6503	6513	6522
45	6532	6542	6551	6561	6571	6580	6590	6599	6609	6618
46	6628	6637	6646	6656	6665	6675	6684	6693	6702	6712
47	6721	6730	6739	6749	6758	6767	6776	6785	6794	6803
48	6812	6821	6830	6839	6848	6857	6866	6875	6884	6893
49	6902	6911	6920	6928	6937	6946	6955	6964	6972	6981
50	6990	6998	7007	7016	7024	7033	7042	7050	7059	7067
51	7076	7084	7093	7101	7110	7118	7126	7135	7143	7152
52	7160	7168	7177	7185	7193	7202	7210	7218	7226	7235
53	7243	7251	7259	7267	7275	7284	7292	7300	7308	7316
54	7324	7332	7340	7348	7356	7364	7372	7380	7388	7396
55	7404	7412	7419	7427	7435	7443	7451	7459	7466	7474
56	7482	7490	7497	7505	7513	7520	7528	7536	7543	7551
57	7559	7566	7574	7582	7589	7597	7604	7612	7619	7627
58	7634	7642	7649	7657	7664	7672	7679	7686	7694	7701
59	7709	7716	7723	7731	7738	7745	7752	7760	7767	7774
60	7782	7789	7796	7803	7810	7818	7825	7832	7839	7846
61	7853	7860	7868	7875	7882	7889	7896	7903	7910	7917
62	7924	7931	7938	7945	7952	7959	7966	7973	7980	7987
63	7993	8000	8007	8014	8021	8028	8035	8041	8048	8055
64	8062	8069	8075	8082	8089	8096	8102	8109	8116	8122
65	8129	8136	8142	8149	8156	8162	8169	8176	8182	8189
66	8195	8202	8209	8215	8222	8228	8235	8241	8248	8254
67	8261	8267	8274	8280	8287	8293	8299	8306	8312	8319
68	8325	8331	8338	8344	8351	8357	8363	8370	8376	8382
69	8388	8395	8401	8407	8414	8420	8426	8432	8439	8445
70	8451	8457	8463	8470	8476	8482	8488	8494	8500	8506
71	8513	8519	9525	8531	8537	8543	8549	8555	8561	8567
72	8573	8579	8585	8591	8597	8603	8609	8615	8621	8627
73	8633	8639	8645	8651	8657	8663	8669	8675	8681	8686
74	8692	8698	8704	8710	8716	8722	8727	8733	8739	8745
75	8751	8756	8762	8768	8774	8779	8785	8791	8797	8802
76	8808	8814	8820	8825	8831	8837	8842	8848	8854	8859
77	8865	8871	8876	8882	8887	8893	8899	8904	8910	8915
78	8921	8927	8932	8938	8943	8949	8954	8960	8965	8971
79	8976	8982	8987	8993	8998	9004	9009	9015	9020	9025
80	9031	9036	9042	9047	9053	9058	9063	9069	9074	9079
81	9085	9090	9096	9101	9106	9112	9117	9122	9128	9133
82	9138	9143	9149	9154	9159	9165	9170	9175	9180	9186
83	9191	9196	9201	9206	9212	9217	9222	9227	9232	9238
84	9243	9248	9253	9258	9263	9269	9274	9279	9284	9289
85	9294	9299	9304	9309	9315	9320	9325	9330	9335	9340
86	9345	9350	9355	9360	9365	9370	9375	9380	9385	9390
87	9395	9400	9405	9410	9415	9420	9425	9430	9435	9440
88	9445	9450	9455	9460	9465	9469	9474	9479	9484	9489
89	9494	9499	9504	9509	9513	9518	9523	9528	9533	9538
90	9542	9547	9552	9557	9562	9566	9571	9576	9581	9586
91	9590	9595	9600	9605	9609	9614	9619	9624	9628	9633
92	9638	9643	9647	9652	9657	9661	9666	9671	9675	9680
93	9685	9689	9694	9699	9703	9708	9713	9717	9722	9727
94	9731	9736	9741	9745	9750	9754	9759	9763	9768	9773
95	9777	9782	9786	9791	9795	9800	9805	9809	9814	9818
96	9823	9827	9832	9836	9841	9845	9850	9854	9859	9863
97	9868	9872	9877	9881	9886	9890	9894	9899	9903	9908
98	9912	9917	9921	9926	9930	9934	9939	9943	9948	9952
99	9956	9961	9965	9969	9974	9978	9983	9987	9991	9996

Table 12
Square Roots of Numbers

This table can be used to extract square roots of numbers. Owing to the nature of a square root, it is important to consider the location of the decimal point. The table provides the square root of numbers, $\sqrt{n}$, directly for numbers between 100 and 999. Also, square roots of ten times these numbers, $\sqrt{10n}$, can be obtained directly from the table. Numbers other than these must be converted into the ones in the table so that the proper answer is obtained. The general rule to follow is that if the decimal point is moved an even number of places use the $\sqrt{n}$ column and if it is moved an odd number of places use the $\sqrt{10n}$ column. To obtain the answer, move the decimal place in the result of the table half the number of places in the opposite direction.

Example In order to find the square root of 185, locate 185 in the n column and the square root in the $\sqrt{n}$ column. The answer is $\sqrt{185} = 13.60147$.

Example In order to find the square root of 1850, which is 10 times 185, we locate 185 and obtain the result from the $\sqrt{10n}$ column. The answer is $\sqrt{1850} = 43.01163$.

Example To find the square root of 1.41 we must move the decimal point two places to the right to obtain 141 which appears in the table. Since this represents an even number of places, we use the result in the $\sqrt{n}$ column which is 11.87434. The answer is obtained by moving the decimal point one place to the left and is $\sqrt{1.41} = 1.187434$.

Example In order to find the square root of 225,000 we move the decimal point three places to the left to obtain 225 which appears in the table and locate the result in the $\sqrt{10n}$ column since this represents an odd number of moves. The value in the table is 47.43416. The answer is found by moving the decimal point one place to the right and is $\sqrt{225{,}000} = 474.3416$.

A number requiring a square root that has more than three non-zero digits must be rounded to a value of n appearing in the table. The resulting square root is an approximate answer.

Table 12. Square Roots of Numbers

n	$\sqrt{n}$	$\sqrt{10n}$	n	$\sqrt{n}$	$\sqrt{10n}$	n	$\sqrt{n}$	$\sqrt{10n}$
100	10.00000	31.62278	150	12.24745	38.72983	200	14.14214	44.72136
101	10.04988	31.78050	151	12.28821	38.85872	201	14.17745	44.83302
102	10.09950	31.93744	152	12.32883	38.98718	202	14.21267	44.94441
103	10.14889	32.09361	153	12.36932	39.11521	203	14.24781	45.05552
104	10.19804	32.24903	154	12.40967	39.24283	204	14.28286	45.16636
105	10.24695	32.40370	155	12.44990	39.37004	205	14.31782	45.27693
106	10.29563	32.55764	156	12.49000	39.49684	206	14.35270	45.38722
107	10.34408	32.71085	157	12.52996	39.62323	207	14.38749	45.49725
108	10.39230	32.86335	158	12.56981	39.74921	208	14.42221	45.60702
109	10.44031	33.01515	159	12.60952	39.87480	209	14.45683	45.71652
110	10.48809	33.16625	160	12.64911	40.00000	210	14.49138	45.82576
111	10.53565	33.31666	161	12.68858	40.12481	211	14.52584	45.93474
112	10.58301	33.46640	162	12.72792	40.24922	212	14.56022	46.04346
113	10.63015	33.61547	163	12.76715	40.37326	213	14.59452	46.15192
114	10.67708	33.76389	164	12.80625	40.49691	214	14.62874	46.26013
115	10.72381	33.91165	165	12.84523	40.62019	215	14.66288	46.36809
116	10.77033	34.05877	166	12.88410	40.74310	216	14.69694	46.47580
117	10.81665	34.20526	167	12.92285	40.86563	217	14.73092	46.58326
118	10.86278	34.35113	168	12.96148	40.98780	218	14.76482	46.69047
119	10.90871	34.49638	169	13.00000	41.10961	219	14.79865	46.79744
120	10.95445	34.64102	170	13.03840	41.23106	220	14.83240	46.90416
121	11.00000	34.78505	171	13.07670	41.35215	221	14.86607	47.01064
122	11.04536	34.92850	172	13.11488	41.47288	222	14.89966	47.11688
123	11.09054	35.07136	173	13.15295	41.59327	223	14.93318	47.22288
124	11.13553	35.21363	174	13.19091	41.71331	224	14.96663	47.32864
125	11.18034	35.35534	175	13.22876	41.83300	225	15.00000	47.43416
126	11.22497	35.49648	176	13.26650	41.95235	226	15.03330	47.53946
127	11.26943	35.63706	177	13.30413	42.07137	227	15.06652	47.64452
128	11.31371	35.77709	178	13.34166	42.19005	228	15.09967	47.74935
129	11.35782	35.91657	179	13.37909	42.30839	229	15.13275	47.85394
130	11.40175	36.05551	180	13.41641	42.42641	230	15.16575	47.95832
131	11.44552	36.19392	181	13.45362	42.54409	231	15.19868	48.06246
132	11.48913	36.33180	182	13.49074	42.66146	232	15.23155	48.16638
133	11.53256	36.46917	183	13.52775	42.77850	233	15.26434	48.27007
134	11.57584	36.60601	184	13.56466	42.89522	234	15.29706	48.37355
135	11.61895	36.74235	185	13.60147	43.01163	235	15.32971	48.47680
136	11.66190	36.87818	186	13.63818	43.12772	236	15.36229	48.57983
137	11.70470	37.01351	187	13.67479	43.24350	237	15.39480	48.68265
138	11.74734	37.14835	188	13.71131	43.35897	238	15.42725	48.78524
139	11.78983	37.28270	189	13.74773	43.47413	239	15.45962	48.88763
140	11.83216	37.41657	190	13.78405	43.58899	240	15.49193	48.98979
141	11.87434	37.54997	191	13.82027	43.70355	241	15.52417	49.09175
142	11.91638	37.68289	192	13.85641	43.81780	242	15.55635	49.19350
143	11.95826	37.81534	193	13.89244	43.93177	243	15.58846	49.29503
144	12.00000	37.94733	194	13.92839	44.04543	244	15.62050	49.39636
145	12.04159	38.07887	195	13.96424	44.15880	245	15.65248	49.49747
146	12.08305	38.20995	196	14.00000	44.27189	246	15.68439	49.59839
147	12.12436	38.34058	197	14.03567	44.38468	247	15.71623	49.69909
148	12.16553	38.47077	198	14.07125	44.49719	248	15.74702	49.79960
149	12.20656	38.60052	199	14.10674	44.60942	249	15.77973	49.89990

Table 12. Square Roots of Numbers

n	$\sqrt{n}$	$\sqrt{10n}$	n	$\sqrt{n}$	$\sqrt{10n}$	n	$\sqrt{n}$	$\sqrt{10n}$
250	15.81139	50.00000	300	17.32051	54.77226	350	18.70829	59.16080
251	15.84298	50.09990	301	17.34935	54.86347	351	18.73499	59.24525
252	15.87451	50.19960	302	17.37815	54.95453	352	18.76166	59.32959
253	15.90597	50.29911	303	17.40690	55.04544	353	18.78829	59.41380
254	15.93738	50.39841	304	17.43560	55.13620	354	18.81489	59.49790
255	15.96872	50.49752	305	17.46425	55.22681	355	18.84144	59.58188
256	16.00000	50.59644	306	17.49286	55.31727	356	18.86796	59.66574
257	16.03122	50.69517	307	17.52142	55.40758	357	18.89444	59.74948
258	16.06238	50.79370	308	17.54993	55.49775	358	18.92089	59.83310
259	16.09348	50.89204	309	17.57840	55.58777	359	18.94730	59.91661
260	16.12452	50.99020	310	17.60682	55.67764	360	18.97367	60.00000
261	16.15549	51.08816	311	17.63519	55.76737	361	19.00000	60.08328
262	16.18641	51.18594	312	17.66352	55.85696	362	19.02630	60.16644
263	16.21727	51.28353	313	17.69181	55.94640	363	19.05256	60.24948
264	16.24808	51.38093	314	17.72005	56.03570	364	19.07878	60.33241
265	16.27882	51.47815	315	17.74824	56.12486	365	19.10497	60.41523
266	16.30951	51.57519	316	17.77639	56.21388	366	19.13113	60.49793
267	16.34013	51.67204	317	17.80449	56.30275	367	19.15724	60.58052
268	16.37071	51.76872	318	17.83255	56.39149	368	19.18333	60.66300
269	16.40122	51.86521	319	17.86057	56.48008	369	19.20937	60.74537
270	16.43168	51.96152	320	17.88854	56.56854	370	19.23538	60.82763
271	16.46208	52.05766	321	17.91647	56.65686	371	19.26136	60.90977
272	16.49242	52.15362	322	17.94436	56.74504	372	19.28730	60.99180
273	16.52271	52.24940	323	17.97220	56.83309	373	19.31321	61.07373
274	16.55295	52.34501	324	18.00000	56.92100	374	19.33908	61.15554
275	16.58312	52.44044	325	18.02776	57.00877	375	19.36492	61.23724
276	16.61325	52.53570	326	18.05547	57.09641	376	19.39072	61.31884
277	16.64332	52.63079	327	18.08314	57.18391	377	19.41649	61.40033
278	16.67333	52.72571	328	18.11077	57.27128	378	19.44222	61.48170
279	16.70329	52.82045	329	18.13836	57.35852	379	19.46792	61.56298
280	16.73320	52.91503	330	18.16590	57.44563	380	19.49359	61.64414
281	16.76305	53.00943	331	18.19341	57.53260	381	19.51922	61.72520
282	16.79286	53.10367	332	18.22087	57.61944	382	19.54482	61.80615
283	16.82260	53.19774	333	18.24829	57.70615	383	19.57039	61.88699
284	16.85230	53.29165	334	18.27567	57.79273	384	19.59592	61.96773
285	16.88194	53.38539	335	18.30301	57.87918	385	19.62142	62.04837
286	16.91153	53.47897	336	18.33030	57.96551	386	19.64688	62.12890
287	16.94107	53.57238	337	18.35756	58.05170	387	19.67232	62.20932
288	16.97056	53.66563	338	18.38478	58.13777	388	19.69772	62.28965
289	17.00000	53.75872	339	18.41195	58.22371	389	19.72308	62.36986
290	17.02939	53.85165	340	18.43909	58.30952	390	19.74842	62.44998
291	17.05872	53.94442	341	18.46619	58.39521	391	19.77372	62.52999
292	17.08801	54.03702	342	18.49324	58.48077	392	19.79899	62.60990
293	17.11724	54.12947	343	18.52026	58.56620	393	19.82423	62.68971
294	17.14643	54.22177	344	18.54724	58.65151	394	19.84943	62.76942
295	17.17556	54.31390	345	18.57418	58.73670	395	19.87461	62.84903
296	17.20465	54.40588	346	18.60108	58.82176	396	19.89975	62.92853
297	17.23369	54.49771	347	18.62794	58.90671	397	19.92486	63.00794
298	17.26268	54.58938	348	18.65476	58.99152	398	19.94994	63.08724
299	17.29162	54.68089	349	18.68154	59.07622	399	19.97498	63.16645

Table 12. Square Roots of Numbers

n	$\sqrt{n}$	$\sqrt{10n}$	n	$\sqrt{n}$	$\sqrt{10n}$	n	$\sqrt{n}$	$\sqrt{10n}$
400	20.00000	63.24555	450	21.21320	67.08204	500	22.36068	70.71068
401	20.02498	63.32456	451	21.23676	67.15653	501	22.38303	70.78135
402	20.04994	63.40347	452	21.26029	67.23095	502	22.40536	70.85196
403	20.07486	63.48228	453	21.28380	67.30527	503	22.42766	70.92249
404	20.09975	63.56099	454	21.30728	67.37952	504	22.44994	70.99296
405	20.12461	63.63961	455	21.33073	67.45369	505	22.47221	71.06335
406	20.14944	63.71813	456	21.35416	67.52777	506	22.49444	71.13368
407	20.17424	63.79655	457	21.37756	67.60178	507	22.51666	71.20393
408	20.19901	63.87488	458	21.40093	67.67570	508	22.53886	71.27412
409	20.22375	63.95311	459	21.42429	67.74954	509	22.56103	71.34424
410	20.24846	64.03124	460	21.44761	67.82330	510	22.58318	71.41428
411	20.27313	64.10928	461	21.47091	67.89698	511	22.60531	71.48426
412	20.29778	64.18723	462	21.49419	67.97058	512	22.62742	71.55418
413	20.32240	64.26508	463	21.51743	68.04410	513	22.64950	71.62402
414	20.34699	64.34283	464	21.54066	68.11755	514	22.67157	71.69379
415	20.37155	64.42049	465	21.56386	68.19091	515	22.69361	71.76350
416	20.39608	64.49806	466	21.58703	68.26419	516	22.71563	71.83314
417	20.42058	64.57554	467	21.61018	68.33740	517	22.73763	71.90271
418	20.44505	64.65292	468	21.63331	68.41053	518	22.75961	71.97222
419	20.46949	64.73021	469	21.65641	68.48357	519	22.78157	72.04165
420	20.49390	64.80741	470	21.67948	68.55655	520	22.80351	72.11103
421	20.51828	64.88451	471	21.70253	68.62944	521	22.82542	72.18033
422	20.54264	64.96153	472	21.72556	68.70226	522	22.84732	72.24957
423	20.56696	65.03845	473	21.74856	68.77500	523	22.86919	72.31874
424	20.59126	65.11528	474	21.77154	68.84766	524	22.89105	72.38784
425	20.61553	65.19202	475	21.79449	68.92024	525	22.91288	72.45688
426	20.63977	65.26868	476	21.81742	68.99275	526	22.93469	72.52586
427	20.66398	65.34524	477	21.84033	69.06519	527	22.95648	72.59477
428	20.68816	65.42171	478	21.86321	69.13754	528	22.97825	72.66361
429	20.71232	65.49809	479	21.88607	69.20983	529	23.00000	72.73239
430	20.73644	65.57439	480	21.90890	69.28203	530	23.02173	72.80110
431	20.76054	65.65059	481	21.93171	69.35416	531	23.04344	72.86975
432	20.78461	65.72671	482	21.95450	69.42622	532	23.06513	72.93833
433	20.80865	65.80274	483	21.97726	69.49820	533	23.08679	73.00685
434	20.83267	65.87868	484	22.00000	69.57011	534	23.10844	73.07530
435	20.85665	65.95453	485	22.02272	69.64194	535	23.13007	73.14369
436	20.88061	66.03030	486	22.04541	69.71370	536	23.15167	73.21202
437	20.90454	66.10598	487	22.06808	69.78539	537	23.17326	73.28028
438	20.92845	66.18157	488	22.09072	69.85700	538	23.19483	73.34848
439	20.95233	66.25708	489	22.11334	69.92853	539	23.21637	73.41662
440	20.97618	66.33250	490	22.13594	70.00000	540	23.23790	73.48469
441	21.00000	66.40783	491	22.15852	70.07139	541	23.25941	73.55270
442	21.02380	66.48308	492	22.18107	70.14271	542	23.28089	73.62065
443	21.04757	66.55825	493	22.20360	70.21396	543	23.30236	73.68853
444	21.07131	66.63332	494	22.22611	70.28513	544	23.32381	73.75636
445	21.09502	66.70832	495	22.24860	70.35624	545	23.34524	73.82412
446	21.11871	66.78323	496	22.27106	70.42727	546	23.36664	73.89181
447	21.14237	66.85806	497	22.29350	70.49823	547	23.38803	73.95945
448	21.16601	66.93280	498	22.31591	70.56912	548	23.40940	74.02702
449	21.18962	67.00746	499	22.33831	70.63993	549	23.43075	74.09453

Table 12. Square Roots of Numbers

n	$\sqrt{n}$	$\sqrt{10n}$	n	$\sqrt{n}$	$\sqrt{10n}$	n	$\sqrt{n}$	$\sqrt{10n}$
550	23.45208	74.16198	600	24.49490	77.45967	650	25.49510	80.62258
551	23.47339	74.22937	601	24.51530	77.52419	651	25.51470	80.68457
552	23.49468	74.29670	602	24.53569	77.58866	652	25.53429	80.74652
553	23.51595	74.36397	603	24.55606	77.65307	653	25.55386	80.80842
554	23.53720	74.43118	604	24.57641	77.71744	654	25.57342	80.87027
555	23.55844	74.49832	605	24.59675	77.78175	655	25.59297	80.93207
556	23.57965	74.56541	606	24.61707	77.84600	656	25.61250	80.99383
557	23.60085	74.63243	607	24.63737	77.91020	657	25.63201	81.05554
558	23.62202	74.69940	608	24.65766	77.97435	658	25.65151	81.11720
559	23.64318	74.76630	609	24.67793	78.03845	659	25.67100	81.17881
560	23.66432	74.83315	610	24.69818	78.10250	660	25.69047	81.24038
561	23.68544	74.89993	611	24.71841	78.16649	661	25.70992	81.30191
562	23.70654	74.96666	612	24.73863	78.23043	662	25.72936	81.36338
563	23.72762	75.03333	613	24.75884	78.29432	663	25.74879	81.42481
564	23.74868	75.09993	614	24.77902	78.35815	664	25.76820	81.48620
565	23.76973	75.16648	615	24.79919	78.42194	665	25.78759	81.54753
566	23.79075	75.23297	616	24.81935	78.48567	666	25.80698	81.60882
567	23.81176	75.29940	617	24.83948	78.54935	667	25.82634	81.67007
568	23.83275	75.36577	618	24.85961	78.61298	668	25.84570	81.73127
569	23.85372	75.43209	619	24.87971	78.67655	669	25.86503	81.79242
570	23.87467	75.49834	620	24.89980	78.74008	670	25.88436	81.85353
571	23.89561	75.56454	621	24.91987	78.80355	671	25.90367	81.91459
572	23.91652	75.63068	622	24.93993	78.86698	672	25.92296	81.97561
573	23.93742	75.69676	623	24.95997	78.93035	673	25.94224	82.03658
574	23.95830	75.76279	624	24.97999	78.99367	674	25.96151	82.09750
575	23.97916	75.82875	625	25.00000	79.05694	675	25.98076	82.15838
576	24.00000	75.89466	626	25.01999	79.12016	676	26.00000	82.21922
577	24.02082	75.96052	627	25.03997	79.18333	677	26.01922	82.28001
578	24.04163	76.02631	628	25.05993	79.24645	678	26.03843	82.34076
579	24.06242	76.09205	629	25.07987	79.30952	679	26.05763	82.40146
580	24.08319	76.15773	630	25.09980	79.37254	680	26.07681	82.46211
581	24.10394	76.22336	631	25.11971	79.43551	681	26.09598	82.52272
582	24.12468	76.28892	632	25.13961	79.49843	682	26.11513	82.58329
583	24.14539	76.35444	633	25.15949	79.56130	683	26.13427	82.64381
584	24.16609	76.41989	634	25.17936	79.62412	684	26.15339	82.70429
585	24.18677	76.48529	635	25.19921	79.68689	685	26.17250	82.76473
586	24.20744	76.55064	636	25.21904	79.74961	686	26.19160	82.82512
587	24.22808	76.61593	637	25.23886	79.81228	687	26.21068	82.88546
588	24.24871	76.68116	638	25.25866	79.87490	688	26.22975	82.94577
589	24.26932	76.74634	639	25.27845	79.93748	689	26.24881	83.00602
590	24.28992	76.81146	640	25.29822	80.00000	690	26.26785	83.06624
591	24.31049	76.87652	641	25.31798	80.06248	691	26.28688	83.12641
592	24.33105	76.94154	642	25.33772	80.12490	692	26.30589	83.18654
593	24.35159	77.00649	643	25.35744	80.18728	693	26.32489	83.24662
594	24.37212	77.07140	644	25.37716	80.24961	694	26.34388	83.30666
595	24.39262	77.13624	645	25.39685	80.31189	695	26.36285	83.36666
596	24.41311	77.20104	646	25.41653	80.37413	696	26.38181	83.42661
597	24.43358	77.26578	647	25.43619	80.43631	697	26.40076	83.48653
598	24.45404	77.33046	648	25.45584	80.49845	698	26.41969	83.54639
599	24.47448	77.39509	649	25.47548	80.56054	699	26.43861	83.60622

Table 12. Square Roots of Numbers

n	$\sqrt{n}$	$\sqrt{10n}$	n	$\sqrt{n}$	$\sqrt{10n}$	n	$\sqrt{n}$	$\sqrt{10n}$
700	26.45751	83.66600	750	27.38613	86.60254	800	28.28427	89.44272
701	26.47640	83.72574	751	27.40438	86.66026	801	28.30194	89.49860
702	26.49528	83.78544	752	27.42262	86.71793	802	28.31960	89.55445
703	26.51415	83.84510	753	27.44085	86.77557	803	28.33725	89.61027
704	26.53300	83.90471	754	27.45906	86.83317	804	28.35489	89.66605
705	26.55184	83.96428	755	27.47726	86.89074	805	28.37252	89.72179
706	26.57066	84.02381	756	27.49545	86.94826	806	28.39014	89.77750
707	26.58947	84.08329	757	27.51363	87.00575	807	28.40775	89.83318
708	26.60827	84.14274	758	27.53180	87.06320	808	28.42534	89.88882
709	26.62705	84.20214	759	27.54995	87.12061	809	28.44293	89.94443
710	26.64583	84.26150	760	27.56810	87.17798	810	28.46050	90.00000
711	26.66458	84.32082	761	27.58623	87.23531	811	28.47806	90.05554
712	26.68333	84.38009	762	27.60435	87.29261	812	28.49561	90.11104
713	26.70206	84.43933	763	27.62245	87.34987	813	28.51315	90.16651
714	26.72078	84.49852	764	27.64055	87.40709	814	28.53069	90.22195
715	26.73948	84.55767	765	27.65863	87.46428	815	28.54820	90.27735
716	26.75818	84.61678	766	27.67671	87.52143	816	28.56571	90.33272
717	26.77686	84.67585	767	27.69476	87.57854	817	28.58321	90.38805
718	26.79552	84.73488	768	27.71281	87.63561	818	28.60070	90.44335
719	26.81418	84.79387	769	27.73085	87.69265	819	28.61818	90.49862
720	26.83282	84.85281	770	27.74887	87.74964	820	28.63564	90.55385
721	26.85144	84.91172	771	27.76689	87.80661	821	28.65310	90.60905
722	26.87006	84.97058	772	27.78489	87.86353	822	28.67054	90.66422
723	26.88866	85.02941	773	27.80288	87.92042	823	28.68798	90.71935
724	26.90725	85.08819	774	27.82086	87.97727	824	28.70540	90.77445
725	26.92582	85.14693	775	27.83882	88.03408	825	28.72281	90.82951
726	26.94439	85.20563	776	27.85678	88.09086	826	28.74022	90.88454
727	26.96294	85.62429	777	27.87472	88.14760	827	28.75761	90.93954
728	26.98148	85.32292	778	27.89265	88.20431	828	28.77499	90.99451
729	27.00000	85.38150	779	27.91057	88.26098	829	28.72936	91.04944
730	27.01851	85.44004	780	27.92848	88.31761	830	28.80972	91.10434
731	27.03701	85.49854	781	27.94638	88.37420	831	28.82707	91.15920
732	27.05550	85.55700	782	27.96426	88.43076	832	28.84441	91.21403
733	27.07397	85.61542	783	27.98214	88.48729	833	28.86174	91.26883
734	27.09243	85.67380	784	28.00000	88.54377	834	28.87906	91.32360
735	27.11088	85.73214	785	28.01785	88.60023	835	28.89637	91.37833
736	27.12932	85.79044	786	28.03569	88.65664	836	28.91366	91.43304
737	27.14774	85.84870	787	28.05352	88.71302	837	28.93095	91.48770
738	27.16616	85.90693	788	28.07134	88.76936	838	28.94823	91.54234
739	27.18455	85.96511	789	28.08914	88.82567	839	28.96550	91.59694
740	27.20294	86.02325	790	28.10694	88.88194	840	28.98275	91.65151
741	27.22132	86.08136	791	28.12472	88.93818	841	29.00000	91.70605
742	27.23968	86.13942	792	28.14249	88.99438	842	29.01724	91.76056
743	27.25803	86.19745	793	28.16026	89.05055	843	29.03446	91.81503
744	27.27636	86.25543	794	28.17801	89.10668	844	29.05168	91.86947
745	27.29469	86.31338	795	28.19574	89.16277	845	29.06888	91.92388
746	27.31300	86.37129	796	28.21347	89.21883	846	29.08608	91.97826
747	27.33130	86.42916	797	28.23119	89.27486	847	29.10326	92.03260
748	27.34959	86.48699	798	28.24889	89.33085	848	29.12044	92.08692
749	27.36786	86.54479	799	28.26659	89.38680	849	29.13760	92.14120

Table 12. Square Roots of Numbers

n	$\sqrt{n}$	$\sqrt{10n}$	n	$\sqrt{n}$	$\sqrt{10n}$	n	$\sqrt{n}$	$\sqrt{10n}$
850	29.15476	92.19544	900	30.00000	94.86833	950	30.82207	97.46794
851	29.17190	92.24966	901	30.01666	94.92102	951	30.83829	97.51923
852	29.18904	92.30385	902	30.03331	94.97368	952	30.85450	97.57049
853	29.20616	92.35800	903	30.04996	95.02631	953	30.87070	97.62172
854	29.22328	92.41212	904	30.06659	95.07891	954	30.88689	97.67292
855	29.24038	92.46621	905	30.08322	95.13149	955	30.90307	97.72410
856	29.25748	92.52027	906	30.09983	95.18403	956	30.91925	97.77525
857	29.27456	92.57429	907	30.11644	95.23655	957	30.93542	97.82638
858	29.29164	92.62829	908	30.13304	95.28903	958	30.95158	97.87747
859	29.30870	92.68225	909	30.14963	95.34149	959	30.96773	97.92855
860	29.32576	92.73618	910	30.16621	95.39392	960	30.98387	97.97959
861	29.34280	92.79009	911	30.18278	95.44632	961	31.00000	98.03061
862	29.35984	92.84396	912	30.19934	95.49869	962	31.01612	98.08160
863	29.37686	92.89779	913	30.21589	95.55103	963	31.03224	98.13256
864	29.39388	92.95160	914	30.23243	95.60335	964	31.04835	98.18350
865	29.41088	93.00538	915	30.24897	95.65563	965	31.06445	98.23441
866	29.42788	93.05912	916	30.26549	95.70789	966	31.08054	98.28530
867	29.44486	93.11283	917	30.28201	95.76012	967	31.09662	98.33616
868	29.46184	93.16652	918	30.29851	95.81232	968	31.11270	98.38699
869	29.47881	93.22017	919	30.31501	95.86449	969	31.12876	98.43780
870	29.49576	93.27379	920	30.33150	95.91663	970	31.14482	98.48858
871	29.51271	93.32738	921	30.34798	95.96874	971	31.16087	98.53933
872	29.52965	93.38094	922	30.36445	96.02083	972	31.17691	98.59006
873	29.54657	93.43447	923	30.38092	96.07289	973	31.19295	98.64076
874	29.56349	93.48797	924	30.39737	96.12492	974	31.20897	98.69144
875	29.58040	93.54143	925	30.41381	96.17692	975	31.22499	98.74209
876	29.59730	93.59487	926	30.43025	96.22889	976	31.24100	98.79271
877	29.61419	93.64828	927	30.44667	96.28084	977	31.25700	98.84331
878	29.63106	93.70165	928	30.46309	96.33276	978	31.27299	98.89388
879	29.64793	93.75500	929	30.47950	96.38465	979	31.28898	98.94443
880	29.66479	93.80832	930	30.49590	96.43651	980	31.30495	98.99495
881	29.68164	93.86160	931	30.51229	96.48834	981	31.32092	99.04544
882	29.69848	93.91486	932	30.52868	96.54015	982	31.33688	99.09591
883	29.71532	93.96808	933	30.54505	96.59193	983	31.35283	99.14636
884	29.73214	94.02127	934	30.56141	96.64368	984	31.36877	99.19677
885	29.74895	94.07444	935	30.57777	96.69540	985	31.38471	99.24717
886	29.76575	94.12757	936	30.59412	96.74709	986	31.40064	99.29753
887	29.78255	94.18068	937	30.61046	96.79876	987	31.41656	99.34787
888	29.79933	94.23375	938	30.62679	06.85040	988	31.43247	99.39819
889	29.81610	94.28680	939	30.64311	96.90201	989	31.44837	99.44848
890	29.83287	94.33981	940	30.65942	96.95360	990	31.46427	99.49874
891	29.84962	94.39280	941	30.67572	97.00515	991	31.48015	99.54898
892	29.86637	94.44575	942	30.69202	97.05668	992	31.49603	99.59920
893	29.88311	94.49868	943	30.70831	97.10819	993	31.51190	99.64939
894	29.89983	94.55157	944	30.72458	97.15966	994	31.52777	99.69955
895	29.91655	94.60444	945	30.74085	97.21111	995	31.54362	99.74969
896	29.93326	94.65728	946	30.75711	97.26253	996	31.55947	99.79980
897	29.94996	94.71008	947	30.77337	97.31393	997	31.57531	99.84989
898	29.96665	94.76286	948	30.78961	97.36529	998	31.59114	99.89995
899	29.98333	94.81561	949	30.80584	97.41663	999	31.60696	99.94999

SELECTED SOLUTIONS AND ANSWERS TO CHAPTER EXERCISES

Chapter 1

6.

x	f
0	6
1	10
2	8
3	4
4	2
	30

9.

Exam Grade	Number of Students
40 but less than 50	1
50 but less than 60	7
60 but less than 70	4
70 but less than 80	8
80 but less than 90	3
90 but less than 100	2
Total	25

13.

Age in Years	*Relative Frequency*	*Cumulative Frequency*
14 & Under	.04	20
15–19	.25	145
20–24	.05	170
25–29	.07	205
30–34	.16	285
35–39	.28	425
40–44	.06	455
45 & Over	.09	500
	1.00	

15. The basic property of data is variability. The general pattern of variability follows a recognizable pattern that is described in many cases by a frequency distribution.

17. Different bodies of data can be described by frequency distributions with different shapes characterized by the different properties of a distribution. The different shapes that result correspond to the way the data are distributed.

21. The class interval size used in Problem 14 is the same for all classes and equals the difference between the upper and lower limit of any class, or 0.2. If the raw data were given to one decimal place, the suggested change could not have an effect on the frequencies. However, if the raw data are given to two decimal places and are not rounded, the change in the limits would alter the frequencies if any of the raw observations actually equalled any of the values of the original upper class limits in Problem 14. Strictly speaking, the suggested change in limits would change the class interval size to 0.19; if the original data are reported to one decimal place it would not really mátter. Generally, the choice of class limits should reflect the precision or decimal place accuracy of the raw data.

22. Rarely will a frequency distribution be perfectly symmetric. Symmetry is definitely not a requirement when constructing a frequency distribution, since the distribution should reflect the underlying shape or behavior of the data.

27. The point of this question is that if there are additional considerations beside describing the overall pattern of variability using a frequency distribution, these considerations should be taken into account when constructing a grouped distribution. In this particular case, if it is desired to report the number of A's, B's, etc., then the numerical equivalents should be used as the class limits.

28. Data are non-homogeneous when they can be subdivided into two or more groups on the basis of some characteristic other than the one measured which yield different results than when the data are considered as one group. For example, the life expectancy of an entire population would be different from the life expectancy of persons within different age groups. Hence, we can say that life expectancy is not homogeneous with respect to age.

29. If a student body is made up of men and women, the group would not be homogeneous with respect to height but would be homogeneous with respect to intelligence.

Chapter 2

4. Mean, $\overline{X} = \frac{\Sigma X}{N} = \frac{148}{20} = 7.4$

   Median, $X_{50} = 7$ since the middle two values in the array are equal to 7. The data in the form of a frequency distribution is given as

X	f
3	5
7	9
9	3
12	2
19	1
	20

   Therefore, the mode equals 7 with the highest frequency of 9.

5. $\overline{X} = 476.9$

   $X_{50} = 400$

   The mode does not exist.

7. The crossproduct terms and cumulative frequencies used to find the mean and median are given as

X Number of Parcels	f Number of Trucks	fX	F
20	10	200	10
25	12	300	22
35	23	805	45
40	9	360	54
55	7	385	61
70	3	210	64
	64	2260	

Mean, $\bar{X} = \frac{\Sigma fX}{N} = \frac{2260}{64} = 35.3$ parcels

The median corresponds to the value which is half the total frequency, N/2 = 64/2 = 32, and equals 35 (i.e., $X_{50} = 35$ parcels). This corresponds to the cumulative frequency of 45 which first exceeds 32.

The mode equals 35 parcels since this corresponds to the highest frequency of 23.

8. Preliminary calculations for the mean and cumulative frequencies for the median are given in the following table. Since the class limits are integers, in this particular case it is satisfactory to use midpoints that are rounded and a class interval size of 100 instead of 99 when calculating the mean.

X	f	m	fm	d	fd	F
100–199	3	150	450	−3	−9	3
200–299	5	250	1250	−2	−10	8
300–399	8	350	2800	−1	−8	16
400–499	15	450	6750	0	0	31
500–599	7	550	3850	1	7	38
600–699	2	650	1300	2	4	40
	40		16400		−16	

Mean (definitional formula), $\bar{X} = \frac{\Sigma fm}{N} = \frac{16{,}400}{40} = 410$

If midpoints equal to 149.5, 249.5, etc., are used, the answer is 409.5.

Mean (shortcut formula), $\bar{X} = m_a + \frac{\Sigma fd}{N} c$

$$= 450 + \frac{-16}{40} \times 100$$

$$= 450 - 0.4 \times 100$$

$$= 450 - 40 = 410$$

If 449.5 is used for m_a and the class interval is set equal to 99, the result is 409.9.

Median, $X_{50} = L_{X_{50}} + \frac{\frac{N}{2} - F_{precede}}{f_{X_{50}}} \times c_{X_{50}}$

$$= 400 + \frac{\frac{40}{2} - 16}{15} \times 100$$

$$= 400 + \frac{4}{15} \times 100$$

$$= 400 + .27 \times 100$$

$$= 400 + 27 = 427$$

The mode can be assumed to equal 449.5 or 450, which corresponds to the midpoint of the class with the highest frequency, 15.

10. $\overline{X} = 5.65$

$X_{50} = 5.0$

Mode = 7.0

11. Since the last class is open-ended, the mean of the last class should be used instead of the midpoint to compute the mean of the distribution.

$\overline{X} = 43.06$

$X_{50} = 39.3$

Mode = 35

12. 1095 million barrels

13. Combined mean = 78.8

15. Typically, averages assume a value that lies somewhere in the vicinity of the center of a distribution corresponding to a set of data. In this sense, they are representative of the entire set of data. The extent to which an average is representative varies and depends upon the average used and the amount of variability and skewness.

20. The concept of a percentile is simple to understand but tricky to determine, owing to the fact that raw data are discrete. This leads to some inconsistencies when finding percentiles. Since there are an even number of observations equal to 20 in Problem 4, the median is found as the value midway between the 10th and 11th observations when arrayed in order of increasing magnitude. The value obtained corresponds to the observation such that 50 percent or half the observations lie on either side. If we consider the data of Problem 4 in the form of an array,

3 3 3 3 3 ↑ 7 7 7 7 7 7 7 7 7 9 ↑ 9 9 12 12 19

the 25th percentile would be the value such that 25 percent or 5 of the 20 observations fall below it, and the 75th percentile would be the value such that 75 percent fall below it. Using the same reasoning to find the median, the 25th percentile can be found as the value midway between the last 3 and the first 7, or the mean of these values. Hence, $X_{25} = \frac{3 + 7}{2} = 5$. In the same way, the 75th percentile lies between the "first 9 and second 9." Since these values are the same, the 75th percentile is 9, or $X_{75} = 9$.

22. $X_{80} = 40$

23. It is necessary to interpolate the median for a grouped distribution because the identity of the raw observations is lost in grouping.

24. $X_{90} = 88.3$

26. (b) The height of the bridge should be determined to accommodate the highest ship expected to pass under it.
(d) Since there is variability in manufactured products, an average is necessary. In this case the average number of miles or tire life would be used. The amount of variation expected from tire to tire for a particular brand would be useful also. The average used would depend on the pattern of variability. If different brands exhibit different patterns of variability, percentiles in addition to the averages may be helpful.
(g) The arithmetic mean is the only average from which a total can be obtained.

Chapter 3

4. Preliminary calculations for the mean and the definitional and shortcut formulas for the variance are given as follows:

X	$X - \overline{X}$	$(X - \overline{X})^2$	X^2
3	−1	1	9
1	−3	9	1
7	3	9	49
2	−2	4	4
3	−1	1	9
12	8	64	144
4	0	0	16
8	4	16	64
1	−3	9	1
5	1	1	25
2	−2	4	4
4	0	0	16
1	−3	9	1
6	2	4	36
4	0	0	16
1	−3	9	1
5	1	1	25
3	−1	1	9
7	3	9	49
1	−3	9	1
80		160	480

$$\text{Mean, } \overline{X} = \frac{\Sigma X}{N} = \frac{80}{20} = 4$$

$$\text{Variance (definitional formula), } S^2 = \frac{\Sigma(X - \overline{X})^2}{N}$$

$$= \frac{160}{20} = 8$$

$$\text{Variance (shortcut formula), } S^2 = \frac{\Sigma X^2}{N} - \overline{X}^2$$

$$= \frac{480}{20} - (4)^2$$

$$= 24 - 16 = 8$$

$$\text{Standard Deviation, } S = \sqrt{S^2} = \sqrt{8} = 2.83$$

5. Preliminary calculations for the mean and the definitional and shortcut formulas for the variance are given as follows:

X	f	fX	$X - \overline{X}$	$(X - \overline{X})^2$	$f(X - \overline{X})^2$	X^2	fX^2
2.0	6	12	−2.6	6.76	40.56	4	24
3.0	10	30	−1.6	2.56	25.60	9	90
5.0	4	20	.4	.16	.64	25	100
9.0	3	27	4.4	19.36	58.08	81	243
13.0	2	26	8.4	70.56	141.12	169	338
	25	115			266.00		795

$$\text{Mean, } \overline{X} = \frac{\Sigma fX}{\Sigma f} = \frac{115}{25} = 4.6$$

$$\text{Variance (definitional formula), } S^2 = \frac{\Sigma f(X - \overline{X})^2}{\Sigma f}$$

$$= \frac{266.00}{25} = 10.64$$

$$\text{Variance (shortcut), } S^2 = \frac{\Sigma fX^2}{\Sigma f} - \overline{X}^2$$

$$= \frac{795}{25} - (4.6)^2$$

$$= 31.8 - 21.16 = 10.64$$

$$\text{Standard Deviation, } S = \sqrt{S^2} = \sqrt{10.64} = 3.26$$

7. $\overline{X} = 26{,}533$
   $S^2 = 214{,}315{,}556$
   $S = 14{,}640$

8. $V^2 = 0.5$
   $V = 0.71$

10. R = 17
    MD = 4.8

13. An average is a measure of the location or point about which a set of observations tend to cluster or concentrate but does not measure the extent to which the observations differ from one another or concentrate about the average. This is accomplished with a measure of variability. A measure of variability does not provide information about the shape of the underlying distribution. It measures the amount of variability, not the pattern of variability.

16. Set 1: $\overline{X} = 9$; $S^2 = 2$

    Set 2: $\overline{X} = 9$; $S^2 = 18$

    A: $\overline{X} = 7$; $S^2 = 7.6$

    B: $\overline{X} = 7$; $S^2 = 14$

18. This question could have been answered based upon the material in Chapter 2. The main point of the question as it relates to this chapter is that variability virtually always exists and that departures from an average by themselves do not indicate unusual behavior. The shape of the underlying distribution and pattern of variability, the amount of variability, and a definition of superiority must be considered before an above average result is considered to be unusual.

20. Since the means for the two funds are identical (equal to 16.4), the variance alone can be used to make the comparison. Variability in Fund 2 is greater than the variability in Fund 1 based upon the variances (i.e., $S_2^2 = 77.59$ vs. $S_1^2 = 4.63$).

22. Since grade point average and SAT score are measured on different scales with greatly differing means, it is more meaningful to compare coefficients of variation rather than standard deviations. The coefficient of variation for grade point averages equals 0.21 and 0.12 for SAT scores. Based upon these results grade point averages exhibit more variability.

Chapter 4

13. (a) $\frac{1}{13}$ or 0.08

    (b) $\frac{1}{52}$ or 0.02

    (c) $\frac{2}{13}$ or 0.15

    (d) $\frac{1}{26}$ or 0.04

    (e) $\frac{4}{13}$ or 0.31

16. 1/8 or 0.125

19. (a) 1/5 or 0.2
    (b) 9/35 or 0.26
    (c) 27/35 or 0.77
    (d) 3/5 or 0.6
    (e) 5/7 or 0.71

22. (a) 2/5 or 0.40
    (b) 7/10 or 0.7
    (c) 1/10 or 0.1
    (d) 5/7 or 0.71

24. (a) P(Disease) $= \frac{250}{10000} = \frac{1}{40}$ or 0.025

    (b) P(No Symptoms and Disease) $= \frac{50}{10000} = \frac{1}{200}$ or 0.005

    (c) P(Disease given No Symptoms) $= \frac{50}{9700} = \frac{1}{194}$ or 0.0052

    (d) P(Disease given Symptoms) $= \frac{200}{300} = \frac{2}{3}$ or 0.67

28. The probability of 0.17 is interpreted to mean that 17 percent of a large number of rolls of a die will result in a six. The probability of 0.25 is interpreted to mean that 25 percent of a large number of rolls will result in a six. The interpretations follow directly from the definition of the probability of an event no matter what the numerical values. The point to be made by this question is that the probability of 1/6 or 0.17 is a theoretical probability based upon the assumption that each possible outcome is equally likely. It is possible that the assumption does not apply to a particular die and that the probability of a six could equal 0.25 or any other number. This value is not determinable theoretically but must be estimated empirically in terms of an observed relative frequency based upon repeated rolls of a particular die.

33. There is not sufficient information to reach the conclusion made.

36. (a) By shuffling a deck of cards, hands received by each player are generated at random. Consequently, every player has the same chance of receiving each of the possible hands and has no control over the hand actually received. Such a method of assigning cards is considered impartial or fair. If the game is played long enough all hands will come to all players.

38. (a) 3 or 3:1; 1/3 or 1:3
    (b) 4; P(W) = 4/5 or 0.8; P(L) = 0.2

40. (a) Marginal, P(D)
    (b) Joint, P(NS & D)
    (c) Conditional, P(D|NS)
    (d) Conditional, P(D|S)

42. (a) P(Acceptable rating) $= \frac{345}{525} = \frac{23}{35}$ or 0.66

This tells us that 66 percent of the salesmen received an acceptable performance rating at the end of one year of service.

(b) P(Medium and Rated acceptable) $= \frac{160}{525} = \frac{32}{105}$ or 0.30

Thirty percent of the salesmen received a medium test score and were rated as acceptable after one year of service.

(c) P(Rated acceptable given Medium score) $= \frac{160}{250} = \frac{16}{25}$ or 0.64

Of the salesmen who received a medium test score, 64 percent were rated as acceptable after one year of service.

44.

	Conditional Probability		**Marginal**
	Symptoms	*No Symptoms*	**Probability**
Disease	0.667	0.005	0.025
No Disease	0.333	0.995	0.975
Total	1.000	1.000	1.000

Since the conditional probabilities in the symptoms and no symptoms columns are not equal (and not equal to the values in the marginal probability column), symptoms and disease are dependent; different information about the probability of disease exists depending upon the presence or absence of symptoms. Given the marginal totals, if the two characteristics were independent the original table would have the following joint frequencies.

NUMBER OF PEOPLE ASSUMING INDEPENDENCE

	Symptoms	*No Symptoms*	*Total*
Disease	7.5	242.5	250
No Disease	292.5	9457.5	9750
Total	300	9700	10000

The frequencies in the table are not rounded to integer values since all would be increased such that they would not sum to the corresponding totals. It is impossible to observe results that indicate "exact" independence with the given totals. The joint frequencies are obtained by multiplying the column totals by the marginal probabilities in the previous table (e.g., 7.5 = .025 × 300). The joint frequencies assuming independence are obtained in such a way that the resulting conditional probabilities between columns are equal and equal to the marginal probabilities.

Chapter 5

11. (a) P(A or B) = 0.32
    (b) P(A|B) = 1/5 or 0.2
    (c) P(B|A) = 0.15
    P(A and B) = 0.03

13. (a) P(A and B) = 0.1
(b) $P(A|B) = 1/3$ or 0.33

15. (a) 0.4
(b) 0.508

16. (a) 0.024
(b) 0.116
(c) 0.3

20. 0.24

22. 0.16

24. 0.77

25. Let D = Democrat wins, R = Republican wins, r = survey indicates Republican will win.

(a)
$$P(D|r) = \frac{P(D)P(r|D)}{P(D)P(r|D) + P(R)P(r|R)}$$
$$= \frac{(.55)(.25)}{(.55)(.25) + (1 - .55)(.75)}$$
$$= \frac{.1375}{.1375 + .45(.75)}$$
$$= \frac{.1375}{.1375 + .3375} = \frac{.1375}{.4750} = 0.29$$

(b)
$$P(R|r) = \frac{P(R)P(r|R)}{P(R)P(r|R) + P(D)P(r|D)}$$
$$= \frac{(.45)(.75)}{(.45)(.75) + (.55)(.25)}$$
$$= 1 - P(D|r)$$
$$= 1 - .29 = 0.71$$

26. 0.28

29. The answer depends upon the nature of the stocks. Stock prices fluctuate for many reasons. If two stocks are affected by the same factors, then the independence assumption should be questioned. There are, however, stocks that are not related and could be assumed to be independent. If two stocks are not independent, it is necessary to know the conditional probability that one will increase given an increase in the price of the other.

32. (a) 8/125 or .064
(b) 1/30 or 0.03

33. 0.189

34. (a) 0.99
(b) 0.97

35. (a) 1/8 or 0.125
(b) 1/4 or 0.25

36. 0.096; 0.769; 0.135

37. (a) 0.727
(c) 0.0674
(d) 6740
(e) 4900

Chapter 6

5. (a) This is a probability distribution since each of the f(x)-values are non-negative numbers between zero and one such that their sum equals one (i.e., .10 + .40 + .30 + .15 + .05 = 1.00)

(b) $P(x = 8.0) = f(8) = .30$

(c) $P(x < 8.0) = P(x = 6.0) + P(x = 7.0)$
$= f(6.0) + f(7.0) = .10 + .40 = .50$

(d) $P(\text{x does not exceed } 8.0) = P(x \leq 8.0)$
$= P(x = 6.0) + P(x = 7.0) + P(x = 8.0)$
$= f(6.0) + f(7.0) + f(8.0)$
$= .10 + .40 + .30 = .80$

$P(\text{x exceeds } 8.0) = P(x > 8.0)$
$= P(x = 9.0) + P(x = 10.0)$
$= .15 + .05 = .20 = 1 - P(x \leq 8.0)$
$= 1 - .80$

7. (a) $P(\text{2 heads}) = P(x = 2) = .375$

(b) $P(\text{at least 1 head}) = P(\text{1 or more heads})$
$= P(x = 1) + P(x = 2) + P(x = 3)$
$= .375 + .375 + .125 = 0.875$

9. (a) 0.03
(b)

x	F(x)
1	1/30 or .03
2	5/30 or .17
3	14/30 or .47
4	30/30 or 1.00

(c) 0.83
(d) 0.44
(e) 3.3

14.

x	f(x)	xf(x)
0	.125	0
1	.375	.375
2	.375	.750
3	.125	.375
		1.5

μ or $E(x) = \Sigma xf(x) = 1.5$ heads

19. (a) The probability that the random variable assumes a value equal to a.
(b) The probability that x is less than a.
(c) The probability that x is less than or equal to a, is at most a, or does not exceed a.
(d) The probability that x is greater than a.
(e) The probability that x is greater than or equal to a, or is at least equal to a.
(f) The probability that x is greater than a and less than b.
(g) The probability that x is greater than or equal to a and less than or equal to b, or that x is at least equal to a and at most equal to b.

The expression in (a) has no meaning for a continuous distribution, since the probability of a single value is defined equal to zero. (b) and (c) are the same if x is continuous, since the equality in (c) contributes nothing because the probability of a single value is defined as equal to zero. The two expressions are different when x is discrete, since the probability of a single value can be non-zero.

Chapter 7

8. $$P(x = 3) = \binom{4}{3}(.35)^3(.65)^{4-3}$$
$$= \frac{4!}{3!1!}(.0429)(.65)^1 = 4(.0429)(.65) = 0.1115$$

10. 0.3840

12. (a) 0.8735
(b) 0.1265
(c) 0.8065
(d) 0.4370

16. $\mu = nP = 50(.2) = 10$
$$\sigma = \sqrt{nPQ} = \sqrt{nP(1 - P)} = \sqrt{50(.2)(.8)}$$
$$= \sqrt{8} = 2.83$$

18. $P(x = 2|n = 10, P = .05) = \binom{10}{2}(.05)^2(.95)^8$

$= 45(.0025)(.6634) = 0.0746$

20. (a) 0.1964
    (b) 0.5982

26. (a) May not be independent if the process possesses some characteristic at the time of production that influences the output.
    (b) Independent
    (c) Since an individual's performance can vary at different times of the day owing to mood, fatigue, etc., consecutive entries probably are not independent.

31. (a) $P(x = 3) = 0.01$ means that 1 percent of a large number of samples drawn from shipments with 5 percent that will not start on the first try will contain 3 motors that do not start on the first try.
    (b) $\mu = 0.5$ means that 0.5 motors per sample will not start on the first try based upon a large number of samples.

Chapter 8

8. (a) $z = \frac{x - \mu}{\sigma} = \frac{16 - 10}{5} = \frac{6}{5} = 1.20$ yields .3849

$P(10 \leq x \leq 16) = 0.3849$

(b) $z_1 = \frac{x_1 - \mu}{\sigma} = \frac{4 - 10}{5} = \frac{-6}{5} = -1.20$ yields .3849

$z_2 = \frac{x_2 - \mu}{\sigma} = \frac{12 - 10}{5} = \frac{2}{5} = .40$ yields .1554

$P(4 \leq x \leq 12) = .3849 + .1554 = 0.5403$

(c) $z_1 = \frac{13 - 10}{5} = \frac{3}{5} = .60$ yields .2257

$z_2 = \frac{14 - 10}{5} = \frac{4}{5} = .80$ yields .2881

$P(13 \leq x \leq 14) = .2881 - .2257 = 0.0624$

(d) $z_1 = \frac{3 - 10}{5} = \frac{-7}{5} = -1.40$ yields .4192

$z_2 = \frac{8 - 10}{5} = \frac{-2}{5} = -.40$ yields .1554

$P(3 \leq x \leq 8) = .4192 - .1554 = 0.2638$

(e) $z = \frac{20 - 10}{5} = \frac{10}{5} = 2.00$ yields .4772

$P(x \geq 20) = 0.5 - .4772 = 0.0228$

(f) $z = \frac{3 - 10}{5} = \frac{-7}{5} = -1.40$ yields .4192

$P(x \leq 3) = .5 - .4192 = 0.0808$

10. (a) 0.1915
(b) 0.1915
(c) 0.3830
(d) 0.6170
(e) 12.2
(f) 31.7
(g) 18.3, 31.7; 5.4, 44.6

11. $x = \mu - z\sigma = 35 - 0.52(5)$
$= 35 - 2.6 = 32.4$

14. 0.0548

16. (a) $z = \frac{(x - \mu)}{\sigma} = \frac{200}{125} = 1.60$ yields 0.4452

$P(|x - \mu| \leq 200) = 2 \times .4452 = 0.8904$

(b) $z = \frac{1200 - 1500}{125} = \frac{-300}{125} = -2.40$ yields 0.4918

$P(x < 1200) = .5 - .4918 = 0.0082$

(c) $z = \frac{1800 - 1500}{125} = \frac{300}{125} = 2.40$ yields 0.4918

$P(x > 1800) = .5 - .4918 = 0.0082$

(d) Lowest 10 percent:
$x = \mu - z\sigma = 1500 - 1.28(125)$
$= 1500 - 160 = 1340$

Highest 5 percent:
$x = \mu + z\sigma = 1500 + 1.65(125)$
$= 1500 + 206 = 1706$

20. The difference in the way probabilities are computed is based upon the fact that the binomial is discrete and the normal is continuous. The z-value is a convenience that allows us to translate any normal distribution into one particular distribution that is tabulated. Otherwise it would be necessary to compile extensive tables of probabilities for different values of μ and σ or to use much more complicated and time consuming methods to find probabilities.

21. The probability of a single point for any continuous distribution is defined as zero, since there is no area under the curve above a point. Areas are assigned to intervals.

22. For any continuous distribution the two probabilities will be identical, since the probability of a single value and therefore the probability of the endpoints of the interval, a and b, is defined as zero.

24. Table 2 does not consider a value of z equal to 4. The highest value in the table is 3.9, which yields a probability of 0.4998. This value is considered to be close enough to 0.5 (area in the corresponding tail equal to 0.0002) that for practical purposes values of z that are greater than 3.9 can be treated as if they correspond to an area of 0.5.

27. This means that approximately 95 percent of a large number of months will have rates between 1 and 3.

Chapter 9

10. Working horizontally from the beginning of the table, one digit at a time, the sequence is

3 8 2 3 3 3 7 1 7 5

An alternative working vertically from the beginning is

3 4 8 4 5 3 2 8 5 3

11. Using two columns of digits working vertically from the beginning of Table 9, a random sequence is

56 51 72 55 51

All two digit numbers not falling between 50 and 75 are ignored.

13. Using Table 9 vertically from the beginning, the sample of 5 families and their incomes is:

Family	*Income*
38	$14174
43	$11865
44	$10500
30	$12141
20	$ 8743

15. The essential difference is that sample data are not complete whereas a census considers all of the data. Hence, one source of error is not present in census data: sampling error.

17. (a) Sample
(b) Census
(c) Sample
(d) Sample
(e) Sample

19. The element of randomness associated with random samples is important because error due to sampling cannot be measured without it.

22. The outcomes of die rolls and results generated from Table 9 are similar in the sense that they are based upon equal probabilities of occurrence. There are, however, six possible outcomes that can result from rolling a die, whereas there are ten possible digits in Table 9; consequently, the probabilities of occurrence in the two cases are not the same.

24. A table of random numbers is one form of a random number generator that is useful

in decision problems where random sequences must be generated. For our purposes here, it is an example of a device that can be used to select random samples. Numbers in a random number table exhibit properties that are similar to outcomes generated by gambling devices. Gambling devices could be used to select random samples; however, they are inefficient to work with.

Chapter 10

15. $\mu_{\bar{x}} = 35$; $\sigma_{\bar{x}}^2 = 1.67$; $\sigma_{\bar{x}} = 1.29$

17. (a) $z = \dfrac{x - \mu_{\bar{x}}}{\sigma_{\bar{x}}} = \dfrac{x - \mu}{\sigma/\sqrt{n}} = \dfrac{60 - 50}{15/\sqrt{9}}$

$= \dfrac{10}{15/3} = \dfrac{10}{5} = 2.00$ yields .4772

$P(50 \leq \bar{x} \leq 60) = 0.4772$

(b) $z = \dfrac{\bar{x} - \mu}{\sigma/\sqrt{n}} = \dfrac{40 - 50}{5} = \dfrac{-10}{5} = -2.00$ yields .4772

$P(40 \leq \bar{x} \leq 50) = 0.4772$

(c) $z_1 = \dfrac{\bar{x}_1 - \mu}{\sigma_{\bar{x}}} = \dfrac{45 - 50}{5} = \dfrac{-5}{5} = -1.00$ yields .3413

$z_2 = \dfrac{\bar{x}_2 - \mu}{\sigma_{\bar{x}}} = \dfrac{55 - 50}{5} = \dfrac{5}{5} = 1.00$ yields .3413

$P(45 \leq \bar{x} \leq 55) = .3413 + .3413 = 2(.3413) = 0.6826$

(d) $z_1 = \dfrac{55 - 50}{5} = \dfrac{5}{5} = 1.00$ yields .3413

$z_2 = \dfrac{65 - 50}{5} = \dfrac{15}{5} = 3.00$ yields .4987

$P(55 \leq \bar{x} \leq 65) = .4987 - .3413 = 0.1574$

(e) $z_1 = \dfrac{35 - 50}{5} = \dfrac{-15}{5} = -3.00$ yields .4987

$z_2 = \dfrac{40 - 50}{5} = \dfrac{-10}{5} = -2.00$ yields .4772

$P(35 \leq \bar{x} \leq 40) = .4987 - .4772 = 0.0215$

19. (1) 54.2
(2) 43.6

22. (1) 138.24, 161.76
(2) 142.32, 157.68

23. (a) $z_1 = \dfrac{\bar{x}_1 - \mu_{\bar{x}}}{\sigma_{\bar{x}}} = \dfrac{\bar{x}_1 - \mu}{\sigma/\sqrt{n}} = \dfrac{16.15 - 16}{.5/\sqrt{9}}$

$$= \frac{.15}{.5/3} = \frac{.15}{.17} = 0.88 \text{ yields } .3106$$

$$z_2 = \frac{16.30 - 16}{.17} = \frac{.30}{.17} = 1.76 \text{ yields } .4608$$

$$P(16.15 \le x \le 16.30) = .4608 - .3106 = 0.1502$$

(b) $$z_1 = \frac{15.7 - 16}{.17} = \frac{-.3}{.17} = -1.76 \text{ yields } .4608$$

$$z_2 = \frac{16.2 - 16}{.17} = \frac{.2}{.17} = 1.18 \text{ yields } .3810$$

$$P(15.7 \le \bar{x} \le 16.2) = .4608 + .3810 = 0.8418$$

(c) $$z = \frac{.3}{.17} = 1.76 \text{ yields } .4608$$

$$P(|\bar{x} - \mu| \ge .3) = 1 - 2(.4608)$$
$$= 1 - .9216 = 0.0784$$
or
$$= 2(.5 - .4608)$$
$$= 2(.0392) = 0.0784$$

(d) $$z = \frac{16.4 - 16}{.17} = \frac{.40}{.17} = 2.35 \text{ yields } .4906$$

$$P(\bar{x} > 16.4) = 0.5 - .4906 = 0.0094$$

(e) $$\bar{x} = \mu_{\bar{x}} - z\sigma_{\bar{x}} = 16 - 0.84(.17)$$
$$= 16 - .1428 = 15.8572 \text{ or } 15.86 \text{ (rounded)}$$

26. Under simple random sampling the mean of the sampling distribution is numerically equal to the mean of the universe, although these represent means of different distributions. Both are considered as parameters. Different samples have different values of the sample mean. The mean of the sampling distribution is the mean of all possible sample arithmetic means. The mean of a particular sample can be considered as a point estimate of the mean of the universe.

27. A universe has a variance, which is a parameter that describes the dispersion or differences among the items in the universe. Different samples have different variances, each of which describes the variability of the items in a particular sample about the mean of the sample. The variance of a particular sample is a statistic and can be considered as a point estimate of the population variance. The variance of the sampling distribution of the sample mean describes the variability among the values of the sample mean among all possible samples that can be selected from the population. It equals the variance of the population divided by the sample size (an adjustment is necessary if sampling is done without replacement). Basically, the answer is the same since standard deviations describe variability similar to variances, but the units are different since a standard deviation is the square root of a variance. The standard error of the mean is another term for the standard deviation of the sampling distribution of the sample mean.

29. When an inference is to be made about a universe, one sample generally is used

based upon a selection from the population. The sample that actually is drawn represents one of many possible samples that exist. The distribution of the sample mean corresponds to the many possible samples rather than the one selected. Although it is possible to continually select samples from a given universe, it is the concept of a distribution of all possible values of the mean that actually is used.

32. If an "exact" answer is required, then a census is the only way to achieve this. If, however, some discrepancy between a sample result and a universe parameter is acceptable, a sample size can be used that will result in a standard error that is as small as required such that there is a high probability that the sample value will fall within specified limits of the true value. In this sense, a required amount of precision can be obtained with near certainty without taking a census. In other words, the error due to sampling can be controlled within acceptable limits and therefore can be ignored for practical purposes.

33. The Central Limit Theorem is an important concept because it enables us to describe the behavior of means (when using large samples) without any knowledge of the nature of the universe from which a sample is selected. Consequently, inferences using the sample mean can be made without knowledge of the universe in many practical applications.

38. (a) 0.8621
    (b) The result in (a) means that 0.8621 or 86.21 percent of all possible samples of 100 monthly charge account balances have means that fall between $435 and $475. Another interpretation is that approximately 86.21 percent of a large number of samples drawn repeatedly from the monthly charge account balances will have a mean between $435 and $475.

39. (a) 0.95
    (c) 0.9882

40. (a) 0.4972
    (b) 0.4972

Chapter 11

11. $\bar{x} \pm z\sigma_{\bar{x}}$ or $\bar{x} \pm z\left(\frac{\sigma}{\sqrt{n}}\right)$

$$12 \pm 1.96\left(\frac{15}{\sqrt{25}}\right)$$

$$12 \pm 1.96\left(\frac{15}{5}\right)$$

$$12 \pm 1.96(3)$$

$$12 \pm 5.88$$

$$6.12-17.88$$

12. 20.68–39.32

13. $\bar{x} \pm ts_{\bar{x}}$ or $\bar{x} \pm t\left(\frac{s}{\sqrt{n-1}}\right)$

$$50 \pm 3.250\left(\frac{24}{\sqrt{10-1}}\right)$$

$$\pm 3.250\left(\frac{24}{\sqrt{9}}\right)$$

$$\pm 3.250\left(\frac{24}{3}\right)$$

$$\pm 3.250(8)$$

$$50 \pm 26$$

$24 - 76$ where $DF = n - 1 = 10 - 1 = 9$

14. (a) 13.3; 4.3
    (b) 10.7–15.9

19. 144 freight cars

24. A confidence interval accounts for sampling variability in terms of the confidence level. The confidence level provides the proportion of repeated samplings that leads to a correct conclusion that the interval contains the universe parameter. In other words, an interval estimate varies from sample to sample but contains the parameter a fixed proportion of the time.

26. The form of the sampling distribution of a statistic that is used to construct an interval estimate is based upon the nature of the universe from which the sample is selected. Consequently, the assumptions made about the universe govern the choice of the sampling distribution that is used.

28. Valid statements: (c) and (h)

29. This can be interpreted to mean that the mean rental is estimated to be between $313 and $337 with 99 percent confidence, or that the mean rental is estimated between $313 and $337.

31. (a) 71–79 seconds
    (b) The interval can be interpreted to mean that the mean reaction time of the smoke detector lies between 71 and 79 seconds with 90 percent confidence.

33. By increasing the confidence level, the width of the interval would increase (i.e., from 71–79 to 73–77) since the confidence coefficient z is larger (1.65 vs. 1.96). If the sample size were increased, the interval would be narrower since the standard error of the mean would be smaller.

40. The t-distribution accounts for the uncertainty by being wider. Consequently, values of t are greater than corresponding values of z, which result in wider confidence intervals. As the sample size increases, the estimate of the standard deviation becomes more reliable and is more likely to be near the true value. Consequently, there is less uncertainty for large samples. This is reflected in values of

t that approach z for a given level of confidence as the sample size and therefore the degrees of freedom increase. Values of z remain constant since they are based upon use of a known σ and reflect the fact that the added uncertainty is not present.

45. Measurements associated with elements in a universe are different, which means that there is variability in the universe. This variability is measured by σ, the universe standard deviation, and is inherent or given for a particular universe. If every observation were the same, there would be no variability in the universe and σ would equal zero. Consequently, every sample would be identical and there would not be variability from sample to sample and no sampling error would exist. The larger the variability in the universe, the more sampling error one can expect in terms of a value of a sample statistic used as an estimate of a universe parameter. This is illustrated with the formula for the standard error of the mean, $\sigma_{\bar{x}} = \sigma/\sqrt{n}$. For fixed n, $\sigma_{\bar{x}}$ equals zero when σ equals zero and becomes larger as σ becomes larger.

Chapter 12

23. $CV = \mu_0 - z_\alpha \sigma_{\bar{x}} = \mu_0 - z_{.05}(\sigma/\sqrt{n})$

$= 35 - 1.65(5/\sqrt{9}) = 35 - 1.65(5/3)$

$= 35 - 1.65(1.67) = 35 - 2.8 = 32.2$

Decision Rule: Accept H_0 if $\bar{x} > 32.2$
Reject H_0 if $\bar{x} \leq 32.2$

$\bar{x} = 29 < CV = 32.2$: Reject H_0

$$\left(\text{z-statistic} = \frac{\bar{x} - \mu_0}{\sigma_{\bar{x}}} = \frac{29 - 35}{1.67} = -3.59\right)$$

24. CV = 22.8; Reject H_0

25. $CV_1 = \mu_0 - z_{\alpha/2}\, \sigma_{\bar{x}} = \mu_0 - z_{.025}(\sigma/\sqrt{n})$

$= 165 - 1.96(40/\sqrt{25}) = 165 - 1.96(40/5)$

$= 165 - 1.96(8) = 165 - 15.7 = 149.3$

$CV_2 = \mu_0 + z_{\alpha/2}\, \sigma_{\bar{x}} = 165 + 15.7 = 180.7$

Decision Rule: Accept H_0 if $149.3 < \bar{x} < 180.7$
Reject H_0 if $\bar{x} \leq 149.7$
or
$\bar{x} > 180.7$

$\bar{x} = 158 > CV_1 = 149.3$

$< CV_2 = 180.7$: Accept H_0

$$\left(\text{z-statistic} = \frac{\bar{x} - \mu_0}{\sigma_{\bar{x}}} = \frac{158 - 165}{8} = -0.8\right)$$

26. Using t-statistic:

$CV = t_\alpha = t_{.01} = 2.821$ where $DF = n - 1 = 10 - 1 = 9$

Decision Rule: Accept H_0 if $t < 2.821$
Reject H_0 if $t \geq 2.821$

$$t = \frac{\bar{x} - \mu_0}{s_{\bar{x}}} = \frac{\bar{x} - \mu_0}{s/\sqrt{n-1}} = \frac{73 - 68}{19/\sqrt{10-1}}$$

$$= \frac{73 - 68}{19/\sqrt{9}} = \frac{5}{19/3} = \frac{5}{6.3} = 0.794$$

$t = 0.794 < t_\alpha = 2.821$: Accept H_0

Alternative in terms of $\bar{x}$:

$$CV = \mu_0 + t_\alpha s_{\bar{x}} = \mu_0 + t_{.01}\left(\frac{s}{\sqrt{n-1}}\right)$$

$$= 68 + 2.821(6.3) = 68 + 17.8 = 85.8$$

$\bar{x} = 73 < CV = 85.8$: Accept H_0

27. $CV = t_{.05} = 2.132$; $t = 3.776$: Reject H_0

28. $CV_1 = -8.21$, $CV_2 = 8.21$: Accept H_0

30. $H_0: \mu \geq 17$; $H_1: \mu < 17$ (Assume $\alpha = 0.02$)

$$CV = \mu_0 - z_\alpha S_{\bar{x}} = 17 - z_{.02}\,(s/\sqrt{n-1})$$

$$= 17 - 2.05\,(6.8/\sqrt{40-1}) = 17 - 2.05\,(6.8/\sqrt{39})$$

$$= 17 - 2.05\,(6.8/6.24) = 17 - 2.05\,(1.09)$$

$$= 17 - 2.2 = 14.8$$

Decision Rule: Accept H_0 if $\bar{x} > 14.8$
Reject H_0 if $\bar{x} \leq 14.8$

$x = 13.6 < CV = 14.8$: Reject H_0 and conclude that the mean number of participants is not at a profitable level.

$$\left(\text{z-statistic} = \frac{\bar{x} - \mu_0}{s_{\bar{x}}} = \frac{13.6 - 17}{1.09} = -3.12\right)$$

31. Reject H_0 and conclude that a change in the average number of accidents has occurred since the new law was introduced.

32. Accept H_0 at all acceptable levels of significance using the t-test: conclude that there is insufficient evidence to justify the inspector's suspicion.

34. $$s_{\bar{x}_1 - x_2} = \sqrt{\frac{s_1^2}{n_1} + \frac{s_2^2}{n_2}} = \sqrt{\frac{15^2}{45} + \frac{19^2}{45}}$$

$$= \sqrt{\frac{1}{45}(225 + 361)} = \sqrt{\frac{586}{45}}$$

$$= \sqrt{13.02} = 3.61$$

$H_0: \mu_1 = \mu_2$; $H_1: \mu_1 \neq \mu_2$

$$CV_1 = (\mu_1 - \mu_2)_0 - z_{\alpha/2}\, s_{\bar{x}_1 - \bar{x}_2} = 0 - z_{.005}\,(3.61)$$
$$= 0 - 2.58\,(3.61) = -9.3$$
$$CV_2 = (\mu_1 - \mu_2)_0 + z_{.005}\, s_{\bar{x}_1 - \bar{x}_2}$$
$$= 0 + 2.58\,(3.61) = 9.3$$

Decision Rule: Accept H_0 if $-9.3 < \bar{x}_1 - \bar{x}_2 < 9.3$
Reject H_0 if $\bar{x}_1 - \bar{x}_2 \leq 9.3$
or
$\bar{x}_1 - \bar{x}_2 \geq 9.3$

$\bar{x}_1 - \bar{x}_2 = 165 - 177 = -12 < CV_1 = -9.3$: Reject H_0 and conclude that the two brands differ in effectiveness (Brand 2 lasts longer on the average).

$$\left(\text{z-statistic} = \frac{(\bar{x}_1 - \bar{x}_2) - (\mu_1 - \mu_2)_0}{s_{\bar{x}_1 - \bar{x}_2}} = \frac{-12 - 0}{3.61} = -3.32\right)$$

35. $\sigma_{\bar{x}} = \sigma/\sqrt{n} = 10/\sqrt{36} = 10/6 = 1.67$

$$z = \frac{CV - \mu_1}{\sigma_{\bar{x}}} = \frac{22.8 - 24}{1.67} = -0.72 \text{ yields } .2642$$

$\beta = 0.5 - .2642 = 0.2358$

$P_w = 1 - \beta = 1 - .2358 = 0.7642$

36. $\sigma_{\bar{x}} = \sigma/\sqrt{n} = 5/\sqrt{9} = 5/3 = 1.67$

$$z_1 = \frac{CV_1 - \mu_1}{\sigma_{\bar{x}}} = \frac{32 - 36}{1.67} = -2.40 \text{ yields } .4918$$

$$z_2 = \frac{CV_2 - \mu_1}{\sigma_{\bar{x}}} = \frac{38 - 36}{1.67} = 1.20 \text{ yields } .3849$$

$\beta = .4918 + .3849 = 0.8767$

$P_w = 1 - \beta = 1 - .8767 = 0.1233$

39.

μ_1	β	P_w
21	.8599	.1401
22.8	.5000	.5000
23	.4522	.5478
25	.0934	.9066

47. (a) CV = 49.6: Reject H_0 and conclude that the collection policy should be changed.
(b) Since the sample mean fell in the region of rejection, a significant difference occurred. Consequently, it can be concluded that this is greater than chance alone would produce if the null hypothesis were true, and therefore it is reasonable to reject it and conclude that it is false.
(c) Assuming $\alpha = 0.05$, we can say that 5 percent of all possible samples of 100 customer accounts would incorrectly lead to the conclusion that the collection policy should be changed when it should not.

51. (a) 0.0062

60. (a) p-value = 0.0082; Reject H_0

63. A test of a hypothesis is performed on the basis of a value of a test statistic by comparing it with a preestablished critical value. The critical value determines the test criterion and the values of β and P_w, which are used primarily to evaluate the test. The value of a test statistic is a function of a particular set of sample observations, whereas β and P_w are a function of the sampling properties of the test statistic treated as a random variable and the corresponding decision rule based upon these sampling properties.

Chapter 13

13. $p = x/n = 45/100 = .45$

$$s_p = \sqrt{\frac{p(1-p)}{n}} = \sqrt{\frac{.45(.55)}{100}} = \sqrt{.002475} = .0497 \text{ or } .05$$

$p \pm zs_p$

$.45 \pm 1.96(.05)$

$\pm .098$

$.342 - .548$ or $.35 - .55$

14. $$CV = P_0 + z_\alpha \sigma_p = .35 + z_{.05}\sqrt{\frac{P_0(1-P_0)}{n}}$$

$$= .35 + 1.65\sqrt{\frac{.35(.65)}{80}} = .35 + 1.65\sqrt{.0028}$$

$$= .35 + 1.65(.053) = .35 + .087 = .437$$

Decision Rule: Accept H_0 if $p < .437$
Reject H_0 if $p \geq .437$

$p = x/n = 34/80 = .425 < CV = .437$: Accept H_0

$$\left(\text{z-statistic} = \frac{p - P_0}{\sigma_p} = \frac{.425 - .35}{.053} = 1.42\right)$$

15. $CV_1 = .393$, $CV_2 = .607$, $p = .44$; Accept H_0

19. $H_0: P \geq .1$; $H_1: P < .1$

$$CV = P_0 - z_\alpha \sigma_p = .1 - z_{.02}\sqrt{\frac{P_0(1-P_0)}{n}}$$

$$= .1 - 2.05\sqrt{\frac{.1(.9)}{75}} = .1 - 2.05\sqrt{.0012}$$

$$= .1 - 2.05(.035) = .1 - .072 = .028$$

Decision rule: Accept H_0 if $p > .028$
Reject H_0 if $p \leq .028$

$p = x/n = 6/75 = .08 > CV = .028$: Accept H_0 and conclude that the assembly line is *not* operating effectively.

(z-statistic $= \frac{p - P_0}{\sigma_p} = \frac{.08 - .1}{.035} = 0.57$)

20. $n = \frac{z^2P(1 - P)}{E^2} = \frac{(2.33)^2(.5)(.5)}{(.1)^2}$

$= \frac{5.4289(.25)}{.01} = 135.7$ or 136 rounded

With no assumptions made about the magnitude of P, the sample size needed equals 136.

$n = \frac{(2.33)^2(.2)(.8)}{(.1)^2} = 86.9$ or 87

Assuming P equals .2, the required sample size is 87.

22. $p_1 = x_1/n_1 = 20/100 = .2$

$p_2 = x_2/n_2 = 35/150 = .23$

$s_{p_1 - p_2} = \sqrt{\frac{p_1(1 - p_1)}{n_1} + \frac{p_2(1 - p_2)}{n_2}} = \sqrt{\frac{.2(.8)}{100} + \frac{.23(.77)}{150}}$

$= \sqrt{.0016 + .0012} = \sqrt{.0028} = .053$

$CV_1 = (P_1 - P_2)_0 - z_{\alpha/2}\, s_{p_1 - p_2} = 0 - 1.96(.053) = -.104$

$CV_2 = (P_1 - P_2)_0 + z_{\alpha/2}\, s_{p_1 - p_2} = 0 + 1.96(.053) = .104$

Decision rule: Accept H_0 if $-.104 < p_1 - p_2 < .104$
Reject H_0 if $p_1 - p_2 \leq -.104$
or
$p_1 - p_2 \geq .104$

$p_1 - p_2 = .2 - .23 = -.03 > CV_1 = -.104$
$< CV_2 = .104$; Accept H_0

(z-statistic $= \frac{(p_1 - p_2) - (P_1 - P_2)_0}{s_{p_1 - p_2}} = \frac{-.03 - 0}{.053} = -0.57$)

26. The sample proportion varies randomly in a way that depends on the population proportion, and this random variation must be taken into account when one makes inferences about the population based on random samples. Knowledge of the distribution of the sample proportion enables one to take the random variation of the samples into account and measure sampling error.

Chapter 14

11. $G = 411$; $GS = 11{,}387$; $\Sigma T^2 = 56{,}441$

$SS_{bg} = 26.8$ $DF_{bg} = 2$ $MS_{bg} = 13.40$

$SS_{wg} = 98.8$ $DF_{wg} = 12$ $MS_{wg} = 8.23$

$F = 1.63 < CV = F_{.05} = 3.89$: Accept H_0 at the 5 percent level of significance and conclude that the means of the three groups are the same.

13. $F = 3.52$: Accept H_0 for $\alpha = .01$ and Reject H_0 for $\alpha = .05$, where $F_{.01} = 4.94$ and $F_{.05} = 3.10$.

15. The within group mean square measures variability due to chance. The between group mean square measures possible differences among the means of the groups. If the samples come from universes with the same mean, both mean squares measure the same thing: variability due to chance. This assumes that the universes have the same variances.

16. The answer to this is based upon the fundamental idea that there is a certain amount of inherent variability among all repeated measurements observed under the same environmental conditions. Consequently, variability among the elements in each group or universe exists about each universe mean. Each of the samples will reflect this inherent variability, which is due to chance.

22. (a) i. 554.5 ii. 184.4 iii. 738.9

Chapter 15

10.

O	R	C	E	O − E	$(O - E)^2$	$(O - E)^2/E$
50	100	90	45	5	25	.5556
50	100	110	55	−5	25	.4545
40	100	90	45	−5	25	.5556
60	100	110	55	5	25	.4545
200			200			$\chi^2 = 2.0202$

Assume $\alpha = .05$; $CV = \chi^2_\alpha = \chi^2_{.05} = 3.841$ where $DF = (r - 1)(c - 1) = (2 - 1)(2 - 1) = 1 \times 1 = 1$.

Decision rule: Accept H_0 if $\chi^2 < 3.841$
Reject H_0 if $\chi^2 \geq 3.841$

$\chi^2 = 2.0202 < CV = 3.841$: Accept H_0 and conclude that there is no relationship between the characteristics.

12. $\chi^2 = 16.3127$ (Significant)

14. $\chi^2 = 89.7712 > CV = \chi^2_{.005} = 28.300$: Reject H_0 and conclude that a relationship exists between accidents and age.

22. (a) $\chi^2 = 15.5052$ (Significant)
(c) $\chi^2 = 0.7873$ (Not significant)

Chapter 16

28. (a) $\Sigma x = 620$; $\Sigma y = 636$; $\Sigma x^2 = 39000$; $\Sigma xy = 33910$.

$$a = \frac{\Sigma x^2 \Sigma y - \Sigma x \Sigma xy}{n\Sigma x^2 - (\Sigma x)^2} = \frac{39000(636) - 620(33910)}{12(39000) - (620)^2}$$

$$= \frac{24804000 - 21024200}{468000 - 384400} = \frac{3779800}{83600}$$

$$= 45.21$$

$$b = \frac{n\Sigma xy - \Sigma x\Sigma y}{n\Sigma x^2 - (\Sigma x)^2} = \frac{12(33910) - 620(636)}{83600}$$

$$= \frac{406920 - 394320}{83600} = \frac{12600}{83600} = 0.15$$

(b) $y' = 45.21 + 0.15(30) = 45.21 + 4.50 - 49.71$ or 50 rounded

30. (a) $y' = 49.6 + 26.5x$ where y = first year sales in $1000 units and x = test score
(b) $175,500

32. (a) $a = 40.699$; $b = 1.290$
(b) $266,449

33. 1.104; 1.05

34. 2.25 − 7.21

35. $n = 11$

$\Sigma x = 290$

$\Sigma y = 4461$

$\Sigma x^2 = 9202$

$\Sigma xy = 140750$

$\Sigma y^2 = 2170565$

(a) $$a = \frac{\Sigma x^2\Sigma y - \Sigma x\Sigma xy}{n\Sigma x^2 - (\Sigma x)^2}$$

$$= \frac{9202(4461) - 290(140750)}{11(9202) - (290)^2}$$

$$= \frac{41050122 - 40817500}{101222 - 84100}$$

$$= \frac{232662}{17122} = 13.586$$

$$b = \frac{n\Sigma xy - \Sigma x\Sigma y}{n\Sigma x^2 - (\Sigma x)^2}$$

$$= \frac{11(140750) - 290(4461)}{17122}$$

$$= \frac{1548250 - 1293690}{17122} = \frac{254560}{17122} = 14.867$$

(b) $$\hat{S}^2_{y \cdot x} = \frac{\Sigma y^2 - a\Sigma y - b\Sigma xy}{n - 2}$$

$$= \frac{2170565 - 13.586(4461) - 14.867(140750)}{11 - 2}$$

$$= \frac{2170565 - 60607.146 - 2092530.25}{9}$$

$$= \frac{17427.604}{9} = 1936.4$$

$$\hat{S}_{y \cdot x} = \sqrt{1936.4} = 44.0$$

38. (a) $a = -6.03;\ b = 1.62$

(b) $\hat{S}^2_{y \cdot x} = 1342.74;\ \hat{S}_{y \cdot x} = 36.64$

39. $$r^2 = \frac{[n\Sigma xy - \Sigma x \Sigma y]^2}{[n\Sigma x^2 - (\Sigma x)^2][n\Sigma y^2 - (\Sigma y)^2]}$$

$$= \frac{[50(6286) - 167(1540)]^2}{[50(693) - (167)^2][50(57460) - (1540)^2]}$$

$$= \frac{[314300 - 257180]^2}{[34650 - 27889][2873000 - 2317600]}$$

$$= \frac{(57120)^2}{(6761)(501400)} = \frac{3262694400}{3389965400}$$

$$= 0.962$$

$r = \sqrt{.962} = 0.981$ where the sign of r is plus since b equal to 8.4 is positive.

40. (a) Since the value of b is not given and it is not necessary to compute it, compute r first using the alternate formula in order to obtain the proper sign.

$$r = \frac{n\Sigma xy - \Sigma x \Sigma y}{\sqrt{n\Sigma x^2 - (\Sigma x)^2}\ \sqrt{n\Sigma y^2 - (\Sigma y)^2}}$$

$$= \frac{20(503) - 83(133)}{\sqrt{20(463) - (83)^2}\ \sqrt{20(960) - (133)^2}}$$

$$= \frac{10060 - 11039}{\sqrt{9260 - 6889}\ \sqrt{19200 - 17689}}$$

$$= \frac{-979}{\sqrt{2371}\ \sqrt{1511}}\ \frac{-979}{(49)(39)}$$

$$= \frac{-979}{1911} = -0.512$$

(b) $H_0: \rho = 0;\ H_1: \rho \neq 0$ (Assume $\alpha = .05$)

$CV_1 = -t_{\alpha/2} = -t_{.025} = -2.101$

$CV_2 = t_{\alpha/2} = t_{.025} = 2.101$ where $DF = n - 2 = 20 - 2 = 18$

Decision rule: Accept H_0 if $-2.101 < t < 2.101$
Reject H_0 if $t \leq -2.101$ or $t \geq 2.101$

$$t = r\sqrt{\frac{n-2}{1-r^2}} = -.512\sqrt{\frac{20-2}{1-.262}} = -.512\sqrt{\frac{18}{.738}}$$

$$= -.512\sqrt{24.390} = -.512(4.939) = -2.529$$

$t = -2.529 < CV_1 = -2.101$: Reject H_0 and conclude that correlation exists.

42. 0.356

43. Let y = loan amount ($1000 units)

    x = current ratio

    n $= 6$

    $\Sigma x = 22.25$

    $\Sigma y = 4650$

    $\Sigma x^2 = 97.6875$

    $\Sigma xy = 21237.5$

    $\Sigma y^2 = 4712500$

    (a) $r = 0.978$; $r^2 = 0.956$
    (b) $t = 9.291$: Reject H_0 and conclude that there is a positive relationship between the current operating ratio and the amount of loans given by the bank.

48. (a) The a-term can be interpreted as the value of y when x equals zero, which is 49.6. In terms of sales, we can say that a trainee with a zero score before going into the field will have $49,600 of sales at the end of one year. The b-term can be interpreted as the change in y for a one unit change in x, which is 26.5. In terms of the problem we can say that two salesmen that differ by one unit in their test score before going into the field will differ in sales by $26,500 at the end of one year.
    (b) The overall point to be made is that the a-term is an extrapolated value since test scores in the sample ranged between 2.37 and 4.91 and do not include a zero score. It is doubtful that a trainee would receive a zero score or that a trainee would be retained with the firm with such a score. Consequently, one really would not know if the same relationship would hold at the zero level, although it is not likely that it would. It is possible, however, that there is some minimum amount that any salesman would make regardless of ability, depending upon the nature of the product sold. If the intercept were negative, it obviously would not have practical meaning. In either case, it is not necessary in order to use the equation to predict trainee sales as long as the relationship holds within the range observed.
    (c) $y' = 49.6 + 26.5(3.5)$

    $= 49.6 + 92.75 = 142.35$ or $142,350

    Like any estimate, this is subject to error or variability and is highly unlikely to exactly equal the trainee's sales. The problem is similar to using a mean to estimate a single value. In this case, however, there is more than one source of error present.

50. (a) $y' = 80.49 + 6.29x$ where y = number of accidents and x = average weight in 100-pound units.

53. Since the standard error of estimate is a measure of variability of the y-values about the line, smaller values of the standard error of estimate indicate that the y-values are closer to the line. A line that goes through all the points is a "perfect fit." The closer the line fits the sample data, the greater the possibility that there actually is a relationship between the variables. Consequently, in such cases the

line may be useful for estimating or predicting values of the dependent variable based upon knowledge of the independent variable. In addition to its use as a measure of error about the line, the standard error of estimate can be incorporated into an interval estimate that explicitly accounts for the variability directly in the estimate itself.

55. 21.59 – 40.41 percent

56. Let y = sales (1,000,000 dollar units)

x = advertising (100,000 dollar units)

$n = 10$

$\Sigma x = 22$

$\Sigma y = 42.9$

$\Sigma x^2 = 54.94$

$\Sigma xy = 108.81$

$\Sigma y^2 = 216.69$

(a) −0.564; 2.206
(b) $3,848,000
(c) $\hat{S}_{y \cdot x} = .326$ or $326,000
(d) $2,903,904 – $4,792,096
A firm spending $200,000 on advertising can expect to have sales between $2,903,904 and $4,792,096 with a probability of 0.98.

58. Poor performer.

66. The value of 0.356 can be interpreted to mean that 35.6 percent of the variance in the percentage change in crime is explained or accounted for by the size of townships' population.

68. (a) $r = 0.755$; $r^2 = 0.57$
(b) The value of r^2 equal to 0.57 is interpreted to mean that 57 percent of the variance in the percentage of students not promoted is explained or accounted for by grade level.
(c) $t = 3.641$: Reject H_0 at the 5 percent level of significance.

81. (a) $y' = 99.67801 + 43.09568x_1 + 1.55566x_2$
(b) $249.19

Chapter 17

6. $N = 12$

$\Sigma t = 66$

$\Sigma t^2 = 506$

$\Sigma Y = 877.9$

$\Sigma tY = 5406.8$

$$A = \frac{\Sigma t^2 \Sigma Y - \Sigma t \Sigma tY}{N\Sigma t^2 - (\Sigma t)^2}$$

$$= \frac{506(877.9) - 66(5406.8)}{12(506) - (66)^2}$$

$$= \frac{444217.4 - 356848.8}{6072 - 4356}$$

$$= \frac{87368.6}{1716} = 50.91$$

$$B = \frac{N\Sigma tY - \Sigma t \Sigma Y}{N\Sigma t^2 - (\Sigma t)^2}$$

$$= \frac{12(5406.7) - 66(877.9)}{1716}$$

$$= \frac{64881.6 - 57941.4}{1716} = \frac{6940.2}{1716} = 4.04$$

$Y_T = 50.91 + 4.04t$

t = 0 in 1967
t is in 1-year intervals

7. (a) $Y_T = 6283 + 313t$

t = 0 in 1968
t is in 1-year intervals

(b) 8474
(c) 11,291

10. $N = 7$

$\Sigma t = 21$

$\Sigma t^2 = 91$

$\Sigma \log Y = 20.20072$

$\Sigma t \log Y = 68.61018$

$\log Y_T = 2.027815 + 0.286001t$

t = 0 in 1970
t is in 1-year intervals

23. The underlying assumption when using r^2 is that the remaining variability about the line is random. With respect to time series as we are treating them there are other systematic components besides the trend present in the data. Consequently, one may describe the trend adequately but obtain a small value of r^2 owing to the presence of a large cyclical movement; in this case the value of r^2 is not informative.

Chapter 18

9.

Quarter	*SI*
1	93.6
2	103.3
3	117.2
4	85.9

10.

Month	*SI*	*Month*	*SI*
Jan	73.5	Jul	84.2
Feb	66.8	Aug	85.8
Mar	87.8	Sep	97.9
Apr	98.6	Oct	100.6
May	103.4	Nov	125.4
Jun	93.7	Dec	182.3

11. Jul–Sep: 405

13. $575,000,000

21. (a) The indexes for the first and fourth quarters indicate seasonal lows, and the indexes for the second and third quarters indicate seasonal highs. The index for the second quarter, for example, can be interpreted to mean that unemployment is 9 percent higher in this quarter than it would be if there were no seasonal effect present, or that owing to seasonal factors unemployment is 9 percent higher than the yearly average. The value for the fourth quarter can be interpreted to mean that unemployment in the fourth quarter is 90 percent of the yearly average, owing to seasonal factors.

Chapter 19

11. An individual cyclical residual or percent of trend demonstrates how much an observation is above or below the trend owing to cyclical and random factors. When viewed for an entire series it is possible in certain cases to determine whether there is an oscillatory or cyclical pattern in the data with either measure.

14. $t = \frac{1974 - 1967}{1} = 7$

$Y_T = 51 + 4(7) = 51 + 28 = 79$ or \$790,000

Absolute cyclical residual $= 930{,}000 - 790{,}000$

$= \$140{,}000$

Percent of trend $= (930{,}000/790{,}000)100$

$= 1.18 \times 100 = 118$ percent

The absolute cyclical residual can be interpreted to mean that cash reserves are $140,000 higher or above trend owing to cyclical (and random) forces. The percent of trend is interpreted to mean that cash reserves are 18 percent higher than the trend owing to cyclical (and random) effects.

15. Cyclical pattern present

17. (a) $Y_T = 362.4 + 22.4t$
t = 0 in January, 1970
t is in 1-quarter intervals
(b) Possible evidence of a minor cyclical effect
(c) Trend = 654
Seasonal = 20
Cycle plus Random = 44

Chapter 20

11.

Item	P_B	P_G	Q_T	Q_TP_B	Q_TP_G
A	10.70	14.50	10	107.00	145.00
B	8.30	6.95	14	116.20	97.30
C	15.47	16.30	6	92.82	97.80
D	2.24	3.90	12	26.88	46.80
				342.90	386.90

$$\frac{\Sigma Q_TP_G}{\Sigma Q_TP_B} \times 100 = \frac{386.90}{342.90} \times 100$$

$$= 1.128 \times 100 = 112.8$$

12.

Item	Q_TP_G	P_G/P_B	$(Q_TP_G)P_G/P_B$
A	145.00	1.36	197.20
B	97.30	.84	81.73
C	97.80	1.05	102.69
D	46.80	1.74	81.43
	386.90		463.05

$$\frac{\Sigma(Q_TP_G)P_G/P_B}{\Sigma Q_TP_G} \times 100 = \frac{463.05}{386.90} \times 100$$

$$= 1.197 \times 100 = 119.7$$

15. Based upon the weighted aggregates index, prices of calculators in 1978 are 23 percent of the prices in 1973. (An average of relatives also could be used but a choice must be made about the typical prices to use in the weights)

18. On the average, prices of the four items increased 13 percent between 1970 and 1978, or prices in 1978 are 112.8 percent of the price of the items in 1970.

INDEX

Q

R

S

T

U

V

Z